Mathematical Models and Applications

Mathematical Models and Applications

WITH EMPHASIS ON THE SOCIAL, LIFE, AND MANAGEMENT SCIENCES

Daniel P. Maki
Maynard Thompson
Indiana University

Prentice-Hall, Inc.
Englewood Cliffs, New Jersey

Library of Congress Cataloging in Publication Data

MAKI, DANIEL P.
Mathematical models and applications.

Includes bibliographies.
1. Mathematical models. I. Thompson, Maynard, joint author. II. Title.
QA402.M32 511'.8 73-4471
ISBN 0-13-561670-0

10 9 8 7 6 5 4 3 2

Printed in the United States of America

PRENTICE-HALL INTERNATIONAL, INC., *London*
PRENTICE-HALL OF AUSTRALIA, PTY. LTD., *Sydney*
PRENTICE-HALL OF CANADA, LTD., *Toronto*
PRENTICE-HALL OF INDIA PRIVATE LIMITED, *New Delhi*
PRENTICE-HALL OF JAPAN, INC., *Tokyo*

(to Judy)^2

Contents

Preface

Mathematics is attractive to some because of the logical consistency and intrinsic beauty of the subject, and it is attractive to others because the ideas and tools of mathematics enable them to better understand the real world. It is primarily for those with the latter interest that this book is intended. Until recently most applications of mathematics dealt with problems and situations arising in the physical sciences and engineering. It has become increasingly apparent, however, that mathematics can also make a significant contribution to the social, life, and management sciences. In these areas it had long been recognized that statistical tools were valuable ones, and now it is clear that ideas from many other fields in the mathematical sciences are similarly useful. An interest in mathematics by scientists is often generated by a desire to form and study abstract models for real situations. This text was written to provide an introduction to the theory and practice of building these models.

This book began as lecture notes developed in connection with a course of the same name given since 1968 at Indiana University. The audience can be loosely grouped as follows: junior and senior mathematics majors, many of whom contemplate graduate work in other fields; undergraduate and graduate students majoring in the social and life sciences and in business; and prospective secondary teachers of mathematics. In addition, portions of the material have been used in NSF institutes for mathematics teachers. The goal of the course has been to provide the student with an appreciation for, an understanding of, and a facility in the use of mathematics in other fields. The role of mathematical models in explaining and predicting phenomena arising in the real world is the central theme.

In keeping with our desire to emphasize model building, we introduce new mathematical ideas only when they are useful in studying the situation at hand, and we have tried not to pursue these ideas past the point of relevancy for that situation. One result of this approach is that many concepts that are intrinsically interesting and important for more sophisticated applications are omitted. We believe that the mathematics discussed, though abbreviated in many cases, is honest. Indeed, we feel that one of the major values of mathematical arguments to the sciences lies in their capability for precision and lack of ambiguity. In this spirit, almost all of the results stated here are proved. A mathematical proof is, after all, the best tool a scientist has for convincing himself and others that an assertion is valid.

The skill of model building is acquired by doing. Simply reading or observing others' efforts is a useful first step, but forms only a part of learning modeling. A crucial activity is that of actually constructing a useful model. The exercises and projects are designed to help develop this creative ability. Exercises, which in general follow each section, provide opportunities to use the concepts and techniques presented in the text. Projects, which follow Chaps. 2–9, require more originality and frequently have many "solutions." Often the projects, as well as some of the exercises, are open-ended. A little or a great deal can be done with them, and there may be no clear conclusions or answers. Many students find such questions unsettling, but in the end they find these activities the most profitable of all.

Most scientists engaged in model building find computers useful or even necessary for their work. While we recognize the fruitful (and sometimes crucial) interaction between computing and model building, work with computers has not been integrated into the text material except in a few instances. This was done because many of our students do not have the training to use a computer effectively. However, the text provides several opportunities for computer work, especially in the exercises and projects; and we urge those with the interest and training to exploit these opportunities.

Selectivity in the choice of topics was essential to keep the book to a manageable size. Our goal was to study "real" situations, that is, those of interest to scientists outside mathematics, and to use widely applicable mathematics. Since our experience is that the text contains more material than can be covered in a standard one-year course, the instructor may also need to make some choices. A course taught in the spirit in which the book was written should include Chap. 1 and most of Chap. 2. These chapters discuss the basis for the use of mathematical models in the sciences; they provide examples which illustrate the model building; and they introduce many of the topics considered in detail in later chapters.

The remainder of the book can be used in a number of different ways, and some sample courses are outlined below. Even within these outlines there is considerable flexibility since most of the chapters are not closely connected. The diagram indicates the dependencies between chapters: a solid line indicates a strong connection; a dashed line means that only a few sections make use of preceding material.

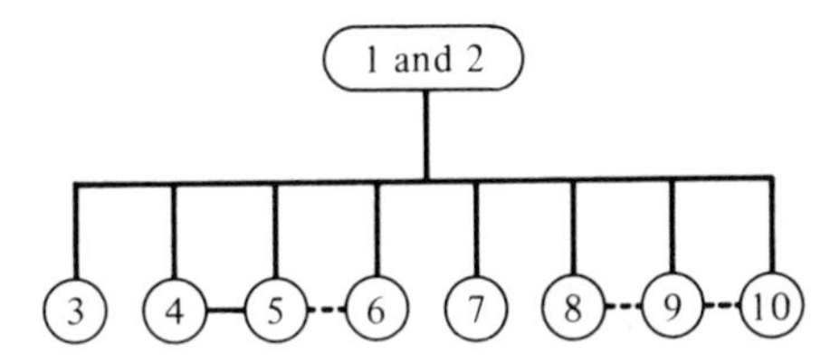

A number of different courses have been taught from this material to various types of students. As a guide to those who wish to tailor a course to their particular needs, we shall outline three courses that we have taught. It is important to recognize that one must select both the topics to be covered and the depth and detail of the mathematical presentation. For convenience we identify the courses as a *survey course*, an *in-depth course*, and a *teacher-preparation course*. We emphasize that these samples are neither exclusive nor exhaustive.

A *survey* course in modeling can be taught to students who have completed courses in finite mathematics and intuitive calculus. In particular, students should have some facility with elementary logic, set theory, probability, matrix notation and manipulations, and calculus. The depth required is that of a knowledge of the definitions and computational techniques. A reasonable one-year course for students with this background consists of Chaps. 1, 2 (omit Sec. 2.5), 3, 4, 6, and 7, and Secs. 8.1–8.3 with Appendix B, and Sec. 10.1. The mathematical details in Chaps. 3, 4, and 7 should be de-emphasized.

Students taking an *in-depth* course should have one semester of elementary probability theory, a standard beginning course in analysis, including an introduction to multivariable calculus, and a course in matrix algebra. Such an in-depth course can be designed to focus on models particularly useful in the management sciences—emphasizing Chaps. 3, 4, 5, 6, 7, Sec. 9.3, and Chap. 10—or the biological sciences—emphasizing Chap. 3, Sec. 7.2, Chap. 8, Secs. 9.1–9.2, and Chap. 10.

It is now recognized that experiences in mathematical model building are an important component in the preparation of secondary school teachers of mathematics (see 1971 AAAS Guidelines for the education of secondary school teachers of sciences and mathematics). We provide these experiences in a *teacher-preparation* course that follows the study of probability, multi-dimensional calculus, and matrix algebra. Our course includes topics from several chapters of the text, and the emphasis is on considering a variety of models and not on the mathematical details of a single model. Thus both in this and in the survey course it is often appropriate to emphasize exercises and projects at the expense of the details of the proofs.

The following conventions have been used for notation and referencing. Vectors are denoted by boldface type, e.g., $\mathbf{x}$, and matrices are denoted by outline letters, e.g., $\mathbb{A}$. Subdivisions of chapters are denoted by two or three digits. The first digit indicates the chapter, the second the section, and the third (if present) the subsection. Theorems and equations are numbered consecutively through each section. Thus, Eq. (2.8) refers to Equation 8 of Section 2 of the chapter in which the reference is made. If an equation in another chapter is referenced, then the other chapter number will also be given. Figures and tables are numbered consecutively through each chapter. There are appendices which provide background material needed in the text—material which is not usually contained in the courses listed as prerequisites for the in-depth course. It is reasonable to expect students to use these appendices, particularly Appendix C which is assumed in Chap. 4 and Appendix B which is assumed in Chaps. 8 and 9.

There is also a short list of references at the end of each chapter. No attempt was made to be comprehensive; and in general the references are selections which (a) are specifically referred to in the text, (b) contain natural extensions of the methods introduced, or (c) may be particularly interesting or useful to the reader.

We are in debt to many for their help and encouragement in the preparation of this text. The Indiana University Mathematics Department actively encouraged and supported our work. We especially acknowledge the encouragement of former department chairmen S. G. Ghurye and George Springer; the assistance of B. E. Rhoades, who among other things arranged financial support for the development of the course; the comments of our colleagues Grahame Bennett, Donald Kerr, Jr., Roger Lynn, and Charles Weaver—all of whom taught from the notes; the aid of former graduate assistants James Burns, Donald Cathcart, and Roger Lautzenheiser; and the reactions of the many students who contributed so much to the evolution of the book. We also recognize the valuable role of informal conversations with faculty members in all departments at Indiana University, especially those with our colleague George Minty. Many of our original ideas were reinforced by conversations with Henry Pollak and with others at Bell Telephone Laboratories in 1968. We have also benefited from the critical reviews of many who have read parts or all of various versions of the text, in particular those of Gerald Thompson of Carnegie-Mellon University. Finally we thank our wives and families without whose cooperation and understanding this writing project would have been terminated (in one way or another) much sooner.

Bloomington, Indiana

Daniel P. Maki
Maynard Thompson

Mathematical Models and Applications

1 Basic Principles

1.1 INTRODUCTION AND PHILOSOPHY

An examination of the origins of any scientific field, be it astronomy or anatomy, physics or psychology, indicates that the discipline began with a mass of observations and experiments. It is natural, then, that the first steps in quantifying the subject should involve the collection, presentation, and treatment of data. Consequently, the study of statistics has played a dominant role in the mathematical preparation of students working in the quantitative areas of the social and life sciences. A statistical treatment of data may be quite elementary, involving little more than listing, sorting, and a few straightforward computations. It may also be quite sophisticated, involving substantial mathematical ideas and delicate problems of experimental design. Once enough data have been collected and adequately analyzed, the researcher tries to imagine a process which accounts for these results. It is this activity, the mental or pencil-and-paper creation of a theoretical system that is the topic of this book. In the scientific literature this activity is commonly known as theory construction and analysis. We shall refer to it as the construction, development, and study of mathematical models.

The original problem almost always arises in the real world, sometimes in the relatively controlled conditions of a laboratory and sometimes in the much less completely understood environment of everyday life. For example, a psychologist observes certain types of behavior in rats running in a maze, a geneticist notes the results of a hybridization experiment, or an economist records the volume of international trade under a specific tariff policy, and then each conjectures certain reasons for his observations. These conjectures may be based completely on intuition, but more often they are the result of detailed study and the recognition of some similarities with other situations which are better understood. This close study of the system, which for the experimenter usually precedes the forming of conjectures, is really the first step in model building. Much of this initial work must be done by a researcher who is familiar with the origin of the problem and the basic biology, psychology, or whatever else is involved.

The next step (after initial study) is an attempt to make the problem as precise as possible. By this we mean arriving at a clear and definite understanding of the words and concepts to be used. This process typically involves making certain idealizations and approximations. One important aspect of this step is the attempt to identify and select those concepts to be considered as basic in the study. The purpose here is to eliminate unnecessary information and to simplify that which is retained as much as possible. For example, with regard to a psychologist studying rats in a maze, the experimenter may decide that it makes no difference that all the rats are gray or that the maze has 17 compartments. On the other hand, he may consider it significant that all the rats are siblings or that one portion of the maze is illuminated more brightly than another. This step of identification, approximation, and idealization will be referred to as constructing a *real model*. This terminology is intended to reflect the fact that the context is still that of real things (animals, apparatus, etc.) but that the situation may no longer be completely realistic. Returning again to the maze, the psychologist may construct a real model which contains rats and compartments, but with the restriction that a rat is always in exactly one compartment. This restriction involves the idealization that rats move instantaneously from compartment to compartment and are never half in one compartment and half in another. Also, he might construct a model in such a way that the rat moves from one compartment to another regularly in time, an approximation which may or may not be appropriate depending on just what behavior is to be investigated.

The third step (after study and formation of a real model) is usually much less well defined and frequently involves a high degree of creativity. One looks at the real model and attempts to identify the operative processes at work. The goal is the expression of the entire situation in symbolic terms. Thus the real model becomes a *mathematical model* in which the real quantities and processes are replaced by symbols and mathematical operations.

Usually, much of the value of the study hinges on this step because an inappropriate identification between the real world and the mathematical world is unlikely to lead to useful results. It should be emphasized that the construction of a mathematical model is highly nonunique. There may be several mathematical models for the same real situation. In such circumstances it may happen that one of the models can be shown to be distinctly better than any of the others as a means of accounting for observations. In fact, it often happens that an elaborate experiment is designed for the purpose of showing that one model is truly better than others. Naturally, if this can be shown, then one usually chooses to use the best model. However, it may also happen that each of a number of models proves to be useful in the study—each model contributing to the understanding of some aspects of the situation, but no one model adequately accounting for all facets of the problem under consideration. Thus there may not be a best model, and the one to be used will depend on the precise questions to be studied.

After the problem has been transformed into symbolic terms, the resulting mathematical system is studied using appropriate mathematical ideas and techniques. The results of the mathematical study are theorems, from a mathematical point of view, and predictions, from the empirical point of view. The motivation for the mathematical study is not to produce new mathematics, i.e., new abstract ideas or new theorems, although this may happen, but more importantly to produce new information about the situation being studied. In fact, it is likely that such information can be obtained by using well-known mathematical concepts and techniques. The important contribution of the study may well be the recognition of the relationship between known mathematical results and the situation being studied.

The final step in the model-building process is the comparison of the results predicted on the basis of the mathematical work with the real world. The happiest situation is that everything actually observed is accounted for by the conclusions of the mathematical study and that other predictions are subsequently verified by experiment. Such agreement is not frequently observed, at least not on the first attempt. A much more typical situation would be that the set of conclusions of the mathematical theory contains some which seem to agree and some which seem to disagree with the outcomes of experiments. In such a case one has to examine every step of the process again. Has there been a significant omission in the step from the real world to the real model? Does the mathematical model reflect all the important aspects of the real model, and does it avoid introducing extraneous behavior not observed in the real world? Is the mathematical work free from error? It usually happens that the model-building process proceeds through several iterations, each a refinement of the preceding, until finally an acceptable one is found. Pictorially, we can represent this process as in Fig. 1–1. The solid lines in Fig. 1–1 indicate the process of building, developing, and

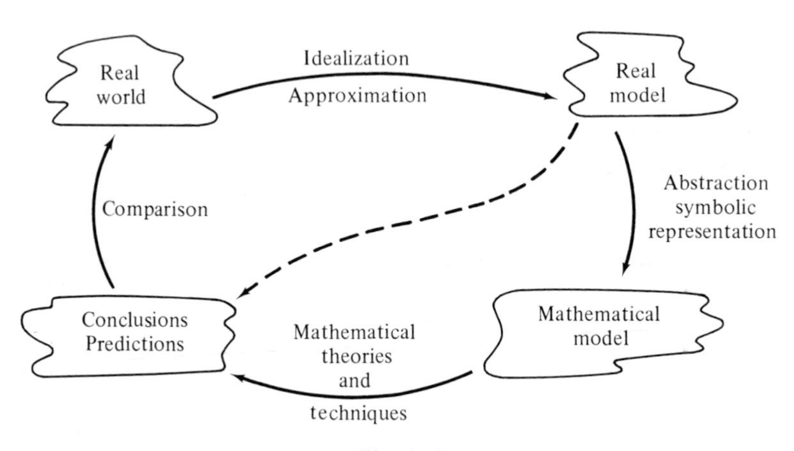

Fig. 1–1

testing a mathematical model as we have outlined it above. The dashed line is used to indicate an abbreviated version of this process which is often used in practice. The shortened version is particularly common in the social and life sciences where mathematization of the concepts may be difficult. In either case, the steps in this process may be exceedingly complex and there may be complicated interactions between them. However, for the purpose of studying the model-building process, such an oversimplification is quite useful. We also note that this distinction between real models and mathematical models is somewhat artificial. It is a convenient way to represent a basic part of the process, but in many cases it is very difficult to decide where the real model ends and the mathematical model begins. In general, research workers often do not worry about drawing such a distinction. Hence in practice one frequently finds that predictions and conclusions are based on a sort of hybrid model, part real and part mathematical, with no clear distinction between the two. There is, however, some danger in this practice. While it may well be appropriate to work with the real model in some cases and the mathematical model in others, one should always keep in mind the setting that is being used. At best, a failure to distinguish between a real model and a mathematical model is confusing; at worst, it may lead directly to incorrect conclusions. Complications may arise because problems in the social, biological, and behavioral sciences often involve concepts, issues, and conditions which are very difficult to quantify. Hence essential aspects of the problem may be lost in the transition from the real model to the mathematical model. In such cases conclusions based on the mathematical model may well not be conclusions about the real world or the real model. Thus there are circumstances in which it is crucial to distinguish the model to which a conclusion refers.

Many skills are needed for successful model building. In particular the ability to recognize patterns and general structures is just as important in model building as it is in the study of pure mathematics. It is not surprising that the same type of mathematics is involved in studying the servicing of automobiles at a turnpike tollbooth and customers in a barbershop. It is less obvious that, from a mathematical point of view, these two situations have much in common with certain models for the propagation of a rumor or epidemic. However, as we shall see later (Chap. 9), very similar mathematical models will serve for both.

One of the first features which should be determined about a situation is whether it is most appropriately modeled in deterministic or stochastic terms. A model is said to be *deterministic* if it predicts that given sufficient information at one instant in time or at one stage, then the entire future behavior of the system being studied can be exactly determined. Thus one might choose to model the growth of a certain population in deterministic terms. The hypothesis, then, is that if the growth rate of the population is known and its size is given at one instant in time, then the size of the population for all future times can be determined exactly. On the other hand, a model is *stochastic* if it incorporates probabilistic behavior. For these models the predictions are such that no matter how much one knows about the system at a given time, it is impossible to determine with absolute certainty the nature of the system for future times. For example, suppose that we are concerned with a concept acquisition experiment in which an unbiased subject is provided stimuli intended to convey, in some explicitly defined manner, the concepts roundness, redness, and fourness. Then one might choose to construct a model in which the knowledge of how quickly the subjects learned the concepts of roundness and redness does not permit an absolute determination of how quickly they will learn the concept of fourness. The strongest statement that can be made in such situations is a probabilistic one such as, "The subject will learn the concept of fourness with fewer than ten presentations of a stimulus with probability 0.8." Such a model might appear very appropriate in light of experimental results. Many of the most useful models in the social and life sciences are of this type, that is, models whose mathematical description involves chance and uncertainty. This is entirely to be expected since the real world gives strong evidence of being a stochastic system. In some instances, one can construct both types of models for the same system, and in some of these cases a comparison serves to check the validity of both (see Chap. 8). The decision as to which type of model should be constructed depends on many factors and is ultimately simply a choice of the investigator. Frequently a deterministic model is taken as a first approximation in a situation when a stochastic model appears more appropriate. For example, when formulated in terms of a real model, a situation involving growth may appear to be best modeled in stochastic terms. However, a stochastic mathematical model may present technical difficulties which are either impossible to overcome or prohibitively time-consuming. It may be desirable to consider a deterministic mathematical model as a first approximation and to compare the conclusions of such an approximation with observations. In general, however, one should not assume that predictions based on one type of model are necessarily better (or worse) than those based on the other type. The relative merits of the two types of models varies from one situation to another.

If a model is to be of practical use, then one must have a means of obtaining results that can be tested or compared with the real world. Consequently, the model builder must keep in mind the necessity for developing realistic computational schemes or algorithms for computing the quantities arising in the study of the mathematical model. There are some particularly important algorithms associated with the topics discussed in this book, and we shall give an introduction to some of them. Most algorithms of practical significance for real problems require computer assistance with the calculations. Another matter related to checking the model against reality is parameter estimation. Many of the models which are considered here lead to mathematical relations involving a parameter. Tracing this parameter back to the real world, one may find that it is related to a learning rate, a probability of a birth, an asset ratio, etc. Thus, in comparing the results of the mathematical study with reality, it may be necessary to give numerical values to these parameters. In general, the estimation of parameters is a real and delicate problem for which each discipline has its own special and refined techniques. Because of the special nature and limited applicability of many of these ideas, and not because of any lack of importance, we shall consider them only briefly (Chap. 10).

The reader will rapidly realize that some of the best models for certain types of problems are simply intractible mathematically. That is, the model leads to mathematical questions which either have no known solution or no solutions which can be reliably and reasonably computed. In such situations, one often turns to computer simulation for assistance. Again the precise nature of the program depends on the specific problem, and we content ourselves here with some examples (Chap. 10).

To conclude this introductory section, we remark that the use of mathematical techniques, in particular the use of mathematical models, is only one method which can be applied to questions arising in the sciences. As noted earlier, many important aspects of a situation in the social or life sciences may be very difficult to quantify. In such cases the use of mathematical models may be limited, and it may be better to study the situation in the context of a real model by nonmathematical means. Indeed, one might legitimately ask what basis we have for expecting mathematical methods to be effective. Our hopes rest on the proven effectiveness of mathematics in the physical sciences and on scattered but significant successes in the social and life sciences. One of the most impressive examples, indeed perhaps *the* most impressive example, of the fruitful use of mathematical models occurs in the study of planetary motion. This is also a fine example of the evolution of a model through several stages. Since this example had such a profound influence on science, it is worthwhile to consider it briefly.

1.2 A CLASSIC EXAMPLE

The creation of a coherent system to explain and predict the apparent motions of the planets and stars as viewed from the earth is certainly a significant triumph of human intellect. That the problem had attracted attention from the most ancient times and that the theory is still undergoing modification in this century give an indication of the enormous time and energy that have gone into its study.

Early views of a fixed and flat earth covered by a spherical celestial dome were studied by the Greeks, who devised a real model in the fourth century B.C. which accounted at least approximately for the rough observations then available. The earth was viewed as fixed with a sphere containing the fixed stars rotating about it. The "seven wanderers" (the sun, moon, and five planets) moved in between. The Greeks' concern was to construct combinations of uniform circular motions centered in the earth by which the movements of the seven wanderers among the stars could be represented. Each body was moved by a set of interconnecting rotating spherical shells. This system was adopted by Aristotle, who introduced 55 shells to account for observed motions. This real model based on geometry was capable of reproducing the apparent motions, at least to a degree consistent with the accuracy of the contemporary observations. However, since it kept each planet a fixed distance from the earth, it could not account for the varying brightness of the planets as they moved.

This system was modified by Ptolemy, the last great astronomer at the famous observatory at Alexandria, in the second century A.D. In its simplest form the Ptolemaic system can be described as follows: Each planet moved in a small circle (epicycle) in the period of its actual motion through the sky, while simultaneously the center of this circle moved around the earth on a larger circle. The basic model was capable of repeated modification to account for new observations, and such modifications in fact took place. The result was that by the thirteenth century the model was extremely complicated, 40–60 epicycles for each planet, without commensurate effectiveness.

By the beginning of the sixteenth century there was widespread dissatisfaction with the Ptolemaic system. Difficulties resulting from more numerous and more refined observations forced repeated and increasingly elaborate revision of the epicycles on which the Ptolemaic system was based. As early as the third century B.C. certain Greek philosophers had proposed the idea of a moving earth, and as the difficulties with the Ptolemaic point of view increased, this alternative appeared more and more attractive. Thus in the first part of the sixteenth century the Polish astronomer Copernicus proposed a heliocentric (sun-centered) theory in which the earth, among the other planets, revolved about the sun. However, he retained the assumption of uniform circular motion—an assumption with a purely philosophical basis—and consequently he was forced to continue the use of epicycles to account for the variation in apparent velocity and brightness of the planets from the earth.

The next step, and a very significant one, was taken by Johannes Kepler. During the years 1576–1596 a Swedish astronomer, Tycho Brahe, had collected masses of observational data on the motion of the planets. Kepler inherited Brahe's records and undertook to modify Copernican theory to fit these observations. He was particularly bothered by the orbit of Mars, whose large eccentricity made it very difficult to fit into circular orbit-epicycle theory. He was eventually led to make a very creative step, a complete break with the circular orbit hypothesis. He posed as a model for the motions of the planets the following three "laws":

1. The planets revolve around the sun in elliptical orbits with the sun at one focus (1609).
2. The radius vector from the sun to the planet sweeps out equal areas in equal times (1609).
3. The squares of the periods of revolution of any two planets are in the same ratio as the cubes of their mean distances to the sun (1619).

These laws are simply statements of observed facts. Nevertheless, they are perceptive and useful formulations of these observations. In addition to discovering these laws, Kepler also attempted to identify a physical mechanism for the motion of the planets. He hypothesized a sort of force emanating from the sun which influenced the planets. This model described very well the accumulated observations and set the stage for the next refinement, due to Isaac Newton.

All models developed up to the middle of the seventeenth century involved geometrical representations with minimal physical interpretation. The fundamental universal law of gravitation provides at once a physical interpretation and a concise and elegant mathematical model for the motion of the planets. Indeed, this law, when combined with the laws of motion, provides a description of the motion of all material particles. The law asserts that every material particle attracts every other material particle with a force which is directly proportional to the product of the masses and inversely proportional to the square of the distance between them. In this framework the motion of a planet could be determined by first considering the system consisting only of the planet and the sun. The latter problem involves only two bodies and is easy to solve. The resulting predictions, the three laws of Kepler, are good first approximations since the sun is the dominant mass in the solar system and the planets are widely separated. However, the law of gravitation asserts that each planet is, in fact, subject to forces due to each of the other planets, and these forces result in perturbations in the predicted elliptical orbits. The mathematical laws proposed by Newton provide such an accurate mathematical model for planetary motion that they led to the disovery of new planets. One could examine the orbit of a specific planet and take into account the influence of all the other planets on this orbit. If discrepancies were observed between the predictions and observations, then one could infer that these discrepancies were due to another planet, and estimates could be obtained on its size and location. The planets Uranus, Neptune, and Pluto were actually discovered in this manner. However, even this remarkable model does not account for all the observations made of the planets. Early in this century small perturbations in the orbit of Mercury, unexplainable in Newtonian terms, provided some motivation for the development of the theory of relativity. The relativistic modification of Newtonian mechanics apparently accounts for these observations. Nevertheless, one should not view this model as ultimate, but rather as the best available at the present time.

The laws of Newton, viewed as a mathematical model, have provided an extremely effective tool to the physical sciences. The concepts of force, mass, velocity, etc., can be made quite precise and the model can be studied from a very abstract point of view. Although the social and life sciences do not yet have their equivalents of Newton's laws, the utility of mathematical models in the physical sciences gives hope that their use may contribute to the development of other sciences as well.

1.3 AXIOM SYSTEMS AND MODELS

In the preceding sections we have surveyed how we intend to use mathematical models to study situations arising outside mathematics, and we considered one example where the use of mathematical models has been particularly rewarding. It is time now for us to begin to make the concept of a model more precise. Since, as we shall see below, a mathematical model can be viewed as an axiom system, it is appropriate to begin with this topic. The following development has been greatly influenced by R. L. Wilder and his book [W].

1.3.1 Axioms

Since the use of the word *axiom* has changed over the years, we begin our discussion of axioms by contrasting the current use of this term with an earlier use. At one time, for example, with Euclid and other Greek mathematicians, the term axiom meant a self-evident truth. It was a universal statement which was obvious to and undebatable by all. An example of such a statement is the proposition "Equals added to equals yield equals." In addition to axioms, mathematicians of the day were also concerned with postulates. These were statements of a more specific character, and they presumably expressed "true facts" about a particular subject, such as geometry. The statement "Through two distinct points there exists one and only one line" qualifies as a postulate. Thus the postulates of geometry use terms such as point and line, which are special to geometry and which are not used in areas such as arithmetic. This separation of basic truths into axioms and postulates has a long history. Euclid used it in his famous *Elements*, calling the axioms "common notions."

The original use of the words axiom and postulate remained relatively unchanged for almost 2000 years. In fact, it is really only in mathematics itself that a second meaning has arisen. In day-to-day conversation one still hears "It is axiomatic," the meaning being that the statement under discussion is a universal truth. In mathematics, however, a change in the meaning of the term axiom (and likewise of the related term postulate) began during the period which saw the development of non-Euclidean geometries. Without going into detail, we simply point out that a number of geometries were developed in which Euclid's fifth postulate failed to be true. (The fifth postulate essentially said that given a line and a point not on the line, there exists one and only one line through this point and parallel to the given line. Gauss, Lobachevski, and Bolyai all used sets of axioms in which this statement is contradicted while all other Euclidean axioms are true.) The development of these geometries demonstrated the fallacy of the earlier belief that the fifth postulate was a consequence of the other postulates and hence that its denial would lead to contradictions. The assumption of an axiom which contradicted the fifth postulate did not lead to an inconsistent system; instead it led to new geometries. This means that there were many geometries and that the postulates in one could contradict the postulates in another. This, in turn, implied that postulates were in some sense a matter of choice and that they were not fundamental truths about geometry which the Greeks had discovered. It also implied that there was now more than one candidate for a mathematical description (i.e., a mathematical model) for physical space. Naturally, the initial reaction was that, although other geometries might be of interest to mathematics as areas for abstract investigations, only Euclidean geometry would be important for providing a description of physical space. This has not been the case, however, as a non-Euclidean geometry plays a prominent role in the theory of relativity.

Along with the change in the meaning of the term postulate, the notion of axiom was also reevaluated. It soon became apparent that these statements also could be denied without necessarily implying a contradiction. Thus the statement "The whole is greater than the part" is not basically different from the fifth postulate. In one context it may be true (in the sense of following logically from other statements), and in another context it may be false, the critical difference being the meaning of the words *whole*, *part*, and *greater than* in each case. For example, we note that in set theory this statement is false when part means "proper subset" and greater than means "has a larger cardinal number." Thus, in mathematics, the distinction between axioms and postulates has essentially disappeared. They both stand for statements which one takes as basic in order to study the logical consequences of such assumptions. They are not, in general, considered to be universal truths, and, in fact, they can be changed and modified to suit different purposes.

As a final comment concerning axioms, we remark that frequently certain of the words which are used in phrasing axioms are left undefined. This is necessary if one is to avoid circular and meaningless definitions. For example, in axiom systems for geometry the words *point* and *line* are often left undefined. In this way one avoids having pseudodefinitions such as "A point is an object having no length, width, or height, while a line has length, but not width or height."

1.3.2 Axiom Systems

An *axiom system* is a collection of undefined terms together with a set of axioms phrased with the use of these common undefined terms. Although it is possible to consider axiom systems with an infinite number of undefined terms and/or an infinite number of axioms, we shall always consider both of these sets to be finite.

Our primary concern with axiom systems is with the logical consequences of the axioms. These logical consequences are called *theorems*, and the collection consisting of all theorems which can be logically deduced from an axiom system Σ is called the *theory determined by* Σ.

Regarding the above definition of a theory, we note that we are using this term as it is used in mathematics and logic. The notion of a theory is used somewhat differently in the social and life sciences. In these sciences, a theory is usually a collection of basic assumptions which is studied in an attempt to explain certain observed phenomena. Below we shall connect these two meanings of the term *theory* by using the concept of a model.

Our notion of axiom system is obviously a very general one, since almost every set of statements now constitutes an axiom system. Naturally, not all these axiom systems are of interest to mathematicians, and hence restrictions are imposed to obtain useful systems. First, since it is the theory of an axiom system that interests us, we want to make sure that the axioms do not contradict each other. This is necessary because it is a consequence of classical logic that assuming the truth of two contradictory statements is logically equivalent to assuming the truth of all statements. Hence the theory determined by a system containing contradictions would contain all statements and would be of no value. This condition about the absence of contradiction can be made precise as follows: An axiom system Σ is said to be *consistent* if the theory determined by Σ does not contain contradictory statements.

This definition of *consistent* is logically sound; however, it is not an especially practical definition. The lack of practicality results from the fact that we usually do not know all the statements in the theory determined by an axiom system Σ. Thus, in general, we cannot tell if the system is consistent because we do not know which statements to check for contradictions. Fortunately, this difficulty can be circumvented by the use of models. Therefore, our next task is to introduce the notion of a model of an axiom system. We shall then show how models can be used to verify the consistency of an axiom system.

As a final comment on axiom systems we note that different branches of mathematics can be characterized by the axiom systems which they use. Thus, group theory is the study of the consequences of the axioms which define a group, topology is the study of the consequences of the axioms for different topological spaces, and geometry is the study of the different axiom systems used to describe geometrical systems.

1.3.3 Models and Formal Model Building

The term *model* is often used in different ways in different contexts. We first consider how it is used in the framework of mathematical logic, since it is here that it is connected to the important concept of the consistency of an axiom system. To this end, suppose that we are given an axiom system Σ consisting of certain undefined terms and certain statements phrased with these terms. Also, suppose that we assign a meaning to each of the undefined terms and that with these meanings the axioms of the system become statements which are known to be true. Here, by "assign a meaning" we mean to associate an object or action of the real world with each of the undefined terms; by the phrase "statements which are known to be true" we mean that the statement is now a meaningful assertion about the real world and that this assertion is an observable and verifiable fact. If this can be done, then we say that this assignment of meanings to the undefined terms of Σ constitutes a *logical model* of the system Σ. We use the adjective *logical* here to distinguish this concept of a model from our main use of the term in the sense of a mathematical model. We also note that a single axiom system may have several different logical models, each associated with a certain assignment of meanings to the terms.

Before proceeding, a word of caution to the reader is in order. In our discussion of logical models we are skirting close to very deep and fundamental matters—matters which are at the very basis of mathematical thought. This brief discussion is certainly an inadequate treatment of these profound concepts. However, since a study of the foundations of mathematics and mathematical logic is clearly beyond the scope of our work, we feel that this short survey is adequate for what follows. The interested reader is referred to the References for more complete discussions of these matters.

Before continuing our discussion of logical models and their relation to real and mathematical models, we illustrate the notion of a logical model by an example. Our example uses a relatively simple axiom system which was chosen so as to avoid any confusion about the nature of the axioms. We also develop a little of the theory of this system to motivate our later comments on the important relationship between the theory of an axiom system and the logical models of the system. As noted earlier, the theory of an axiom system is the collection of statements which can be logically deduced from the axioms of the system. The development of such a theory (i.e., the discovering of the statements) can proceed in many ways, and it can be expressed in many forms. The most economical form, and at the same time the form least likely to be misinterpreted, is in terms of formal theorems. Thus, if a statement A is to be in the theory determined by Σ, then the statement $\Sigma \Rightarrow A$ (read "Σ implies A") must be logically true. To show that $\Sigma \Rightarrow A$ is logically true in any specific instance, we use the traditional methods of classical logic.

We now proceed with our example.

Example. Let Σ be the axiom system whose undefined terms are *tree* and *fence* and whose axioms are the following:

- A_1: Every fence is a set of trees containing at least two elements.
- A_2: There exist at least three trees.
- A_3: Given any two trees T_1 and T_2, there exists one and only one fence containing them.
- A_4: Given any fence F and a tree T not in F, then there exists one and only one fence F' containing T and disjoint from F (i.e., $F \cap F' = \varnothing$).

Models of Σ

1. Tree will have its usual real-world meaning. A single clump of three trees will then form a logical model of Σ. The set whose three elements are the three trees is defined to be a fence. There is only one fence.
2. Tree again has its usual meaning. This time we have a grove of four trees which we call A, B, C, and D. We define six fences in the following way: Each is a set of two and only two trees, $F_1 = \{A, B\}$,

$F_2 = \{A, C\}$, $F_3 = \{A, D\}$, $F_4 = \{B, C\}$, $F_5 = \{B, D\}$, $F_6 = \{C, D\}$. With these definitions of fence and tree it is easy to see that the axioms are true. To check this, the reader may find it useful to draw some pictures.

3. Tree is now defined to be a point of Euclidean geometry, while a fence is a line. Then the axioms of Σ follow from the axioms of Euclidean geometry.

Model 3 is somewhat different from models 1 and 2 because it uses another axiom system which itself has undefined terms (i.e., the axiom system for Euclidean geometry). Strictly speaking, it is not a model of Σ in the sense of our definition. However, the word model is often used in this way, and we introduced this interpretation of Σ to illustrate this common stretching of our definition. Henceforth, we shall try to stay within the strict framework of our definition and shall use only logical models such as 1 and 2 above.

One might be inclined to guess that the theory associated with such a simple system is trivial and uninteresting. This is not the case, and we shall provide two examples of theorems which are not a priori obvious.

Theorem 1: If there are two distinct fences, then there are three distinct fences.

Note. Two fences F_1 and F_2 are said to be *distinct* if $F_1 \neq F_2$ in the set theoretic sense.

Proof: Let F_1 and F_2 be distinct fences. By Axioms A_1 and A_3 and the meaning of set equality, there exists a tree in each one which is not in the other. Thus suppose that $T_1 \in F_1$, $T_1 \notin F_2$ and $T_2 \in F_2$, $T_2 \notin F_1$. Next, by A_3 there is a unique fence containing T_1 and T_2. Clearly, $F_1 \neq F_3$ since $T_2 \notin F_1$, and $F_2 \neq F_3$ since $T_1 \notin F_2$. Therefore F_3 is a distinct fence.

Q.E.D.

The reader is invited to test his understanding of the proof of Theorem 1 by providing a proof of the following result (Exercise 2).

Theorem 2: If there are three distinct fences, then there are four distinct fences.

It is natural at this point to consider the relationship between the theory (i.e., the set of theorems) of any axiom system and any logical model of that system. This relationship is based on certain general assumptions about models. These assumptions have been called *principles of applied logic* by R. L. Wilder, and they will be basic to our future work. A more complete discussion of these principles can be found in Chap. II of [W].

Principle 1: All the theorems in the theory determined by an axiom system Σ are true in every model of Σ.

Principle 2: The law of contradiction holds for all meaningful statements about each model of an axiom system Σ.

For the reader who is not familiar with the laws of classical logic, the law of contradiction says that a statement S and its negation $\sim S$ cannot both be true.

Using Principles 1 and 2, we are now able to provide an easier method for checking the consistency of an axiom system. Recall that an axiom system is consistent if it does not imply contradictory statements. Suppose that Σ is an axiom system with a model. By Principle 1, if Σ implies two contradictory statements, these statements will both be true about the model. But Principle 2 says that the law of contradiction holds for statements about the model. Hence contradictory statements cannot be true about the model, and Σ must not imply contradictory statements. Summarizing these remarks, we have the important result that *every axiom system which has a logical model is a consistent system.* In particular, the axiom system Σ of the example on trees and fences is a consistent system.

In addition to the result on consistency, the principles of applied logic also yield a result of central importance in the use of axiom systems and models. Once a theorem in the theory determined by a system Σ has been established, then this theorem is a true statement about every model of Σ. Thus in regard to the axiom system Σ of the above example, we see that no logical model of this system can have exactly two fences or exactly three fences. This, of course, is an aid in the search for models of the system since it is clear that any candidate for a model should be rejected if it has either exactly two fences or exactly three fences. A similar situation holds for a general axiom system. In the search for models of the system, one can use as a guide the knowledge that each theorem must be a true statement about each model.

As the reader is no doubt aware, the axiom system Σ involving trees and fences was created with the intention of forming an axiom system that had certain predetermined logical models. It is common to do this, that is, to begin with a specific real situation and to set down statements which describe this situation as accurately as possible. We have discussed this process in Sec. 1.1, and at that time we called the process model building. This brings us to the second (and for us the main) use of the word *model.*

The use of the term *mathematical model* in regard to the idea that we are about to discuss first became common outside of mathematics. It is likely that if mathematicians and logicians had had the prerogative of introducing terminology, then it would have been somewhat different. However, the usage is now well established and we shall make no attempt to change it.

When an investigator forms statements which he feels express basic principles in an area of observation and study, then it is often said that he has formed a model. Actually he has made a major step in forming an axiom system, and it is his hope that this axiom system has a logical model in his area of study. The process of forming this axiom system is called *model building*. The model builder experiments and observes facts about the real world in his area of specialization. He then tries to explain and describe the phenomena that he is studying. He usually does this by proposing certain statements as the ones which are basic and most important. Frequently these statements contain terms which are undefined and which are created to aid in explaining certain observed facts. Together these terms and statements constitute an axiom system. It is this axiom system which we choose to call a mathematical model. The next step in this process is the testing and modifying of the model. This is done by considering the consequences of the axioms which constitute the model and checking whether these consequences are true statements about that part of the real world under study. If they are not true statements, then the model should be changed. Thus model-building is frequently an evolutionary process.

At this point the word *model* has been used in a number of different ways, and it is now time for us to review and relate the different uses of this term. First, a *real model* is a collection of statements about real objects which is obtained by a process of observation, identification, and approximation. A *mathematical model* is an axiom system consisting of undefined terms and axioms which are obtained by abstracting and quantifying the essential ideas of a real model. A *logical model* is an association of real objects with the undefined terms of an axiom system so that the axioms become verifiable statements about the real world. Pictorially, we represent the different uses of the term model in Fig. 1–2. In this figure the solid lines indicate the usual relationships between the different models: The real model is obtained from the real world, the mathematical model is obtained from the real model, and the logical model (if one exists) is again in the real world. The dashed line

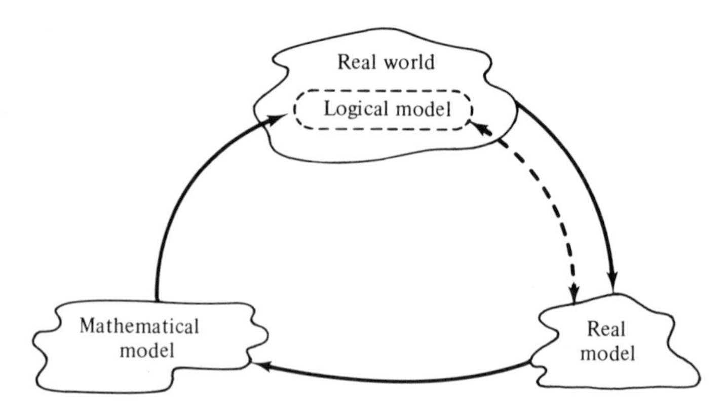

Fig. 1–2

between the real and logical models indicates that often the real world setting that led to the real model is itself a logical model. Thus, the real world setting which motivated the study is a natural place to look for a logical model.

1.3.4 Independence and Equivalence

In their study of general axiom systems (and hence, in particular, of mathematical models) logicians and mathematicians have invented many properties which are of interest in the study of axiom systems. We have already introduced one of these properties, namely the property of consistency. We now consider another such property, and in addition we discuss some relationships which may hold between axiom systems.

Let Γ be an axiom system. We would like to be able to say that every axiom in the system Γ is making a contribution to the theory determined by Γ. Thus we would not want some proper subset of the axioms of Γ to determine the same theory that is determined by Γ. This would certainly happen if, for example, one of the axioms could be derived as a theorem from the others. Hence we are led to consider a property of axiom systems which guarantees that none of the axioms of the system logically follow from the others. Our definition requires the use of models and the notion of the negation of a statement, and it is stated as follows:

Definition: Let A be one of the axioms in the axiom system Γ. We say that A is *independent* in Γ if there exists a logical model of Γ and a logical model of $(\Gamma \setminus A) + (\sim A)$.

With regard to this notation, we remark that $(\Gamma \setminus A) + (\sim A)$ represents the axiom system obtained by deleting axiom A from Γ and adding the negation of A as a new axiom.

Thus, if A is to be shown to be independent in Γ, then two logical models must be produced. In one of them, A and all the other axioms of Γ are true. Hence the other axioms of Γ could not possibly imply $\sim A$. If they did, then Principles 1 and 2 would be contradicted because both A and $\sim A$ would be true in the model. In the second model, all axioms except A are true, and also the statement $\sim A$ is true. Again using Principles 1 and 2, it follows that the other axioms of Γ do not imply A. Therefore, if A is independent in Γ, we know that $\Gamma \setminus A$ does not logically imply A or $\sim A$. In other words, A is making a definite contribution to the theory determined by Γ.

As an example of the property of independence, we return to the axiom system involving trees and fences. Suppose that we add to this system an axiom which says that there are exactly four trees. Then this new axiom is independent in the enlarged axiom system. The independence of this new axiom follows from the fact that we can find a model of the old system which has exactly four trees and we can find another model which does not have four trees.

We conclude this short discussion of independence by pointing out that the independence of the axioms of a system is by no means necessary for the system to be a useful one. Removing a nonindependent axiom from an axiom system does not change the theory of the system. Thus, if the theory is interesting and useful with the axioms all independent, this is still true if one adds a nonindependent axiom, because the theory does not change. One reason for seeking independence is clarity. If all the axioms of a system are independent, then one knows that each axiom is actually contributing to the theory, and changing or deleting any axiom will change the theory.

As a final topic dealing with general axiom systems, we mention a method for comparing certain axiom systems. As usual, we start with a definition so that we may be absolutely clear about the idea that we are considering.

Definition: Let Σ_1 and Σ_2 be two axiom systems with the same undefined terms. We say that Σ_1 is *stronger than* Σ_2 (denoted by $\Sigma_1 \geq \Sigma_2$) if the axioms of Σ_2 are contained in the theory determined by Σ_1. If both $\Sigma_1 \geq \Sigma_2$ and $\Sigma_2 \geq \Sigma_1$, we say that Σ_1 and Σ_2 are *equivalent* (denoted by $\Sigma_1 \cong \Sigma_2$). Finally, if $\Sigma_1 \geq \Sigma_2$ but Σ_2 is not stronger than Σ_1, then we say that Σ_1 is *strictly stronger than* Σ_2 (denoted by $\Sigma_1 > \Sigma_2$).

The utility of these concepts in the study of mathematical models is due to the following observation. Suppose that two investigators each form an axiom system about the same area of study. Suppose further that the undefined terms in one system can be renamed in such a way that both systems involve the same set of undefined terms. This rephrasing may be a difficult process and there are no absolute rules for accomplishing it. However, once it is completed, the concepts of "equivalence," "stronger," and "strictly stronger" may be relevant for comparing the two systems. If it can be shown that the systems are equivalent, then they both yield the same theory, and in this sense they are completely interchangeable. Also, they both have the same logical models. On the other hand, if one system is strictly stronger than the other, then there are two conclusions that the investigators should keep in mind. First, the stronger system has a richer structure, and hence it will be more difficult to find logical models of this system. Second, the stronger system will contain all the theorems of the weaker system, plus some special theorems which do not hold in the weaker system, and hence the theory itself may be richer and more precise.

The following example may help to clarify the notions of strength and equivalence.

Example. Consider two axiom systems each using the undefined terms *movement* and *coordination*. Thus a comparison can be made immediately without any need to rename the undefined terms.

The first system which we denote by Γ consists of four axioms.

A_1: There exists at least one movement.

Definition: Let $\mathfrak{M}$ be the set of all movements.

A_2: Coordination is a function c defined on $\mathfrak{M} \times \mathfrak{M}$ with range in $\mathfrak{M}$. [We write $c(m_1, m_2) = m_1 \, c \, m_2$.]
A_3: For each $m \in \mathfrak{M}$ there exists $i(m) \in \mathfrak{M}$ such that $m \, c \, i(m) = m$.
A_4: For each $m \in \mathfrak{M}$ there exists $m^* \in \mathfrak{M}$ such that $m^* \, c \, m = i(m)$.

The expression $c(m_1, m_2)$ (or $m_1 \, c \, m_2$) may be read "the coordination of m_1 and m_2."

We exhibit a logical model of Γ. Let a and b be two physical locations. For example, a might be New York and b Los Angeles. The movements are the three statements

S_1: Go from a to b.
S_2: Go from b to a.
S_3: Stay where you are.

The coordination function combines these movements in the following manner:

$$\begin{array}{lll} S_1 \, c \, S_1 = S_1 & S_2 \, c \, S_1 = S_3 & S_3 \, c \, S_1 = S_1 \\ S_1 \, c \, S_2 = S_3 & S_2 \, c \, S_2 = S_2 & S_3 \, c \, S_2 = S_2 \\ S_1 \, c \, S_3 = S_1 & S_2 \, c \, S_3 = S_2 & S_3 \, c \, S_3 = S_3. \end{array}$$

The reader can check that with $\mathfrak{M}$ and c so defined the axioms of Γ are true statements, and thus this system provides a logical model for Γ (Exercise 7). For example, to check that axiom A_3 is satisfied, it is necessary to find $i(S_1)$, $i(S_2)$, $i(S_3)$. Inspection of the definition of c shows that one can take $i(S_1) = S_1$ $(S_1 \, c \, S_1 = S_1)$. Note that one could also take $i(S_1) = S_3$. Thus there may be more than one element of $\mathfrak{M}$ which will do for $i(S_1)$, and this is perfectly acceptable.

The second axiom system Δ is defined by five axioms. Two of these, A_1 and A_2, are the same as for Γ; the remaining three are different.

A_1: Same as above.
A_2: Same as above.
A'_3: There exists $i \in \mathfrak{M}$ such that $m \, c \, i = i \, c \, m = m$ for every $m \in \mathfrak{M}$.
A'_4: For each $m \in \mathfrak{M}$ there exists a $m^* \in \mathfrak{M}$ such that $m^* \, c \, m = m \, c \, m^* = i$.
A_5: $m_1 \, c \, (m_2 \, c \, m_3) = (m_1 \, c \, m_2) \, c \, m_3$, for every $m_1, m_2, m_3 \in \mathfrak{M}$.

It is clear that if A'_3 is true, then A_3 is true. Indeed, one can take $i(m) = i$

for all $m \in \mathfrak{M}$. Likewise, if A'_4 is true, then so is A_4, with $i(m) = i$ as above. Thus the axioms of Γ are theorems in the theory of Δ. Using the terminology of the definition, we say that Δ is stronger than Γ. It is natural to ask whether Δ is strictly stronger than Γ or if they are equivalent. To show equivalence, one must prove that $\Gamma \geq \Delta$. As a consequence of equivalence, it would follow that each model of Γ is also a model of Δ. In particular, the model of Γ given above would be a model for Δ. But this model is not a model of Δ (Exercise 9), and consequently the statement $\Gamma \geq \Delta$ is false. Therefore, the axiom system Δ is strictly stronger than Γ.

As a striking and important example of axiom systems for which comparison is a question of considerable significance, we cite the systems commonly known as Newtonian mechanics and relativistic mechanics. Newtonian mechanics (referred to briefly in Sec. 1.2) can be put in the form of a formal axiom system, and the results of nineteenth-century mechanics then appear as theorems in the theory of this system. The changes necessary in Newtonian mechanics due to relativity are not superficial ones dealing only with isolated phenomena, but rather they are fundamental ones dealing with the most basic concepts. One can view the axioms of Newtonian mechanics as in a certain sense being included in those of relativistic mechanics. (Inclusion here does not refer to the ordinary notion. Instead it means that the axioms of Newtonian mechanics are a sort of limit of the axioms of relativistic mechanics.) Also, there are theorems in the theory of the latter system which do not occur in the theory of the former. Thus in our terminology the system of relativistic mechanics is strictly stronger than that of Newtonian mechanics.

EXERCISES

In these exercises Σ is the axiom system involving trees and fences and Γ and Δ are the axiom systems involving movement and coordination.

1. Show that models 1 and 2 of axiom system Σ are actually models of this system. That is, show that each of the axioms is satisfied.
2. Prove Theorem 2 of the example on trees and fences.
3. Add to axiom system Σ the additional axiom that there are exactly five trees. How many different models does this new system possess? First discuss what the term *different* should mean in this case. Perhaps a new term such as *essentially different* should be used instead.
4. Consider the axiom system Σ' consisting of Axioms A_1, A_2, and A_3 of Σ and A'_4: Given any fence F and a tree T not in F, there exists *at least one* fence F' containing T disjoint from F.
   (a) Show that model 2 is a model of Σ'.
   (b) How many different models of Σ' have exactly five trees? Define your use of the term different.

5. Show that if there are no more than eight trees in a model for Σ, then this model contains either one fence or six fences.
6. Find a model for Σ which contains nine trees and twelve fences.
7. Show that the model proposed for Γ is, in fact, a logical model for this axiom system.
8. Show that the axioms of Γ are theorems in the theory of Δ.
9. Show that the model given for Γ is not a logical model for Δ.
10. Find a logical model for Δ.

1.4 A GROWTH EXAMPLE

The reader no doubt recognizes that Sec. 1.3 presents a rather idealistic view of the model-building process. The examples introduced there were selected because of their simplicity and because it was easy to see the ways in which the concepts are related. In practice it is often difficult to decide which are the best choices for undefined terms and axioms, and the distinction between the various kinds of models is blurred. We now consider an example which is intended to convey more about how the notions of Sec. 1.3 are used in practice. It is a short example and still quite simple—indeed, much of the remainder of the book can be thought of as a collection of more significant examples—and it also illustrates how a mathematical model may depend on other axiom systems that are not explicitly listed as a part of the system under study. We shall return to a much more thorough study of the questions raised here in Chaps. 8 and 9.

Example: Cell Growth. A beginning biologist (BB for short) is assigned the task of studying the reproduction of certain types of abnormal cells. She is asked to provide information on two questions about each type of cell. First she is to determine how long it takes a cluster with a given number, N, of cells of a certain type to grow to a prescribed size. Second, she is to determine the rate of growth for the total number of cells of each type.

As a first step in obtaining the desired information, BB examines a laboratory culture which contains cells of a particular type, say type D. She observes that they multiply by dividing and that the process is as follows: Each cell grows for a certain period of time and then divides into two separate cells. Then each of these parts grows to become a complete cell and itself divides into two parts. After observing a number of these growth cycles, BB determines that the time between successive divisions is almost always the same and that the average time between successive divisions is T seconds. She now decides to make an approximation, and her reasoning goes something like this: It will be very difficult for me to deal with the "almost always" and the "average time" of these observations; therefore, I will *assume* that the behavior of the dividing cells is much more regular. She summarizes her approximation in the following way.

Fundamental Law of Type D Cell Division: There is a lapse of T seconds between successive divisions of type D cells.

After having stated her approximations in this precise manner, she proceeds to the next step. She seeks to translate this assumption into numerical terms, and in particular she seeks to determine a function which gives the total number of cells. Since the number of cells changes with the passage of time, the function she seeks will give a number of cells for each time. She denotes this function and its argument by f and t respectively, where t is the elapsed time since the initial observation, measured in seconds. She now tries to express the statement of the Fundamental Law in terms of the function f. She claims that if N is the number present at time 0, which is taken to be immediately after division, then

$$f(0) = N,$$
$$f(T) = 2 \cdot N,$$
$$f(2T) = 4 \cdot N,$$

and, in general,

$$f(kT) = 2^k \cdot N.$$

This gives her the total number of cells at certain times, namely integral multiples of T. Being a clever girl, she recognizes that the last formula can be used to extend the domain of definition of the function f to nonintegral multiples of T as well. Thus she sets $t = kT$ and obtains the following formal expression for f in terms of t:

$$f(t) = 2^{t/T} N.$$

This formal expression has the correct values for all integral multiples of T. She now asserts that her task is essentially complete and that the questions can be answered using this definition of f and elementary calculus.

At this point we stop for a moment and consider the work that BB has done. She has studied a problem, made a simplification, proposed a basic principle, and drawn some initial conclusions. In our terminology she has formed a real model, has started (but not completed) the formulation of a mathematical model, and has indicated how certain conclusions may be drawn. A critical examination of her work will raise a number of grounds for concern about the validity of her results. We invite the reader to criticize her use of the function f to arrive at the answers to the questions asked of her (Exercise 3). However, it is not the specific conclusions of BB that are of interest to us at this point; instead, we are interested in the setting from which the conclusions are drawn. Rather than work in a setting which is partly a real model and partly a mathematical model, we believe that the deductions should be made in the formal setting of a true mathematical model. One such model is the following:

Axiom System Λ

Undefined term: Cell.

Definitions: N is a positive integer.

T is a positive real number.

Time (denoted by t) is any nonnegative real number.

Axioms: A_1: The number of cells present at time t depends on t, T, and N. If this dependence is denoted by the function f, $f(N, T, t) =$ number of cells at time t, then $f(N, T, 0) = N$ and $f(N, T, t + T) = 2f(N, T, t)$ for all t.

A_2: The function $f(N, T, t)$ is differentiable with respect to time.

A_3: There exists a constant k, depending on N and T, such that $df/dt = kf$.

Remarks

1. Axiom system Λ contains much more than might be apparent at first glance. In addition to the obvious statements, other assumptions are included in the definitions where axiom system Λ is tacitly based on properties of real numbers and integers. Since these numbers and their properties are themselves rigorously defined by means of axiom systems, system Λ depends on these other systems and is not completely defined without a knowledge of them. This is a common feature of model building. Frequently one does not state all the basic axioms that will be used in the model. Instead, one often assumes that certain entire axiom systems are well enough known so that it is not necessary to explicitly state the axioms of these systems. Axiom systems which are often taken as known are systems within mathematics. For example, it is common to use facts from set theory, Euclidean geometry, or probability theory or facts about real numbers without explicitly listing the respective axioms.

2. Axiom system Λ contains certain mathematical assumptions, Axioms A_2 and A_3, which were not included in the real model developed by BB. Thus any conclusions which follow from system Λ must be checked to see if they are relevant statements about the original problem. Alternatively, one can try to verify that Axioms A_2 and A_3 are actually valid statements

about the function which describes the growth of the cells. We shall discuss this further after developing a little of the theory of system Λ. The theory developed here consists of three theorems which have been chosen for their possible relationship to the questions asked of BB.

Theorem 1: The function f is given by the formula

$$f(N, T, t) = N \cdot 2^{t/T}. \tag{4.1}$$

Proof: From Axiom A_3 we know that there is a constant k such that $df/dt = kf$. Also, from the axioms we see that f is never zero and that, in fact, f is always positive. Hence, we obtain $(df/dt)/f = k$, and since this holds for all nonnegative values of t, we may integrate both sides of the equality between the limits 0 and t. This gives the relationship

$$\log f(N, T, t) - \log f(N, T, 0) = kt. \tag{4.2}$$

Note. Here and throughout the text the notation $\log x$ is used to denote the natural logarithm of x (i.e., base e). If the logarithm is to be taken to another base, b for example, we shall write $\log_b x$.

Next, in (4.2) we set $t = T$ and recall that $f(N, T, T) = 2N$ and that $f(N, T, 0) = N$. This shows that $k = (\log 2)/T$. Using this value of k in (4.2), we have

$$\log \frac{f(N, T, t)}{N} = \frac{t}{T}(\log 2) = \log 2^{t/T}.$$

This is the same as $f(N, T, t) = N \cdot 2^{t/T}$. Q.E.D.

The important aspect of this theorem is that we have proceeded from the knowledge that the function is reasonably nice (A_2), is related in a simple way to its derivative (A_3), and has some prescribed values for certain times to the explicit form of the function. Such a deduction is quite typical, and we shall encounter similar arguments again. We can now use Eq. (4.1) to compute the amount of time necessary for the number of cells to reach a prescribed size.

Theorem 2: The time required for the total number of cells to increase from N to n is given by

$$t = T \cdot \log_2 \left(\frac{n}{N}\right).$$

Proof: Using Theorem 1 and Axiom A_1, we know that we must find a value of t such that $f(N, T, t) = n = N \cdot 2^{t/T}$. Taking logarithms to base 2 of both sides gives $\log_2(n) = \log_2(N) + \log_2(2^{t/T})$. By the basic properties of logarithms, this is the same as $t = T \cdot \log_2(n/N)$. Q.E.D.

Since we know the values of the function $f(N, T, t)$ for all values of t, Axiom A_3 guarantees that we know the rate of growth of the number of cells for all times.

Theorem 3: At time t, the instantaneous rate of growth of the total number of cells is given by

$$\left(\frac{N}{T}\right)(2^{t/T})(\log 2)$$

in units of cells per second.

Proof: We recall from calculus that the derivative of a function gives the rate of change of that function at the point where the derivative is calculated. Thus the instantaneous rate of growth of the total number of cells at time t is given by

$$\frac{df(N, T, t)}{dt} = \frac{d[N \cdot 2^{t/T}]}{dt} = \left(\frac{N}{T}\right)(2^{t/T})(\log 2). \qquad \text{Q.E.D.}$$

If we recall the discussion preceding this example, it would certainly seem that Theorems 1, 2, and 3 should shed some light on the questions that BB was to answer. In fact, there seems to be a good chance that these theorems answer the questions since the theorems are phrased in a way to provide the appropriate information. However, this appearance is somewhat deceptive, and one must take care in drawing conclusions about the actual growth of type D cells from these theorems. We have already noted that some of the axioms of system Λ are not known to be true for the function which gives the actual number of cells at time t. Hence there is no reason to believe in advance that Theorems 1, 2, and 3 are accurate statements about this function. These theorems must be checked and verified before they can be used with confidence to predict the growth of type D cells.

As an alternative to checking the theorems for accuracy, one can instead check the axioms. If Axioms A_1, A_2, and A_3 can all be shown to be true statements about type D cell growth, then type D cell growth is a logical model for system Λ, and Theorems 1, 2, and 3 are correct statements about this growth. Intuitively, it might seem that there is no chance that Axioms A_2 and A_3 are correct statements about the function which actually describes type D growth since these axioms state that the function f is differentiable. This, in turn, means that f is continuous because every differentiable function is continuous. But how can f be continuous when it measures the total number of cells and this number must change suddenly whenever a cell divides? The answer lies in the phrase "total number of cells." Perhaps the total number of cells is determined by weighing the culture which contains the cells and dividing by an average cell weight. In this way, it would make sense to have a total number of cells which was not an integer. Also, it is

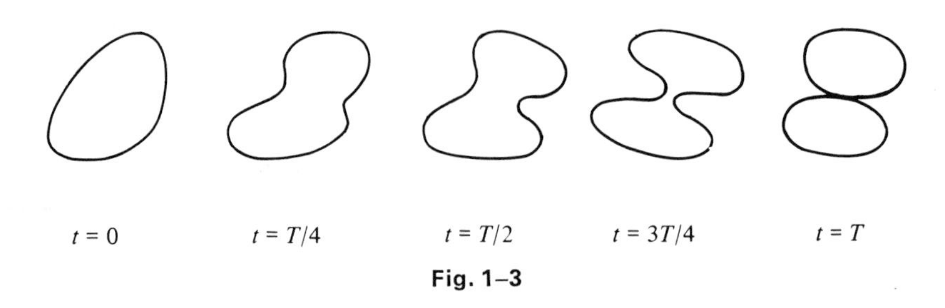

Fig. 1–3

very likely that the process of cell division is not really as sudden as it has been depicted. Perhaps the cells begin to divide well before the actual split takes place and pictorially the division looks like Fig. 1–3. In this case the process seems continuous, and a continuous function f is not unreasonable. Similarly, it is possible to justify Axiom A_3 on the basis of measurements, and hence there is hope that system Λ is a mathematical model for type D cell growth. Of course, this does not make it a good model. Many assumptions have been made and these would have to be checked. This checking must ultimately be done by comparing the results predicted by the theory of Λ with actual observations. If the model does not make accurate predictions about the cell growth, then the assumptions of the model must be questioned. Perhaps too great an approximation has been made in Axiom A_1. It may be that variations in the dividing times must be considered in the model. One method for doing this is considered in the Project of Sec. 2.7.1, and the discussion is continued in Chap. 8, Part B.

Remarks. We conclude this section and the chapter with a few remarks intended to put the formal and practical aspects of mathematical model building into proper perspective. We have been quite formal in our introduction to mathematical models, and one could conclude from our discussion that either a mathematical model must be perfect (i.e., it has the appropriate logical model in the real world) or else it is useless. This is not at all the case. It often happens that a mathematical model is very useful even though it does not have an appropriate logical model in the real world. Instead, it *almost* has an appropriate logical model. Here *almost* can mean many things depending on the model being considered. It may mean that the mathematical model is true in most real cases but not in all, or it may mean that even though the mathematical model is never precisely correct, it is almost correct all the time. As an illustration of these vague ideas we consider again the cell-counting problem of this section. It is quite likely that the main axiom on cell growth (cell divisions are T seconds apart) is not really true about every cell. However, it may be true for most cells, and for the other cells the difference between the actual dividing time and T may be small enough so that the predictions of the model are quite accurate and useful. Also, it may be that the axiom is never really true (no cell divides after exactly T seconds) but instead that the dividing time of every cell is very close to T, and again the predictions of the model are reasonably close to the true situation. Thus a model may be quite useful even though it is known that the model is not perfect.

The problem of evaluating mathematical models is a complex one, and it is not the main topic of this text. However, it is relevant to our principle theme, and we do consider this very important aspect of model building in Chap. 10.

EXERCISES

In Exercises 1–3, f is the function obtained by BB in her study of cell growth.

1. What is the value of t for which $f(N, T, t)$ equals 10^5? Express your answer in terms of N and T and evaluate this answer for $N = 10^2$ and $T = 10$ minutes.
2. What is the smallest integer N such that $f(N, T, 10^2)$ exceeds 10^5? Express your answer in terms of T and evaluate this answer for $T = 10$. Assume that T and t are measured in the same units.
3. Give reasons the function f may not yield the correct answers to the questions asked. In particular, criticize the derivation of the function f.
4. Find and describe mathematical models which are used in the following fields: (a) biology, (b) chemistry, (c) psychology, and (d) business. State the undefined terms and the axioms as clearly as possible.
5. Find an area in which two contradictory models are used in an attempt to understand a single phenomenon. (Two models are said to be contradictory if one of them implies a statement A while the other implies $\sim A$.)

REFERENCES

General comments on the role of model building in the social and life sciences can be found in Chap. I of [KS] and in several articles in [SW]. In the same vein, although occasionally more mathematically demanding, are several articles in [COSRIMS], particularly those by S. Ulam, J. G. Kemeny, L. R. Klein, Z. Harris, and H. Cohen. A discipline-oriented (mathematical psychology) discussion containing additional comments on parameter estimation can be found in the introduction of [ABC].

Chapter II of [W] and Chaps. 1 and 3 of [S] are very relevant to the topics that we have discussed in this chapter. For a general survey of the nature of mathematics, [BP] is appropriate. This contains a series of survey articles by prominent mathematicians and logicians concerning their views of mathematics. From a historical point of view, [N] is a very important work. These four volumes reproduce many fundamental papers by leading mathematicians and scientists, and many of the papers involve the construction of a mathematical model. A more specialized work concerned with this topic is [F], another collection of articles; we especially recommend to the reader the article by Suppes.

[ABC] ATKINSON, R. C., G. H. BOWER, and E. J. CROTHERS, *An Introduction to Mathematical Learning Theory.* New York: Wiley, 1965.

[BP] BENAURRAF, P., and H. PUTNAM, *Philosophy of Mathematics, Selected Readings.* Englewood Cliffs, N. J.: Prentice-Hall, 1964.

[COSRIMS] The National Research Council's Committee on Support of Research in the Mathematical Sciences, ed., *The Mathematical Sciences.* Cambridge, Mass.: M.I.T. Press, 1969.

[F] FREUDENTHAL, H., ed., *The Concept and the Role of the Model in Mathematics and Natural and Social Sciences.* New York: Gordon & Breach, 1961.

[KS] KEMENY, J. G., and J. L. SNELL, *Mathematical Models in the Social Sciences.* Waltham, Mass.: Ginn/Blaisdell, 1962.

[N] NEWMAN, J. R., *The World of Mathematics.* New York: Simon and Schuster, 1956.

[S] SAATY, T. L., *Mathematical Methods of Operations Research.* New York: McGraw-Hill, 1959.

[SW] SAATY, T. L., and F. J. WEYL, *The Spirit and the Uses of the Mathematical Sciences.* New York: McGraw-Hill, 1969.

[W] WILDER, R. L., *Introduction to the Foundations of Mathematics*, 2nd ed. New York: Wiley, 1965.

2 Model Construction: Selected Case Studies

2.0 INTRODUCTION

We repeat here a remark made in the preface: Skill in constructing and interpreting mathematical models is gained only by actively participating in their creation. There is no easy formula or well-defined algorithm which tells you how to construct a model for a given situation. Indeed, the ideas and techniques which are useful for investigating one problem may be useless for studying the next. Also, the very nature of the process may seem somewhat mysterious to one who has not thought of it in exactly this way before. Therefore, in this chapter the reader is presented with several instances in which the model-building process has proved to be useful. The examples have been selected to give an indication of the range of application of the concept and also to illustrate the various mathematical ideas which may enter. Although reading accounts of model building is not sufficient, it is a reasonable beginning. There are exercises, ranging from very easy to quite difficult, which give the reader an opportunity to get his feet wet in the subject.

This chapter presents a number of short case studies which are examples of model building in the spirit of Chap. 1. In the early sections we pay close attention to identifying concepts to be taken as undefined terms, to constructing axioms, and to the other aspects of formal model building. However, as we mentioned in the preceding chapter, the literature contains many instances of model building which are not formalized. In fact this informal approach is traditionally used to study some of the models of this text, and we adopt this approach where it seems appropriate. Consequently in the later sections of this chapter we pay less attention to formalizing the process in the text and we leave this to the reader in the exercises.

Also, it will become apparent to the reader that the models constructed here serve two somewhat different purposes. Some models are primarily designed to assist in understanding the basic processes at work in the situation being investigated. Sec. 2.1 and 2.4 are of this type. Other models are intended to aid in studying rather specific problems, as in Secs. 2.2 and 2.6. Still others serve a combination of both purposes.

Finally, the material of this chapter may be considered as motivation for much of what follows. In later chapters we shall frequently return to ideas introduced here. At that time we shall provide more detail and, frequently, much more generality. Also, there are questions introduced here to which we do not return but for which one can proceed in much the same manner as is done with the others. We shall provide references and we encourage the reader to try his hand.

2.1 PEAS AND PROBABILITIES

In 1866 an Augustinian monk, who had previously failed the biology portion of the examination for a teaching license, published the results of his research into plant hybridization. The fundamental nature of his results went unnoticed for more than 30 years, and it was not until they had been rediscovered several times that their significance was appreciated. The monk was Gregor Mendel and his results, Mendel's laws, are basic to the science of genetics. Our interest in his studies rests on his use of mathematics to explain and predict observations. His was one of the earliest uses of the concept of a mathematical model in the biological sciences. We shall not, however, simply survey or summarize Mendel's work, although it is a fascinating story. It is our intention to stress a certain point of view and to attribute goals to the experimenter that cannot be historically supported. Also, for our purposes it is convenient to consider a somewhat simpler experiment than that conducted by Mendel. Thus we shall introduce a fictitious experiment conducted by an equally fictitious scientist to illustrate those aspects of Mendel's work which are important for us.

2.1.1 The Problem

A certain biologist (we refer to him as Dr. Monk) was concerned with the problem of accurately predicting the reproductive properties of a certain plant, say the pea plant. The very simple observation which motivated his question was the following. He selected two apparently identical peas from plants which appeared identical, or even from the same plant, and, after they matured, planted them and observed the results. Even though all external conditions were the same for the two plants, they did not produce identical peas. Dr. Monk wished to understand this phenomenon.

2.1.2 A Real Model

As a first step in simplifying the problem, Dr. Monk decided to concentrate on only two aspects of the peas, color and texture. His decision was to separate the peas into two groups according to color, green and yellow, and into two groups according to texture, smooth and wrinkled. He idealized the situation by assuming that he could always determine exactly which characteristics a specific pea exhibited. Therefore each pea could be assigned to exactly one of the following classifications, which we shall refer to as *types*. We shall use the abbreviations as indicated.

Green–Smooth (G–S)	Green–Wrinkled (G–W)
Yellow–Smooth (Y–S)	Yellow–Wrinkled (Y–W)

He next began a series of experiments involving successive generations of plants. His procedure was as follows. A seed of a certain type was planted and the plant that grew from it was permitted to reproduce by self-fertilization (this is biologically possible). The resulting peas were collected, identified as to type, planted, and the process continued. The data collected at each stage regarding the peas of each generation can be arranged in a set of sequences of types, each sequence consisting of the types of a line of direct descendants of the original pea. If a pea is such that the sequences of types arising in this manner all contain only a single type, then the pea is said to belong to a *pure line*. That is, if a pea of a certain type is a member of a pure line, then all its descendants (remember that reproduction is by self-fertilization) are of the same type.

Dr. Monk continued his study by considering cross fertilization between different pure lines. His experiments indicated that if the original seeds differed in some characteristic, then all first-generation peas were of the same type. However, the first-generation seeds, when allowed to reproduce by self-fertilization, did not reproduce in a consistent manner. For example, if the original parents were of types G–S and G–W, then all first-generation descendants were of type G–S. However, self-fertilization of the first-generation descendants produced second-generation peas of both types, G–S and G–W. Moreover, the number of descendants of G–S type outnumbered those of G–W type by approximately 3 to 1.

One of the conclusions of this experiment, that G-S seeds could have either G–S or G–W descendants, indicated that superficial appearance did not completely determine the results of reproduction by self-fertilization. This led Dr. Monk to believe that his notion of type based on appearance only was inadequate. Accordingly, he decided to introduce the term *phenotype*, to refer to what had previously been known simply as type, and to also consider the possibility of further notions of type.

It appeared that the seeds carried with them some undetectable units (on a microscopic scale) which also entered into the reproductive process. Whatever behavior these units exhibited, they should account for the 3 to 1 distribution observed in the above experiment. He felt that further experimentation would be useful, and he continued the cross-fertilization studies previously mentioned. He had already observed that the first-generation seeds with phenotype G–S yielded seeds with phenotypes G–S and G–W under self-fertilization. The additional experiments permitted only self-fertilization. He observed that the G–W seeds produced in the first generation reproduced as a pure line with G–W phenotype. Also, of the G–S seeds produced in the first generation, approximately one third reproduced as a pure line with G–S phenotype, and the remaining two thirds reproduced as if a pure-line G–S seed had been cross-fertilized with a pure-line G–W seed. That is, three fourths of the seeds were of phenotype G–S and one fourth of phenotype G–W. Note that throughout these experiments he was dealing only with green peas, so that texture was the only variable. He now made a major creative step. Concentrating on the texture characteristic, he imagined that each cell of each pea, other than those actually involved in reproduction, carried along with it a fundamental unit which determined the texture of its descendants. The fundamental unit consisted of a pair of *genes*, if we adopt standard terminology. Cells actually involved in reproduction carry only one gene, but more of this later. He assumed that, since there were two alternative forms for the texture, there were two alternative forms for the associated genes. During reproduction, the pair of genes is split and one of them is contributed to the offspring. Stated with somewhat more detail, the assumption is that the reproductive cells, known as *gametes*, each carry only one gene (remember we are talking about a single characteristic) and that that gene is selected arbitrarily from the pair of the respective parent. The cell created during reproduction receives two gametes, one selected arbitrarily from each parent. Each cell could then be classified according to its genetic composition, and since every cell of a plant contains exactly the same gene structure, the entire plant can be so classified. The genetic nature of a plant is known as its *genotype*.

To connect this conjecture with observation, it was necessary to propose a connection between genotype and phenotype. On the basis of the experimental evidence, he proposed that a plant would bear smooth peas if its seeds acquired a gene associated with this form of the characteristic from either the male or female gamete, while it would bear wrinkled peas only if its seed acquired a gene associated with this form from both gametes. This phenomenon of an individual exhibiting that form of a characteristic associated with one form of a gene even though both forms are present is known as *dominance*. In this case we would say that smoothness is the *dominant* form of texture. The remaining one of each pair of alternative forms of a characteristic is known as *recessive*. Finally, we remark that with this understanding of the words, a knowledge of the genotype of an individual and the dominance relations is sufficient to determine its phenotype. The converse does not hold.

The biological situation presented here is admittedly somewhat oversimplified for purposes of illustration. However, in this context, the assumptions and approximations do amount to the formation of a real model. We shall now try to make the whole process more precise and more abstract. Our goal is to identify those aspects which seem to shed the most light on the underlying processes.

2.1.3 A Mathematical Model

The discussion given above in connection with the formation of a real model leads us to believe that the notion of gene is an important one. Also, since for the moment we are considering a single characteristic (texture) which has two alternative forms (smooth and wrinkled), the gene associated with texture can take on one of two alternative forms. These two forms are known as *alleles*, and we denote the alleles associated with the characteristic of texture by A and a. Dr. Monk conjectured that genes occur in pairs, and thus we are led to consider the set of pairs AA, Aa, aa. If we agree that the alleles denoted by A and a correspond to smooth and wrinkled forms, respectively, then (using the conjectured dominance relation) peas with genes AA or Aa will be smooth and those with genes aa will be wrinkled.

We now consider the relation between the genetic makeup of the parents and that of the offspring. Each reproductive cell or gamete receives one gene selected at random from the two available from the appropriate parent. Thus, if the parents have genes AA and aa, then each cell resulting from the mating will have genes Aa. Also, if the parents have genes AA and Aa, then half of the offspring will have genotype AA and half will have Aa. Let us now make this more precise.

We may continue to think of a gene as a fundamental unit which determines heredity, but formally we take it to be an undefined term.

Axiom 1: Each gene occurs in two forms (alleles), denoted by A and a.

Definition: $U = \{AA, Aa, aa\}$.

Thus U is a set consisting of three elements, and each element of U is a symbol consisting of a pair of letters. The letters used are those which represent the alternative forms of genes.

Definition: Reproduction is a function from $U \times U$ to R^3.

Notation. Let r denote the reproduction function. Then $r: (u_1, u_2) \longrightarrow (r_1, r_2, r_3) \in R^3$ or $r(u_1, u_2) = (r_1, r_2, r_3)$. The real numbers r_1, r_2, r_3 are the values of the coordinate functions associated with r.

Definition: For each $u \in U$ and $\alpha \in \{A, a\}$, let $p(\alpha; u)$ be the probability that α is selected if a random selection is made between the two letters which make up the symbol u, each choice assumed equally likely.

It is clear that $p(\alpha, u)$ is given by the following table:

	$p(\alpha, u)$	u: AA	Aa	aa
α	A	1	$\frac{1}{2}$	0
	a	0	$\frac{1}{2}$	1

We are now able to state an axiom which contains the essential assumptions regarding reproduction. This axiom forms the heart of our model.

Axiom 2: The reproduction function satisfies

$$r_1(u_1, u_2) = p(A; u_1)p(A; u_2)$$
$$r_2(u_1, u_2) = p(A; u_1)p(a, u_2) + p(a; u_1)p(A; u_2)$$
$$r_3(u_1, u_2) = p(a; u_1)p(a; u_2).$$

We now have enough structure to begin to develop a theory. The first result is the following.

Theorem 1: The range of the reproduction function is a set of probability vectors.

Recall that a vector $\mathbf{x} = (\xi_1, \ldots, \xi_n)$ in R^n is a probability vector if $0 \leq \xi_i \leq 1$, $i = 1, 2, \ldots, n$ and $\sum_{i=1}^{n} \xi_i = 1$. The proof of the theorem is a straightforward computation. Using the definition of reproduction function, the associated notation, and Axiom 2, we have

$$\begin{aligned} r(u_1, u_2) = (&p(A; u_1)p(A; u_2), p(A; u_1)p(a; u_2) \\ &+ p(a; u_1)p(A; u_2), p(a; u_1)p(a; u_2)). \end{aligned}$$

Also, by definition $p(\alpha, u)$ is a probability and $p(a; u) + p(A; u) = 1$ for each $u \in U$. Therefore each coordinate of $r(u_1, u_2)$ is nonnegative, and the sum of the coordinates is

$$\begin{aligned} p(A; u_1)(p(A; u_2) + p(a; u_2)) + p(a; u_1)(p(A; u_2) + p(a; u_2)) \\ = p(A; u_1) + p(a; u_1) = 1. \end{aligned}$$

The coordinates must therefore each be no larger than 1. The proof is complete.

We now connect this system to the real model discussed above. Recall that if the parents are of genotype AA and Aa, respectively, then the gametes of one parent carry gene A with probability 1, and the gametes of the other parent carry gene A with probability $\frac{1}{2}$ and gene a with probability $\frac{1}{2}$. Thus the offspring will be of genotype AA with probability $\frac{1}{2}$ and genotype Aa with probability $\frac{1}{2}$. One method of indicating this fact is to assign to the element (AA, Aa) of $U \times U$ the vector $(\frac{1}{2}, \frac{1}{2}, 0)$. The general interpretation of the reproduction function follows from this example. Namely, the first coordinate $r_1 = r_1(u_1, u_2)$ of $r = r(u_1, u_2)$ is to be interpreted as the probability that the genotype of an offspring of parents with genotypes u_1 and u_2 will be AA. The second and third coordinates, r_2 and r_3, are to be interpreted as the probability that its genotype will be Aa and aa, respectively.

For this interpretation of reproduction to be reasonable, it is necessary that the genetic character of the offspring be unchanged under a permutation of the parents. This is indeed the case and we state the result formally.

Theorem 2: If $u_1, u_2 \in U$, then $r(u_1, u_2) = r(u_2, u_1)$.

The proof is again straightforward and the details are left as an exercise (Exercise 1).

We return to the interpretation introduced above and compare the line of descendants generated by the model with the actual sequence of descendants observed in the laboratory. The results of such a comparison indicate the validity of the model as an abstraction of the reproduction process. Consider the case discussed above in which two pure lines were crossed initially and the descendants reproduced by self-fertilization. The parents have genotypes AA and aa, and $r(AA, aa) = (0, 1, 0)$. Thus the first-generation descendants all have genotype Aa. If the alleles denoted by A and a correspond to smooth and wrinkled forms, respectively, and if the smooth form dominates, then all first-generation descendants will be smooth. Recall that this was observed. Next consider reproduction of the first generation by self-fertilization. Using the model, we compute $r(Aa, Aa) = (\frac{1}{4}, \frac{1}{2}, \frac{1}{4})$.

The information obtained so far with our interpretation of the terms can be summarized as follows:

Parent generation	AA, aa.
First-generation descendants	All Aa.
Second-generation descendants	$\begin{cases} \frac{1}{4} \text{ of population will be } AA. \\ \frac{1}{2} \text{ of population will be } Aa. \\ \frac{1}{4} \text{ of population will be } aa. \end{cases}$

Alternatively, this information can be presented as in Fig. 2–1, where the numbers indicate the probability that in a specific case an offspring with the given genotype results.

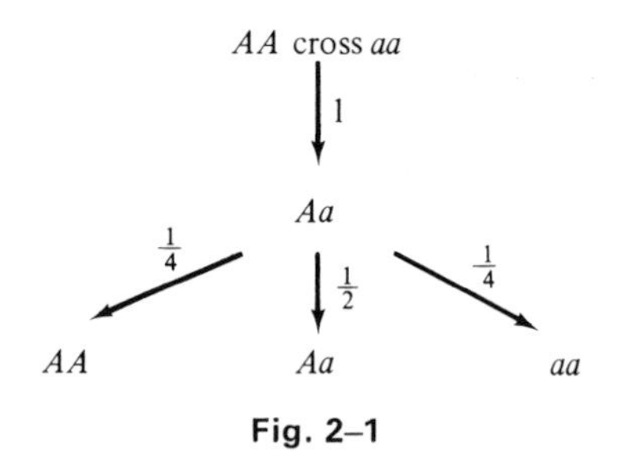

Fig. 2–1

The results predicted above seem to agree with the observations of Dr. Monk, and they have, in fact, been verified many times. The statistics resulting from Mendel's original work can be found in the books cited in the References. Another way of saying this is that the real model constructed above is, insofar as we have proceeded, a logical model for our axiom system.

It should be remembered throughout the entire discussion that any results derived from the axioms and definitions using legitimate mathematical arguments are valid results. Whether they are useful, or suitable, when interpreted in the real model is an independent question. It is, of course, related to the question of whether the real model is a logical model for the axiom system.

Finally, let us return to the original problem where Dr. Monk was concerned with two characteristics, texture and color. Suppose that the genes associated with texture are denoted as above and that those associated with color, which also occur in two alternative forms, are denoted by B and b. Suppose that B is associated with yellow peas and that b is associated with green ones. Assume that yellow is the dominant form. Now, instead of three possible genotypes for any individual, there are nine different ones. The development of the appropriate definitions and axioms for this system is left to the reader.

EXERCISES

1. Prove Theorem 2.
2. Consider the case of one characteristic with initial cross fertilization between individuals with genotypes *Aa* and *aa*. Find a table and diagram similar to Fig. 2–1 in this case.
3. Continue the table and diagram in Fig. 2–1 for one more generation (self-fertilization).
4. Formulate appropriate definitions and axioms for the case of two distinct characteristics, each with two forms.
5. Let the alleles for the two characteristics of Exercise 4 be denoted by A, a and B, b, respectively. Suppose that plants with genotypes (*AA*, *Bb*) and (*Aa*, *BB*) are crossed and that the first-generation descendants reproduce by self-fertilization. Find a table and diagram analogous to that of Fig. 2–1 for this situation.
6. (Hardy-Weinberg law) Consider a large population in which individuals mate completely randomly and a characteristic of these individuals which occurs in two forms. Let the alleles associated with this characteristic be denoted by A and a. Suppose that in the ith generation, $i = 0, 1, 2, \ldots$, the proportions of the three genotypes *AA*, *Aa*, and *aa* are p_i, $2q_i$, and r_i, respectively; $p_i + 2q_i + r_i = 1$. Using the model developed in this chapter, show that

   (a) $p_1 = (p_0 + q_0)^2$, $2q_1 = 2(p_0 + q_0)(q_0 + r_0)$, $r_1 = (q_0 + r_0)^2$.

   (b) $p_2 = p_1$, $q_2 = q_1$, $r_2 = r_1$.

   Thus the distribution of genotypes established in the first filial generation is preserved in the second filial generation and consequently in all following generations. That is, in the absence of external factors, there is no inherent tendency for genetic differences to disappear; in fact, the genetic composition of the population is stable after at most two generations.

2.2 MOVING MOBILE HOMES

In this section we shall consider a situation in which the basic concern is with scheduling or allocation. A general theory of such decision processes, which will include a consideration of problems similar to this one as special cases, will be developed in Chaps. 4, 5, and 7. Although it will be necessary to develop some new mathematical ideas for the general study, we shall be able to handle the necessary technical details of this problem using only basic algebra.

2.2.1 The Situation

Mr. Wheeler owns two factories at which he assembles mobile homes, and in addition he operates three regional distribution centers at which the homes are sold. Over a period of time he has kept careful records, and consequently he has a good idea of about how many homes will be sold each month at each of the centers. Also, he is able to estimate production rates, and therefore he knows (at least approximately) how many homes will be available each month at the factories. Finally, he is able to obtain commitments from mobile home movers to ship homes from his factories to his lots at a specific rate. The rate, of course, varies with the origin and destination of the home. Mr. Wheeler's concern is to determine the most economical allocation of homes from his factories to the distribution centers. This is to be interpreted as minimizing the shipping costs while fulfilling demand.

2.2.2 A Real Model

As a first step in his study of the situation, Mr. Wheeler examines the data which he has collected, and he concludes that he can estimate the supply and demand accurately enough to assume they are known quantities. Of course, he does not know exactly what supply and demand will be, and if the range of possibilities is large, then it might be an oversimplification to assume that they are known exactly (see Sec. 4.4). Similar comments apply to the shipping costs, which for this model we assume to be known constants. Thus we take the following information as known:

d_i = demand at distribution center i, $i = 1, 2, 3$.
s_i = supply at factory i, $i = 1, 2$.
c_{ij} = cost of moving one mobile home from factory i to distribution center j, $i = 1, 2; j = 1, 2, 3$.

The next matter to be considered is the cost of moving several homes. Although there may be minor efficiencies in moving several homes over the same route, it seems to be a reasonable approximation to assume that the cost of moving M mobile homes over a route is M times the cost of moving one home over that route. The total cost of the project is taken to be the sum of the costs of moving homes over each route.

In this work Mr. Wheeler has created a real model. He has simplified his problem by ignoring certain features of the real world, e.g., the uncertainties of supply and demand, and by making assumptions regarding others, e.g., the way in which the total transportation cost was obtained. He hopes that with these simplifications he can determine the least expensive shipping schedule. The merits of his conclusions ultimately rest on the validity of the assumptions.

Let us now consider the situation in more detail. The problem involves six unknown quantities, the number of homes to be shipped from each of the two factories to each of the three stores. If we let f_{ij} be the number of homes to be moved from factory i to distribution center j, then $f_{11}, f_{12}, f_{13}, f_{21}, f_{22}$, and f_{23} are to be determined. The total number of homes sent to center j is $f_{1j} + f_{2j}$. Since it is required that the number of homes sent to each center satisfy the demand there, we have the inequalities

$$\begin{aligned} f_{11} + f_{21} &\geq d_1 \\ f_{12} + f_{22} &\geq d_2 \\ f_{13} + f_{23} &\geq d_3. \end{aligned} \tag{2.1}$$

Also, the total number sent from each factory must not exceed the number produced there; hence we have

$$\begin{aligned} f_{11} + f_{12} + f_{13} &\leq s_1 \\ f_{21} + f_{22} + f_{23} &\leq s_2. \end{aligned} \tag{2.2}$$

At this point it is clear that the problem has a natural side condition. If the demand at each center is to be met, then there must be at least $d_1 + d_2 + d_3$ homes available at the factories. Thus the condition $d_1 + d_2 + d_3 \leq s_1 + s_2$ is a necessary one for the problem to be solvable. In this section we actually choose to assume that $d_1 + d_2 + d_3 = s_1 + s_2$. In Chaps. 4 and 7 we shall drop this assumption. In those chapters we shall also consider larger numbers of factories and centers.

Finally, we consider the quantity in which Mr. Wheeler is most interested, namely the cost. With each set of f_{ij} one can associate a total moving cost C. For notational convenience we let $\mathbf{y} = (f_{11}, f_{12}, f_{13}, f_{21}, f_{22}, f_{23})$. Note that $\mathbf{y}$ is a vector and therefore the order of the f_{ij} is crucial. In view of the assumptions regarding moving costs, we have

$$C(\mathbf{y}) = c_{11}f_{11} + c_{12}f_{12} + c_{13}f_{13} + c_{21}f_{21} + c_{22}f_{22} + c_{23}f_{23}. \tag{2.3}$$

The problem can now be phrased in precise terms in the following way:

Find numbers $f_{11}, f_{12}, f_{13}, f_{21}, f_{22}, f_{23}$ such that the inequalities (2.1) and (2.2) are satisfied and so that the value of the function C defined by (2.3) is as small as possible. A problem of this form is known as a *two-by-three* transportation problem.

In general, the systems (2.1) and (2.2) are true inequalities. However, here we made a special assumption that supply is equal to demand, and using this, we can show that (2.1) and (2.2) are actually equalities. Indeed, adding the inequalities in system (2.1) and those in (2.2), we obtain

$$f_{11} + f_{12} + f_{13} + f_{21} + f_{22} + f_{23} \geq d_1 + d_2 + d_3$$

and

$$f_{11} + f_{12} + f_{13} + f_{21} + f_{22} + f_{23} \leq s_1 + s_2$$

Now, $s_1 + s_2 = d_1 + d_2 + d_3$, and thus each of these inequalities must be an equality, and consequently each of the inequalities in (2.1) and (2.2) must be an equality as well.

As a final side condition on our problem, note that each of the f_{ij} must be a nonnegative integer. That each must be an integer is clear; we do not want to break a mobile home into pieces. That each must be nonnegative reflects the fact that homes are to move from factories to distribution centers and not in the other direction.

2.2.3 A Mathematical Model

At this point, we could proceed to study the situation in its present formulation. However, we prefer first to formulate a mathematical model in quite precise terms. This model will be generalized and studied in some depth in subsequent chapters.

A mathematical model for this situation should incorporate the concepts identified by Mr. Wheeler as being important ones, and it should also incorporate the assumptions and simplifications introduced in the real model. The quantities and concepts appearing in the mathematical model will be abstractions of their counterparts in the real model. Thus we now replace the commonplace notions of supply and demand by an undefined term called a data vector. Similarly, the shipping schedule is replaced by an allocation vector, and moving costs are replaced by a cost vector. These terms and the relationships between them are governed by certain axioms, and the theory is obtained from them by using standard mathematical arguments. The object of the study is to obtain information on a quantity which represents the total cost of moving mobile homes. One such mathematical model is the following:

Undefined terms: data vector, allocation vector, cost vector.

Axioms

A_1: A data vector is an ordered set of five nonnegative integers $\mathbf{d} = (s_1, s_2, d_1, d_2, d_3)$.

A_2: $s_1 + s_2 = d_1 + d_2 + d_3$.

A_3: An allocation vector is an ordered set of six nonnegative integers $\mathbf{y} = (f_{11}, f_{12}, f_{13}, f_{21}, f_{22}, f_{23})$.

A_4: A cost vector is an ordered set of six nonnegative real numbers $\mathbf{c} = (c_{11}, c_{12}, c_{13}, c_{21}, c_{22}, c_{23})$.

A_5: A data vector and an allocation vector satisfy the following equalities:

$$f_{11} + f_{21} = d_1, \quad f_{12} + f_{22} = d_2, \quad f_{13} + f_{23} = d_3,$$
$$f_{11} + f_{12} + f_{13} = s_1, \quad f_{21} + f_{22} + f_{23} = s_2.$$

Definition: The *total transportation cost* associated with the cost vector $\mathbf{c}$ and the allocation vector $\mathbf{y}$ is the number $\mathbf{y} \cdot \mathbf{c}$ defined by

$$\mathbf{y} \cdot \mathbf{c} = c_{11}f_{11} + c_{12}f_{12} + c_{13}f_{13} + c_{21}f_{21} + c_{22}f_{22} + c_{23}f_{23}.$$

Using this mathematical model of the situation, the question facing Mr. Wheeler can be stated as follows: Given a data vector $\mathbf{d}$ and a cost vector $\mathbf{c}$, find an allocation vector $\mathbf{y}$ which makes the total transporation cost as small as possible.

We now solve Mr. Wheeler's problem in the context of the model which he developed. To begin our search for the desired allocation vector $\mathbf{y}$, we examine the equations that the coordinates of $\mathbf{y}$ must satisfy. These equations can be written in the form

$$\begin{array}{llllllll}
f_{11} & & & + f_{21} & & & = d_1 \\
& f_{12} & & & + f_{22} & & = d_2 \\
& & f_{13} & & & + f_{23} & = d_3 \\
f_{11} & + f_{12} & + f_{13} & & & & = s_1 \\
& & & f_{21} & + f_{22} & + f_{23} & = s_2.
\end{array}$$

Axiom A_2 can be used to show that these equations are not independent. In fact, it is clear that the last equation is the sum of the first three minus the fourth. This means that it is sufficient to solve only the first four equations as a system. With this point of view we now have two extra unknowns. We consider f_{22} and f_{23} as parameters, and we solve for the other unknowns in terms of f_{22} and f_{23}. If we carry out the necessary algebra, we obtain

$$\begin{aligned} f_{11} &= d_1 - s_2 + f_{22} + f_{23} \\ f_{12} &= d_2 - f_{22} \\ f_{13} &= d_3 - f_{23} \\ f_{21} &= s_2 - f_{22} - f_{23}. \end{aligned} \tag{2.4}$$

Next, we use the information that each of the f_{ij} must be nonnegative to obtain the following constraints on f_{22} and f_{23}:

$$\begin{aligned} 0 &\leq f_{22} \leq d_2 \\ 0 &\leq f_{23} \leq d_3 \\ s_2 - d_1 &\leq f_{22} + f_{23} \leq s_2. \end{aligned} \tag{2.5}$$

Next, we express the cost function C in terms of the parameters f_{22} and f_{23}. We obtain

$$C = A + Bf_{22} + Df_{23},$$

where A, B, and D are constants given by

$$\begin{aligned} A &= [c_{11}(d_1 - s_2) + c_{12}d_2 + c_{13}d_3 + c_{21}s_2] \\ B &= [c_{11} + c_{22} - c_{12} - c_{21}] \\ D &= [c_{11} + c_{23} - c_{13} - c_{21}]. \end{aligned}$$

Thus this two-by-three transportation problem can be phrased as the problem of finding the proper choice of the parameters f_{22} and f_{23} to satisfy conditions (2.5) and make C as small as possible. The specific choices for f_{22} and f_{23} which should be made in a given problem are determined by the relative magnitudes of the constants B and D. For example, if B is negative and D is positive, then one should choose f_{22} large and f_{23} small, since these choices make C small. There are many such cases to consider. We work out one numerical example in detail, and others are considered in Exercise 6.

Let the demands, supplies, and moving costs be as follows:

$$d_1 = 6, \quad d_2 = 9, \quad d_3 = 10, \quad s_1 = 15, \quad s_2 = 10, \quad c_{11} = 70,$$
$$c_{12} = 70, \quad c_{13} = 60, \quad c_{21} = 80, \quad c_{22} = 60, \quad c_{23} = 80.$$

The constants A, B, and D have the values $A = 1750$, $B = -20$, and

$D = 10$, and the constraints (2.5) have the form

$$0 \le f_{22} \le 9$$
$$0 \le f_{23} \le 10$$
$$4 \le f_{22} + f_{23} \le 10.$$

Thus, in this case, to make C as small as possible, we choose f_{22} as large as possible, $f_{22} = 9$, and f_{23} as small as possible, $f_{23} = 0$. This gives

$$C = A + Bf_{22} + Df_{23} = 1750 - 20(9) = 1570.$$

Hence the minimum shipping cost, \$1570, is achieved by the allocation vector whose coordinates are

$$f_{11} = 5 \qquad f_{21} = 1$$
$$f_{12} = 0 \qquad f_{22} = 9$$
$$f_{13} = 10 \qquad f_{23} = 0$$

We note that in general a problem of this sort need not have a unique solution. However, in this case the solution is in fact unique (Exercise 3).

It is interesting to examine the special case which corresponds to the condition $d_3 = 0$. Thus there are actually only two distribution centers, and the problem reduces to a two-by-two transportation problem. It is easily shown that conditions (2.4) and (2.5) become, respectively,

$$f_{11} = s_1 - d_2 + f_{22}$$
$$f_{12} = d_2 - f_{22} \tag{2.6}$$
$$f_{21} = s_2 - f_{22}$$

and

$$\max[0, d_2 - s_1] \le f_{22} \le \min[d_2, s_2]. \tag{2.7}$$

The cost function C now has the form

$$C = A + Bf_{22},$$

where A and B are constants (Exercise 1). In this case the solution to the problem of finding a least cost shipping schedule is given by the following result.

Theorem 1: The allocation vector which minimizes the cost function C is the vector determined by (2.6) and the choice of f_{22} specified by the following conditions:

1. If $B > 0$, then $f_{22} = \max[0, d_2 - s_1]$.
2. If $B < 0$, then $f_{22} = \min[d_2, s_2]$.

3. If $B = 0$, then f_{22} is any value such that

$$\max[0, d_2 - s_1] \leq f_{22} \leq \min[d_2, s_2].$$

Proof: Exercise 2.

As an application of Theorem 1, consider the two-by-two problem with the data

$$d_1 = 8, \quad d_2 = 2, \quad s_1 = 4, \quad s_2 = 6, \quad c_{11} = 8,$$
$$c_{12} = 4, \quad c_{21} = 6, \quad c_{22} = 3.$$

Then $B = 1$, and hence $f_{22} = \max[0, -2] = 0$. Therefore, the shipping schedule is given by

$$f_{11} = 2, \quad f_{12} = 2, \quad \text{and} \quad f_{21} = 6.$$

It is interesting to note that even though the least expensive shipping rate is between factory 2 and distribution center 2, there are no shipments along this route, i.e., $f_{22} = 0$.

EXERCISES

1. Formulate the two-by-two transportation problem and show that this is the same problem as the two-by-three problem with $d_3 = 0$.

2. Prove Theorem 1.

3. Show that the solution of the sample two-by-three problem worked in the text is unique. *Hint:* Show that for all least-cost schedules f_{22} and f_{23} are as given.

4. Formulate an m-by-n transportation problem.

5. Discuss the usefulness of Theorem 1 for problems modeled as two-by-two transportation problems. Under what conditions will this theorem give useful information, and when might it give misleading results? More precisely, give three aspects of the real-world situation which seem to be adequately incorporated in the model and three which are not.

6. Solve the two-by-three transportation problems with the following data:

(a) $d_1 = 8, d_2 = 12, d_3 = 10, s_1 = 15 = s_2, c_{11} = 50, c_{12} = 80, c_{13} = 40,$
$c_{21} = 50, c_{22} = 40, c_{23} = 40.$
(b) Same as (a) with $c_{23} = 30$.
(c) Same as (a) with $c_{22} = 90$ and $c_{23} = 50$.
(d) $s_1 = 4, s_2 = 5, d_1 = 2, d_2 = 4, d_3 = 3, c_{11} = 3, c_{12} = 5, c_{13} = 1,$
$c_{21} = 6, c_{22} = 7, c_{23} = 6.$
(e) Same as (d) with $c_{13} = 4$.

7. Discuss the two-by-three transportation model with Axiom A_2 replaced by Axiom A_2': $d_1 + d_2 + d_3 \leq s_1 + s_2$. How might the introduction of a fictitious distribution center be used to convert this to the previous model?

8. How should the transportation model of this section be modified to account for a partial shipping strike which limits the total number of homes that can be sent over certain routes?

2.3 WATER AND MIRRORS

The discussion of this section is intended to illustrate two common aspects of the model-building process. First, there may be another entire axiom system subsumed in the one under construction. Here, for example, the basic structure of Euclidean geometry is assumed in the axioms given for these models. Since the difficulties of axiomitizing geometry are well known, it is clear that this is not a trivial assumption. However, our concern is not with the foundations of geometry, and consequently we do not hesitate to use whatever geometrical facts we need without additional comment. The same point of view is taken later in connection with the real number system, probability theory, etc. Second, we shall see how a model which is constructed to account for certain observed facts may fail to account for others. In such situations our model must be modified if it is to be useful in explaining and predicting reality.

As in Sec. 2.1, it is convenient to introduce a fictitious experiment to motivate our model building. The basic questions were considered long ago—certainly the Greeks used at least portions of the models constructed here. Although the models are relatively simple and many of the observed facts are accounted for, there are experiments not considered here which require further modification of the models.

2.3.1 The Problem

A certain scientist—we shall refer to him as Dr. Greek—is interested in the properties of light. He is particularly concerned with the path of a ray of light in a single medium, at the interface of two different media, and at a reflecting surface.

2.3.2 A Real Model

Observations of shafts of sunlight or a study of shadows cast by an obstruction in a ray of light indicate that light travels in a straight line. Here, we are supposing that the entire path lies in the same medium and that the medium has the same physical characteristics throughout; that is, it is *homogeneous*. This belief can be strengthened by simple laboratory experiments involving a light source, a fixed pinhole and a movable one. See Fig. 2–2.

To discuss the notion of reflection, some additional terminology is required. Suppose that a ray of light strikes a plane-reflecting surface as

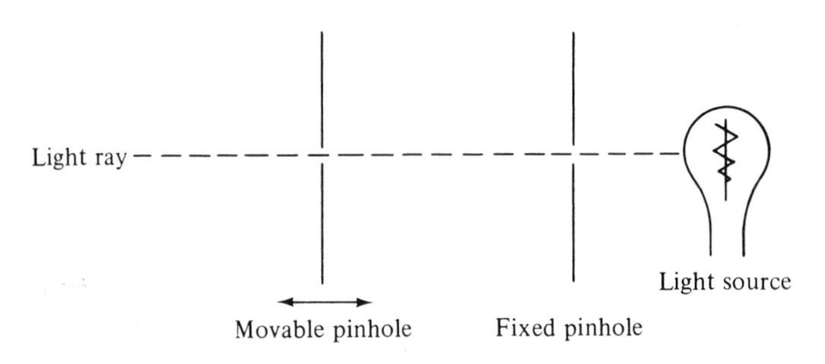

Fig. 2–2

indicated in Fig. 2–3. Here l_1 indicates the incoming ray, known as the *incident* ray, and l_2 the outgoing ray, known as the *reflected* ray. It will be useful for us to identify certain angles, and, to make the identification in the usual way, we introduce the normal n to the surface at the point of reflection of the ray. The lines l_1 and n determine a plane, and we define θ, the angle of incidence, as the angle between them. Likewise we define φ, the angle of reflection, as the angle between l_2 and n. We are concerned only with the magnitude of θ and φ and not their sense. With these conventions we can state the results of observations easily: The incident ray, the normal at the point of reflection, and the reflected ray are coplanar, and the angle of incidence is equal to the angle of reflection. All these observations are the sort that can be checked with relatively simple equipment. Of course, as with

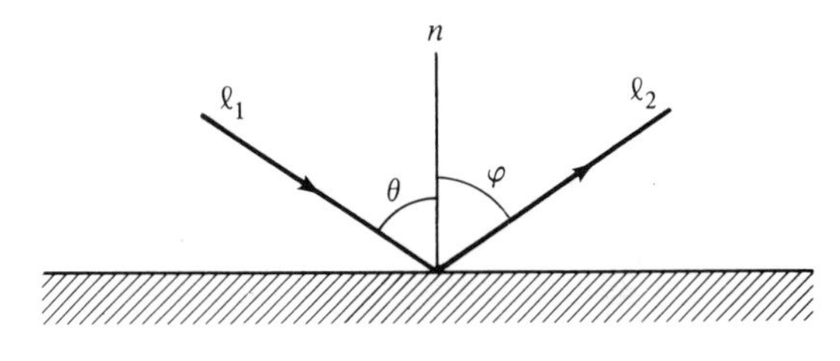

Fig. 2–3

any experiment, one expects approximate verification, the degree of approximation depending on the nature of the apparatus involved. Many of the optical phenomena occurring in a single medium can be accounted for in terms of these observations. There are difficulties at the interface of different media, and we shall return to this question later.

2.3.3 A Mathematical Model

One of the goals of model building is to account for as many observations as possible with a single concept. One example of such a concept is a *minimum principle*, that is, a principle which asserts that the real world behaves in such a way that a certain quantity is a minimum. It turns out that we can account for the observations cited above with a single axiom stated in the form of a minimum principle. We take light ray and medium as undefined terms.

Definition: The *source* and *observer* are two points in Euclidean 3-space.

It is useful to interpret the observer as either a person with one eye closed or else as a pinhole camera. The geometric notions will have their usual meanings. In particular, by a *path* between two points we mean a curve connecting the two points. This concept can be made quite precise, but it is unnecessary for us to do so here. It is understood for this part of our discussion that we are considering only the propagation of light in a single medium.

Axiom 1: In a single medium, a light ray travels the shortest possible path between source and observer.

We note for future reference that this axiom makes no mention of plane-reflecting surfaces. Thus, there is the possibility of making use of this axiom to predict the behavior of light in systems involving curved mirrors.

We now turn to the theory. We phrase the results in such a way that the connections with the real model introduced above are clear.

Theorem 1: In a single medium a light ray travels in a straight line between source and observer.

This follows at once from the axiom system and the basic facts of Euclidean geometry; i.e., the shortest distance in Euclidean 3-space is a straight line.

The next definition is clearly motivated by the observation referred to above.

Definition: Let π be a given plane in 3-space and let a source and observer be given, both on the same side of π. A light ray from the source to the observer is said to be *reflected* in π if there is a point $P \in \pi$ such that the ray travels from the source to P to the observer. A point P satisfying these conditions is known as a *point of reflection*. See Fig. 2–4.

In most applications of this sort, where a quantity is defined that is clearly intended to be analogous to something observable and the observable is unique, then we are obliged to prove that the definition identifies a unique entity as well. This is the point of our next result.

Theorem 2: Let source, observer, plane, and reflected light ray satisfy the conditions of the above definition. Then there is exactly one point of reflection.

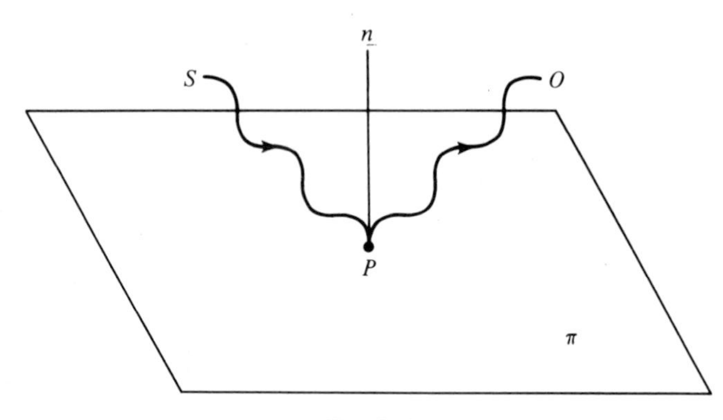

Fig. 2–4

Proof: For the proof, let us denote the source by S and the observer by O. Select any point of reflection and call it P. Think of the path of the light ray from S to O as consisting of two parts, the first from S to P and the second from P to O. Then we claim that the path from S to P and the path from P to O must both be straight lines. Otherwise, either the path from S to P would not be the shortest possible or the path from P to O would not be, or both. Then, we could find a shorter path from S to O via P using the straight lines. But this contradicts the axiom, and therefore both paths are straight lines. Finally, $P \in \pi$ and neither S nor O is in π, so that clearly no other points on the path from S to O can be points of reflection. Q.E.D.

We have actually provided a portion of the proof of a much stronger result. Namely,

Theorem 3: Let a source S, an observer O, a plane π, and a reflected light ray satisfy the conditions of the above definition. If P is the point of reflection, then

1. The paths of a light ray from S to P and from P to O are straight lines.
2. The plane determined by the straight lines in 1 is perpendicular to π.
3. If n is the normal to the plane π at P directed into the half-space containing S and O, then the angle between n and the path from S to P is equal to the angle between the path from P to O and n.

Proof: The proof of Theorem 2 contains the proof of part 1. For part 2 we observe that if P is not in the plane perpendicular to π containing S and O, then a shorter path can be found from S to O via a point on π. This contradicts Axiom 1, and it follows that part 2 holds.

Finally, we consider part 3. It follows from part 2 that we may assume that S, O, and P are coplanar and lie in a plane π^* perpendicular to π. Thus n lies in π^* as well. Let O' be the point symmetric to O with respect to the plane π, note that O' is also contained in the plane π^*. Figure. 2–5 depicts

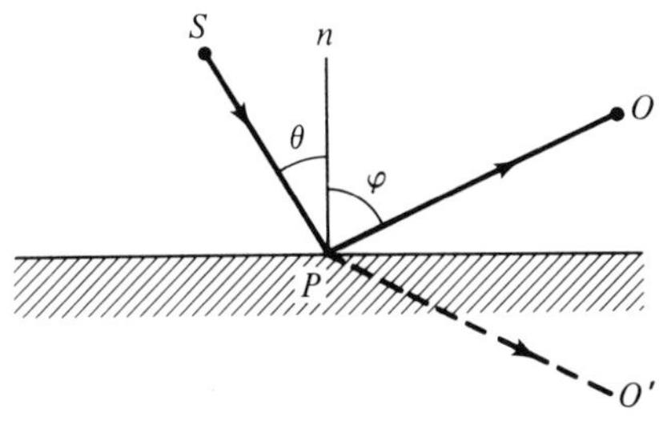

Fig. 2–5

the situation in the plane π^*. The horizontal line in this figure represents the intersection of π and π^*. Then it follows that the length of the path from S to O, via P, is equal to the length of the path from S to O', via P. But the path from S to O' is clearly a minimum when SO' is a straight line. A simple argument involving similar triangles then completes the proof that $\theta = \varphi$. Q.E.D.

It should be remarked that this proof also shows that given a source S, an observer O, and a reflecting plane π, there is a unique path for a light ray from S to O which is reflected in π.

We have now shown that the mathematical model given by the axiom system of this section accounts for the observations referred to in Sec. 2.3.2. However, it was pointed out there that difficulties arise when one considers more than one media. It is now time to consider these difficulties.

2.3.4 Refinements of the Models

In our discussion thus far we have restricted ourselves to a consideration of the behavior of light rays in a homogeneous medium. We shall consider now the effects observed if this restriction is relaxed. It is useful to return to Dr. Greek.

In observing the behavior of light as it passes from one medium to another, Dr. Greek noted that it no longer traveled in a straight line. Indeed, the simple act of inserting a pencil halfway into a pan of water shows that something is different: The pencil appears to be bent at the surface of the water. With additional careful work, Dr. Greek was able to give a rule which accounted in a quantitative way for all the observations. Again, we require some new terminology. Suppose that we have a plane interface between two

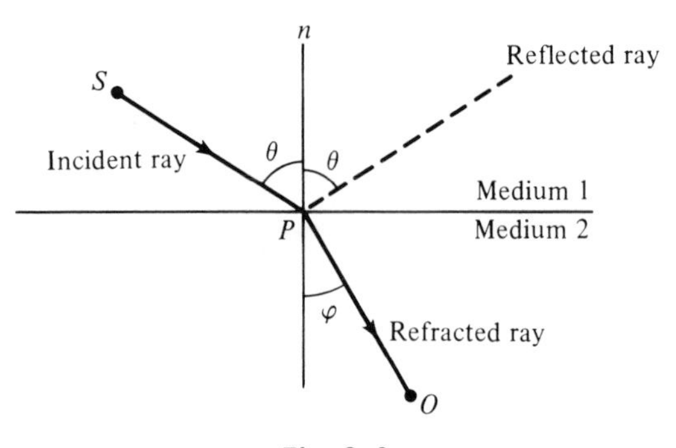

Fig. 2–6

transparent media, and a light ray, denoted by SP (see Fig. 2–6), incident upon this interface. In addition to a ray reflected back into the same medium as the source, there will be a ray propagated into the other medium. This ray, denoted by PO, is known as the *refracted* ray. With reference to Fig. 2–6, the angle φ between the refracted ray and the normal to the surface at the point P of refraction is known as the *angle of refraction*. With this terminology, the law illustrated by Dr. Greek's experiments can be summarized by stating that the incident ray, the normal to the plane interface at the point of refraction, and the refracted rays are coplanar. Moreover, there is a constant μ, which depends only on the media, such that $\sin\theta = \mu \sin\varphi$, where θ is the angle of incidence and φ is the angle of refraction. The constant μ is known as the relative index of refraction of the media and depends on their physical properties.

If one of the media is an agreed upon standard one which is kept fixed, say medium 1 in Fig. 2–6, and if medium 2 is successively replaced by several different ones, then it is found that with each of the samples used as medium 2 there is a physical constant known as the *index of refraction*. Then for any choice of media 1 and 2 with respective indices of refraction μ_1 and μ_2, we have

$$\mu_1 \sin\theta = \mu_2 \sin\varphi. \tag{3.1}$$

Finally, other experiments indicated that the velocity of light in a fixed homogeneous medium is inversely proportional to the index of refraction. That is, if v denotes the velocity of light and μ the index of refraction for a specific medium, then $v = k/\mu$, where k is a physical constant which is independent of the medium. Assuming the validity of this result, we can rewrite Eq. (3.1) in the form

$$\frac{\sin\theta}{v_1} = \frac{\sin\varphi}{v_2}, \tag{3.2}$$

where v_1 and v_2 are the velocities of light in media 1 and 2, respectively.

All attempts to account for this behavior in the mathematical model determined by the axiom system of Sec. 2.3.3 failed. It was concluded, therefore, that the real model constructed from these observations was not a logical model for that axiom system and that a new system must be devised. The formulation of an axiom in terms of a minimum principle was found useful before, and we shall find it so again. Before stating the new axiom, we first turn our attention to the notions of a light ray and a medium. These were undefined terms previously, but now we shall give them definitions. It should be remarked that this is possible in general and frequently useful. Obviously all results which hold for light ray as an undefined term also hold with it defined in any way which is consistent with the axiom using the term. Our definition will have the desired consistency.

From an experimental point of view, we can think of the apparatus used as a source consisting of a continuously illuminated bulb or other light-producing device and a screen with a pinhole. The light ray is then that which emanates from the pinhole. Suppose that the bulb is not continuously illuminated but instead is illuminated in pulses. Then we can think of a pulse of light being emitted through the pinhole. These pulses can be made shorter and shorter until we imagine an elementary packet or a point of light coming through the pinhole. The velocity of light is then the velocity of this point. Here, velocity is used in the usual sense, time rate of change of position. With this interpretation, a light ray can be thought of as a continuous string of pulses, or even as a continuous string of points. We now make these notions precise.

Definition: The time axis is the nonnegative real numbers with their usual order. A *point of light* is a piecewise continuously differentiable function λ, defined on some interval of the time axis. A *ray of light* is a family of points, Λ, with a common domain of definition, such that for each fixed t in this common domain the set $\{\lambda(t)\}_{\lambda \in \Lambda}$ is a single arc on a path. The velocity of the point λ is $d\lambda/dt$.

Definition: A homogeneous medium is one in which the velocity of light is constant. In particular, for each such medium m there is a constant c such that if λ is a point of light and $\lambda(t) \in m$, then $d\lambda/dt(t) = c$.

The essential axiom for this model is

Axiom 2: A light ray travels that path between source and observer which minimizes the time of travel.

In a homogeneous medium the velocity of light is constant so that the path of minimum time of travel is also the shortest path. We have

Theorem 4: In a homogeneous medium, Axiom 2 implies Axiom 1.

In particular, therefore, Axiom 2 accounts for the observations regarding reflection. To show that it also accounts for the experimental results concerning refraction, we must make that concept precise. Again, the definition is clearly made with experiments in mind.

Definition: Let π be a plane in 3-space with a homogeneous medium on each side and P a point on π. Also, let S be a source and O an observer, each on different sides of π. A light ray from the source to the observer is said to be *refracted* at P if the ray travels from S to P and from P to O. A point P satisfying these conditions is known as a *point of refraction.*

The following result can be proved in a manner similar to Theorem 2.

Theorem 5: Let a source, observer, plane, and light ray satisfying the conditions of the above definition be given. Then there is exactly one point of refraction.

The next theorem is essentially the statement that the observations involving refraction can be accounted for in this system.

Theorem 6: Let a source S, an observer O, and a plane π be given. Consider a light ray satisfying the conditions of the above definition. If P is the point of refraction of this light ray, then

a. The paths traveled by a light ray from S to P and from P to O are straight lines.
b. The paths of *a* lie in a plane π', which is perpendicular to π.
c. If the source is in medium 1 in which the velocity of light is v_1, the observer is in medium 2 in which the velocity of light is v_2, and θ and φ are the angles of incidence and refraction, respectively, then

$$\frac{\sin \theta}{v_1} = \frac{\sin \varphi}{v_2}. \tag{3.2}$$

Proof: Parts *a* and *b* have proofs which differ only in detail from those given above, and thus only the proof of part *c* will be provided. The technique of proof involves some elementary calculus. Consider the plane π' which contains S, O, and P and in which the situation appears as in Fig. 2–7. The horizontal line represents the intersection of π and π' and therefore that part of π' above this intersection is in one medium (medium 1) and the part below in another (medium 2). Let the path from the source S to the point P of refraction be denoted by l_1 and that from P to the observer O by l_2. If the respective velocities of light in the two media are v_1 and v_2, then the times of travel of

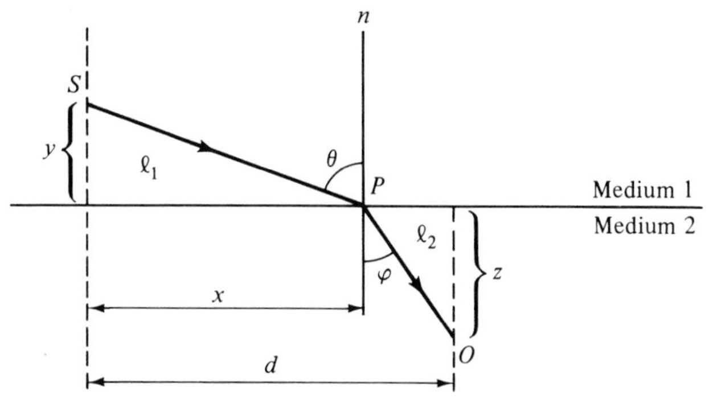

Fig. 2–7

a point of light from S to P and from P to O, denoted by t_1 and t_2, respectively, are

$$t_1 = \frac{\text{length of } l_1}{v_1}, \qquad t_2 = \frac{\text{length of } l_2}{v_2}.$$

According to Axiom 2, the total time $T = t_1 + t_2$ is a minimum for the path traveled by the point. Thus, to find the point of refraction P, it is sufficient to find the minimum for the expression

$$T(x) = \frac{\sqrt{y^2 + x^2}}{v_1} + \frac{\sqrt{z^2 + (d - x)^2}}{v_2}.$$

Standard techniques of differential calculus give

$$\frac{x}{v_1\sqrt{y^2 + x^2}} - \frac{d - x}{v_2\sqrt{z^2 + (d - x)^2}} = 0,$$

or

$$\frac{\sin\theta}{v_1} - \frac{\sin\varphi}{v_2} = 0,$$

as the condition for minimum T. Since this is clearly Eq. (3.2), the proof of the theorem is complete.

Thus, the revised model yields theorems which are consistent with all the observations mentioned here and in a sense "explains" them. There are, however, other observations which require additional modifications of the model. For example, even the second version is inadequate to account for some of the observed phenomena in the study of diffraction. In fact, it is very difficult to construct a single model which accounts for all the many observations involving optical phenomena.

EXERCISES

1. Consider the phenomena of reflection discussed in Secs. 2.3.2 and 2.3.3. Find an expression for the length of the path *SPO* in analytic terms and use the calculus to prove that if this path has minimum length, then $\theta = \varphi$. *Hint:* Consider Fig. 2–8 and express the length of the path *SPO* as a function of x.

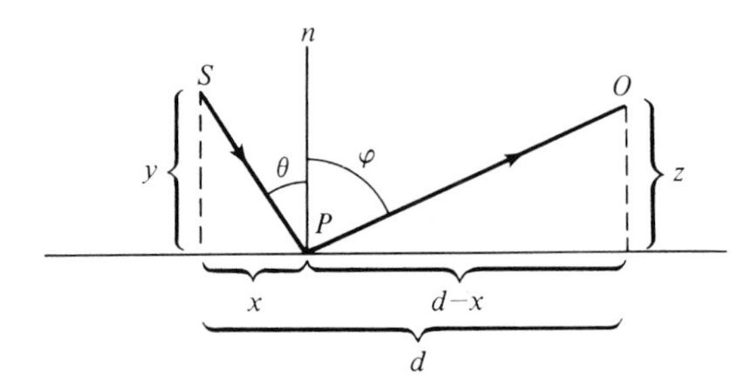

Fig. 2–8

2. Discuss the relative strength of Axioms 1 and 2.
3. Consider the phenomena of refraction with a fixed source, observer, and plane interface. If medium 1 is kept fixed and medium 2 is varied, how does the point *P* of refraction vary? In particular, how does *P* vary with v_2, the velocity of light in medium 2?
4. Provide an argument by which Eq. (3.1) follows from the preceding discussion. What additional assumptions are necessary?

2.4 INSTANT INTELLIGENCE

Several interesting mathematical models have been studied in connection with problems arising in psychology and, in particular, learning theory. In this section we shall discuss one of the models which has been proposed to explain the learning of simple tasks, such as a single word in a vocabulary list. The learning process is complex, and we cannot hope to do justice to it here. However, our aim is not the detailed study of learning. Instead, we are concerned with illustrating the construction of models, and thus we feel free to concentrate on only one aspect of this complex subject. In the References at the end of the chapter we list several sources which offer a more complete analysis. In particular, [RG] provides information on the development, testing, and comparison of a number of mathematical models for learning.

2.4.1 The Problem

How does a person learn a simple task? Is it a slow process of incremental learning, or is it a sudden transition which takes the subject from the completely unlearned state to the completely learned state? Mathematical models have been proposed for both of these conjectures, and a good deal of study has been carried out using and testing these models. The incremental theory seems to have been developed first and to conform to everyday experience. While practice may not always make perfect, it usually does bring about improvement. One of the most influential of the theories which used the incremental assumption was given by Clark Hull in 1943 in *Principles of Behavior*. We shall not study this work; instead we shall consider a special case of the competing all-or-none model.

One of the earliest proposals of the all-or-none model was made by E. R. Guthrie in 1935. He assumed that the connection between a stimulus and a response was not something that varied in strength and grew stronger but rather was an all-or-none connecting bond. He believed that his all-or-none assumption did not contradict the observed fact that people improved their performance on complicated tasks by practicing. He simply felt that complicated tasks were made up of many simple tasks and that each of these was learned in an all-or-none fashion. The simple tasks are not all learned at the same trial; however, more and more of them are learned at each trial, and thus there is improvement on the complicated task. This assumption by Guthrie was not extensively studied until the early 1950s. Then a substantial amount of experimental evidence accumulated which seemed to reinforce Guthrie's assumption and to refute the incremental assumption. These experiments were carried out by G. A. Kimble and by I. Rock and later by W. K. Estes. We shall consider one model based on the assumption that simple tasks are learned in an all-or-none manner.

2.4.2 A Real Model

We consider an experiment consisting of subjects learning the correct responses to certain simple stimuli such as a single pair of letters. The response is either the number 1 or the number 2, and each subject is to learn the correct responses for ten stimuli. The correct responses are randomly chosen for each subject, so it is reasonable to assume that initially the subject does not know the correct responses. After each stimulus is shown and the subject makes a response, he is told whether or not his response was correct. At this point we encounter one of the difficulties in studying learning. Namely, learning is not directly observed in a subject. Instead, one observes changes in behavior, and then an assumption is made to relate these behavior changes with learning. The particular form of this assumption, that is, the criterion used to connect observed changes in behavior with learning, will vary with the experimental design, the goals of the investigator, and the model being considered. For example, in the experiment described here an appropriate criterion for concluding that a subject learns might be that twice in succession he responds correctly to all ten stimuli.

The main assumption on which our model is based is that all subjects learn each correct response in an all-or-none manner. Once they learn it, they remember it during all future trials, and until they learn it, they are simply guessing. The term *trial* is used to indicate the presentation of a single stimulus, the subject's response, and the evaluation (right or wrong) of his response. In quantifying this assumption of all-or-none learning, we assume that there is a number c, $0 < c < 1$, which gives the probability that a subject who does not know the correct response to a stimulus learns it on any given trial. This number c is to be independent of trial number and stimulus. It may, however, vary from one subject to another. As a final technical assumption, we suppose that each subject is a "good" subject. That is, he pays attention to every stimulus and to the correct responses and thus no trials are wasted.

Our approach will be based on these assumptions, and we shall find that it is relatively easy to derive predictions about the learning behavior of the subjects. These predictions are obtained by introducing some concepts from elementary probability theory, and they give the probability that a subject learns the correct response by a certain trial. Naturally, these probabilities depend on the constant c which is associated with the subject. We shall briefly indicate how such predictions can be obtained. Finally, we shall give a discussion of a mathematical model for this real model.

Let L_n be the probability that the subject has learned the correct response to a certain stimulus prior to the start of the nth trial for that stimulus. Then $1 - L_n$ is the probability that he has not learned the correct response at the start of trial n. From our assumptions, $L_1 = 0$, $L_2 = c$, and $L_3 = c + (1 - c)c = 2c - c^2$. We note at this point that L_n is *not* the probability that the subject makes a correct response at trial n. It is possible to make a correct response by guessing, even though the subject has not yet learned the correct response. In fact, since this experiment has only two responses, the subject has at least a 50: 50 chance at all times. Since it is responses that are recorded, it is also of interest to know the probability that the subject will make a correct response on trial n. We denote this quantity by C_n. It follows from the laws of probability theory that

$$C_1 = \tfrac{1}{2}, \qquad C_2 = 1 \cdot c + \tfrac{1}{2} \cdot (1 - c) = \tfrac{1}{2}(1 + c).$$

Indeed, we can compute both C_n and L_n for all values of n, and we state these results as theorems. These are theorems in the real model since they are stated in terms of subjects and responses.

Theorem 1: The probability that the subject knows the correct response at the start of the nth trial is given by

$$L_n = [1 - (1 - c)^{n-1}], \qquad n = 1, 2, \ldots .$$

Proof: The event "the subject *does not* know the correct response at the start of the nth trial" can occur only if the subject fails to learn the correct response at each of the first $n - 1$ trials. This event has probability $(1 - c)^{n-1}$ because this model assumes that the probability of learning (and hence of not learning) is independent of the trial, and the probability of not learning is $1 - c$. Hence, $L_n = [1 - (1 - c)^{n-1}]$, $n = 1, 2, \ldots$. Q.E.D.

Theorem 2: The probability that the subject makes a correct response on the nth trial is given by

$$C_n = 1 - \tfrac{1}{2}(1 - c)^{n-1}.$$

Proof: Using Theorem 1 and the laws of probability theory we have

$C_n =$ Prob [subject knows correct response] · Prob [subject gives correct response | he knows correct response] + Prob [subject does not know correct response] · Prob [subject gives correct response | he does not know correct response]

$$= [1 - (1 - c)^{n-1}] \cdot 1 + (1 - c)^{n-1} \cdot \tfrac{1}{2}$$

$$= 1 - \tfrac{1}{2}(1 - c)^{n-1}.$$

Q.E.D.

Rather than prove additional theorems at this point, we turn instead to the development of a mathematical model for this real model. In this way we shall be able to obtain some additional information and insight by using well-known mathematical tools.

2.4.3 A Mathematical Model

It is a relatively simple task to form a mathematical model for the real model introduced above. In fact, this real model is a concrete example of a special concept in probability theory known as a Markov chain. In Chap. 3 we shall develop some of the elementary theory of Markov chains. At this point we shall show how some of those topics are related to this model in learning theory.

In the study of Markov chains one has an undefined term called a process and a set of undefined objects called states. One also has certain real numbers, called transition probabilities. To make this formal, let $S = \{s_1, s_2, \ldots, s_r\}$ be the set of states, and let p_{ij} be the probability that the process moves from state s_i to state s_j in one step. The numbers p_{ij} are defined for every ordered pair (i, j), and they are referred to as the transition probabilities. They form a square matrix $\mathbb{P} = (p_{ij})$, which is known as the transition matrix. There is an extensive theory of Markov chains and some of the theorems are useful for the special case that we are considering. We have only two states, which will be called L and U, for learned and unlearned.

The transition matrix is

$$\begin{array}{c} \\ L \\ U \end{array}\begin{array}{c} \begin{array}{cc} L & U \end{array} \\ \begin{bmatrix} 1 & 0 \\ c & 1-c \end{bmatrix} \end{array}, \qquad \text{for} \qquad 0 < c < 1.$$

The form of the transition matrix shows that the process belongs to a special class of Markov chains whose properties will be investigated in the next chapter. The following theorem is a special case of a more general result, but because of the simplicity of the process, it is easy and worthwhile to give an independent proof. The result is that essentially all subjects eventually reach the learned state. The precise result is a probabilistic statement to the effect that if the process is allowed to continue through enough trials, then the probability that the subject is in state L can be made arbitrarily close to 1. Naturally, the number of trials necessary to achieve a preassigned probability varies with the subject.

The symbol L_n was given a meaning above in the real model. Since it is necessary to use L_n in the theorem, it must be given a meaning in the context of this model. We define L_n to be the probability that a process which starts out in state U reaches state L before the nth transition.

Theorem 3: Given any $\epsilon > 0$, there exists an integer $N = N(\epsilon)$ such that $L_n > 1 - \epsilon$ for $n \geq N$.

Proof: Just as before, we have $L_n = [1 - (1 - c)^{n-1}]$. Since $0 < c < 1$, it follows that $(1 - c)^{n-1} \longrightarrow 0$ and $L_n \longrightarrow 1$ as $n \longrightarrow \infty$. This is just the statement of the theorem. Q.E.D.

To check the validity of the model, it is useful to have theorems which predict the behavior of quantities which can be measured in the laboratory. Frequently the theorems predict expected values which can then be compared with the results of experiments. The next two theorems are of this sort.

Theorem 4: The expected number of times that the subject is in state U before he enters state L is $1/c$.

Proof: Let n be the random variable which assigns to each sequence of trials for a subject the number of trials that he is in state U. If $E[n]$ denotes the expected value of n, then we have

$$\begin{aligned} E[n] &= \sum_{k=1}^{\infty} k(L_{k+1} - L_k) \\ &= \sum_{k=1}^{\infty} k[1 - (1 - c)^k - 1 + (1 - c)^{k-1}] \\ &= \sum_{k=1}^{\infty} k \cdot c(1 - c)^{k-1} \\ &= c \cdot \frac{1}{[1 - (1 - c)]^2} = \frac{1}{c}. \end{aligned}$$

Q.E.D.

The next theorem is stated in terms of subject and responses for convenience.

Theorem 5: The expected total number of incorrect responses by the subject is $1/2c$.

Proof: We use Theorem 4 and the fact that the subject makes incorrect responses only if he is in state U; then, on the average, he makes an incorrect response half the time. Since he is in state U an average of $1/c$ times, we have

$$\text{Expected number of incorrect responses} = \frac{1}{c} \cdot \frac{1}{2} = \frac{1}{2c}. \qquad \text{Q.E.D.}$$

Further results in the theory of this model could be given; however, the flavor of the theory has now been established. These theorems can be considered to be predictions about the learning behavior of the subjects, and as such they can be tested. As a first step in this testing one must determine the constant c. This is done by examining the data from past experiments, and it involves the science and techniques of parameter estimation. Using the value obtained for c, one can check future experimental results against those predicted by the theorems. This has been done for experiments similar to the one discussed here, and the predictions have been quite accurate. In particular, the work of Bower is relevant here (see [Bo]). Naturally, the fact that the predictions of the theory have proved reliable provides a stimulus for continued study of this model. It does not prove that this mathematical model completely describes learning; however, it indicates that the assumptions of this model should probably not be ignored.

EXERCISES

1. Find the probability that the subject will make an incorrect response on the nth trial.

2. (a) Find the smallest value of c such that 95% of the time the subject is in state L after three trials.
(b) Let $c = 2/3$. Find the smallest value of n such that the subject is in state L 95% of the time after n trials.

3. Construct a model for simple learning based on the assumption that learning is a three-state process: all-none-halfway.

4. Construct a model for simple learning based on the assumption that learning is an incremental process with K stages.

5. Use the transition matrix denoted by $\mathbb{P}$ and the definition of L_n to show that

$$\mathbb{P}^{n-1} = \begin{bmatrix} 1 & 0 \\ * & 1 - L_n \end{bmatrix}.$$

Also find the missing term $*$.

2.5 WHEN TO BROADCAST AND WHEN TO WAIT

The need for models for decision making arises frequently in the social and management sciences and occasionally in the other sciences as well. There are many types of such decision problems, and we shall consider here an example of a class of *timing problems*. In these situations the decision is that of selecting a time to carry out an action. Common features of these problems are the desirability of postponing the action as long as possible and the presence of a penalty for waiting too long. An example with these features as well as competition and uncertainty is given here.

2.5.1 The Problem

The situation involves two political parties, the Scoundrels (the "in" party) and the Redeemers (the "out" party), and an approaching election. Each party plans to use the various media—newspapers, magazines, radio, and television—to attempt to convince the voters of the merits of their candidates. In particular, each party feels that they can afford exactly one nationwide television broadcast, and they wish to determine when to schedule it. Political strategists in both camps agree that a broadcast becomes more effective as the election approaches and that if it is scheduled too soon, then its effects will be dissipated before the election. On the other hand, if it is scheduled too late and if the other party goes on the air previously, it may be that many of the voters will be so thoroughly convinced by the arguments of the opposition that they will remain unmoved by the other arguments presented, and the effect will be small. Given these considerations, when is the best time to broadcast?

2.5.2 Models

We shall now make the situation more precise. It is assumed by both parties that only broadcasts made within a one-month period prior to the election can be effective, and therefore each party decides to broadcast in that period. Arrangements are made so that neither party knows in advance when the other plans to broadcast. Moreover, by means of public opinion polls it is possible to measure immediately after the broadcast whether it was effective. Each party has its own criteria for effectiveness, but each decides that if the other party broadcasts first and if their program is effective, then they will cancel their own program and use the funds to provide for door-to-door canvasing and newspaper advertisements. The only exception to this is if both schedule broadcasts on the same day, in which case they both proceed as planned.

Neither party is able to predict in advance whether their broadcast will be effective or not, but they can estimate the probability that it will be. It is observed that these probabilities vary with time and increase as the date of the election approaches. Let $P_R(t)$ denote the probability that a broadcast by the Redeemer Party t days before the election will be effective, and let $P_S(t)$ be defined similarly for the Scoundrels. In terms of these functions, the above observations can be expressed as

$$P_R(t) = 0, \qquad P_S(t) = 0, \qquad t > 30.$$

$$P_R(t), P_S(t) \quad \text{increase as } t \text{ decreases to } 0.$$

Finally, we assume that the probability of a broadcast being effective is 1 on the day before the election. A consequence of this assumption is the fact that if one party broadcasts and it is not effective, then the other party will wait until the day before the election to broadcast, at which time its effectiveness is assured.

In constructing a mathematical model for this situation, it is natural to take Redeemer Party, Scoundrel Party, and effectiveness as undefined terms and P_R, P_S as undefined symbols. We introduce a model in which the comments made above are included in a formal way.

Definition: N^+ is the set of positive integers.

Definition: A function f defined on N^+ is in class $\mathfrak{U}$ if

1. $f(t) = 0$ for $t \geq 30$; $f(1) = 1$.
2. $f(t_1) \geq f(t_2)$ for $t_1 \leq t_2$, $t_1, t_2 \in N^+$.

Thus, $\mathfrak{U}$ is a class of *nonincreasing* functions.

Axiom 1: P_R and P_S are functions in class $\mathfrak{U}$.

Axiom 2: The effectiveness of the Redeemer Party and the Scoundrel Party at time t is given by $P_R(t)$ and $P_S(t)$, respectively.

We continue our discussion in real terms. Suppose that there is 1 unit of weakly committed votes in each party and these votes may be influenced by the broadcast. A unit of votes may be 10 or 10 million votes, but it is a fixed quantity for the study. If one of the broadcasts is effective and the other is not, then the party making the effective broadcast retains its own weakly committed votes and picks up those of the opposing party. If both are effective or both are not effective, then each party retains its own weakly committed votes. Since it is not known with certainty just what the outcome will be, the proper measure to introduce is the expected gain. Consider the situation from the point of view of the Redeemer Party. The expected gain for this party is a function of two variables, the times at which the two parties plan to broadcast. If we let these times be t_R and t_S, where the subscripts refer to the party, then the expected gain for the Redeemer Party depends on the relative sizes of t_R and t_S. If $t_R > t_S$, then the Redeemer Party will broadcast first. If their broadcast is effective, then the party gains 1 unit of votes. This occurs with probability $P_R(t_R)$. If they are unsuccessful, then they lose 1 unit of votes since the Scoundrel Party will surely wait until the day before the election to broadcast. Clearly this occurs with probability $1 - P_R(t_R)$. We are tacitly assuming that the Scoundrel Party behaves rationally, i.e., they act so as to minimize the number of votes cast for the Redeemer Party. In a similar manner we can consider the cases $t_R = t_S$ and $t_R < t_S$. Combining these results, we obtain the expected gain for the Redeemer Party to be

$$E_R(t_R, t_S) = \begin{cases} -1 + 2P_R(t_R), & t_R > t_S, \\ P_R(t_R) - P_S(t_S), & t_R = t_S, \\ 1 - 2P_S(t_S), & t_R < t_S. \end{cases} \tag{5.1}$$

The problem for the Redeemer Party is to choose t_R so as to maximize E_R. Naturally, the Scoundrel Party wants to choose t_S so as to minimize E_R. The theory which is needed for a general study of problems of this type is known as game theory, and we shall consider it in some detail in Chap. 6. However, it is interesting to continue the present discussion with specific choices for P_R and P_S. For simplicity, we shall assume that P_R increases steadily during the month preceding the election. In particular,

$$P_R(t) = \begin{cases} 0, & t > 30, \\ \dfrac{31 - t}{30}, & t = 1, 2, \ldots, 30. \end{cases}$$

Also, we suppose that the function P_S is always smaller than P_R. The voters are naturally dissatisfied with the party in power. But also P_S increases more rapidly than P_R as the election approaches and the voters begin to believe that the present government is not so bad after all. In particular, we assume that

$$P_S(t) = \begin{cases} 0, & t > 30, \\ \left(\dfrac{31 - t}{30}\right)^2, & t = 1, 2, \ldots, 30. \end{cases}$$

With these choices for P_R and P_S, and with the implicit restriction that always $0 \leq t_R \leq 30$, $0 \leq t_S \leq 30$, we have

$$E_R(t_R, t_S) = \begin{cases} \dfrac{16 - t_R}{15}, & t_R > t_S, \\ \dfrac{-31 + 32t_R - t_R^2}{900}, & t_R = t_S, \\ \dfrac{-511 + 62t_S - t_S^2}{450}, & t_R < t_S. \end{cases}$$

Strategists in the Redeemer Party reason as follows: For any specific t_R the Scoundrel Party will try to minimize E_R and hence will choose t_S to satisfy

$$E_R(t_R, t_S) = \underset{\substack{1 \le \sigma \le 30 \\ \sigma \in N^+}}{\text{minimum}}\ E_R(t_R, \sigma).$$

Therefore the Redeemer Party should choose t_R so that it satisfies

$$E_R(t_R, t_S) = \underset{\tau \in \Lambda}{\text{maximum}}\ \underset{\sigma \in \Lambda}{\text{minimum}}\ E_R(\tau, \sigma),$$

where $\Lambda = \{t : t \in N^+, 1 \le t \le 30\}$. We have

$$E_R(\tau, \sigma) = \begin{cases} \dfrac{16 - \tau}{15}, & \tau > \sigma, \\ \dfrac{-31 + 32\tau - \tau^2}{900}, & \tau = \sigma, \\ \dfrac{-511 + 62\sigma - \sigma^2}{450}, & \tau < \sigma, \end{cases} \tag{5.2}$$

and consequently (Exercise 2)

$$\underset{\tau \in \Lambda}{\text{maximum}}\ \underset{\sigma \in \Lambda}{\text{minimum}}\ E_R(\tau, \sigma) =$$

$$\underset{\tau \in \Lambda}{\text{maximum}}\left[\text{minimum}\left\{\frac{16 - \tau}{15}, \frac{-31 + 32\tau - \tau^2}{900}, \frac{-450 + 60\tau - \tau^2}{450}\right\}\right].$$

The last term in the expression on the right-hand side comes from the term for $\tau < \sigma$ in (5.2). Remember that $1 \le \tau < \sigma$ so $\sigma \ge 2$, and therefore this term has a minimum for $\sigma = \tau + 1$. To compare the functions

$$\gamma_1(\tau) = \frac{16 - \tau}{15},$$

$$\gamma_2(\tau) = \frac{-31 + 32\tau - \tau^2}{900},$$

$$\gamma_3(\tau) = \frac{-450 + 60\tau - \tau^2}{450},$$

it is useful to graph them. Since these functions are defined for $\tau = 1, 2, \ldots, 30$, the graph of each will consist of a set of 30 points. For the moment let us assume that the functions are extended to the interval $1 \le \tau \le 30$ by using the same formulas. The graphs of these extended functions are given in Fig. 2-9. It is clear from this figure that the minimum of the three functions is γ_3 for $1 \le \tau \le \tau_1$, where τ_1 is the solution of the equation $\gamma_3(\tau) = \gamma_2(\tau)$; it is γ_2 for $\tau_1 \le \tau \le \tau_2$, where τ_2 is the solution of $\gamma_1(\tau) = \gamma_2(\tau)$; and it is γ_1 for $\tau_2 \le \tau \le 30$. Thus the function $\gamma(\tau) = \min[\gamma_1(\tau), \gamma_2(\tau), \gamma_3(\tau)]$ has

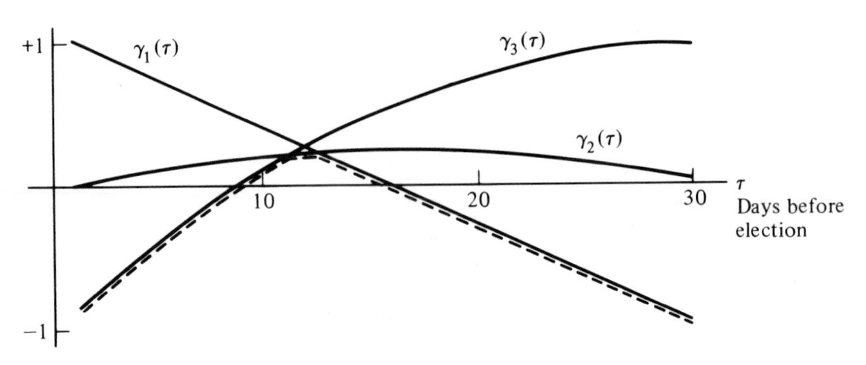

Fig. 2–9

the graph given by the dashed line in Fig. 2–9. Restricting our attention to the values of these functions on the integers, it follows from the graph and a little computation that

$$\begin{aligned} \gamma(\tau) &= \gamma_3(\tau), & \tau &= 1, 2, \dots, 11, \\ \gamma(\tau) &= \gamma_2(\tau), & \tau &= 12, \\ \gamma(\tau) &= \gamma_1(\tau), & \tau &= 13, 14, \dots, 30, \end{aligned} \tag{5.3}$$

and that

$$\underset{\tau \in \Lambda}{\text{maximum}} \ \underset{\sigma \in \Lambda}{\text{minimum}} \ E_R(\tau, \sigma) = \gamma_2(12). \tag{5.4}$$

The best strategy for the Redeemer Party is now clear. They should broadcast 12 days before the election. Also, since the value of the right-hand side of (5.4) is $209/900 \cong 0.232$, the Redeemer Party can expect to gain 0.232 unit of votes even if the Scoundrel Party utilizes its best strategy. This result is determined by the relative behavior of the functions P_R and P_S.

We emphasize once again that it is not the answer which is of primary interest here, but instead the model-building process.

EXERCISES

1. What is the best strategy for the Scoundrel Party in the situation considered in the text?
2. Use the expression of $E_R(\tau, \sigma)$ in (5.2) to show that
$$\max_{\tau \in \Lambda} \min_{\sigma \in \Lambda} E_R(\tau, \sigma) = \max_{\tau \in \Lambda} \left[\min \left\{ \frac{16 - \tau}{15}, \frac{-31 + 32\tau - \tau^2}{900}, \frac{-450 + 60\tau - \tau^2}{450} \right\} \right].$$
3. Show that γ is as given by (5.3).

4. Show that (5.4) is true.

5. Using the undefined terms, definitions, and Axioms 1 and 2 as a start, continue the development of the mathematical model for this problem. Your axioms should include something similar to (5.1) and a criterion for the selection of t_R.

6. Consider the situation described in this section with the following modification. The Scoundrel Party has A units of weakly committed votes and the Redeemer Partly has B units of such votes. Find an expression analogous to (5.1) in this situation.

7. Suppose in contrast to the situation discussed in this section that all votes presently committed to either party are firmly committed and will not be influenced by a broadcast of either party. However, there are undecided votes which prior to the broadcast are not committed to either party. Construct a model and find the expression analogous to (5.1) in the case of a campaign with 2 units of undecided votes 30 days prior to the election.

8. Formulate a model for the following situation, known in the literature as a *noisy duel*. Mr. Honorable and Mr. Rogue decide to settle a dispute on the field by the river at sunrise. Each has a loaded pistol with one bullet, and we assume that the pistols are noisy This means that if one person fires his pistol the other will know it. They begin at a certain distance apart and they walk toward one another. We assume that they become more accurate as the distance between them decreases. Indeed, if either fires at the original distance he is certain to miss, and when they are face to face each is certain to hit. The rules of the duel do not permit retreat. State your definitions and assumptions carefully.

9. Formulate a model for the *silent duel*. The situation is the same as in Exercise 8 except that if one person fires and misses, his opponent does not know he has been fired upon.

2.6 THE BEST MAN FOR THE JOB

There are problems in every branch of science which are essentially combinatorial in nature. Certain of these problems arising in the management sciences, especially allocation and search questions, have relatively well-developed theories associated with them. Other questions, particularly in the life sciences, are almost completely open. Many of these questions can be phrased using everyday language and yet involve interesting and frequently very deep combinatorial problems. It is possible to give the flavor of the type of questions which arise without embarking on a systematic study, and we shall give here an example of an assignment problem which is typical. This example also illustrates a practical concern which is common in these problems, namely, the need of finding an effective means of handling large-scale situations. That is, the task of obtaining a useful algorithm is both important and nontrivial. Most problems actually arising in practice require the use of computers, and an increase in the speed of computation has a very real effect on the types of algorithms which are feasible. We shall return to the discussion of algorithms for assignment problems in Chap. 7.

2.6.1 The Problem

The well-known firm of T.A. and A. finds itself in the position of having many available positions at its worldwide facilities. It decides to use its scientific personnel in a large-scale recruiting drive. These men will go out and interview applicants and then report back to the main office concerning the qualifications of the men interviewed. The personnel staff at the main office will review these reports and decide who should be hired and what job they should be given. The personnel department must decide when enough people have been recruited so that all jobs can be filled. Since these recruiting activities obviously take skilled scientists away from their regular work, it is important that the recruiting be as limited as possible.

2.6.2 A Real Model

Here we shall identify some specific problems and look more closely at some special cases. The problem can be illustrated by means of a simple diagram. Let points $J_1, \ldots, J_n$ represent available jobs, the jobs need not be distinct, and points $A_1, \ldots, A_m$ represent applicants. Let the fact that applicant A_i is qualified for job J_k be indicated by connecting A_i to J_k with a line. In this manner the qualifications of the applicants, both singly and as a group, can be easily evaluated and compared. Figure 2–10 is an example of such a diagram or graph with $m = 3$ and $n = 4$. In a situation

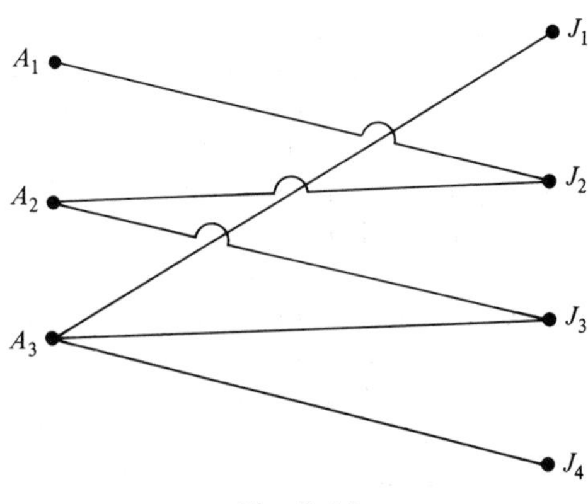

Fig. 2–10

involving many jobs and many applicants, a diagram such as this becomes quite complicated, but it may still be of some use.

There are certain approximations which go into a treatment of this problem. First it is difficult to decide when an applicant is really qualified for a job. In the general case, it would probably be best to rate people for jobs, and thus someone could be "well qualified" as opposed to "barely qualified." However, in this section, we shall assume that either an applicant is qualified for a job or he is not. There are no intermediate stages. In Chap. 7 we shall discuss some generalizations.

As a second approximation, we assume that each applicant will remain available while the firm considers his credentials. In fact, it may be that a number of applicants will accept other jobs or for other reasons will become unavailable.

Subject to these approximations, an initial examination would seem to indicate that the problem can be easily handled by trial and error. One simply tries all the possible ways of assigning applicants to jobs, and then a check is made to see if each job is filled by a qualified applicant under one of these assignments. In principle this is true. If a suitable assignment of applicants exists, then certainly it can be found by trial and error. However, the number of trials may be extremely large. The number of possible assignments increases very quickly with an increase in the numbers m and n, and it soon becomes virtually impossible to check all of them. For example, with $m = n$, there are $n!$ ways of assigning the applicants to the jobs. Even for n as small as 10, this gives more than 3 million possible assignments. Since it is likely that both m and n exceed 100 in practice, even the use of a computer does not make the procedure realistic. A method which is more efficient than trial and error must be found.

2.6.3 Comments on the Model and Its Theory

There are really two separate problems here. First, there is the problem of deciding when you have enough qualified applicants to fill the available jobs. Second, there is the problem of actually assigning the applicants to the jobs. For large numbers m and n, this is a nontrivial problem, and some thought is necessary to decide how to proceed. This situation is a good example of the difference between proving that a problem has a solution (giving an existence proof that there are enough applicants) and actually constructing a solution (giving an algorithm for assigning the applicants to the jobs). In this section, we shall concentrate on giving necessary and sufficient conditions that a given set of applicants be capable of filling all the available jobs. Certain aspects of the algorithm will be considered in Chap. 7.

An obvious necessary condition is that there be as many applicants as there are jobs. Clearly, this is not sufficient since all applicants might be qualified for the same job and only for that one. It seems, therefore, that we need a condition that the applicants are as a group qualified for all jobs. However, even this is not sufficient. For example, consider the situation depicted in Fig. 2–11.

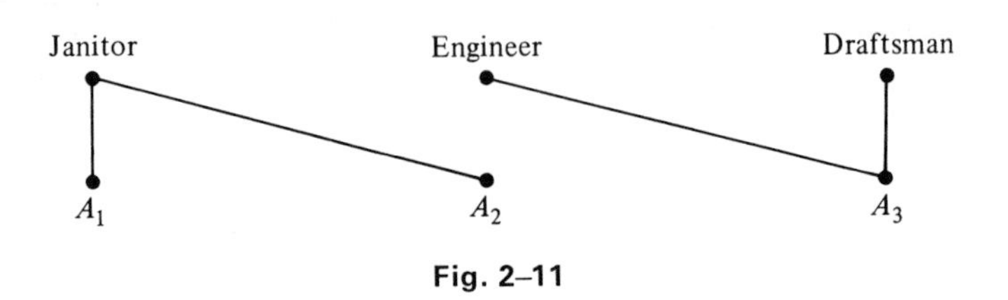

Fig. 2–11

This situation raises another point. Although applicant A_3 is (probably) qualified for the position as janitor, it is unlikely that he would accept the position if it were offered. This motivates the following definition.

Definition: An applicant is *appropriate* for a job if he is qualified for the job and willing to accept it if it is offered. A set of applicants is *complete* if for each subset of k jobs there are at least k applicants who are appropriate for the jobs, in the sense that each applicant is appropriate for at least one of the jobs, $k = 1, 2, \ldots, n$.

We can now state the main result of this section as a theorem.

Theorem 1: A necessary and sufficient condition that a job assignment problem be solvable is that the set of applicants be complete.

Proof: The necessity part of the theorem is quite simple. It is enough to point out that if we can assign all the jobs, then corresponding to each subset of k of the vacant jobs, we automatically have a set of k appropriate applicants. We just take the k applicants who were assigned to these jobs.

The sufficiency part is much harder. Our proof is given by an induction on n, the total number of jobs to be filled. We suppose that the set of applicants is complete.

$n = 1$. In this case there is only one job to be filled. Since the set of applicants is complete, there is an applicant who is appropriate for this job. We assign him, and the problem is solved.

$n = 2$. There are two vacant jobs, J_1 and J_2. Since the set of applicants is complete, we know that there are at least two applicants, each of whom is appropriate for at least one of the two jobs J_1 and J_2. Also, there is at least one applicant appropriate for J_1 and at least one appropriate for J_2. We conclude that there are two applicants A_1 and A_2, for which one of the following statements is true:

(a) Both A_1 and A_2 are appropriate for both jobs.
(b) One applicant is appropriate for both jobs, and the other is appropriate for only one job.
(c) One applicant is appropriate only for J_1, and the other is appropriate only for J_2.

No matter which case holds, it is clear that job assignment is possible.

Now we assume that the theorem holds for $n = 1, 2, \ldots, k$ and prove that it holds for $n = k + 1$. There are two cases to consider.

Case 1. We suppose that the set of applicants is complete with "room to spare." Thus, given any l jobs, $1 \leq l \leq k$, there are always *more* than l applicants who are appropriate for these jobs. In this case we begin by choosing a specific vacant job. By the completeness condition we know that some applicant is appropriate for this job. Assign this applicant to the chosen job. Now consider the remaining k jobs. For each l, $1 \leq l \leq k$, there must be at least l applicants who find these jobs appropriate since originally there were more than l and only one applicant has been removed. By the induction hypothesis we can fill these jobs with appropriate applicants.

Case 2. Suppose that the set of applicants is not complete with room to spare. Thus there is a subset B of jobs containing l jobs, $1 \leq l \leq k$, and there are exactly l applicants appropriate for these jobs; i.e., no applicant outside of this set of l applicants is appropriate for any job in B. Now $1 \leq l \leq k$ and the set of applicants for the jobs in B is complete (Exercise 3), and therefore by the induction hypothesis we can assign jobs in B. Consider the remaining $k + 1 - l$ jobs. Since $1 \leq k + 1 - l \leq k$, the proof of the assertion will be complete if we can show that the remaining applicants are complete for the remaining jobs. If not, there would exist a subset D of m jobs, $1 \leq m \leq k + 1 - l$, for which there does not exist m applicants (not assigned to jobs in B) each of which is appropriate for some job in D. But then there does not exist a set of at least $m + l$ applicants each of which is appropriate for a job in $D \cup B$. This contradicts the completeness of the original set of applicants and therefore no such set D can exist.

The proof is now complete by induction.

EXERCISES

1. Form a mathematical model for the situation discussed in this section. Be careful to identify your undefined terms, definitions, axioms, etc.
2. Find an algorithm for efficiently assigning five applicants to five jobs. Estimate

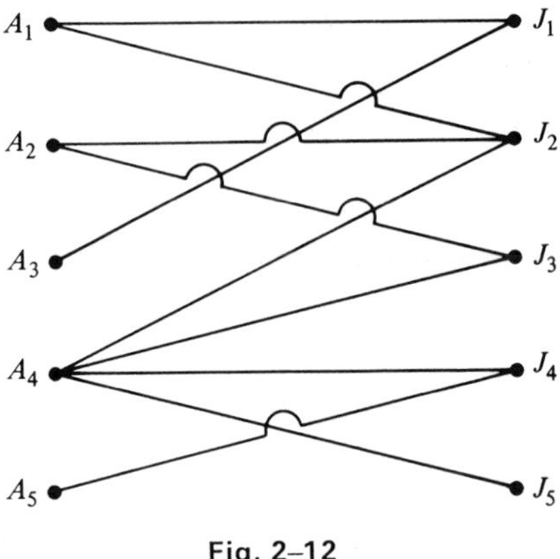

Fig. 2–12

the efficiency of your method, and use your method on the set of applicants given in Fig. 2–12. (A_i is connected to J_k if and only if A_i is appropriate for J_k.)

3. Establish the assertion in Case 2 of the induction proof of Theorem 1. That is, show that the set of applicants for the jobs in B is complete.

4. Discuss the situation when each applicant is either unqualified, minimally qualified, or highly qualified for each job. Find an analog to Theorem 1 in this case.

2.7 PROJECTS

The first six sections of this chapter have provided examples of model building and some exercises for the reader to test his understanding of the discussion. We now provide an opportunity for deeper study and more creative work. The projects provided here, and throughout the book, are of different types. Some include a rather complete description of the model and the reader is asked to develop some of the theory. In others, it is necessary to develop a model more or less from first principles. We emphasize that in general there will be several models, and the conclusions resulting from the theory may not always be the same. The reader is encouraged to think about those aspects of the theory which seem to go against his intuition or to be incompatible with reality. Those assumptions in the model which give rise to such conclusions must be considered carefully.

2.7.1 Another Look at Cell Growth

In Sec. 1.4 we studied a model which arose in the consideration of a cell growth problem. That model is of a type known as *deterministic*.

Recall that this name results from the fact that these models lead to a prediction of the precise nature of future behavior. Thus in the cell model the total number of cells at time t is completely determined by the number of cells at time $t = 0$ and the time T between divisions. It was pointed out in Chap. 1 that this deterministic model may not give correct predictions about the actual growth of the cells. In this project a second model for cell growth will be developed. This model is of a type called *probabilistic* or *stochastic*. In this case the name reflects the fact that future behavior is not completely determined but that instead there are a number of possible results and that each has a certain probability of occurring.

The change from the deterministic model to the stochastic one is brought about by changing the axiom dealing with the time distribution of cell divisions. Instead of stating that a cell always divides after T seconds, a new axiom is chosen to allow for the possibility that a cell may divide at almost any time. Naturally, it is more likely to divide at certain times than at other times, and this fact must be expressed mathematically. This is done by considering the function f which gives the probability that a cell which divided at time $t = 0$ will divide again before time t.

Problem 1. Discuss the nature of the function f. What can you say about the graph of f?

When one is dealing with a specific type of cell, then the function f can be estimated from observable data. The observer records the different times between successive divisions and uses them to estimate the desired probabilities. It might be assumed, for example, that the function f has the form $f(t) = 1 - \alpha^t$ for some constant α, $0 < \alpha < 1$. This function has the properties which one would expect of f based on its definition. However, since it is a transcendental function, it might (and indeed will) lead to rather complicated expressions later in the development. Thus we look for a simpler function which gives an acceptable approximation to f, at least for certain times. It is natural to consider polynomial approximations.

Problem 2. Let $\lambda = -\log \alpha$, and note that $\lambda > 0$. Show that the first-order polynomial λt agrees with f and its derivative agrees with f' at $t = 0$, and that it is the only first-order polynomial for which this is true.

Recall that $f(t)$ gives the probability that a cell which divided at time 0 divides again before time t. Thus the probability that a cell which divided at time t_1 will divide again before time $t_1 + \Delta t$ is, by Problem 2, approximately equal to $\lambda \, \Delta t$. This provides some motivation for the following:

Axiom: In a short time Δt, a cell has probability $\lambda \, \Delta t$ of dividing exactly once, and the probability of dividing more than once in time Δt is small in comparison to Δt.

Next, let $P_n(t)$ be the probability that there are exactly n cells at time t.

Problem 3. Explain why the following is a true statement:

$$P_n(t + \Delta t) = P_n(t)(1 - n\lambda \Delta t) + P_{n-1}(t)(n - 1)\lambda \Delta t + h(\Delta t),$$

where $\lim_{\Delta t \to 0}[h(\Delta t)/\Delta t] = 0$.

Hint. Consider the various ways in which a collection of n cells can arise at time $t + \Delta t$. Use elementary probability theory and then the binomial theorem to expand terms of the form $(1 - a)^n$.

Problem 4. Use the definition of the derivative to show that

$$P_n'(t) = -n\lambda P_n(t) + (n - 1)\lambda P_{n-1}(t).$$

In Problem 4 we have derived a differential equation for the unknown function P_n. It is possible to solve this equation and obtain P_n. However, this requires a knowledge of how to solve such equations, and, moreover, in general one is not really interested in the individual probabilities P_n. Instead, one usually would like some idea of the total number of cells to be expected to exist at time t. Such a situation often occurs in problems where many outcomes are possible and a numerical value of some sort changes with the outcome. The quantity which is usually used in such cases is the *expected value* of the numerical quantity. From the definition of expected value, the expected number of cells at time t is given by the formula

$$E(t) = \sum_{n=0}^{\infty} nP_n(t)$$

Problem 5. Show that $E'(t) = \lambda E(t)$ and that if the initial size of the population is N, then $E(t) = Ne^{\lambda t}$. Let $\lambda = (\log 2)/T$ and compare this result with Sec. 1.4.

2.7.2 To ABM or Not to ABM

Formulate a mathematical model for the defense of cities against attack by enemy missiles. Assume that the offense has a stockpile of missiles each of which can be launched against any city. Also, assume that the defense has radar, computers, and antimissile missiles to counter any attacking force and that the offense knows this. Consider the different possible strategies for both the offense and defense and show how these will depend on the number and accuracy of the missiles.

2.7.3 Return to Simple Learning

Consider an experiment where a rat is placed at the bottom of a T-maze and allowed to run free for a certain period of time. Suppose that a food pellet is placed on the right-hand side of the maze before each trial and that the left-hand side is constructed so that if the rat turns left, it is confined. Each trial terminates when the rat discovers the food pellet, is confined on the left, or a specified amount of time elaspes. We wish to construct a model which describes the rat's presumed tendency to favor the right as trials progress.

We shall assume that there are only two modes of behavior for the rat—turning to the right and not turning to the right. In other words, we shall not distinguish between the response when the rat turns left and, say, when the rat just stays in one position until the time for the trial runs out. Furthermore, we shall assume that the probability of the rat turning to the right on a given trial depends on the probability that it turned to the right on the preceding trial. In particular, we assume that this dependence is a linear one. That is, if P_n is the probability that the rat turns to the right on trial n, then

$$P_{n+1} = \alpha + (1 - \alpha)P_n,$$

$0 < \alpha < 1$, $n = 1, 2, 3, \ldots$. It follows from this that

$$P_{n+1} - P_n = \alpha(1 - P_n).$$

This relation seems to be plausible, since it means the increase in probability from one trial to the next is proportional to the maximum possible increase.

Problem 1. Construct a mathematical model for this real model. Take *trial* and *alternative* as undefined terms.

Problem 2. Prove that the following facts are true in the theory of the model:

1. The sequence $P_1, P_2, \ldots$ is increasing.
2. $P_{n+1} = 1 - (1 - \alpha)^n(1 - P_1)$ for $n = 0, 1, 2, \ldots$.
3. $P_n \longrightarrow 1$ as $n \longrightarrow \infty$.

Next, define an error to be any behavior other than a turn to the right, and let T_n denote the total number of errors made by the rat in the first n trials.

Problem 3. Show that the expected value of the quantity T_n is given by $E[T_n] = (1 - P_1)[1 - (1 - \alpha)^n]/\alpha$.

Hint. Define a number x_n in the following manner:

$$x_n = \begin{cases} 0, & \text{if a turn to the right is made on trial } n, \\ 1, & \text{if an error is made on trial } n. \end{cases}$$

Then

$$T_n = \sum_{i=1}^{n} x_i \qquad \text{and} \qquad E[T_n] = \sum_{i=1}^{n} E[x_i].$$

Returning to our fundamental assumption, we have studied the situation in which $P_{n+1} = \alpha_0 + \alpha_1 P_n$, where $\alpha_1 = 1 - \alpha_0$. A more general model can be obtained if we let $\alpha_0 = \alpha$ and $\alpha_1 = 1 - \alpha - \beta$, where $0 < \alpha < 1$ and $0 \leq \beta \leq 1$. Then

$$P_{n+1} = \alpha + (1 - \alpha - \beta)P_n = P_n + \alpha(1 - P_n) - \beta P_n.$$

The last relation may be taken as a description of the situation when there is punishment as well as reward; that is, an increase in the parameter β is associated with a decrease in the chance of a positive response (right turn), and an increase in the parameter α is associated with an increase in the chance of a correct response. Regarding $\alpha(1 - P_n)$ as the increment in correct response probability, we see that this increment is proportional to the maximum possible increase, $1 - P_n$. Also βP_n can be interpreted as the increment in error probability, and this is proportional to the maximum decrease in correct response probability.

Problem 4. Establish the following facts about P_n under the new assumptions:

1. $P_{n+1} = [\alpha/(\alpha + \beta)](1 - (1 - \alpha - \beta)^n) + (1 - \alpha - \beta)^n P_1$.
2. $P_n \longrightarrow \alpha/(\alpha + \beta)$ as $n \longrightarrow \infty$.

Problem 5. Compare this model for learning with that of Sec. 2.4. How are they different and how are they similar?

2.7.4 Getting There Is Half the Fun

Formulate a mathematical model for the flow of intercity traffic within a collection of cities. Assume that there are different modes of transportation and that these have various capacities and costs. Consider ways of making the traffic flow "efficient."

2.7.5 More on Mirrors

In Sec. 2.3.2 it was pointed out that Axiom 1 could be used to construct a model for reflection in curved mirrors. With this in mind, discuss reflection in spherical, elliptical, and parabolic mirrors. Your discussion should include consideration of the following problems.

Problem 1. Show that if light is propagated from a source located at one focus of an elliptical mirror, then rays reflected from the mirror pass through the other focus.

Problem 2. Show that if light is propagated from a source located at the focus of a parabolic mirror, then rays reflected from the mirror are parallel.

Be sure to state your assumptions clearly and comment on whatever approximations you make. For example, it is necessary to make precise what is meant by a mirror and reflection. Can you use your work on Problem 1 to shed light on Problem 2?

In connection with Problem 2, it should be mentioned that since light propagation can be reversed, a parabolic mirror will focus parallel rays at one point. This fact has important applications both in everyday life and in astronomical research. In view of this importance, the answer to the following problem is interesting.

Problem 3. Characterize as completely as possible those mirrors for which parallel light rays are reflected through a single point.

Remembering our admonition that it is sometimes necessary to simplify the situation, at least initially, in order to make progress, it is helpful to consider the following. Suppose that we have an axially symmetric mirror with light rays parallel to the axis of symmetry incident upon the mirror. Because of the symmetry we consider a plane section containing the axis of symmetry. Introduce a rectangular coordinate system with the y axis as the axis of symmetry and the origin at the intersection of this axis with the surface of the mirror. The situation is pictured in Fig. 2–13, where the plane of the

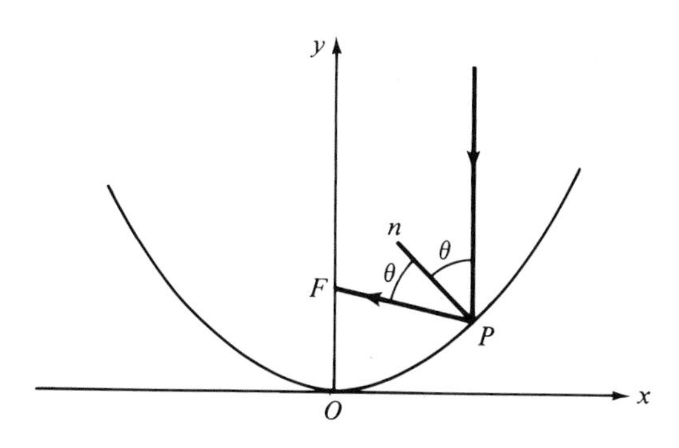

Fig. 2–13

paper is the plane section. Suppose that the equation of the plane section of the mirror is $y = f(x)$. The essential assumption is that if a ray is incident upon the mirror and is reflected at P (see Fig. 2–13), then the reflected ray passes through a fixed point F, *independent of which incident ray is considered.*

Problem 3a. If n is the normal to the surface of the mirror at P and if θ is as indicated in Fig. 2–13 show that

$$f(x) - x\tan\left(\frac{\pi}{2} + 2\theta\right) = k,$$

where k is a constant which is independent of P.

Problem 3b. If $f' = df/dx$, show that the equation obtained in Problem 3a leads to the following differential equation for f:

$$f + \frac{x}{2f'} - \frac{1}{2}xf' = k.$$

Problem 3c. Solve the differential equation for f.

Hint. Differentiate the equation, set $f' = g$, and obtain a homogeneous differential equation involving g. Note that this equation can be factored, and obtain a first-order differential equation for g. Solve for g and recall that $g = f'$, so that the result is a differential equation for f which can be solved by a simple integration.

2.7.6 How Is the Patient Today, Nurse?

The town of Healthville, U.S.A., has an acute shortage of doctors at its hospital. As a result, incoming patients can usually obtain only one quick examination. Since doctors have been known to make diagnostic mistakes, it is desired to have a check on each diagnosis by means of tests carried out by nurses and technicians. Moreover, information is desired about the probability that the results of the tests are correct, so that if they do not confirm the doctor's diagnosis, then there is justification for requiring that a second doctor examine the patient.

Construct a mathematical model for the diagnosis of a disease. Use undefined terms of *health state* and *test results*.

REFERENCES

This chapter is organized somewhat differently than the others in that it consists of six more or less independent sections. Although the point of view is the same in each section, the subject matter is not. Indeed, a reference which is very useful for one section may have nothing at all to contribute to the topic of another section. Therefore we have separated the references by section and provided comments for each section individually. If a reference is pertinent for more than one section, we give complete details the first time it is listed, and in subsequent listings we only refer back to the section containing the details. As is our custom, we make no attempt to provide a comprehensive bibliography. However, the reader will, of course, find other sources listed in those provided here.

2.1 Peas and Probabilities. Elementary probability textbooks often contain at least a short discussion of the use of probability theory in genetics. The last section of Chap. 2 in [Go] and Secs. 5 and 6 of Chap. V in [F] are at an appropriate level. Both derive the Hardy-Weinberg law on stationary genotype distributions. A mathematical account of population genetics is contained in [E], and some discussion of the experimental situation can be found in Chap. II of [N]. Also, the reader may find general works on genetics helpful, for example, [L]. A detailed account of Mendel's experiments, including statistics, can be found in [SS].

[E] EWENS, W. J., *Population Genetics*. London: Methuen, 1968.

[F] FELLER, W., *An Introduction to Probability Theory and Its Applications*, Vol. 1, 3rd ed. New York: Wiley, 1968.

[Go] GOLDBERG, S., *Probability, An Introduction*. Englewood Cliffs, N.J.: Prentice-Hall, 1960.

[L] LI, C. C., *Population Genetics*. Chicago: University of Chicago Press, 1955.

[N] NEYMAN, J., *Mathematical Statistics and Probability*. Washington, D.C.: U.S. Department of Agriculture, 1952.

[SS] STERN, C., and E. R. SHERWOOD, *The Origin of Genetics, A Mendel Sourcebook*. San Francisco: Freeman, Cooper, 1966.

2.2 Moving Mobile Homes. There are many good sources for a detailed study of transportation problems. In general, one should look under the topics of linear programming, integer programming, and networks and flows. [CC], [D], and [W] are appropriate basic references for the application of the techniques of linear programming to problems in business and economics. Elementary linear programming from a mathematical point of view is presented in [Sp]. Other examples of transportation problems are given in [KF].

[CC] CHARNES, A., and W. W. COOPER, *Management Models and Industrial Applications of Linear Programming*, Vols. I and II. New York: Wiley, 1961.

[D] DANTZIG, G., *Linear Programming and Extensions*. Princeton, N.J.: Princeton University Press, 1963.

[KF] KAUFMANN, A., and R. FAURE, *Introduction to Operations Research*. New York: Academic Press, 1968.

[Sp] SPIVEY, W. ALLEN, *Linear Programming: An Introduction*. New York: Macmillan, 1963.

[W] WAGNER, H. M., *Principles of Management Science*. Englewood Cliffs, N.J.: Prentice-Hall, 1970.

2.3 Water and Mirrors. There is a qualitative discussion of the nature of light and the historical development of the concepts in [Gl]. A more quantitative discussion of the notions of Sec. 2.3 can be found in the beginning chapters of [M]. This reference proceeds much further, and in Chap. 10 there is a discussion of the nature of light which compares the corpuscular and wave theories. Although written primarily for high school teachers of mathematics, Chap. 3 of [Sc] is much in the spirit of the approach taken here.

[Gl] GLUCK, I. D., *Optics*. New York: Holt, Rinehart and Winston, 1964.

[M] MORGAN, J., *Geometrical and Physical Optics*. New York: McGraw-Hill, 1953.

[Sc] SCHIFFER, M. M., *Studies in Mathematics*, Vol. X, *Applied Mathematics in the High School*. School Mathematics Study Group, Stanford, 1963.

2.4 Instant Intelligence. Models in learning theory can be found in many books and articles in the literature on mathematical psychology. The experiment considered here is known as an experiment in paired-associate learning. Much of this is motivated by the work reported in [Bo]. [ABC] presents a broad survey at the level of this book; there is a historical review in Sec. 1.5. [BBY] is a problem book designed to accompany [ABC], and most problems involve model construction in some way. [R] gives an application of the all-none learning theory model to education, and [RG] discusses and compares a number of models for learning.

[ABC] ATKINSON, R. C., G. H. BOWER, and E. J. CROTHERS, *An Introduction to Mathematical Learning Theory*. New York: Wiley, 1966.

[BBY] BATCHELDER, W. H., R. A. BJORK, and J. I. YELLOTT, JR., *Problems in Mathematical Learning Theory*. New York: Wiley, 1966.

[Bo] BOWER, G. H., "Application of a Model to Paired-Associate Learning," *Psychometrika*, **26** (1961), 255–280.

[BE] BUSH, R. R., and W. K. ESTES, *Studies in Mathematical Learning Theory*. Stanford, Calif.: Stanford University Press, 1959.

[E] ESTES, W. K., "Learning Theory and the New Mental Chemistry," *Psychological Review*, **67** (1960) V, 207–223.

[R] RESTLE, F., "The Relevance of Mathematical Models for Education," in *Theories of Learning and Instruction*, Sixty-third Yearbook of the National Society for the Study of Education, part I. Chicago: University of Chicago Press, 1964.

[RG] RESTLE, F., and J. GREENO, *Introduction to Mathematical Psychology*. Reading, Mass.: Addison-Wesley, 1970.

2.5 When To Broadcast and When To Wait. There are many references which indicate the relevance of game theory for problems arising in business and economics but fewer for applications to the behavioral sciences. However, [LR] discusses the relationship between the theory of games and the social sciences, and there are several articles in [BuN] which have a similar viewpoint to that taken here. [D] presents both the mathematical theory and applications and includes a chapter on games of timing (Chap. 9).

[BuN] BUCHLER, I. R., and H. G. NOTINI, eds., *Game Theory in the Behavioral Sciences*. Pittsburgh: University of Pittsburgh Press, 1969.

[D] DRESHER, M., *Games of Strategy: Theory and Applications.* Englewood Cliffs, N.J.: Prentice-Hall, 1961.

[LR] LUCE, R. D., and H. RAIFFA, *Games and Decisions.* New York: Wiley, 1957.

2.6 The Best Man for the Job. Assignment problems are usually studied in the general setting of networks and flows. [BeGH] and [Hu] present this material from a mathematical point of view. Some special cases are considered in [Ga]. Additional examples can be found in [KF].

[BeGH] BERGE, C., and A. GHOUILA-HOURI, *Programming, Games and Transportation Networks.* London: Methuen, 1965.

[Ga] GALE, D., *The Theory of Linear Economic Models.* New York: McGraw-Hill, 1960.

[Hu] HU, T. C., *Integer Programming and Network Flows.* Reading, Mass.: Addison-Wesley, 1969.

[KF] KAUFMANN, A., and R. FAURE, see 2.2.

3 Markov Chain Models

3.0 INTRODUCTION

There are many models which are based on the mathematical notion of a Markov chain and which have proved to be quite useful in the study of problems arising in the social and life sciences. Section 2.4 contains one example of a situation which can be modeled in these terms, and another is presented in Sec. 3.1. The situation described in Sec. 3.1 was selected for inclusion because it provides an application of Markov chain models to a question of current research interest in psychology and because the model building can be so clearly illustrated. Markov chains are introduced formally in Sec. 3.2, and the properties of some special types are discussed in Secs. 3.3 and 3.4. These discussions are continued and other situations in which Markov chains prove to be useful as models are introduced in the projects.

3.1 SMALL-GROUP DECISION MAKING

The decisions made by small groups, e.g., juries, panels of contest judges, and evaluation teams, have affected all of us to some extent. In most cases the decisions reached by a group of individuals depend on many factors. The results of the discussions of a team of automobile designers, for example, may have been determined by an extremely complex process. Not only are the merits of the alternatives involved, but also the personal interrelations of the members of the team play a role. If the corporation president is known to have a strong preference, this is likely to influence the thinking of the team. It may also happen that the structure of the group itself plays a role in the decision-making process. Once a group of six people reaches a division of five to one in favor of some alternative, the mere fact of this division exerts some influence on the dissenting member. In this section we shall explore the question of the effect of the structure of the group on the decision-making process.

3.1.1 An Experiment

We shall describe an experiment which has contributed to an understanding of the role of group structure in the decision-making process. To study only this aspect, the experiment must be designed so that other influences on the decision-making process are minimized. In particular, care must be taken to ensure that the alternative choices appear to be equally attractive and that no individual in the group assumes a leadership position. Such an experiment might be constructed along the following lines, and in fact more sophisticated versions of this experiment have been conducted [G]. A small group—we shall take a group of four members—performs a sequence of trials. Each trial consists of the presentation of a stimulus to be evaluated and a discussion on the merits of the various alternatives which continues until consensus is reached. A stimulus consists of a set of three pictures of geometrical designs which are to be evaluated according to some criteria. Each member of the group is able to convey his preference to the investigator without the other members of the group knowing what it is. Also, each member can change his preference at will. The subjects are instructed to express a preference as soon as the stimulus is displayed to the group and then to begin discussions seeking to reach a consensus. Each subject is to inform the group of each preference change as near as possible to the time it is conveyed to the investigator. After a consensus has been reached, the group is told which of the set of the three pictures is "best" in the sense of the criteria. The group is led to believe that there is a system in the assignment of values to each of the three pictures in each set, but actually the experimenter ranks each set in a random manner. Thus, insofar as the group is concerned, each of the three pictures in a set is equally preferable. Also, a technique of selective reinforcement is utilized to discourage the emergence of a group leader. In particular, the investigator will manipulate the selection of the best picture so that each member of the group makes the "correct" initial selection about the same percentage of the time. Thus there is no possibility that any member of the group appears to be a better evaluator than the others. Subjects are to be selected who have no previous acquaintance with one another.

Preference selections are monitored and recorded by the investigator. The process continues either for a certain number of trials or for a certain period of time. The experiment is repeated with several different groups, each consisting of four individuals. One might choose to discard the first few trials of each experiment, since during this period extraneous factors might enter in. For example, during this period the subjects are becoming familiar with the experimental procedure.

3.1.2 A Model

Since there are four individuals and three choices for each, there are $3^4 = 81$ possible distributions of preferences. However, the experiment has been designed so that the alternatives appear to be equally attractive and hence one ought not to distinguish among them. For example, if the alternatives are a, b, and c and if the choices of the respective members are a, b, b, b in one case and b, a, a, a in another, then these choices should be viewed as equivalent from the standpoint of the structure of the group. Also, there is no reason to distinguish between members of the group. That is, three votes for a and one vote for b is to be considered the same, independent of which group member casts the vote for b.

It follows that the only important information is the number of subjects who voted for the most popular alternative, the number which voted for the second most popular, and the number which voted for the third. That is, the relevant information concerning the preferences of the group at any instant is contained in a set of three integers (x, y, z), where x is the number voting for the most popular alternative, y is the number voting for the second most popular alternative, z is the number voting for the least popular alternative, and $x + y + z = 4$. The possible triples are $(4, 0, 0)$, $(3, 1, 0)$, $(2, 2, 0)$, and $(2, 1, 1)$.

The investigator is monitoring the preferences of the subjects continuously, and he records every change. A change of preferences of the group occurs whenever any subject changes his vote. If we refer to the above triples as *group compositions* and write them simply as xyz, then each preference change is equivalent to a change from one triple to another. It is sufficient to assume that only one vote changes at a time. In the rare event that two votes change simultaneously, one is selected arbitrarily as having been changed first. Thus preference changes are transitions between group compositions that can be effected by the change of a single vote. For example, $211 \rightarrow 220$ is an admissible transition, but $220 \rightarrow 400$ is not. The permissible shifts are

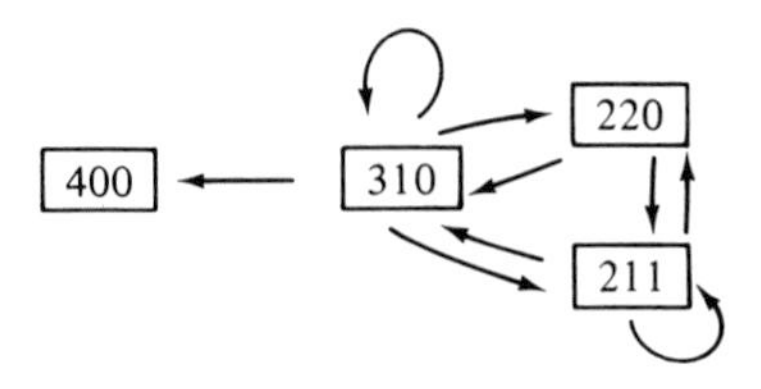

Fig. 3–1

indicated by arrows in Fig. 3–1. Note that it is possible for a single vote to change and a group with composition 211 to shift to a group with the same composition. This is also true for a group with composition 310. Since a new stimulus is presented as soon as consensus is reached, there are no possible shifts from the composition 400. Each experiment can be viewed as a sequence of transitions between various group compositions, terminating when the composition 400 is reached. Each transition corresponds to the change of a single vote on the part of a subject. Obviously, it is only the voting position of the subjects that is being measured, and all other aspects of what transpires during the experiment are neglected. Such a study might shed light on the relationship between the number of members voting for a certain alternative and the probability that others will change their votes to agree. Intuitively, one might expect that the size of a majority is important in the decision making of minority members. In some cases, these expectations have been supported by experimental studies [A].

The probabilities of the various transitions can be conveniently summarized in tabular form. The entries in the table will depend, of course, on just what is being assumed about how the subjects change their votes. For example, if each of the subjects is equally likely to change his vote to each of the other alternatives, then the probabilities of the shifts are given in the accompanying table. In developing the mathematical theory of Markov chain models, it is useful to write entries such as those occurring in this table in matrix form. Such matrices, known as transitions matrices, will be discussed in the next section. There we shall see how the entries in this table, when written in matrix form, play a fundamental role in our study of the

	Probability of shift	Composition after shift			
		400	310	220	211
Composition before shift	310	1/8	1/8	3/8	3/8
	220	0	1/2	0	1/2
	211	0	1/4	1/4	1/2

experiment. When writing the table as a transition matrix it will be necessary to add a row corresponding to transitions from a group with composition 400. In the real model, such transitions are meaningless and the row is omitted from the table.

EXERCISES

1. Consider an experiment analogous to that discussed in this section in which groups of five individuals select from four alternatives. Find the possible shifts in group composition and display them as in Fig. 3–1. Find a table similar to the one worked out above in this case.
2. Consider the experiment discussed in the text and suppose that the votes change in such a way that each of the possible shifts from a given group composition is equally likely. Find a table similar to the one in the text for this case.
3. In the experiment discussed in this section, define a shift toward consensus as one of the following: 211 $\longrightarrow$ 310, 220 $\longrightarrow$ 310, 310 $\longrightarrow$ 400. Find a table similar to that constructed in the text with the assumption that shifts toward consensus are twice as likely to occur as any other shift from the same group composition. All shifts other than those toward consensus are equally likely to occur.
4. What sort of assumptions regarding shifts in group compositions in the example of the text might account for a table of the form shown below.

	Probability of shift	Composition after shift 400	310	220	211
Composition before shift	310	1	0	0	0
	220	0	1	0	0
	211	0	1	0	0

5. Suppose that the probability of transition from one group composition to another is as given by the table in the text. If the initial group composition is 220, find the probability that after two vote changes the composition will be 400. (*Hint:* Use a tree diagram.)
6. Same as Exercise 5 with the table of Exercise 1.

3.2 BASIC PROPERTIES

3.2.1 Definitions and Notations

With the examples of Secs. 3.1 and 2.4 as a guide, we turn to a discussion of more general sequential processes. We consider a system which can be in any of N possible states, and we observe the system at n successive times. The concepts of *state* and of a system being *in a state* or *occupying a state* are taken as undefined terms. The reader may wish to assign these terms specific meanings and keep an example in mind as we proceed. We let $S = \{s_1, \ldots, s_N\}$ be the set of states, and when there is no danger of confusion, we shall refer to the states simply by their subscripts $1, 2, \ldots, N$. If the system is in state i at the kth observation and in state j at the $(k + 1)$th observation, then we say that the system has made a *transition* from state i to state j at the kth *trial*, *step*, or *stage* of the process.

At this point it is helpful to introduce another example. Consider a maze consisting of four connecting compartments as shown in Fig. 3–2. The compartments are identified both by color and by number. Suppose that a

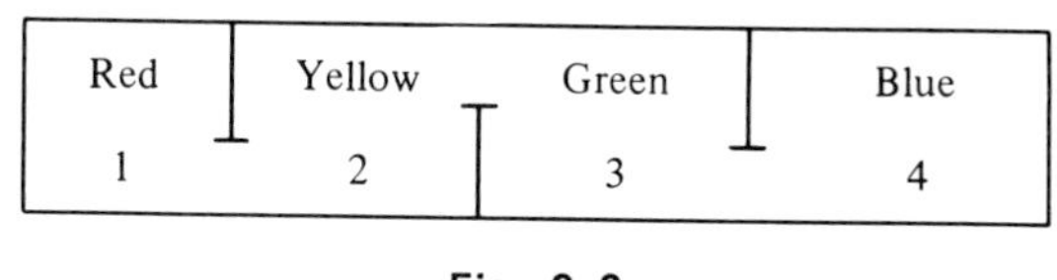

Fig. 3–2

mouse is released in the maze and that his behavior is observed. The set consisting of the mouse and the maze is the system to be investigated. The system is said to be in state i if the mouse is in compartment number i, $i = 1, 2, 3, 4$. Observations are to be made and the state of the system recorded at regular intervals and also each time the mouse moves from one compartment to another, necessarily an adjacent one. We assume that it is always possible to determine exactly which compartment the mouse occupies. To illustrate the possible transitions, suppose that on one observation the mouse is in compartment 2. Then, on the next observation it might be in compartment 1, 2, or 3; according to our conventions, it is not possible for it to be in compartment 4. We refer to this example as a one-dimensional four-state maze. One-dimensional mazes with N states are defined in the obvious way.

We assume that the mouse moves in a somewhat unpredictable way, and we describe its movements in probabilistic terms. Thus we are interested in the probabilities of the several possible movements. It is possible that the behavior of the mouse at a specific instant depends on where it is and how it arrived there. It might be, for example, that if the mouse is in compartment 3 on one observation, then the probability that it will be in compartment 4 on the next observation depends on which compartment it occupied at the preceding observation. The probability of a transition from 3 to 4 might be 1/2 if the preceding transition was from 2 to 3, and 1/4 if it was from 3 to 3 or 4 to 3. In this chapter we shall consider only the simplest possible situation, namely, the one in which the probabilities of the various movements from a specific compartment are independent of all the previous moves of the mouse. One might interpret this as an assumption that the mouse has no memory.

We make this idea precise in terms of the general processes under study.

Definition: Let p_{ij} denote the conditional probability that a system in state i on the kth observation is in state j on the $(k + 1)$th observation, $i, j = 1, 2, \ldots, N$. These probabilities are known as *transition probabilities.*

The process is said to be a *Markov chain* if the following fundamental condition is satisfied:

Basic Assumption: The transition probabilities p_{ij} depend only on i (the state occupied at the kth observation) and on j (the state occupied at the $(k + 1)$th observation). This condition includes the assumption that p_{ij} does not depend on the observation k. That is, p_{ij} is the same as if the system began in state i initially and consequently is completely independent of its previous moves.

The $N \times N$ matrix $\mathbb{P} = (p_{ij})$ (see Appendix C for notation) is known as the *transition matrix* for the Markov chain. If the system is in state i on the kth observation, then it must be in some state on the $(k + 1)$th observation, and therefore $\sum_{j=1}^{N} p_{ij} = 1$. Thus $\mathbf{p}_i = (p_{i1}, \ldots, p_{iN})$ is a probability vector, and since this holds for each $i, i = 1, 2, \ldots, N$, we conclude that each row of $\mathbb{P}$ is a probability vector.

The entries in the matrix $\mathbb{P}$ give the probabilities of transitions between states in one step. The probabilities of transitions between states in several steps are also of interest.

Definition: Let $p_{ij}(m)$ denote the conditional probability that a system in state i initially is in state j on the mth observation.

In the example of a one-dimensional four-state maze introduced above, suppose that there is a constant probability p that the mouse moves toward the red compartment on any trial, independent of its previous moves. Also, suppose that the probability q of the mouse moving toward the blue compartment on any trial is independent of its previous moves. Then $t = 1 - p - q$ is the probability that on any trial it remains in the same compartment. With these assumptions this process can be modeled as a Markov chain. The transition matrix for such a model is

$$\begin{array}{c} \\ 1 \\ 2 \\ 3 \\ 4 \end{array} \begin{array}{c} \begin{array}{cccc} 1 & 2 & 3 & 4 \end{array} \\ \begin{bmatrix} 1-q & q & 0 & 0 \\ p & t & q & 0 \\ 0 & p & t & q \\ 0 & 0 & p & 1-p \end{bmatrix} \end{array}. \tag{2.1}$$

The numbers above and to the left of the matrix refer to the states. Thus the entry in row 2 and column 3 is p_{23}. The reader should keep such identifications in mind even though we do not usually include them. An exception occurs in Sec. 3.4, where such identification is crucial. The matrix (2.1) reflects several things. First, note that $p_{21} = p_{32} = p_{43} = p$ and that $p_{12} = p_{23} = p_{34} = q$. That is, the probability of moving toward the red compartment is the same whether the mouse is in the yellow, green, or blue one. A similar observation holds for the probability of moving toward the blue compartment. This might be interpreted as an assumption concerning the tendency of the mouse to move toward colors with certain characteristics. Also, the form of p_{11} and p_{44} reflects the fact that if the mouse is in one of the end compartments, then there are only two options available: Move toward the opposite end or remain in the same compartment.

As another example, consider the small-group decision-making experiment (Sec. 3.1) with a group of four individuals selecting from among three alternatives. In the notation of this section, we identify states as follows:

State	1	2	3	4
Group composition	400	310	220	211

Recall that the experiment ends when state 1 is reached. This can be interpreted as meaning that the probability of a transition from state 1 to another state is 0. Therefore, the probability of transition from state 1 to state 1 is 1. Also, of the remaining states 2, 3, 4, both states 2 and 4 admit transitions to themselves. The general form of the transition matrix for this example is

$$\begin{array}{cc} & \begin{array}{cccc} 1 & 2 & 3 & 4 \end{array} \\ \begin{array}{cc} 400 & 1 \\ 310 & 2 \\ 220 & 3 \\ 211 & 4 \end{array} & \begin{bmatrix} 1 & 0 & 0 & 0 \\ p & t & q & s \\ 0 & p' & 0 & q' \\ 0 & p'' & q'' & t'' \end{bmatrix} \end{array}, \tag{2.2}$$

where $p + q + s + t = 1$, $p' + q' = 1$, and $p'' + q'' + t'' = 1$. Notice that this form includes all the special cases introduced in Sec. 3.1. In particular, for the situation described in detail and summarized in the table at the end of Sec. 3.1 the transition matrix is

$$\begin{bmatrix} 1 & 0 & 0 & 0 \\ 1/8 & 1/8 & 3/8 & 3/8 \\ 0 & 1/2 & 0 & 1/2 \\ 0 & 1/4 & 1/4 & 1/2 \end{bmatrix}.$$

Note the presence of the first row reflecting transitions from state 1 (group composition 400), and recall that this was unnecessary when considering the real model.

3.2.2 Transitions in More than One Step

The matrix $\mathbb{P}(m)$ whose entries are $p_{ij}(m)$ is another transition matrix. The task of writing its entries is our present concern. Consider the four-state one-dimensional maze whose transition matrix is (2.1). For definiteness take $p = 2/3$, $q = 1/3$, and $t = 0$. Suppose that the mouse is initially released in compartment 2. If we think of each observation of the system as an experiment, then the tree diagram for the first three observations is given in Fig. 3–3. The numbers in the circles refer to the states, and the branch weights are the appropriate transition probabilities. These branch weights provide a well-defined tree measure. From this tree diagram, we deduce, for example, that a system which begins in state 2 will be in state 1 after three transitions with probability $(2/3)^3 + (2/3)^2(1/3) + (2/3)^2(1/3) = 16/27$. It will be in state 2 with probability $(2/3)^2(1/3) = 4/27$, in state 3

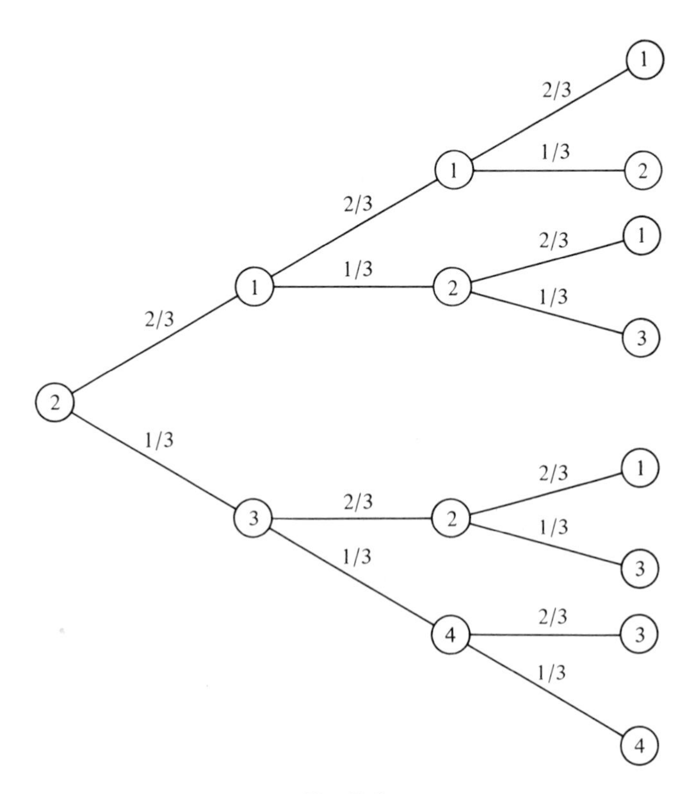

Fig. 3–3

with probability $(2/3)(1/3)^2 + (2/3)(1/3)^2 + (2/3)(1/3)^2 = 6/27$, and in state 4 with probability $(1/3)^3 = 1/27$. Using our usual notation, we have $p_{21}(3) = 16/27$, $p_{22}(3) = 4/27$, $p_{23}(3) = 6/27$, and $p_{24}(3) = 1/27$. It is clear that one can use tree diagrams to compute $p_{ij}(m)$ for $i, j = 1, 2, 3, 4$ and any m, but it is also clear that this will be a cumbersome technical process with many opportunities for error. Another method of a simpler nature is preferable. It is a fundamental property of Markov chains that one can obtain the higher-order transition probabilities $p_{ij}(m)$, $m > 1$, in a straightforward manner.

Lemma: If $\mathbb{P} = (p_{ij})$ is the transition matrix of a Markov chain, then $p_{ij}(m)$ is the i–j entry of $\mathbb{P}^m$, the mth power of the matrix $\mathbb{P}$.

Proof: According to the definition, $p_{ij}(m)$ is the conditional probability that a system in state i will be in state j after m transitions. Using the concept of conditional probability, we have

$$p_{ij}(m) = \sum_{k=1}^{N} \begin{bmatrix}\text{Conditional probability that a system in state} \\ i \text{ will be in state } k \text{ after } m-1 \text{ transitions}\end{bmatrix} \times \begin{bmatrix}\text{Conditional probability that a system in state} \\ k \text{ will be in state } j \text{ after one transition}\end{bmatrix}$$

$$= \sum_{k=1}^{N} p_{ik}(m-1) \cdot p_{kj}.$$

If we let $(p_{ij}(m)) = \mathbb{P}(m)$, then using the definition of matrix multiplication (Appendix C), we have $\mathbb{P}(m) = \mathbb{P}(m-1)\mathbb{P}$. Applying the same reasoning to $\mathbb{P}(m-1)$, we find that $\mathbb{P}(m-1) = \mathbb{P}(m-2)\mathbb{P}$. Continuing the argument, we obtain $\mathbb{P}(m) = \mathbb{P}(m-1)\mathbb{P} = \mathbb{P}(m-2)\mathbb{P}^2 = \ldots = \mathbb{P}(1)\mathbb{P}^{m-1}$. But $\mathbb{P}(1)$ is clearly equal to $\mathbb{P}$, and consequently $\mathbb{P}(m) = \mathbb{P}^m$ (see Exercise 5).

Q.E.D.

Example. Consider the four-state one-dimensional maze with $p = 2/3$, $q = 1/3$, and $t = 0$. Using (2.1) we have

$$\mathbb{P} = \begin{bmatrix} 2/3 & 1/3 & 0 & 0 \\ 2/3 & 0 & 1/3 & 0 \\ 0 & 2/3 & 0 & 1/3 \\ 0 & 0 & 2/3 & 1/3 \end{bmatrix} \tag{2.3}$$

$$\mathbb{P}^2 = \begin{bmatrix} 6/9 & 2/9 & 1/9 & 0 \\ 4/9 & 4/9 & 0 & 1/9 \\ 4/9 & 0 & 4/9 & 1/9 \\ 0 & 4/9 & 2/9 & 3/9 \end{bmatrix} \tag{2.4}$$

$$\mathbb{P}^3 = \begin{bmatrix} 16/27 & 8/27 & 2/27 & 1/27 \\ 16/27 & 4/27 & 6/27 & 1/27 \\ 8/27 & 12/27 & 2/27 & 5/27 \\ 8/27 & 4/27 & 10/27 & 5/27 \end{bmatrix} \tag{2.5}$$

Note that the second rows of $\mathbb{P}^2$ and $\mathbb{P}^3$ give the transition probabilities from state 2 in two and three steps, respectively. Thus the probability vector (4/9, 4/9, 0, 1/9) gives the probabilities that after two steps a system which is initially in state 2 will be in states 1, 2, 3, 4, respectively. Likewise, (16/27, 4/27, 6/27, 1/27) gives the probabilities for the system to be in these states after three transitions from state 2. A comparison with the tree diagram shows that this agrees with the corresponding probabilities calculated from the tree measure.

EXERCISES

1. Find the transition matrix for a five-state one-dimensional maze. Assume that p, q, and t have the meanings of this section.
2. Compute $\mathbb{P}^2$ and $\mathbb{P}^3$ for the transition matrix $\mathbb{P}$ defined by (2.2).
3. Let $\mathbb{P}$ be the transition matrix of a Markov chain. Prove directly, using the definition of matrix multiplication, that the rows of $\mathbb{P}^m$ are probability vectors.
4. Consider a four-state one-dimensional maze and suppose that a mouse moves according to the following rules:
   (a) Its first move is random, to an adjacent cell or the same one, each possibility equally likely.
   (b) For moves after the first, the mouse does not change direction. Precisely,
      (1) If the last previous move which involved a change of compartment was toward red, then the current move is toward red with probability 1/2, and is to the same compartment with probability 1/2.
      (2) If the last previous move which involved a change of compartment was toward blue, then the current move is toward blue with probability 2/3, and is to the same compartment with probability 1/3.
      (3) If no previous move involved a change of compartment, then the current move is toward red with probability 1/6 and toward blue with probability 1/3 and is to the same compartment with probability 1/2.

   Find the transition matrices $\mathbb{P} = \mathbb{P}(1)$, $\mathbb{P}(2)$, and $\mathbb{P}(3)$. Is this process a Markov chain (support your answer)?
5. The proof of the lemma given in the text is actually an induction argument. Write the proof in this form; i.e., use mathematical induction explicity.

3.3 REGULAR AND ERGODIC MARKOV CHAINS

The first two sections of this chapter contain several examples of the use of Markov chain models, and we now turn to a study of these processes in more detail. All Markov chains have certain features in common; however, they do not all behave in exactly the same way. Indeed, if $\mathbb{P}$ is the transition matrix (2.3) of a four-state one-dimensional maze, then by (2.5) every element of $\mathbb{P}^3$ is positive: $p_{ij}(3) > 0$ for all i and j. Consequently, for every pair of states i and j there is a positive probability that if the system is in state i on the initial observation, then it will be in state j on the third observation. It is easy to see that if $\mathbb{P}^3$ contains only positive entries, then the same is true of $\mathbb{P}^k$ for all $k > 3$ (Exercise 1). On the other hand, if $\mathbb{P}$ is the transition matrix (2.2) for the small-group decision-making experiment, then the first row of every power of $\mathbb{P}$ is (1, 0, 0, 0) (Exercise 2). Thus, if the system is ever in state 1, then it continues in state 1 for all subsequent trials. In particular, if on the first observation the system is in state 1, then this is the only state ever occupied by the system.

These examples give some indication of the possible types of behavior of Markov chains. In this and the next section we shall investigate certain aspects of the behavior of two quite different classes of Markov chains. These classes are generalizations of the examples discussed above. Our discussion is elementary and very much oriented toward the specific results which provide information on the models considered here.

3.3.1 Regular Markov Chains

Consider the one-dimensional maze with transition matrix $\mathbb{P}$ given by (2.1). A reasonable question is whether the mouse exhibits any preference for the blue compartment over the red one. In studying this question, one might consider the probabilities $p_{i1}(m)$ and $p_{i4}(m)$, $m \geq 1$. Near the beginning of an experiment, i.e., for the first few transitions, these probabilities may depend strongly on i, the compartment into which the mouse is released. However, as m becomes larger, it sometimes happens that $p_{i1}(m)$ tends to a limit which is independent of i. The meaning of this is, of course, that after many observations (or trials) the probability that the mouse is in the red compartment, state 1, is independent of the compartment into which it was released. Concern for the long-run behavior of the process is typical of many sequential experiments which can be modeled using Markov chains. The investigator hopes that no matter how the process begins, it will settle down to some stable and hopefully predictable behavior. Stable long-run behavior of a process which is independent of the initial state is, of course, not always to be expected. In the one-dimensional maze with $p = q = 0$, $t = 1$, the transition matrix $\mathbb{P}$ is the identity matrix $\mathbb{I}$ (1s on the main diagonal and 0s elsewhere), and consequently the system never leaves its initial state, whatever that may be. Thus, although the long-run behavior is highly stable, it is certainly not independent of the initial state.

We consider first a particularly nice type of process, one in which the long-run behavior is very regular. By the *long-run* or *asymptotic* behavior of the process we mean the behavior of $p_{ij}(m)$ for large m, that is, the behavior of the m-step transition probabilities for a large number m of trials. It was shown above that $p_{ij}(m)$ is the i–j entry in the matrix $\mathbb{P}^m$. Thus it is natural to single out those processes to be studied by specifying certain behavior of $\mathbb{P}^m$ for large m. Our definition is given in terms of behavior which is of interest to those doing experimental work.

Definition: A Markov chain with transition matrix $\mathbb{P}$ is said to be *regular* if

1. $\lim_{m \to \infty} p_{ij}(m)$ exists and is positive, $1 \leq i, j \leq N$, and
2. For each j, $1 \leq j \leq N$, the limit $\lim_{m \to \infty} p_{ij}(m) = \sigma_j$ is independent of i.

Since $\sum_{j=1}^{N} p_{ij}(m) = 1$ for $i = 1, 2, \ldots, N$, it follows that $\sum_{j=1}^{N} \sigma_j = 1$. Therefore the vector $\mathbf{s} = (\sigma_1, \ldots, \sigma_N)$ is a probability vector and its coordinates will be called the *stable probabilities* of the process. Using this terminology, a process is regular if for large m the m-step transition probability $p_{ij}(m)$ is close to the stable probability σ_j for all i.

If a Markov chain is regular, then by part 1 of the definition there is an integer $r \geq 1$ such that all entries in $\mathbb{P}^r$ are positive. Our first result shows that the converse of this is also true. Thus we have a characterization of regular Markov chains which does not require the evaluation of limits as does the definition, and consequently it is frequently a useful test for regularity.

Theorem 1: Let $\mathbb{P}$ be the transition matrix of a Markov chain. If there is an integer $r \geq 1$ such that $\mathbb{P}^r$ contains only positive entries, then the chain is regular.

We recall our convention regarding terminology: Positive means strictly positive; i.e., every entry in $\mathbb{P}^r$ is greater than 0. Also, we note that one can estimate the magnitude of r in terms of the number of states (see Exercise 12).

To demonstrate the utility of this theorem, we point out that it guarantees the regularity of the process based on a one-dimensional maze with transition matrix (2.1) and any p, $0 < p < 1$, $q = 1 - p$, $t = 0$. For this process we may take $r = 3$. It should be noted, however, that the theorem provides a test for regularity and is not a means of determining the stable probabilities. We return to the latter question in the next subsection.

The proof of Theorem 1 requires some preparation. First, it is convenient to restate conditions 1 and 2 of the definition of a regular chain. These conditions involve the convergence of the entries of a sequence of $N \times N$ matrices, and we turn to this notion. Let $\{\mathbb{A}(k)\}_{k=1}^{\infty}$ be a sequence of $N \times N$ matrices, $\mathbb{A}(k) = (a_{ij}(k))$. We say that $\{\mathbb{A}(k)\}$ converges and has limit

$\mathbb{A} = (a_{ij})$, written $\lim_{k\to\infty} \mathbb{A}(k) = \mathbb{A}$, if $\lim_{k\to\infty} a_{ij}(k) = a_{ij}$, $1 \le i, j \le N$. Thus the convergence of a sequence of $N \times N$ matrices is equivalent to the convergence of N^2 sequences of real numbers, the sequences of entries in the matrices. For example, the sequence $\{\mathbb{A}(k)\}$ with

$$\mathbb{A}(k) = \begin{bmatrix} 1 & e^{-k} \\ k^{-2} & k^{1/k} \end{bmatrix}, \qquad k = 1, 2, \ldots,$$

converges to the 2×2 identity matrix, while the sequence obtained by replacing $a_{22}(k)$ by $(-1)^k$, $k = 1, 2, \ldots$, does not converge at all. In terms of convergence of matrices, it is clear that the definition of a regular chain is equivalent to the following:

Definition: A Markov chain with transition matrix $\mathbb{P}$ is regular if there is a probability vector $\mathbf{s}$ with positive coordinates such that

$$\lim_{m\to\infty} \mathbb{P}^m = \begin{bmatrix} \mathbf{s} \\ \mathbf{s} \\ \cdot \\ \cdot \\ \cdot \\ \mathbf{s} \end{bmatrix}$$

The coordinates of $\mathbf{s}$ are the σ_i's of part 2 of the previous definition.

Another way of stating the last limiting relation is that the ith column of $\mathbb{P}^m$ tends to a limit vector all of whose coordinates are equal and positive. We approach Theorem 1 in this way.

We make use of a special notation for the rows and columns of $\mathbb{P}^m$ (Appendix C). Let $\mathbf{p}^i(m)$ and $\mathbf{p}_i(m)$ denote the ith column and row of $\mathbb{P}^m$, respectively, $i = 1, 2, \ldots, N$. Also, let $\mathbf{u}$ denote the vector in R^N all of whose coordinates are 1. We are to prove that there are positive constants $\sigma_1, \ldots, \sigma_N$ such that

$$\lim_{m\to\infty} \mathbf{p}^i(m) = \sigma_i \mathbf{u}, \qquad i = 1, 2, \ldots, N. \tag{3.1}$$

Let r be an integer such that all entries in $\mathbb{P}^r$ are positive, say all are greater than $\rho > 0$. It follows that all the entries of $\mathbb{P}^m$ are greater than ρ for all $m \ge r$ (Exercise 1). To begin our proof, we consider the first column $\mathbf{p}^1(m)$ of $\mathbb{P}^m$. Set $s(m) = \min\{p_{i1}(m), i = 1, 2, \ldots, N\}$ and $l(m) = \max\{p_{i1}(m), i = 1, 2, \ldots, N\}$. In terms of the model, $s(m)$ is the smallest probability that the system is in state 1 on the mth observation, and $l(m)$ is the largest such probability. The monotonicity of the sequences $\{s(m)\}_{m=1}^{\infty}$ and $\{l(m)\}_{m=1}^{\infty}$ is a useful fact.

Lemma: $s(m+1) \ge s(m)$ and $l(m+1) \le l(m)$ for $m = 1, 2, \ldots$.

Proof: We have

$$s(m+1) = \min\{p_{i1}(m+1), i = 1, 2, \ldots, N\}$$
$$= \min\{\mathbf{p}_i \cdot \mathbf{p}^1(m), i = 1, 2, \ldots, N\}.$$

Also, since $\mathbf{p}_i$ is a probability vector,

$$\mathbf{p}_i \cdot \mathbf{p}^1(m) = \sum_{j=1}^{N} p_{ij} p_{j1}(m) \geq s(m) \sum_{j=1}^{N} p_{ij} = s(m),$$

and consequently the minimum of these inner products must also be $\geq s(m)$. The proof of $l(m+1) \leq l(m)$ is similar. Q.E.D.

Returning to the proof of the theorem, we have $s(m) \leq l(m) \leq l(1)$ for all m, and hence $\{s(m)\}$ is an increasing sequence of real numbers which is bounded above. Therefore, by a fundamental property of the real numbers the limit $\lim_{m \to \infty} s(m)$ exists, and we denote this limit by s. Also, $l(m) \geq s(m) \geq s(1)$, and consequently $\lim_{m \to \infty} l(m) = l$ exists. Since $s(m) \leq l(m)$ for all m, we conclude that $s \leq l$. If $s = l$, then the proof of (3.1), for $i = 1$, is complete. We show that the remaining possibility, namely $s < l$, is impossible. In this case, the sequences $\{s(m)\}$, $\{l(m)\}$ and the limits s and l are as in Fig. 3–4. Set

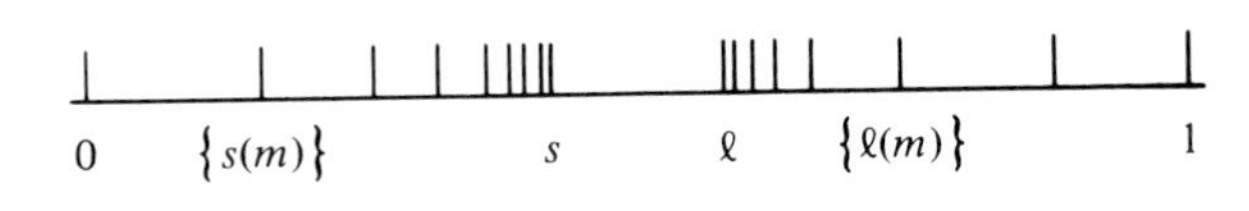

Fig. 3–4

$d = l - s$, $d > 0$. To deduce a contradiction to the assumption $s < l$, we show that in this case $s(m)$ increases more than a certain fixed amount (depending on ρ and d) in each sequence of r trials. In the notation of the present discussion we make the following assertion.

Assertion: If $p_{ij}(n) \geq \rho$ for $1 \leq i, j \leq N$, and $n \geq r$, if $s(m) \leq s < l \leq l(m)$ for all m, and if $l - s = d$, then $s(m+r) \geq s(m) + \rho d$, $m \geq 1$.

This assertion can be verified as follows. Let k be an index for which $s(m+r) = p_{k1}(m+r)$. We have

$$s(m+r) = \mathbf{p}_k(r) \cdot \mathbf{p}^1(m)$$
$$= \mathbf{p}_k(r) \cdot (\mathbf{p}^1(m) - s(m)\mathbf{u}) + s(m)(\mathbf{p}_k(r) \cdot \mathbf{u})$$
$$\geq s(m) + \max\{p_{kj}(r)(p_{j1}(m) - s(m)), j = 1, 2, \ldots, N\}$$
$$\geq s(m) + \rho d.$$

The second inequality results from the fact that at least one of the terms $p_{j1}(m) - s(m)$ is as large as d, and all the $p_{kj}(r)$ are at least as large as ρ.

It is now easy to show that the assumption $s < l$ leads to a contradiction. Indeed, since $s(m) \longrightarrow s$ as $m \longrightarrow \infty$, we can find an integer m_0 such that $s(m_0) > s - (\rho d/2)$. But the above assertion then implies that $s(m_0 + r) \geq s(m_0) + \rho d > s + (\rho d/2)$, which is obviously impossible in view of the definition of s. The proof of (3.1) for $i = 1$ is complete, and we have shown that the entries in the first column of $\mathbb{P}^m$ all tend to the same limit. Since $s(r) > 0$, this limit must be positive.

The same proof can be utilized with the column vectors $\mathbf{p}^i(m)$, $i = 2, 3, \ldots, N$, instead of $\mathbf{p}^1(m)$ to show that the entries in the ith column of $\mathbb{P}^m$ all tend toward the same limit. The limit will, in general, depend on i, the column index. The proof of Theorem 1 is now complete.

There is a generalization of Theorem 1 whose proof does not differ significantly from that given above. The result is that if a Markov chain has a transition matrix $\mathbb{P}$ such that $\mathbb{P}^r$ has some (strictly) positive columns for an integer r, then this Markov chain is *partially regular*. The meaning of partially regular is contained in the following precise statement of this more general result.

Theorem 1′: Let $\mathbb{P}$ be an $N \times N$ transition matrix for a Markov chain with the property that there is an integer r and a nonempty set $J \subset \{1, 2, \ldots, N\}$ such that every entry in $\mathbf{p}^j(r)$ (the jth column of $\mathbb{P}^r$) is no smaller than $\delta > 0$ for all $j \in J$. Then $\lim_{n\to\infty} \mathbb{P}^n = \mathbb{Q}$, and the matrix $\mathbb{Q}$ has the form

$$\mathbb{Q} = \begin{bmatrix} \mathbf{w} \\ \cdot \\ \cdot \\ \cdot \\ \mathbf{w} \end{bmatrix}, \qquad \mathbf{w} = (\omega_1, \ldots, \omega_N), \qquad \omega_j \geq \delta \quad \text{for} \quad j \in J.$$

3.3.2 Computation of Stable Probabilities

Theorem 1 provides a useful method of determining whether a chain is regular. It does not, however, give any information on the stable probabilities σ_i, $i = 1, 2, \ldots, N$. The next two results show how these probabilities may be obtained without explicitly evaluating the limit of $\mathbb{P}^m$.

Lemma: If $\mathbb{P}$ is a transition matrix and $\mathbb{P}^m$ converges to a matrix $\mathbb{Q}$ all of whose rows are the same vector $\mathbf{s}$, then $\mathbf{z}\mathbb{P}^m$ converges to $\mathbf{s}$ for any probability vector $\mathbf{z}$.

Proof: The condition $\lim_{m\to\infty} \mathbb{P}^m = \mathbb{Q}$ implies that $\mathbf{z}\mathbb{P}^m$ converges to $\mathbf{z}\mathbb{Q}$. Since $\mathbf{z}$ is a probability vector and all the rows of $\mathbb{Q}$ are the same, $\mathbf{z}\mathbb{Q} = \mathbf{s}$ (Exercise 4).

Theorem 2: If $\mathbb{P}$ is the transition matrix of a regular Markov chain, then there is a unique probability vector $\mathbf{s}$ which has positive coordinates and which satisfies $\mathbf{s}\mathbb{P} = \mathbf{s}$. Moreover, $\lim_{m\to\infty} \mathbf{p}_i(m) = \mathbf{s}$.

The vector $\mathbf{s}$ is therefore the vector whose coordinates give the stable probabilities. A vector $\mathbf{x}$ which satisfies $\mathbf{x} = \mathbf{x}\mathbb{P}$ is known as a *stationary vector* for the matrix $\mathbb{P}$. Thus Theorem 2 can be interpreted as saying that the stable probabilities of a regular Markov chain are the coordinates of the (unique) vector which is both a probability vector and a stationary vector for the transition matrix.

Proof: If $\mathbb{P}$ is the transition matrix of a regular Markov chain, then there is a probability vector $\mathbf{s}$ with positive coordinates such that $\lim_{m\to\infty} \mathbf{p}_i(m) = \mathbf{s}$, $i = 1, 2, \ldots, N$. Let $\mathbb{Q}$ be the $N \times N$ matrix all of whose rows are $\mathbf{s}$. We have both $\mathbb{P}^{m+1} \longrightarrow \mathbb{Q}$ and $\mathbb{P}^{m+1} = \mathbb{P}^m\mathbb{P} \longrightarrow \mathbb{Q}\mathbb{P}$, and therefore $\mathbb{Q}\mathbb{P} = \mathbb{Q}$. This is equivalent to $\mathbf{s}\mathbb{P} = \mathbf{s}$.

Finally, we prove uniqueness. That is, we show that if $\mathbf{x}$ is any stationary probability vector for $\mathbb{P}$, then $\mathbf{x}$ must be equal to $\mathbf{s}$. Using the relation $\mathbf{x}\mathbb{P} = \mathbf{x}$, we deduce $\mathbf{x}\mathbb{P}^2 = (\mathbf{x}\mathbb{P})\mathbb{P} = \mathbf{x}\mathbb{P} = \mathbf{x}$, and in general $\mathbf{x}\mathbb{P}^m = \mathbf{x}$ for $m \geq 1$. But the lemma asserts that if $\mathbf{x}$ is any probability vector, then $\mathbf{x}\mathbb{P}^m \longrightarrow \mathbf{s}$ as $m \longrightarrow \infty$. Thus $\mathbf{x} = \mathbf{x}\mathbb{P}^m \longrightarrow \mathbf{s}$ and $\mathbf{x}$ must be equal to $\mathbf{s}$. Q.E.D.

Theorem 2 can also be generalized to include the transition matrices of partially regular Markov chains (see Theorem 1′) and even more general transition matrices. The proof, however, is substantially more complicated in the general cases, and we do not consider it here. The reader is referred to [D], pp. 172–184, for the details.

To illustrate the applicability of this theorem, we consider the problem of finding the stable probabilities for the four-state one-dimensional maze with $p = 2/3$, $q = 1/3$, $t = 0$. The transition matrix $\mathbb{P}$ for this process is given by (2.3), and we are looking for a probability vector $\mathbf{s}$ such that $\mathbf{s}\mathbb{P} = \mathbf{s}$. Finding $\mathbf{s}$ is equivalent to solving a system of four linear equations in four unknowns and taking that solution which is a probability vector. If $\mathbf{s} = (\sigma_1, \sigma_2, \sigma_3, \sigma_4)$, then the equation $\mathbf{s}\mathbb{P} = \mathbf{s}$ is

$$\begin{aligned}
\tfrac{2}{3}\sigma_1 + \tfrac{2}{3}\sigma_2 \qquad\qquad\qquad\qquad &= \sigma_1, \\
\tfrac{1}{3}\sigma_1 \qquad\quad + \tfrac{2}{3}\sigma_3 \qquad\qquad &= \sigma_2, \\
\tfrac{1}{3}\sigma_2 \qquad\quad + \tfrac{2}{3}\sigma_4 &= \sigma_3, \\
\tfrac{1}{3}\sigma_3 + \tfrac{1}{3}\sigma_4 &= \sigma_4,
\end{aligned}$$

or

$$\begin{aligned}
-\tfrac{1}{3}\sigma_1 + \tfrac{2}{3}\sigma_2 \qquad\qquad\qquad &= 0,\\
\tfrac{1}{3}\sigma_1 - \sigma_2 + \tfrac{2}{3}\sigma_3 \qquad\quad &= 0,\\
+ \tfrac{1}{3}\sigma_2 - \sigma_3 + \tfrac{2}{3}\sigma_4 &= 0,\\
\tfrac{1}{3}\sigma_3 - \tfrac{2}{3}\sigma_4 &= 0.
\end{aligned}$$

Replacing the second equation by the sum of the first and second and the third equation by the sum of the third and fourth and eliminating the unnecessary equation, we have

$$\begin{aligned}
-\tfrac{1}{3}\sigma_1 + \tfrac{2}{3}\sigma_2 \qquad\qquad &= 0,\\
- \tfrac{1}{3}\sigma_2 + \tfrac{2}{3}\sigma_3 \qquad &= 0,\\
\tfrac{1}{3}\sigma_3 - \tfrac{2}{3}\sigma_4 &= 0.
\end{aligned}$$

We solve for σ_2, σ_3, σ_4 in terms of σ_1. If we set $\sigma_1 = \sigma$, then we find that $\sigma_2 = \sigma/2$, $\sigma_3 = \sigma/4$, $\sigma_4 = \sigma/8$. Thus the vector **s** must be $(\sigma, \sigma/2, \sigma/4, \sigma/8)$, and since it must be a probability vector, $\sigma + \sigma/2 + \sigma/4 + \sigma/8 = 1$, or $\sigma = 8/15$. Therefore, $\mathbf{s} = (8/15, 4/15, 2/15, 1/15)$. We conclude that after a sufficiently large number m of trials the system will be in state 1 about 8/15 of the time, in state 2 about 4/15 of the time, in state 3 about 2/15 of the time, and in state 4 about 1/15 of the time, independent of the initial state. It is interesting to compare this technique for obtaining **s** with a straightforward computation. Such a computation gives

$$\mathbb{P}^3 = \begin{bmatrix} 0.593 & 0.296 & 0.074 & 0.037 \\ 0.593 & 0.148 & 0.222 & 0.037 \\ 0.297 & 0.444 & 0.074 & 0.185 \\ 0.297 & 0.148 & 0.370 & 0.185 \end{bmatrix},$$

$$\mathbb{P}^6 = \begin{bmatrix} 0.568 & 0.258 & 0.129 & 0.053 \\ 0.516 & 0.302 & 0.107 & 0.075 \\ 0.516 & 0.214 & 0.194 & 0.075 \\ 0.429 & 0.302 & 0.151 & 0.119 \end{bmatrix},$$

$$\mathbb{P}^{18} = \begin{bmatrix} 0.534 & 0.266 & 0.133 & 0.066 \\ 0.534 & 0.267 & 0.133 & 0.067 \\ 0.534 & 0.266 & 0.134 & 0.067 \\ 0.533 & 0.267 & 0.133 & 0.067 \end{bmatrix},$$

where the entries are correct to three decimal places.

3.3.3 Ergodic Markov Chains

It was pointed out above that one important characteristic of regular Markov chains is that no matter what the initial state may be there is a positive probability that the system will be in any other state after sufficiently many trials. This sort of behavior is exhibited by some Markov chains which are not regular, and it is useful to define a class of chains having this property.

Definition: A Markov chain is *ergodic* if for every pair of states i and j there is an integer m, which in general depends on i and j, such that $p_{ij}(m) > 0$.

The class of ergodic chains contains the class of regular chains. Indeed, for each regular chain there is an integer r such that $p_{ij}(r) > 0$ for all i, j. It is not so obvious that the class of regular chains is a smaller class than the class of ergodic ones. To show this, we exhibit a chain which is ergodic but not regular. Consider the two-dimensional maze pictured in Fig. 3–5. The com-

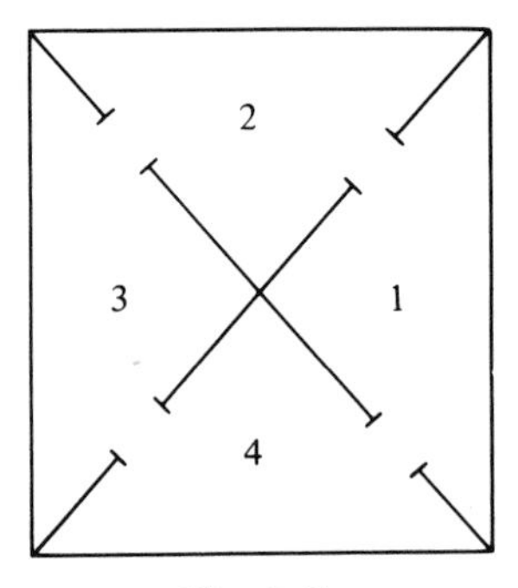

Fig. 3–5

partment numbers also identify the states. A mouse is released into a compartment and we assume that it moves to the adjacent compartment in the counterclockwise direction with probability q, into the adjacent compartment in the clockwise direction with probability p, and remains in the same compartment with probability t, $p + q + t = 1$. The transition matrix for this chain is

$$\mathbb{P} = \begin{bmatrix} t & q & 0 & p \\ p & t & q & 0 \\ 0 & p & t & q \\ q & 0 & p & t \end{bmatrix}. \tag{3.2}$$

Consider the special case $t = 0$. In this situation the mouse must move from a compartment with an even state number to one with an odd state number or vice versa. It follows that the odd and even powers of the transition matrix have the forms

$$\mathbb{P}^{2k+1} = \begin{bmatrix} 0 & X & 0 & X \\ X & 0 & X & 0 \\ 0 & X & 0 & X \\ X & 0 & X & 0 \end{bmatrix}$$

$$\mathbb{P}^{2k} = \begin{bmatrix} X & 0 & X & 0 \\ 0 & X & 0 & X \\ X & 0 & X & 0 \\ 0 & X & 0 & X \end{bmatrix},$$

where the Xs indicate nonzero entries. Every power of $\mathbb{P}$ contains some zeros, and consequently this chain is not regular. However, the chain is clearly ergodic. Either $p_{ij}(1) > 0$ or $p_{ij}(2) > 0$ for every i and j. Markov chains which are ergodic but not regular possess some of the properties of regular chains but not others. For example, if $\mathbb{P}$ is the transition matrix of an ergodic chain, then there is a unique stationary probability vector $\mathbf{s}$, $\mathbf{s}\mathbb{P} = \mathbf{s}$. However, the coordinates of $\mathbf{s}$ need not be stable probabilities in the sense of this section. Indeed, in general the powers $\mathbb{P}^m$ of an ergodic transition matrix do not tend toward a limit as m tends toward ∞. Nevertheless, the entries in this stationary vector are important in discussing the asymptotic behavior of the ergodic chain. In fact, these entries are the long-run average probabilities of being in the respective states. Thus, if $\mathbf{s} = (\sigma_1, \sigma_2, \ldots, \sigma_N)$ is the fixed probability vector for the transition matrix of an ergodic chain, then the long-run average probability of being in state i is σ_i, $i = 1, \ldots, N$. The proof of this result is somewhat involved, and we do not give it here. The interested reader should see [D], pp. 172–184, where this notion, among others, is made precise.

In concluding this section, it is appropriate to comment on terminology. Definitions of the various types of Markov chains are not uniform in the literature. The reader is urged to check carefully the usage of each author before beginning a study of particular results. Our terminology is consistent with [KMST] but somewhat different from that of [Fe] and [Fi]. Our Theorem 1 is sometimes called an *ergodic theorem* (e.g., Theorem 7.4.1 of [Fi]).

EXERCISES

In exercises consisting of a declarative statement, the reader is to prove the assertion.

1. If $\mathbb{P}$ is a transition matrix and r is an integer such that all entries of $\mathbb{P}^r$ are $\geq \rho$, then all entries of $\mathbb{P}^m$ are $\geq \rho$ for all $m \geq r$.

2. If $\mathbb{P}$ is the transition matrix (2.2) for the example of Sec. 3.1, then the first row of every power of $\mathbb{P}$ is $(1, 0, 0, 0)$.

3. If $\mathbb{P}$ is a transition matrix for a Markov chain and $\{\mathbb{P}^m\}$ converges to a matrix $\mathbb{Q}$ then the rows of $\mathbb{Q}$ are probability vectors. If some power of $\mathbb{P}$ has only positive entries, then $\mathbb{Q}$ has only positive entries.

4. If $\mathbb{Q}$ is a matrix all of whose rows are the same vector $\mathbf{s}$ and $\mathbf{z}$ is any probability vector, then $\mathbf{z}\mathbb{Q} = \mathbf{s}$.

5. Decide which of the following transition matrices are regular and find the stationary probability vector for each regular matrix.

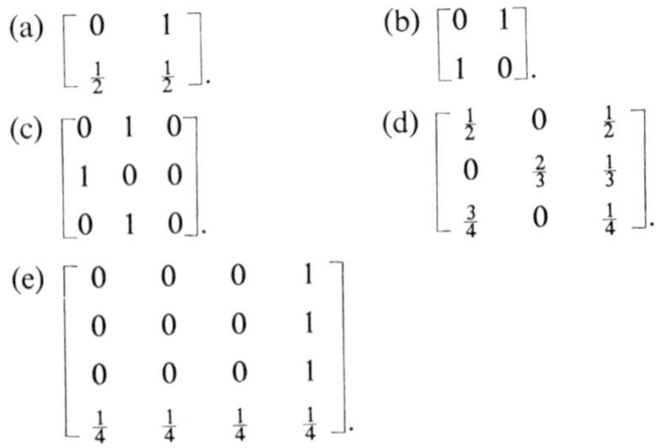

(a) $\begin{bmatrix} 0 & 1 \\ \frac{1}{2} & \frac{1}{2} \end{bmatrix}$. (b) $\begin{bmatrix} 0 & 1 \\ 1 & 0 \end{bmatrix}$.

(c) $\begin{bmatrix} 0 & 1 & 0 \\ 1 & 0 & 0 \\ 0 & 1 & 0 \end{bmatrix}$. (d) $\begin{bmatrix} \frac{1}{2} & 0 & \frac{1}{2} \\ 0 & \frac{2}{3} & \frac{1}{3} \\ \frac{3}{4} & 0 & \frac{1}{4} \end{bmatrix}$.

(e) $\begin{bmatrix} 0 & 0 & 0 & 1 \\ 0 & 0 & 0 & 1 \\ 0 & 0 & 0 & 1 \\ \frac{1}{4} & \frac{1}{4} & \frac{1}{4} & \frac{1}{4} \end{bmatrix}$.

6. Let $\{\mathbb{P}^{(k)}\}$ be a sequence of transition matrices ($\mathbb{P}^{(k)}$ is not necessarily a power of a fixed matrix) with the property that every entry in every matrix is $\geq \rho > 0$. Let $\mathbf{x} = \mathbf{x}^0$ be an arbitrary vector, and define $\mathbf{x}^{(m)} = \mathbb{P}^{(m)}\mathbf{x}^{(m-1)}$, $m = 1, 2, \ldots$. Prove that there is a constant γ such that $\lim_{m\to\infty} \mathbf{x}^{(m)} = \gamma\mathbf{u}$.

7. Find the stable probabilities for the four-state one-dimensional maze in the general case $p > 0$, $q > 0$, $t > 0$. Note that the stable probabilities depend only on the ratio p/q. Can you account for this in the model?

8. Prove or disprove (by giving a counterexample) that the transition matrix of an N-state regular chain must have at least $2N$ nonzero entries.

9. Let $\mathbb{P}$ be the transition matrix for the ergodic chain of the maze example of Sec. 3.3.3. Prove that if $t = 0$, then there is no probability vector $\mathbf{s}$ for which $\lim_{m\to\infty} \mathbf{p}_i(m) = \mathbf{s}$, $i = 1, 2, \ldots, N$. Is this true if $t \neq 0$?

10. Let $\mathbb{P}$ be the transition matrix of an N-state ergodic chain. Prove that for each $i, j, i \neq j$, there is an integer $m \leq N - 1$ such that $p_{ij}(m) > 0$. That is, there is a positive probability that the system will go between any two different states in not more than $N - 1$ steps. Use this to show that if the diagonal entries in $\mathbb{P}$ are positive ($p_{ii} > 0$, $i = 1, 2, \ldots, N$), then the chain is in fact regular.

11. (Continuation of 10) Let $\mathbb{P}$ be the transition matrix of an N-state ergodic chain and suppose that *some* diagonal entry of $\mathbb{P}$ is positive (there is *some* i such that $p_{ii} > 0$). Prove that the chain is regular.

12. Let $\mathbb{P}$ be the transition matrix of an N-state regular chain. Let r be the smallest integer for which $\mathbb{P}^r$ contains only positive entries.

(a) Show by example that in general $r \leq N$ is not true.
(b) It is easy to show $r \leq 2N(N-1)$. Do so.
(c) It is somewhat more difficult to show that $r \leq N^2 - N + 1$. Do so.
It is true, but still more difficult to show, that $r \leq (N-1)^2 + 1$ and that there are Markov chains for which $r = (N-1)^2 + 1$.

3.4 ABSORBING MARKOV CHAINS

3.4.1 Definitions

The basic property of an ergodic (and *a fortiori* of a regular) Markov chain is that for any states i and j it is possible for the system to pass from state i to state j. That is, if the system is in state i on a certain observation, then there is a positive probability that it will be in state j on the mth succeeding observation for some m. If direct transition from state i to state j is possible, then we can take $m = 1$, but in general it will be necessary to make the transition in several steps, i.e., $m > 1$. This possibility of transition between any two states can fail to hold in several ways. It may be that there is only a single pair i, j such that transition from state i to state j is impossible. Or it may be that the set S of states can be divided into two subsets S_1 and S_2 such that no state in S_2 can be reached from any state in S_1 and conversely. In the latter case one can consider the original process as consisting of two quite independent subprocesses. We shall consider here a class of Markov chains which fail to have the basic property of ergodic chains in a strong sense. We assume not only the existence of states i and j such that transition from state i to state j is impossible but also that transition from state i to any other state is impossible. The precise formulation of this notion is the following:

Definition: The ith state of a Markov chain is said to be *absorbing* if $p_{ii} = 1$.

If the ith state of a Markov chain is absorbing, then it follows immediately that $p_{ij} = 0, j \neq i$. The set of conditions $p_{ii} = 1$, $p_{ij} = 0, j \neq i$, is the mathematical statement of the fact that once the system reaches state i, then it never leaves that state. The existence of an absorbing state is one characteristic of the chains to be studied in this section; however, we choose to require even more of those chains which we call absorbing.

Definition: A Markov chain is said to be *absorbing* if

1. There is at least one absorbing state, and
2. Transition from each nonabsorbing state to some absorbing state is possible.

Condition 2 is equivalent to the statement that if state j is nonabsorbing, then there is an absorbing state, say state i, and an integer m much that $p_{ji}(m) > 0$.

Some examples will be useful.

Example 1. Consider a Markov chain whose transition matrix $\mathbb{P}$ is

$$\begin{bmatrix} 1 & 0 & 0 \\ \frac{1}{3} & \frac{1}{3} & \frac{1}{3} \\ 0 & \frac{1}{2} & \frac{1}{2} \end{bmatrix}.$$

The first state of this chain is an absorbing one and the others are nonabsorbing. It is easy to check that $p_{21}(1) > 0$, $p_{31}(2) > 0$ and consequently that passage from states 2 and 3 to state 1, an absorbing state, is possible. Thus the chain with this transition matrix is an absorbing chain.

Example 2. Let the transition matrix for a Markov chain be

$$\begin{bmatrix} 1 & 0 & 0 \\ 0 & \frac{2}{3} & \frac{1}{3} \\ 0 & \frac{1}{2} & \frac{1}{2} \end{bmatrix}.$$

The first state is absorbing and the others are not. Since $p_{21}(m) = 0$ and $p_{31}(m) = 0$ for all integers m, this is not the transition matrix of an absorbing chain.

Example 3. Consider the four-state one-dimensional maze introduced in Sec. 3.2. Suppose that the maze is modified so that the doors to the red and blue compartments permit entrance to these compartments but prevent exit from them. The transition matrix for this process is

$$\begin{bmatrix} 1 & 0 & 0 & 0 \\ p & 0 & q & 0 \\ 0 & p & 0 & q \\ 0 & 0 & 0 & 1 \end{bmatrix}, \tag{4.1}$$

where p and q have the same meaning as previously. That is, q is the probability that the mouse moves toward the blue compartment (#4) and p is the probability that it moves toward the red compartment (#1) when such moves are possible, $p + q = 1$. This process is an absorbing chain with two absorbing states (1 and 4) and two nonabsorbing ones (2 and 3).

3.4.2 Properties of Absorbing Chains

In studying absorbing chains, it is convenient to adopt certain conventions regarding the labeling of the states. Let us agree that the states are to be numbered in such a way that the absorbing states come first. Thus, if there are N states and k of them are absorbing, then we suppose that states numbered 1 through k are absorbing and those numbered $k + 1$ through N are not. With this convention the transition matrix $\mathbb{P}$ has the form

$$\mathbb{P} = \begin{bmatrix} \mathbb{I}_{k \times k} & \mathbb{O}_{k \times (N-k)} \\ \mathbb{R}_{(N-k) \times k} & \mathbb{Q}_{(N-k) \times (N-k)} \end{bmatrix}.$$

The transition matrix $\mathbb{P}$ is decomposed into block submatrices $\mathbb{I}$, $\mathbb{O}$, $\mathbb{R}$, and $\mathbb{Q}$, and these matrices have the dimensions shown. Recall that $\mathbb{I}$ is an identity matrix and $\mathbb{O}$ is a matrix containing only 0 entries. If the states are labeled so that the transition matrix of an absorbing chain has this form, then the transition matrix is said to be in *canonical form*.

One of the reasons for writing the transition matrix in canonical form is that the definition of matrix multiplication gives particularly nice results when used to compute the powers of matrices having this form. In particular, we have

$$\mathbb{P}^2 = \begin{bmatrix} \mathbb{I} & \mathbb{O} \\ \mathbb{R}_2 & \mathbb{Q}^2 \end{bmatrix},$$

where $\mathbb{R}_2 = \mathbb{R} + \mathbb{Q}\mathbb{R}$. In general for any integer m, $m \geq 2$,

$$\mathbb{P}^m = \begin{bmatrix} \mathbb{I} & \mathbb{O} \\ \mathbb{R}_m & \mathbb{Q}^m \end{bmatrix},$$

where $\mathbb{R}_m$ can be computed successively as $\mathbb{R}_m = \mathbb{R} + \mathbb{Q}\mathbb{R}_{m-1} = \mathbb{R}_{m-1} + \mathbb{Q}^{m-1}\mathbb{R}$, $\mathbb{R}_1 = \mathbb{R}$ (Exercise 3). The matrix $\mathbb{P}^m$ contains the m-step transition probabilities; hence $\mathbb{Q}^m$ contains the m-step transition probabilities from one nonabsorbing state to another. Intuitively, as m increases, the entries in $\mathbb{Q}^m$ should be decreasing, because as m becomes large, it is increasingly likely that the system enters an absorbing state. We make this idea precise.

Lemma: Let

$$\mathbb{P} = \begin{bmatrix} \mathbb{I} & \mathbb{O} \\ \mathbb{R} & \mathbb{Q} \end{bmatrix}$$

be the canonical form of the transition matrix for an absorbing chain. Then

$$\lim_{m \to \infty} \mathbb{Q}^m = \mathbb{O}.$$

Proof: This lemma plays an important role in our work, and we shall give a proof before proceeding. Although the proof given here is not particularly difficult, it is rather lengthy and the notation is complicated. We give an outline and then the details.

Outline. We show the following facts:

1. For sufficiently large values of m, the matrix $\mathbb{R}_m$ has a positive entry in each row.
2. For all sufficiently large m the sums of the entries in each row of $\mathbb{Q}^m$ are all smaller than a constant $\eta < 1$.
3. There is an integer n and a constant $\eta < 1$ such that each entry of the matrix $\mathbb{Q}^{rn}$ is less than η^r, $r = 1, 2, 3, \ldots$.
4. $\mathbb{Q}^m \longrightarrow \mathbb{O}$ as $m \longrightarrow \infty$.

Details. Since $\mathbb{P}$ is the transition matrix of an absorbing chain, it follows that for each nonabsorbing state i there is an absorbing state j and an integer m such that $p_{ij}(m) > 0$. If there are k absorbing states in a total of N states and if $\mathbb{P}$ is in canonical form, then it is possible for the system to go from state $k + 1$ to an absorbing state. Thus there is an absorbing state j_1 and an integer m_1 such that $p_{k+1,j_1}(m_1) > 0$. A similar argument holds for states $k + 2, \ldots, N$, and we conclude that there are pairs of integers $(j_2, m_2), \ldots, (j_{N-k}, m_{N-k})$ with the properties $p_{k+i,j_i}(m_i) > 0$, $i = 1, 2, \ldots, N - k$. Each integer j_i is the number of an absorbing state and each integer m_i is positive. Next, $\mathbb{R}_m = \mathbb{R}_{m-1} + \mathbb{Q}^{m-1}\mathbb{R}$, and since the entries in all these matrices are nonnegative, it follows that the entries in $\mathbb{R}_m$ are at least as large as those in $\mathbb{R}_{m-1}$. Therefore, if we set $n = \max\{m_1, \ldots, m_{N-k}\}$, then $\mathbb{R}_n$ has a positive entry in each row. Moreover, $\mathbb{R}_m$ has a positive entry in each row for every $m \geq n$.

Define $\delta(m) = \min\{\sum_{j=1}^{k} r_{ij}(m), i = 1, 2, \ldots, N - k\}$, where $r_{ij}(m)$ is the i–j entry in the $(N - k) \times k$ matrix $\mathbb{R}_m$. Since the entries in $\mathbb{R}_m$ are nondecreasing with increasing m, it follows that $\delta(m) \geq \delta(n)$ for $m \geq n$. Also, writing $(q_{ij}(m)) = \mathbb{Q}^m$, we have

$$\begin{aligned}\max\left\{\sum_{j=1}^{N-k} q_{ij}(m), \qquad i = 1, 2, \ldots, N - k\right\} &= 1 - \delta(m)\\ &\leq 1 - \delta(n)\\ &= \eta < 1.\end{aligned}$$

Therefore, for large enough m, in fact for all $m \geq n$, the rows of $\mathbb{Q}^m$ have entries whose sum is no larger than $\eta < 1$. In particular, each element must be less than or equal to η.

Next, consider $\mathbb{Q}^{2n}$. Each entry in $\mathbb{Q}^{2n}$ is of the form

$$\sum_{l=1}^{N-k} q_{il}(n)q_{lj}(n) \leq \eta\left[\sum_{l=1}^{N-k} q_{il}(n)\right] \leq \eta^2.$$

Similarly, each entry in $\mathbb{Q}^{3n}$ is less than or equal to η^3, and in general each entry in $\mathbb{Q}^{rn}$ is less than or equal to η^r, $r = 1, 2, 3, \ldots$. Also, each entry in the matrices $\mathbb{Q}^{rn+1}, \mathbb{Q}^{rn+2}, \ldots, \mathbb{Q}^{(r+1)n-1}$ is less than or equal to η^r (Exercise 4). Since $\eta < 1$, $\eta^r \longrightarrow 0$ as $r \longrightarrow \infty$, and the proof of the lemma is complete.

The lemma contains the major part of the proof of our first theorem on the behavior of absorbing chains. The result is that as the number of trials becomes large, it is increasingly likely that the system reaches an absorbing state.

Theorem 1: Given any absorbing chain and any $\epsilon > 0$, there is an integer n such that for every $m \geq n$ the probability that the system is in an absorbing state after m trials is at least $1 - \epsilon$.

Proof: Let $\mathbb{P}$ be the transition matrix for the chain. We suppose that $\mathbb{P}$ is written in canonical form so

$$\mathbb{P}^m = \begin{bmatrix} \mathbb{I} & \mathbb{O} \\ \mathbb{R}_m & \mathbb{Q}^m \end{bmatrix}.$$

According to the lemma, $\mathbb{Q}^m \longrightarrow \mathbb{O}$ as $m \longrightarrow \infty$. Thus for any $\epsilon > 0$ there is an integer n such that

$$\max \left\{ \sum_{j=1}^{N-k} q_{ij}(m), \qquad i = 1, 2, \ldots, N-k \right\} < \epsilon$$

for all $m \geq n$. The conditional probability that a system which was initially in state i is in an absorbing state after the mth transition is $\sum_{j=1}^{k} p_{ij}(m)$. If $i \leq k$, then certainly this sum is 1. If $k + 1 \leq i \leq N$, then the sum is $\geq 1 - \epsilon$ for all $m \geq n$. It follows that no matter what the initial state happened to be, after n transitions the system is in an absorbing state with probability at least $1 - \epsilon$. Q.E.D.

Paraphrasing the above theorem, we say that in the long run every absorbing chain will be absorbed with probability 1. The theorem gives a precise formulation of the rather vague assertion that every absorbing chain is eventually absorbed. It is natural to ask for the number of trials necessary for absorption and also to study the behavior of the process prior to absorption. Obviously, since we are investigating a stochastic process, the situation cannot be described with specific numbers that are exact for every sequence of trials. Appropriate descriptions involve averages or expected values. One can determine the expected number of transitions before absorption and the expected number of times the system will be in each nonabsorbing state before it is absorbed.

The next lemma involves the notion of an invertible matrix and an explicit formula for the inverse (see Appendix C for the definitions of these terms).

Lemma: If

$$\mathbb{P} = \begin{bmatrix} \mathbb{I} & \mathbb{O} \\ \mathbb{R} & \mathbb{Q} \end{bmatrix}$$

is the canonical form for the transition matrix of an absorbing chain, then $\mathbb{I} - \mathbb{Q}$ is invertible and $(\mathbb{I} - \mathbb{Q})^{-1} = \mathbb{I} + \mathbb{Q} + \mathbb{Q}^2 + \cdots$.

Proof: For each m

$$(\mathbb{I} - \mathbb{Q})(\mathbb{I} + \mathbb{Q} + \mathbb{Q}^2 + \cdots + \mathbb{Q}^m) = \mathbb{I} - \mathbb{Q}^{m+1}.$$

Since $\mathbb{Q}^{m+1} \longrightarrow \mathbb{O}$ as $m \longrightarrow \infty$, it follows that

$$\lim_{m \to \infty} (\mathbb{I} - \mathbb{Q})(\mathbb{I} + \mathbb{Q} + \mathbb{Q}^2 + \cdots + \mathbb{Q}^m) = \mathbb{I}.$$

Likewise,

$$\lim_{m \to \infty} (\mathbb{I} + \mathbb{Q} + \mathbb{Q}^2 + \cdots + \mathbb{Q}^m)(\mathbb{I} - \mathbb{Q}) = \mathbb{I}.$$

The proof is completed by showing that the series $\mathbb{I} + \mathbb{Q} + \mathbb{Q}^2 + \ldots$ actually converges (in the sense of convergence of matrices). This follows from our earlier estimates on the size of the entries in $\mathbb{Q}^n$. Indeed, as shown earlier, there is an integer n and a constant η, $0 < \eta < 1$, such that the entries of $\mathbb{Q}^{rn}$ are all no larger than η^r and such that the entries of $\mathbb{Q}^{rn+1}$, $\mathbb{Q}^{rn+2}, \ldots, \mathbb{Q}^{(r+1)n-1}$ are no larger than η^r, $r = 1, 2, \ldots$. Thus, each entry in the sum $\sum_{m=n}^{\infty} \mathbb{Q}^m$ is bounded by $n \sum_{1}^{\infty} \eta^r = n\eta/(1 - \eta)$. Since each matrix $\mathbb{Q}^m$ has only nonnegative entries, the sequence $\{\sum_{m=1}^{M} q_{ij}(m)\}_{M=1}^{\infty}$ is increasing and bounded above and therefore it is convergent, $i, j = 1, \ldots, N-k$. Q.E.D.

Definition: The matrix $\mathbb{N} = (\mathbb{I} - \mathbb{Q})^{-1} = \sum_{i=0}^{\infty} \mathbb{Q}^i$ is said to be the *fundamental matrix* for the Markov chain whose transition matrix is $\mathbb{P}$.

The fundamental matrix is very useful in studying the long-run behavior of absorbing chains. Indeed, the limit of $\mathbb{P}^n$ as n increases is determined by $\mathbb{N}$ (Exercise 9), and also $\mathbb{N}$ can be used to find how long an absorbing process will proceed before absorption. The next theorem is directed toward the latter question.

Theorem 2: The entry n_{ij} in the fundamental matrix $\mathbb{N}$ is the expected number of times that the system is in state $k + j$ given that it began in state $k + i$ and it ran until absorbed.

Proof: Suppose that the process begins in state $k + i$, and let $E_m(i, j)$ be the expected number of times the system is in state $k + j$ in the first m transitions. For the moment assume $i \neq j$. The probability that the system is in state $k + j$ after one transition is q_{ij}, and hence $E_1(i, j) = q_{ij}$. Also, from the laws of probability theory

$$\begin{aligned} E_2(i, j) &= 1 \cdot q_{ij}(1 - q_{jj}) + 1 \cdot \sum_{l \neq j} q_{il} q_{lj} + 2 \cdot q_{ij} q_{jj} \\ &= q_{ij} + q_{ij}(2). \end{aligned}$$

An induction proof (Exercise 5) verifies the relation

$$E_m(i, j) = q_{ij} + q_{ij}(2) + \cdots + q_{ij}(m), \qquad i \neq j.$$

If $i = j$, then the system is in state j initially and consequently

$$E_m(i, i) = 1 + q_{ii} + \cdots + q_{ii}(m).$$

Therefore

$$\lim_{m \to \infty} E_m(i, j) = \begin{cases} \sum_{m=1}^{\infty} q_{ij}(m), & i \neq j, \\ 1 + \sum_{m=1}^{\infty} q_{ii}(m), & i = j, \end{cases}$$

and the proof of the theorem is complete.

To illustrate this theorem, we return to the maze described in Sec. 3.4.1 with transition matrix (4.1). Reordering the states in the order 1, 4, 2, 3 gives a transition matrix in canonical form

$$\begin{bmatrix} 1 & 0 & 0 & 0 \\ 0 & 1 & 0 & 0 \\ p & 0 & 0 & q \\ 0 & q & p & 0 \end{bmatrix} = \begin{bmatrix} \mathbb{I} & \mathbb{O} \\ \mathbb{R} & \mathbb{Q} \end{bmatrix}.$$

In this example

$$\mathbb{Q} = \begin{bmatrix} 0 & q \\ p & 0 \end{bmatrix}$$

and

$$\mathbb{I} - \mathbb{Q} = \begin{bmatrix} 1 & -q \\ -p & 1 \end{bmatrix}.$$

The inverse of $\mathbb{I} - \mathbb{Q}$ can be obtained by solving the system of four equations in four unknowns given by

$$(\mathbb{I} - \mathbb{Q})\begin{bmatrix} x & y \\ z & w \end{bmatrix} = \mathbb{I}.$$

The solution of this system is $x = \alpha$, $y = \alpha q$, $z = \alpha p$, $w = \alpha$, with $\alpha = 1/(1 - pq)$. Therefore

$$\mathbb{N} = (\mathbb{I} - \mathbb{Q})^{-1}$$
$$= \begin{bmatrix} \dfrac{1}{1 - pq} & \dfrac{q}{1 - pq} \\ \dfrac{p}{1 - pq} & \dfrac{1}{1 - pq} \end{bmatrix}.$$

Alternatively, the inverse of $\mathbb{I} - \mathbb{Q}$ can be obtained by computing the entries in the series $\mathbb{I} + \mathbb{Q} + \mathbb{Q}^2 + \cdots$. The four infinite series which occur are easily summed by using the formula for the sum of a geometric series (Appendix A). We leave the details to the reader (Exercise 7).

Suppose that $p = 1/4$ and $q = 3/4$ in this example; then

$$\mathbb{N} = \begin{bmatrix} \frac{16}{13} & \frac{12}{13} \\ \frac{4}{13} & \frac{16}{13} \end{bmatrix},$$

and we conclude from the theorem that if the mouse begins in the green compartment (state 4 after relabeling), then on the average it is in the yellow compartment (state 3 after relabeling) 4/13 times before it enters one of the end compartments. The other entries in $\mathbb{N}$ carry similar information.

The entries in the ith row of $\mathbb{N}$ give the expected number of times that the process is in each nonabsorbing state given that it began in state $k + i$. Thus, the sum of the entries in this row is the expected number of trials during which the system is in a nonabsorbing state, given that it started in state $k + i$ and continued until absorbed. Returning to the above example, if the system is initially in state 3 (yellow compartment), then the expected number of trials before absorptions is $\frac{16}{13} + \frac{12}{13} = \frac{28}{13}$.

3.4.3 Applications to Ergodic Chains

We have seen that it is possible to determine a number of facts about the expected behavior of an absorbing chain. It is natural to ask if the same sort of information can be obtained for ergodic chains. Such information can be obtained, and the techniques make use of the results already obtained for absorbing chains.

Consider an ergodic chain with transition matrix $\mathbb{P}$. It is possible for the system to pass from state i to state j for any i and j. What is the expected number of trials necessary for the system to go from state i to state j? To answer this question, we temporarily replace the ergodic chain by an absorbing one with state j replaced by an absorbing state. Denote the transition matrix for this new process by $\mathbb{P}[j]$ to indicate that only the jth row of $\mathbb{P}$ has been altered:

$$\text{Row } k \text{ of } \mathbb{P}[j] = \begin{cases} \text{Row } k \text{ of } \mathbb{P} & \text{if } k \neq j, \\ \mathbf{u}_j & \text{if } k = j. \end{cases}$$

Here and in what follows, $\mathbf{u}_j$ denotes the jth unit vector. That is, the vector all of whose coordinates are zero expect for the jth which is one. The modified chain behaves as follows:

1. If the system begins in state j, then it remains there.
2. If the system begins in state i, $i \neq j$, then it proceeds just as in the original ergodic process until it reaches state j for the first time. Once in state j, it remains there.

Since the original process was ergodic, it is possible to reach state j from every other state, and the new process satisfies the conditions for an absorbing chain. It follows that state j is certain to be reached; i.e., it will be reached with probability 1, regardless of the initial state.

Write the transition matrix for the new chain in canonical form and let $\mathbb{N}$ be the fundamental matrix. By Theorem 2 the entries of $\mathbb{N}$ give the expected number of times the system is in each state prior to entering an absorbing state. Interpreting this result in terms of the original process, we see that these numbers give the expected number of times the system is in each state i, $i \neq j$, before it reaches state j. Let us denote the states of the original ergodic chain by $S_1, \ldots, S_N$ and the relabeled states of the absorbing chain by $\tilde{S}_1, \ldots, \tilde{S}_N$. In particular, $S_j = \tilde{S}_1$. The sum of the entries in the ith row of the fundamental matrix gives the expected number of times the process is in any of the nonabsorbing states before reaching $\tilde{S}_1 = S_j$ given that it began in $\tilde{S}_{i+1}$. One can think of this as the expected number of trials before the system reaches state S_j from the initial state $\tilde{S}_{i+1}$.

As an example, consider the ergodic chain of the two-dimensional maze example of Sec. 3.3.3. The transition matrix of this chain is given by Eq. (3.2), and we shall study the special case $p = \frac{1}{3}$, $q = \frac{2}{3}$, $t = 0$. The transition matrix is

$$\mathbb{P} = \begin{bmatrix} 0 & \frac{2}{3} & 0 & \frac{1}{3} \\ \frac{1}{3} & 0 & \frac{2}{3} & 0 \\ 0 & \frac{1}{3} & 0 & \frac{2}{3} \\ \frac{2}{3} & 0 & \frac{1}{3} & 0 \end{bmatrix}.$$

Suppose that we are interested in how long, on the average, it takes for the system to reach state 3 from state 1. We first form an absorbing chain by replacing the third row of $\mathbb{P}$ by $\mathbf{u}_3$. We renumber the states so that $\tilde{S}_1 = S_3$, $\tilde{S}_2 = S_1$, $\tilde{S}_3 = S_2$, $\tilde{S}_4 = S_4$. The transition matrix (in canonical form) for the absorbing process is

$$\begin{bmatrix} 1 & 0 & 0 & 0 \\ 0 & 0 & \frac{2}{3} & \frac{1}{3} \\ \frac{2}{3} & \frac{1}{3} & 0 & 0 \\ \frac{1}{3} & \frac{2}{3} & 0 & 0 \end{bmatrix}.$$

The matrix $\mathbb{Q}$ is

$$\begin{bmatrix} 0 & \frac{2}{3} & \frac{1}{3} \\ \frac{1}{3} & 0 & 0 \\ \frac{2}{3} & 0 & 0 \end{bmatrix},$$

and the fundamental matrix is $\mathbb{N} = (\mathbb{I} - \mathbb{Q})^{-1}$. Since we are interested only in the expected number of transitions from state 1 to state 3, we are interested only in the first row of $\mathbb{N}$. Recall that $S_1 = \tilde{S}_2$. The first row of $\mathbb{N}$, call it (x, y, z), can be determined from the system

$$\begin{aligned} x - \tfrac{1}{3}y - \tfrac{2}{3}z &= 1, \\ -\tfrac{2}{3}x + y \qquad &= 0, \\ -\tfrac{1}{3}x \qquad + z &= 0. \end{aligned}$$

We obtain $x = \frac{9}{5}$, $y = \frac{6}{5}$, $z = \frac{3}{5}$, and the first row of $\mathbb{N}$ is $(\frac{9}{5}, \frac{6}{5}, \frac{3}{5})$. We conclude that if the system is initially in state $S_1 = \tilde{S}_2$, then on the average it requires $\frac{9}{5} + \frac{6}{5} + \frac{3}{5} = \frac{18}{5}$ trials before the system reaches state $S_3 = \tilde{S}_1$.

As a final item on ergodic chains we note that there is a simple formula for the mean recurrence time of the states in an ergodic chain. If the process begins in state i, then the average number of steps until the process returns to state i is $1/\sigma_i$, where $\mathbf{s} = (\sigma_1, \ldots, \sigma_N)$ is the stationary probability vector for the transition matrix of the process. See Exercise 11 for a discussion of the proof of this fact.

EXERCISES

1. Give an example of a four-state Markov chain which is neither ergodic nor absorbing. Characterize as completely as possible all four-state Markov chains.
2. Each of the following matrices is a transition matrix for a Markov chain. Decide which of these chains are ergodic, which are absorbing, and which are neither. For each absorbing chain compute the fundamental matrix.

(a) $\begin{bmatrix} 0 & 0 & 1 \\ 1 & 0 & 0 \\ 0 & 1 & 0 \end{bmatrix}$. (b) $\begin{bmatrix} 0 & 0 & 1 \\ 0 & 1 & 0 \\ 0 & \frac{1}{2} & \frac{1}{2} \end{bmatrix}$.

(c) $\begin{bmatrix} \frac{3}{4} & 0 & \frac{1}{4} & 0 \\ 0 & 0 & 1 & 0 \\ \frac{1}{3} & 0 & \frac{2}{3} & 0 \\ 0 & 0 & 0 & 1 \end{bmatrix}$. (d) $\begin{bmatrix} \frac{1}{4} & \frac{1}{4} & \frac{1}{4} & \frac{1}{4} \\ 0 & \frac{1}{3} & \frac{1}{3} & \frac{1}{3} \\ 0 & 0 & \frac{1}{2} & \frac{1}{2} \\ 0 & 0 & 0 & 1 \end{bmatrix}$. (e) $\begin{bmatrix} 0 & 0 & 0 & 1 \\ 0 & 0 & \frac{1}{2} & \frac{1}{2} \\ 0 & \frac{1}{3} & \frac{1}{3} & \frac{1}{3} \\ \frac{1}{4} & \frac{1}{4} & \frac{1}{4} & \frac{1}{4} \end{bmatrix}$.

3. Show that if $\mathbb{P}$ is the transition matrix written in canonical form of an absorbing Markov chain, then

$$\mathbb{P}^m = \begin{bmatrix} \mathbb{I} & \mathbb{O} \\ \mathbb{R}_m & \mathbb{Q}^m \end{bmatrix}$$

where $\mathbb{R}_m = \mathbb{R} + \mathbb{Q}\mathbb{R}_{m-1} = \mathbb{R}_{m-1} + \mathbb{Q}^{m-1}\mathbb{R}$, $\mathbb{R}_1 = \mathbb{R}$.

4. Let $\mathbb{Q}$ be a square matrix with the following property: The sum of the entries in each row of $\mathbb{Q}$ is no larger than η. Prove that for every integer r the entries in $\mathbb{Q}^r$ are no larger than η^r.

5. Let i and j, $i \neq j$, be two nonabsorbing states of an absorbing Markov chain. Prove that if the system is initially in state i, then the expected number of times it is in state j in m trials is

$$E_m(i, j) = q_{ij} + q_{ij}(2) + \cdots + q_{ij}(m).$$

6. In the notation of Exercise 3, prove that $\lim_{m \to \infty} \mathbb{R}_m$ exists and characterize the limit matrix as completely as possible.

7. Let

$$\mathbb{Q} = \begin{bmatrix} 0 & q \\ p & 0 \end{bmatrix},$$

$p + q = 1$, and find $(\mathbb{I} - \mathbb{Q})^{-1}$ by summing the series $\mathbb{I} + \mathbb{Q} + \mathbb{Q}^2 + \cdots$.

8. Consider the four-state one-dimensional maze whose transition matrix is given by Eq. (2.1). For what values of p, q, and t is the associated Markov chain regular, ergodic (but not regular), absorbing? Also answer the same question for the Markov chain arising from the two-dimensional maze with transition matrix (3.2).

9. Let

$$\mathbb{P} = \begin{bmatrix} \mathbb{I} & \mathbb{O} \\ \mathbb{R} & \mathbb{Q} \end{bmatrix}$$

be the canonical form of a transition matrix for an absorbing Markov chain. Show that the fundamental matrix $\mathbb{N} = (\mathbb{I} - \mathbb{Q})^{-1}$ determines the limiting behavior of $\mathbb{P}^n$ as $n \longrightarrow \infty$. *Hint:* Let

$$\lim_{n \to \infty} \mathbb{P}^n = \begin{bmatrix} \mathbb{I} & \mathbb{O} \\ \mathbb{A} & \mathbb{O} \end{bmatrix}$$

and use Exercise 3 to show that $\mathbb{A} = \mathbb{N}\mathbb{R}$.

10. Frequently the entries in a transition matrix are obtained by experimentation and observation. Thus, they are subject to experimental error, and it is important to consider the effect (if any) of such errors on conclusions based on a Markov chain model. In particular, answer the following questions (give reasons for your answers).

(a) If small changes are made in the entries of a transition matrix for an absorbing Markov chain, does the chain remain absorbing?

(b) Same as (a) for regular chains.

(c) Same as (a) for ergodic but not regular chains.

(d) What qualifications (if any) should be made in conclusions about a Markov chain for which there is likely to be some error in the entries in the transition matrix.

11. Let $\mathbb{P} = (p_{ij})$ be the transition matrix for an ergodic Markov chain, and suppose that $\mathbf{s} = (\sigma_1, \sigma_2, \ldots, \sigma_N)$ is a stationary probability vector for $\mathbb{P}$. Define m_{ij} to be the mean number of steps for this process to go from state i to state j, $i, j = 1, \ldots, N$. Carry out the following sequence of steps to show that $m_{ii} = 1/\sigma_i$, $i = 1, \ldots, N$.

(a) Show that

$$\begin{aligned} m_{ij} &= p_{ij} + \sum_{k \neq j} p_{ik} \cdot (m_{kj} + 1) \\ &= \sum_{k \neq j} p_{ik} m_{kj} + 1. \end{aligned} \qquad (*)$$

(b) Let $\mathbb{M} = (m_{ij})$, let $\mathbb{D}$ be the matrix obtained by setting all nondiagonal entries in $\mathbb{M}$ equal to 0, and let $\mathbb{C}$ be an $N \times N$ matrix with all entries equal to 1. Show that Eq. (*) has the form

$$\mathbb{M} = \mathbb{P}(\mathbb{M} - \mathbb{D}) + \mathbb{C}. \qquad (**)$$

(c) Multiply both sides of Eq. (**) by $\mathbf{s}$, and thus show that $\mathbf{s}\mathbb{M} = \mathbf{s}(\mathbb{M} - \mathbb{D}) + (1, 1, \ldots, 1)$.

(d) Use the result of (c) to show that $m_{ii} = 1/\sigma_i$.

3.5 PROJECTS

3.5.1 Competition on the Campaign Trail

We study the competitive aspect of a primary election in which the candidate of the Redeemer Party for High Public Office is to be selected. The three candidates vying for the nomination represent the three dominant factions of the party: Lucius of the Libertine Left (LLL for short), Reginald of the Righteous Right (RRR), and Manfred of the Mild Middle (MMM). We focus on a single phase of the race, namely the use of the media to influence votes. Suppose that each week each candidate places an advertisement in the Sunday newspaper. The advertising is an effort to attract votes, and each faction can prepare its copy in such a way that its primary effect is on one of the other candidates. If the advertising of one faction has its primary effect on another faction, then we say that the advertising of the first is *directed against* the second. Also, a specific advertisement may or may not be effective. If the advertising is effective, then we suppose that the candidate at which it was directed refrains from further competition for votes by newspaper advertising. Assume that the probabilities of being effective depend only on the talents of the writing staff of each faction and do not vary from week to week. The probability that LLL is effective is assumed to be 1/2, that RRR is effective is 1/3, and that MMM is effective is 1/6. The advertising of each candidate is directed against the strongest of the remaining competitors. Here strength is measured in terms of the probability of being effective. Thus LLL is the strongest, RRR is the second strongest, and MMM is the weakest of the three candidates.

Problem 1. Construct a Markov chain model for this advertising campaign. *Hint:* Define a system which has as states the several possible combinations of candidates remaining in the campaign. The campaign begins with all three candidates participating. The newspaper campaign is said to end when there is at most one candidate remaining.

Problem 2. Show that the model constructed in Problem 1 is an absorbing chain.

Problem 3. Find the expected number of advertisements to appear before the campaign ends.

Problem 4. Determine as accurately as possible the probability that MMM is the only surviving candidate.

3.5.2 To Tell a Tale

Consider a community consisting of n individuals and suppose that insofar as repeating rumors is concerned, the set of individuals can be divided into three groups. The first consists of those individuals who will always repeat a rumor, no matter how many times they have heard it. The second consists of those individuals who will never repeat a rumor. The third group consists of those individuals who will repeat a rumor about half the time. Suppose that in these three groups there are n_1, n_2, and n_3 individuals, respectively. Also assume that the rumor is spread through personal contacts and that meetings between individuals occur randomly. Formulate a Markov chain model for the spread of a rumor through the community. Assume that initially a single person knows the tale.

Problem 1. Find the transition matrix for your model in the special case $n_1 = 1$, $n_2 = 1$, and $n_3 = 1$.

Problem 2. Find the transition matrix for the special case $n_1 = n_3 = 2$, $n_2 = 1$.

Problem 3. In general, how does the size of the transition matrix depend on n_1, n_2, and n_3?

3.5.3 Peas and Probabilities Revisited

The object of this project is the construction of a Markov chain model for the passage of genetic characteristics from parents to offspring.

To begin, suppose that we are concerned with a single gene which can occur in two alternative forms A and a (see Sec. 2.1 for definitions and notation). Consider the sequential experiment in which a trial consists of planting a pea, allowing it to reproduce by self-fertilization, and selecting one of the peas so produced at random.

Problem 1. Formulate a Markov chain model for this experiment and find the transition matrix.

Problem 2. Suppose that the genotype of the original pea is unknown but that it is known that it resulted from a cross between parents with genotypes AA and Aa. What is the probability that the pea selected at the end of the second trial has genotype Aa? Genotype AA?

Next, consider a similar situation in which we are concerned with two genes, each with two forms. Denote the alleles by A, a and B, b. An experiment identical to that described above is performed.

Problem 3. Formulate a Markov chain model for this experiment and find the transition matrix. Note that in this case the genotype of each pea is described by four letters.

Problem 4. Suppose that A and B are dominant and that the original pea resulted from a cross between parents of genotypes $AAbb$ and $AaBb$. What is the probability that the phenotype of a pea selected at random at the end of the first trial is that of the a allele with regard to the respective gene.

Next consider the so-called brother-sister mating between descendants and suppose that we are concerned with a single gene which occurs in two alternative forms. In more detail, we suppose that two plants are crossed and that among the resulting peas two are selected at random. They are again crossed and the process continues.

Problem 5. Formulate a Markov chain model for this process and find the transition matrix.

Problem 6. What sort of Markov chain is this (regular, ergodic, absorbing, or none of these)? Support your answer.

REFERENCES

Several other experiments in small-group decision making are discussed in [A], and details of an experiment similar to that described in Sec. 3.1 are contained in [G]. Other presentations of the mathematical theory of Markov chains which include some applications can be found in [KMST]—and at a more advanced level in [Fe] and [Fi]. The monograph [R] contains a discussion of Markov chain models for learning.

[A] ASCH, S. E., "Effects of Group Pressures upon the Modification and Distortion of Judgment," in H. Guetzkow, ed., *Groups, Leadership, and Men*. Pittsburgh: Carnegie Press, 1951.

[D] DOOB, J. L., *Stochastic Processes*. New York: Wiley, 1953.

[Fe] FELLER, W., *An Introduction to Probability Theory and Its Applications*, Vol. I, 3rd ed. New York: Wiley, 1968.

[Fi] FISZ, M., *Probability Theory and Mathematical Statistics*, 3rd ed. New York: Wiley, 1963.

[G] GODWIN, W. F., *Subqroup Pressures in Small Group Consensus Processes*, Doctoral Thesis. Bloomington: Indiana University, 1970.

[KMST] KEMENY, J. G., H. MIRKIL, J. L. SNELL, and G. L. THOMPSON, *Finite Mathematical Structures*. Englewood Cliffs, N.J.: Prentice-Hall, 1959.

[R] RESTLE, F., *Mathematical Models in Psychology*. Baltimore: Penguin, 1971.

4 The Theory of Models for Linear Optimization

4.0 INTRODUCTION

In the fields we consider in this book one frequently encounters situations in which the primary concern is making something as large or as small as possible. Such questions will be referred to as questions of optimization. Basically they have the property that one is concerned with a function depending on several variables, and the object is to select values of the variables in such a manner that the function takes on the largest (or smallest) possible value. In the cases of interest, one is not usually free to select arbitrary values of the variables. Indeed, it is common for each variable to be subject to constraints of one type or another. An illustration taken from elementary calculus is the problem of finding the rectangular field of maximum area which is enclosed by a fence of given length.

The reader is already familiar with at least one instance of such a problem in the social sciences, namely the two-by-three transportation problem considered in Sec. 2.2. We begin this chapter by discussing two situations, one in health care and one in business, which indicate the breadth of usefulness of the concepts. Also, they lead to the main mathematical notions of the chapter, systems of linear inequalities. We digress for a brief discussion of the theory of such systems before considering models for linear optimization. Our main objective is a consideration of problems of maximizing or minimizing a linear function of several variables over a domain defined by a system of linear inequalities. Such a problem is known as a problem in linear programming, and there is a vast literature on the subject. The reader is encouraged to consult the works cited in the References for a more complete treatment. Many of the notions are most easily understood when presented in geometric language, and we recommend [Go] and [GoT], particularly from this point of view. We shall consider the geometric aspect of the question further in Chap. 5.

At the very outset it should be mentioned that for our purposes it is sufficient to consider only a very restricted class of optimization questions. It follows that the reader should not be surprised if the problems he encounters in formulating models for his own questions deviate somewhat from those discussed in this book. It is rare in real applications to find the known theory both appropriate and applicable. We shall indicate some of the possibilities for refined models in Chap. 7. However, there is little comprehensive general theory for situations which in some essential way fall outside of the scope of this chapter. Our hope is that by carefully considering what follows and by using some ingenuity the reader will be better prepared to cope with what nature turns up.

4.1 OPTIMIZATION IN HEALTH CARE AND BUSINESS

Questions of optimization and in particular questions of linear optimization arise naturally in many different contexts. A *linear process* is defined as one in which a specific input yields a specific output and which is characterized by the following two properties:

1. If input x yields output X, then input λx yields output λX for all real numbers λ.
2. If input x yields output X and input y yields output Y, then input $x + y$ yields output $X + Y$. It is tacitly assumed here that $x + y$ and $X + Y$ have a well-defined meaning.

Linear processes have several properties that make them a useful starting point for our discussion. First, such processes are very common in nature, although not as common as would be convenient. Second, there is a well-developed and relatively complete theory of such processes, so that the appropriate mathematics is likely to be available. And finally, even if the assumption of linearity is not completely valid, it can be used as a first approximation, and in fact it may be quite good in describing certain aspects of the situation. We turn now to some examples, and we shall attempt to point out where the various assumptions enter.

Let us consider how one might go about selecting foods for consumption from a supply known to be contaminated. The contamination might arise from the presence of bacteria due to inadequate care in processing, from the use of insecticides and weed killers in the growing of vegetables, from industrial pollution of water and air, and from natural radioactivity, radioactive wastes, or even fallout resulting from nuclear explosions. We suppose that only the effects of one such source are to be considered at a time. Our basic assumption is that if one unit of food of a certain type contains A units of contamination, then λ units of that food contain λA units of contamination, and if two different types of food contain A and B units of contamination per unit of food, respectively, then the consumption of one unit of each introduces $A + B$ units of contamination into the system. That is, the amount of contamination is not modified by taking foods in combination.

It is recognized that an adequate diet must contain a variety of foods each of which contributes certain essential nutrients. For these nutrients a specific minimum daily requirement has been established. It is reasonable to assume that the nutrient content of a number of different foods is known. We consider a set of n foods and m nutrients. Let $\{\beta_i\}_{i=1}^m$ be the set of minimum daily requirements for the m nutrients, and let a_{ij}, $1 \leq i \leq m$, $1 \leq j \leq n$, denote the concentration of the ith nutrient in the jth food. We are concerned with determining a *diet* or *consumption* vector $\mathbf{x} = (\xi_1, \ldots, \xi_n)$, where ξ_j is the amount of the jth food to be consumed each day. Clearly, $\xi_j \geq 0$, $1 \leq j \leq n$, and if the minimum daily requirement is to be met, then

$$\sum_{j=1}^{n} a_{ij}\xi_j \geq \beta_i, \qquad 1 \leq i \leq m.$$

We now introduce the effect of contamination. Let γ_j, $1 \leq j \leq n$, be the concentration of contamination in the jth food. Then the total daily consumption of contaminants associated with the diet given by $\mathbf{x}$ is $\sum_{j=1}^{n} \gamma_j\xi_j$. Thus one might define a "best" diet as one which minimizes the consumption of contaminants. With this convention, the best diet $\mathbf{x}$ is one which satisfies

$$\xi_j \geq 0, \qquad 1 \leq j \leq n, \tag{1.1}$$

$$\sum_{j=1}^{n} a_{ij}\xi_j \geq \beta_i, \qquad 1 \leq i \leq m, \tag{1.2}$$

and which minimizes $\sum_{j=1}^{n} \gamma_j\xi_j$ over all such diets.

It is frequently the case that more than one type of contamination is present. To include this possiblity, we suppose that there are p different contaminants to be considered, and let γ_{kj}, $1 \leq k \leq p$, $1 \leq j \leq n$, denote the concentration of the kth contaminant in the jth food. The total consumption of the kth contaminant in the diet given by $\mathbf{x}$ is therefore equal to $\sum_{j=1}^{n} \gamma_{kj}\xi_j$. To proceed further, it is necessary to assume something about the interaction of contaminants. In general different contaminants have different effects on the human system. We assume that there are constants w_k, $1 \leq k \leq p$, which we refer to as *weights*, such that the total effect of the p different contaminants is

$$\sum_{k=1}^{p} w_k \cdot [\text{Amount of } k\text{th contaminant consumed}].$$

Thus the total effect of contamination in the diet given by $\mathbf{x}$ is

$$\sum_{k=1}^{p} w_k \sum_{j=1}^{n} \gamma_{kj}\xi_j = \sum_{j=1}^{n} \left(\sum_{k=1}^{p} w_k\gamma_{kj}\right)\xi_j. \tag{1.3}$$

Again it is reasonable to try to minimize the total effect of contamination by selecting a diet $\mathbf{x}$ which satisfies (1.1) and (1.2) and minimizes (1.3). Notice that if we set $\gamma'_j = \sum_{k=1}^{p} w_k\gamma_{kj}$, $1 \leq j \leq n$, then this problem is the same as the original one, with $\sum_{j=1}^{n} \gamma'_j\xi_j$ to be minimized instead of $\sum_{j=1}^{n} \gamma_j\xi_j$.

Turning to another setting, we present a situation which is superficially quite different from that just considered. However, in stating the problem precisely, it will become clear that it is actually quite similar in mathematical nature to the one just discussed. As the setting for this discussion, we consider a company which refines oil at a number of locations and transports it via pipelines to a number of terminals where it is sold and delivered to the purchaser. The company wishes to set up a shipping schedule which meets the natural conditions imposed by the problem and minimizes the transportation costs to be paid by the company. The conditions resulting from the nature of the problem are that each refinery has a certain capacity which cannot be exceeded and that the demand at each terminal must be met. The first is purely a physical condition, the size of the plant, while the second is more an administrative decision, dictated by competitive factors, for example. We suppose that the pipeline is capable of carrying any amount of oil asked of it. This assumption will be modified in Chap. 7, where the notion of capacity is introduced.

The data relevant to the problem consist of information on production at the refineries, demand at the terminals, and transportation costs. The essential assumption which we shall make with regard to the last of these is that the transportation cost depends linearly on the amount shipped. That is, if the cost is k dollars for one barrel, then it is λk dollars for λ barrels, for any (positive) real number λ. Also if the cost for a shipment from one refinery to one terminal is A and the cost for a shipment from another refinery to another terminal is B, then the cost for both shipments is $A + B$. These assumptions are essentially that no discounts are available for either bulk shipments or total business, and they may or may not hold in a specific instance.

Let us now specify the situation in more detail. We assume that the company refines oil at m different refineries $R_1, R_2, \ldots, R_m$ and that the oil is delivered to n different terminals $T_1, \ldots, T_n$. Let s_i be the capacity measured in barrels per day of refinery R_i, $i = 1, 2, \ldots, m$, and let d_j be the demand measured in barrels per day at terminal T_j, $j = 1, 2, \ldots, n$. Using the data on the pipeline facilities available for use by the company and the assumption that transportation costs are linear, the company can compute the cost of transporting any given amount of oil over any specific route. This is easy once the company knows the cost c_{ij} to move one barrel of oil from refinery i to terminal j, $i = 1, 2, \ldots, m$; $j = 1, 2, \ldots, n$. The concern of the company is that of devising a shipping schedule which satisfies the following conditions:

1. The schedule is such that there is always a sufficient amount of oil allocated to each terminal; i.e., the demand is satisfied.
2. The schedule is such that no more oil is required from any refinery than it is able to supply; the supply is sufficient.
3. The total shipping cost incurred by the company is no more than the cost associated with any other schedule which satisfies conditions 1 and 2 above.

Before proceeding further in our considerations, it is necessary to be more specific with regard to exactly what is meant by a shipping schedule. Intuitively, a shipping schedule is simply a listing of how much oil to ship from each refinery to each terminal. Precisely, a shipping schedule F is a set of nonnegative real numbers, $\{f_{ij}\}_{i=1}^{m}{}_{j=1}^{n}$, where f_{ij} is the quantity of oil measured in barrels to be shipped from refinery R_i to terminal T_j. Using this definition of a shipping schedule and the assumptions concerning transportation costs, we can assign to each schedule F a transportation cost $C[F]$ by

$$C[F] = \sum_{i=1}^{m} \sum_{j=1}^{n} c_{ij} f_{ij}.$$

Also, conditions 1 and 2 can be stated in quantitative terms using the notion of a schedule. Indeed, condition 1 becomes

$$\sum_{i=1}^{m} f_{ij} \geq d_j, \qquad j = 1, 2, \ldots, n, \tag{1.4}$$

and condition 2 becomes

$$\sum_{j=1}^{n} f_{ij} \leq s_i, \qquad i = 1, 2, \ldots, m. \tag{1.5}$$

Since it is among schedules which satisfy (1.4) and (1.5) that we must search for a schedule satisfying condition 3, let us call such schedules *admissible*. Then a schedule F^* satisfying condition 3 satisfies

$$C[F^*] \leq C[F],$$

where F is any admissible schedule.

Clearly, our study divides into two parts. First, one must determine whether or not there are admissible schedules. A related question of practical as well as theoretical importance is that of characterizing those situations in which admissible schedules exist. Next, one must select from the set of all admissible schedules that one giving a minimum value for $C[F]$. Since admissible schedules are determined by (1.4) and (1.5), the question of existence of admissible schedules is equivalent to the question of existence of solutions of sets of linear inequalities. It is to this question that we shall turn next.

EXERCISES

1. Consider a situation in which the effect of p different contaminants in a food supply is to be studied. Suppose that the total effect of contaminants in a diet is assumed to be

$$\sum_{k=1}^{p} v_k \cdot [\text{Amount of } k\text{th contaminant consumed}]^2,$$

where v_k, $1 \leq k \leq p$, are constants, State a problem analogous to those discussed in this section for this situation.

2. Consider the contamination of a food supply by p different contaminants as discussed in this section. Suppose that one arrives at bounds on the consumption of each contaminant which may not be exceeded. It might be, for example, that consumption above these bounds will result in irreparable damage to essential organs. If these bounds are τ_k for the kth contaminant, $1 \leq k \leq p$, find a system of inequalities which must be satisfied by the diet.

3. (Continuation of 2) It may be that no diet satisfies the constraints of Exercise 2 and also provides the minimum daily requirement of each nutrient. Define the total nutrition of a diet by

$$\sum_{i=1}^{m} w_i \cdot [\text{Amount of } i\text{th nutrient consumed}],$$

where w_i, $1 \leq i \leq m$, are constants. Formulate a problem similar to those of this section which asks for a diet which satisfies the constraints on consumption of contaminants and maximizes the total nutrition.

4. How does the model for the shipping problem change if there are discounts for bulk shipments.

5. Discuss from a qualitative point of view the features which would have to be incorporated in a model for a shipping problem where demand is not precisely known.

6. Consider the following refinement of the oil distribution problem. Suppose that after the oil reaches the terminals it must be further distributed to retail outlets. Let there be p outlets, and let the cost of shipping from terminal i to outlet k be h_{ik}. Formulate a mathematical model for the expanded problem.

4.2 SYSTEMS OF LINEAR INEQUALITIES

Although the notion of a linear inequality is more general, and in a sense more fundamental, than that of a linear equation, this material is much less likely to be a part of the reader's background. Consequently, we digress here for a brief and (together with Appendix C) self-contained discussion of linear inequalities. The reader who is familiar with this material may proceed directly to Sec. 4.3. The goal of our study of systems of linear inequalities is to establish criteria which guarantee the existence of solutions to such systems. We prefer to give an unhurried account and to proceed in small steps. The main results are summarized in the theorems which appear toward the end of the section. However, many of the lemmas are of independent interest and worthy of consideration in their own right.

This subject is one of those where notation is particularly important. Unless one adopts an appropriate notation, the essential simplicity of the results can become lost in a maze of symbols. Consequently, we make systematic use of vector notation, and the reader is referred to Appendix C for the relevant definitions and conventions. We write the system

$$\begin{array}{c} a_{11}\xi_1 + a_{12}\xi_2 + \cdots + a_{1n}\xi_n \leq \beta_1 \\ \hdashline a_{m1}\xi_1 + a_{m2}\xi_2 + \cdots + a_{mn}\xi_n \leq \beta_m \end{array},$$

where $\{a_{ij} : i = 1, \ldots, m; j = 1, \ldots, n\}$ and $\{\beta_i : i = 1, \ldots, m\}$ are given sets of real numbers, in vector notation as

$$\mathbb{A}\mathbf{x} \leq \mathbf{b}. \tag{2.1}$$

Here $\mathbf{x}$ is the vector $(\xi_1, \ldots, \xi_n)$ written as a column vector, and $\mathbf{b}$ is the vector $(\beta_1, \ldots, \beta_m)$ also written as a column vector. $\mathbb{A}$ is the matrix whose i–j entry is a_{ij}. Finally, so that all the symbols in (2.1) make sense, it is necessary to specify what is meant by the use of the inequality sign. The meaning of $\leq$ when used between two real numbers is obvious. We use the symbol between two vectors to indicate that the relation holds between each corresponding pair of coordinates. That is, if $\mathbf{z} = (\zeta_1, \ldots, \zeta_n)$ and $\mathbf{w} = (\omega_1, \ldots, \omega_n)$ are two vectors in R^n, then $\mathbf{w} \leq \mathbf{z}$ is equivalent to the n inequalities $\omega_1 \leq \zeta_1, \ldots, \omega_n \leq \zeta_n$. We use $\geq$, $>$, and $<$ similarly. If $\mathbf{x} \geq \mathbf{0}$ or $\mathbf{x} > \mathbf{0}$, we say that $\mathbf{x}$ is nonnegative or positive, respectively.

The reader should note that inequalities between vectors display many but not all of the properties of ordinary inequality of real numbers. For example, it is not true that given $\mathbf{x}$ and $\mathbf{y}$ one must have either $\mathbf{x} \geq \mathbf{y}$ or $\mathbf{y} \geq \mathbf{x}$. Likewise, $\mathbf{x} \geq \mathbf{0}$ and $\mathbf{x} \neq \mathbf{0}$ does not imply that $\mathbf{x} > \mathbf{0}$. However, the following assertions and the corresponding statements for $\geq$ are valid. Certain of them also hold for $<$ and $>$. The proofs follow directly from the definitions and are left to the reader.

1. If $\mathbf{x} \leq \mathbf{y}$ and $\mathbf{y} \leq \mathbf{z}$, then $\mathbf{x} \leq \mathbf{z}$.
2. If $\mathbf{x} \leq \mathbf{y}$ and $\mathbf{z} \leq \mathbf{w}$, then $\mathbf{x} + \mathbf{z} \leq \mathbf{y} + \mathbf{w}$.
3. If $\mathbf{x} \leq \mathbf{y}$ and λ is a nonnegative real number, then $\lambda\mathbf{x} \leq \lambda\mathbf{y}$; if $\mathbf{x} \leq \mathbf{y}$ and μ is a negative real number, then $\mu\mathbf{x} \geq \mu\mathbf{y}$.

Since we are dealing with inequalities, the connection between order relations and vector operations is important. One connection between the order relations and inner products is contained in

Lemma 1: Let $\mathbf{x}$, $\mathbf{y}$, $\mathbf{z}$ be vectors in R^n, $\mathbf{x} > \mathbf{0}$, $\mathbf{y} \geq \mathbf{z}$, and $\mathbf{y} \neq \mathbf{z}$. Then $\mathbf{x} \cdot \mathbf{y} > \mathbf{x} \cdot \mathbf{z}$. Similarly, if $\mathbf{x} \geq \mathbf{0}$, $\mathbf{x} \neq \mathbf{0}$, and $\mathbf{y} > \mathbf{z}$, then $\mathbf{x} \cdot \mathbf{y} > \mathbf{x} \cdot \mathbf{z}$.

We prove the first statement; the proof of the second is similar. Here and in what follows $\mathbf{x} = (\xi_1, \ldots, \xi_n)$, $\mathbf{y} = (\eta_1, \ldots, \eta_n)$, and $\mathbf{z} = (\zeta_1, \ldots, \zeta_n)$. We have $\xi_i > 0$, $i = 1, \ldots, n$, and $\eta_i \geq \zeta_i$, $i = 1, \ldots, n$. Also, there is some j, $1 \leq j \leq n$, for which $\eta_j > \zeta_j$. Therefore, $\xi_j\eta_j > \xi_j\zeta_j$, and $\xi_i\eta_i \geq \xi_i\zeta_i$ for $i \neq j$. Adding the terms on each side, we obtain the desired result. Q.E.D.

The above notation and preliminaries will be adequate for the moment, and we can now begin our work. There is a well-developed and elegant theory of linear inequalities, both from a geometric and an algebraic point of view. However, in light of the purposes of this book, we do not propose to give a broad survey of the subject. Instead, we proceed rather directly to a result which serves our purposes adequately. The subject is rich in structure and there are several developments which would serve in place of the one selected.

The presentation given here naturally reflects the prejudices of the authors, although we have been significantly influenced by [Ga] and [T]. The reader is referred to these works for further details.

It is clear from the examples given in Sec. 4.1 that we ultimately need to consider nonhomogeneous inequalities, that is, inequalities of the form

$$\mathbb{A}\mathbf{x} \leq \mathbf{b}, \qquad \mathbf{b} \neq \mathbf{0}.$$

However, it is useful to first consider the homogeneous case, that is, $\mathbb{A}\mathbf{x} \leq \mathbf{0}$. One of the most interesting and important features of the theory is the interplay between relations involving $\mathbb{A}\mathbf{x}, \mathbf{x} \in R^n$, and those involving $\mathbb{A}^T\mathbf{y}, y \in R^m$. Here $\mathbb{A}^T$ denotes the transpose matrix (see Appendix C) of $\mathbb{A}$. To be more specific, we shall begin our study by considering an $m \times n$ matrix $\mathbb{A}$ and the following problems:

1. Find $\mathbf{y} \in R^m$ such that $\mathbb{A}^T\mathbf{y} \geq \mathbf{0}$.
2. Find $\mathbf{x} \in R^n$, $\mathbf{x} \geq \mathbf{0}$ such that $\mathbb{A}\mathbf{x} = \mathbf{0}$.

We are interested in pairs of vectors $\mathbf{x}$ and $\mathbf{y}$, at least one of which is not the zero vector.

Let $\mathbf{a}^1, \ldots, \mathbf{a}^n$ denote the columns of $\mathbb{A}$. Each $\mathbf{a}^i$ is a vector in R^m. As is our custom in this book, we think of them as being written in either column or row form according to our needs. Unless there is a possible ambiguity we shall not comment on the form of the vectors which appear in the following discussion. In each of the several possible approaches to the study of systems of linear inequalities, one must eventually prove a somewhat complicated result. We elect to do so immediately. Thus the next lemma has a rather involved proof, a portion of which is deferred to the exercises. The details will be clearer if the reader keeps the geometry of the situation in mind.

Lemma 2: Let $\mathbb{A}$ be an $m \times n$ matrix. Then there are vectors $\mathbf{x} \in R^n$, $\mathbf{y} \in R^m$ such that the following conditions are satisfied:

1. $\mathbf{x} \geq \mathbf{0}$ and $\mathbb{A}\mathbf{x} = \mathbf{0}$.
2. $\mathbb{A}^T\mathbf{y} \geq \mathbf{0}$.
3. $\mathbf{a}^1 \cdot \mathbf{y} + \xi_1 > 0$.

Proof: Consider first the case with $n = 1$. If $\mathbf{a}^1$ is the zero vector, then $\mathbf{y}$ can be selected arbitrarily and $\mathbf{x} = (\xi_1)$ can be taken as positive. If $\mathbf{a}^1$ is not the zero vector, then $\mathbf{a}^1 \cdot \mathbf{a}^1 > \mathbf{0}$ and therefore we may take $\mathbf{y} = \mathbf{a}^1$ and $\mathbf{x} = \mathbf{0}$.

To continue, suppose that $n = 2$. We may suppose that $\mathbf{a}^2 \neq \mathbf{0}$ since otherwise the situation is the same as in the case considered above. Now, either $\mathbf{a}^1$ and $\mathbf{a}^2$ are linearly dependent or they are not. If they are linearly independent, then there is a vector $\mathbf{v} \in R^m$ such that $\mathbf{v} \cdot \mathbf{a}^1 = 1$, $\mathbf{v} \cdot \mathbf{a}^2 = 0$ (Theorem C.5 in Appendix C). Therefore, in this case the vectors $\mathbf{x} = \mathbf{0}$, $\mathbf{y} = \mathbf{v}$ have the desired properties. On the other hand, if $\mathbf{a}^1$ and $\mathbf{a}^2$ are dependent, then there are constants λ_1, λ_2, not both zero, such that

$$\lambda_1\mathbf{a}^1 + \lambda_2\mathbf{a}^2 = \mathbf{0}. \tag{2.2}$$

Since $\mathbf{a}^2 \neq \mathbf{0}$ and since at least one of the two constants λ_1, λ_2 is nonzero, it follows that $\lambda_1 \neq 0$. It is no restriction to assume that $\lambda_1 > 0$. There are now two possibilities to consider. If $\lambda_2 \geq 0$, then the vectors $\mathbf{x} = (\lambda_1, \lambda_2)$, $\mathbf{y} = \mathbf{0}$ have the desired properties. If $\lambda_2 < 0$, then it follows from (2.2) that

$$\mathbf{a}^1 = \mu \mathbf{a}^2, \qquad \mu = -\frac{\lambda_2}{\lambda_1} > 0.$$

In this case the vectors $\mathbf{x} = \mathbf{0}$ and $\mathbf{y} = \mathbf{a}^1$ have the desired properties.

It is useful to consider the geometry of the case $n = 2$. There is a plane π containing $\mathbf{a}^1$, $\mathbf{a}^2$ and the origin $\mathbf{0}$, and we study the situation in that plane. Therefore we can think of $\mathbf{a}^1$ and $\mathbf{a}^2$ as elements of R^2. If $\mathbf{a}^1$ and $\mathbf{a}^2$ are linearly independent, then they do not lie on the same straight line through the origin, and there is a vector $\mathbf{v}$ which is perpendicular to $\mathbf{a}^2$ but not perpendicular to $\mathbf{a}^1$. By multiplying $\mathbf{v}$ by a constant, if necessary, we can assume that $\mathbf{v} \cdot \mathbf{a}^1 = 1$. This situation is pictured in Fig. 4–1(a). If $\mathbf{a}^1$ and $\mathbf{a}^2$ are depen-

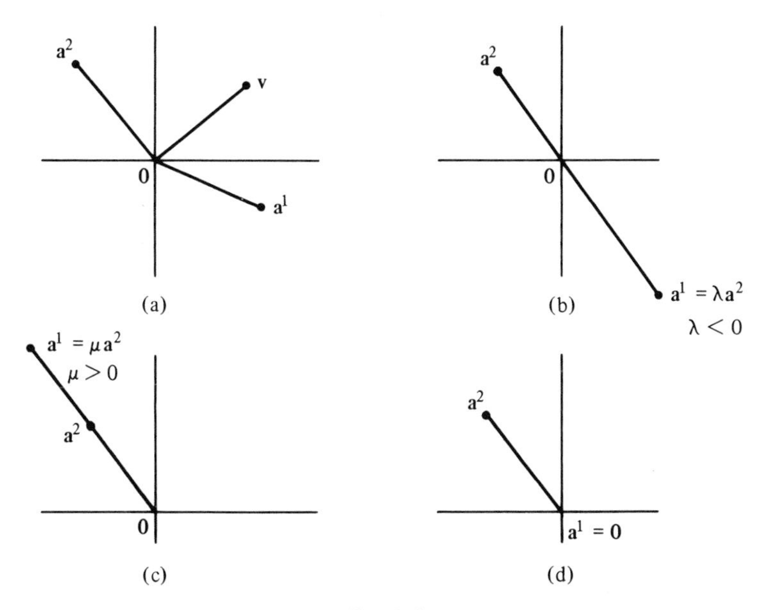

Fig. 4–1

dent, then they lie on the same straight line through the origin. If they lie on opposite sides of the origin, then (remembering that $\mathbf{a}^2 \neq \mathbf{0}$) $\mathbf{a}^1 = \lambda \mathbf{a}^2$ with $\lambda < 0$, as in Fig. 4–1(b). Thus we can represent the zero vector as a linear combination of $\mathbf{a}^1$ and $\mathbf{a}^2$ with nonnegative coefficients: $\mathbf{a}^1 + (-\lambda)\mathbf{a}^2 = \mathbf{0}$. We take $\mathbf{y} = \mathbf{0}$, $\mathbf{x} = (1, -\lambda)$, and then conditions 1–3 are satisfied. If $\mathbf{a}^1$ and $\mathbf{a}^2$ lie on the same side of the origin, then $\mathbf{a}^1 = \mu \mathbf{a}^2$ with $\mu > 0$, as in

Fig. 4–1(c). In this case the only linear combination of $\mathbf{a}^1$, $\mathbf{a}^2$ with non-negative coefficients which gives the zero vector must be the trivial one with zero coefficients. Thus $\mathbf{x} = \mathbf{0}$. But in this case $\mathbf{a}^1 \cdot \mathbf{a}^2 = \mu\mathbf{a}^2 \cdot \mathbf{a}^2 > 0$ and condition 3 is satisfied with $\mathbf{y} = \mathbf{a}^2$. If $\mathbf{a}^1 = \mathbf{0}$, Fig. 4–1(d), then the representation $1 \cdot \mathbf{a}^1 + 0 \cdot \mathbf{a}^2 = \mathbf{0}$ gives condition 1 with $\mathbf{x} = (1, 0)$, and one can take $\mathbf{y}$ to be the zero vector.

If the matrix $\mathbb{A}$ contains more than two columns, a proof along these same lines can again be given. However, as the reader may guess by looking at the geometry of the situation, the details become quite complicated. Instead, the proof can be completed by induction (Exercise 5).

This lemma includes as a special case a classical result of J. Farkas [F] on linear inequalities. Although Farkas derived this lemma in connection with problems in mechanics, the result has had its most fruitful applications in the study of questions similar to those considered here.

Farkas' Lemma: Let $\mathbb{A}$ be an $m \times n$ matrix and $\mathbf{b} \in R^m$. If for all solutions $\mathbf{y}$ of the inequality $\mathbb{A}^T\mathbf{y} \geq \mathbf{0}$, it follows that $\mathbf{b} \cdot \mathbf{y} \geq 0$, then there are non-negative constants $\lambda_1, \ldots, \lambda_n$ such that

$$\mathbf{b} = \sum_{j=1}^{n} \lambda_j \mathbf{a}^j.$$

Proof: Consider a matrix $\hat{\mathbb{A}}$ which has columns $-\mathbf{b}, \mathbf{a}^1, \ldots, \mathbf{a}^n$, and apply Lemma 2 to this matrix. The conclusion is that one can find vectors $\hat{\mathbf{x}} = (\xi_0, \xi_1, \ldots, \xi_n)$ and $\mathbf{y} = (\eta_1, \ldots, \eta_m)$ such that

$$\begin{aligned} \hat{\mathbf{x}} &\geq \mathbf{0}, \\ \hat{\mathbb{A}}\hat{\mathbf{x}} &= \mathbf{0}, \\ \hat{\mathbb{A}}^T\mathbf{y} &\geq \mathbf{0}, \\ (-\mathbf{b}) \cdot \mathbf{y} + \xi_0 &> 0. \end{aligned}$$

The equation $\hat{\mathbb{A}}^T\mathbf{y} \geq \mathbf{0}$ implies in particular that $\mathbb{A}^T\mathbf{y} \geq \mathbf{0}$, so that by the hypothesis of the lemma we conclude that $\mathbf{b} \cdot \mathbf{y} \geq 0$. From this and $(-\mathbf{b}) \cdot \mathbf{y} + \xi_0 > 0$ we conclude that $\xi_0 > 0$. Next, consider $\hat{\mathbb{A}}\hat{\mathbf{x}} = \mathbf{0}$ and note that this can be written as

$$-\mathbf{b}\xi_0 + \sum_{j=1}^{n} \mathbf{a}^j \xi_j = \mathbf{0},$$

with $\xi_j \geq 0$, $j = 1, 2, \ldots, n$. Since $\xi_0 > 0$, we have

$$\mathbf{b} = \sum_{j=1}^{n} \frac{\xi_j}{\xi_0} \mathbf{a}^j,$$

where

$$\frac{\xi_j}{\xi_0} \geq 0 \quad \text{for} \quad 1 \leq j \leq n.$$

EXERCISES

1. Given an $m \times 2$ matrix $\mathbb{A}$ and a vector $\mathbf{b} \neq \mathbf{0}$, prove that exactly one of the following two statements holds:
   (a) $\mathbf{b} = \sum_{i=1}^{2} \lambda_i \mathbf{a}^i$, $\lambda_i \geq 0$, or
   (b) There is a vector $\mathbf{y} \in R^m$ such that $\mathbf{y} \cdot \mathbf{a}^i \geq 0$, $i = 1, 2$, and $\mathbf{b} \cdot \mathbf{y} < 0$.
2. Discuss in detail the geometry of Exercise 1. State the corresponding result in geometric terms for an arbitrary $m \times n$ matrix $\mathbb{A}$ and $\mathbf{b} \neq \mathbf{0}$ in R^m.
3. Given an $m \times n$ matrix $\mathbb{A}$, prove that there are vectors $\mathbf{x} \in R^n$, $\mathbf{y} \in R^m$ such that the following conditions are satisfied:
   (a) $\mathbf{x} \geq 0$ and $\mathbb{A}\mathbf{x} = \mathbf{0}$.
   (b) $\mathbb{A}^T\mathbf{y} \geq \mathbf{0}$.
   (c) $\mathbb{A}^T\mathbf{y} + \mathbf{x} > \mathbf{0}$.
4. For each of the following matrices $\mathbb{A}$ find vectors $\mathbf{x}$, $\mathbf{y}$ satisfying the conclusions of Lemma 2:

$$\begin{bmatrix} 1 & 0 \\ -2 & 1 \\ 3 & -1 \end{bmatrix} \quad \begin{bmatrix} 1 & 0 & -1 \\ -2 & -1 & 0 \end{bmatrix} \quad \begin{bmatrix} 1 & 0 & -1 \\ -2 & 1 & 1 \\ 3 & -1 & 0 \end{bmatrix}.$$

5. Use an induction argument on the number of columns of $\mathbb{A}$ to complete the proof of Lemma 2. *Hint:* One way to proceed is as follows: Let $\mathbf{a}^1, \ldots, \mathbf{a}^n$ be the columns of $\mathbb{A}$, and let $\tilde{\mathbb{A}}$ denote the $m \times (n-1)$ matrix obtained from $\mathbb{A}$ by omitting the last column. The lemma has been shown to be true for $n = 1$ (and in fact also for $n = 2$). Assume that it is true for a matrix with $n - 1$ columns. Next,
   (a) Show that there are vectors $\tilde{\mathbf{x}}$ and $\tilde{\mathbf{y}}$ such that $\tilde{\mathbf{x}} \geq \mathbf{0}$, $\tilde{\mathbb{A}}\tilde{\mathbf{x}} = \mathbf{0}$, $\tilde{\mathbb{A}}^T\tilde{\mathbf{y}} \geq \mathbf{0}$, and $\mathbf{a}^1 \cdot \tilde{\mathbf{y}} + \tilde{\xi}_1 > 0$ $(\tilde{\mathbf{x}} = (\tilde{\xi}_1, \ldots, \tilde{\xi}_n))$.
   (b) If $\tilde{\mathbf{y}} \cdot \mathbf{a}^n \geq 0$, then the proof is complete. Find $\mathbf{y}$ and $\mathbf{x}$ satisfying the conclusions of the lemma. If $\tilde{\mathbf{y}} \cdot \mathbf{a}^n < 0$, then additional work is required. Define an $m \times (n-1)$ matrix $\mathbb{B}$ whose columns $\mathbf{b}^j$ are

$$\mathbf{b}^j = \mathbf{a}^j - \left(\frac{\mathbf{a}^j \cdot \tilde{\mathbf{y}}}{\mathbf{a}^n \cdot \tilde{\mathbf{y}}}\right)\mathbf{a}^n, \qquad 1 \leq j \leq n-1.$$

   (c) Prove that $\mathbb{B}^T\tilde{\mathbf{y}} = \mathbf{0}$, and that there are vectors $\mathbf{x}^*$ and $\mathbf{y}^*$ with the following properties:

$$\mathbf{x}^* = (\xi_1^*, \ldots, \xi_{n-1}^*) \geq 0,$$
$$\mathbb{B}\mathbf{x}^* = \mathbf{0},$$
$$\mathbb{B}^T\mathbf{y}^* \geq \mathbf{0},$$
$$\mathbf{b}^1 \cdot \mathbf{y}^* + \xi_1^* > 0.$$

Define

$$\mathbf{x} = \left(\xi_1^*, \ldots, \xi_{n-1}^*, -\sum_{j=1}^{n-1}\left(\frac{\mathbf{a}^j \cdot \tilde{\mathbf{y}}}{\mathbf{a}^n \cdot \tilde{\mathbf{y}}}\right)\xi_j^*\right) \in R^n.$$

(d) Prove that $\mathbf{x} \geq \mathbf{0}$, $\mathbb{A}\mathbf{x} = \mathbf{0}$.
Define

$$\mathbf{y} = \mathbf{y}^* - \left(\frac{\mathbf{a}^n \cdot \mathbf{y}^*}{\mathbf{a}^n \cdot \tilde{\mathbf{y}}}\right)\tilde{\mathbf{y}}.$$

(e) Prove that $\mathbb{A}^T\mathbf{y} \geq \mathbf{0}$ and, if ξ_1 is the first coordinate of the vector $\mathbf{x}$, that $\mathbf{a}^1 \cdot \mathbf{y} + \xi_1 > 0$.

(f) The vectors $\mathbf{x}$ and $\mathbf{y}$ satisfy the conclusions of the lemma, and the proof by induction is complete.

4.3 LINEAR PROGRAMMING

In Sec. 4.1 it was asserted that the examples presented there were particular instances of problems frequently encountered in optimization. The type of problem of which these are special cases is known broadly as a problem in linear programming. The field of linear programming first received widespread attention in the 1940s in connection with allocation problems. Considerable effort has been devoted to the subject in recent years, and the basic theory is now well understood. It has proved to be an extraordinarily fruitful concept, both from the point of view of model construction and the practical solution of specific problems.

4.3.1 Primal Problems

It is our object here to pose a problem which is sufficiently general so that it will include those instances which are commonly encountered as special cases, and yet is sufficiently special so that we can develop an adequate theory without excessive technical complications. We begin by introducing what we shall refer to as the *standard problem*.

Definition: The *standard maximum problem* of linear programming is the problem of finding nonnegative real numbers $\xi_1, \ldots, \xi_n$ which satisfy a given set of inequalities

$$\sum_{j=1}^{n} \xi_j a_{ij} \leq \beta_i, \qquad i = 1, 2, \ldots, m,$$

and which maximize a certain linear function

$$L(\xi_1, \ldots, \xi_n) = \sum_{i=1}^{n} \gamma_i \xi_i.$$

As with our discussion of systems of linear inequalities, there is much to be gained by phrasing this definition in vector terminology. To this end, we introduce the $m \times n$ matrix $\mathbb{A}$ whose i–j element is a_{ij}, the vector $\mathbf{c} = (\gamma_1, \ldots, \gamma_n)$, and the column vector $\mathbf{b}$ whose coordinates are $\beta_1, \ldots, \beta_m$, respectively. The above definition can be rephrased as follows.

Definition: Let $\mathbb{A}$ be an $m \times n$ matrix, $\mathbf{b} \in R^m$, $\mathbf{c} \in R^n$. The feasible set $\mathfrak{F}$ for the pair $(\mathbb{A}, \mathbf{b})$ is

$$\mathfrak{F} = \{\mathbf{x} \in R^n : \mathbf{x} \geq \mathbf{0},\ \mathbb{A}\mathbf{x} \leq \mathbf{b}\}.$$

The *standard maximum problem* is that of finding $\mathbf{x}^* \in \mathfrak{F}$ such that

$$\mathbf{x}^* \cdot \mathbf{c} = \max_{\mathbf{x} \in \mathfrak{F}} [\mathbf{x} \cdot \mathbf{c}]. \tag{3.1}$$

In both cases the maximum is to be a finite maximum. For notational convenience we frequently write $L(\mathbf{x}) = \mathbf{x} \cdot \mathbf{c}$.

It is clear from either form of this definition that the matrix $\mathbb{A}$ and the vectors $\mathbf{b}$ and $\mathbf{c}$ characterize the problem. We refer to the standard maximum problem defined above as the *primal* problem $[\mathbb{A}, \mathbf{b}, \mathbf{c}]$ determined by these quantities. The adjective *primal* is used to refer to the original problem, and to distinguish it from another related problem. The details will become clear in the next subsection. We omit the $[\mathbb{A}, \mathbf{b}, \mathbf{c}]$ if it is clear from the context which problem is under discussion.

One sometimes encounters problems of a slightly different form. Let $\mathbb{A}'$ be an $m \times n$ matrix, $\mathbf{c}' \in R^n$, and $\mathbf{b}' \in R^m$. Find $\mathbf{x}^* \in R^n$ such that

$$\mathbf{x}^* \in \mathfrak{F}' = \{\mathbf{x} \in R^n : \mathbf{x} \geq \mathbf{0},\ \mathbb{A}'\mathbf{x} \geq \mathbf{b}'\}$$

and

$$\mathbf{x}^* \cdot \mathbf{c}' = \min_{\mathbf{x} \in \mathfrak{F}'} [\mathbf{x} \cdot \mathbf{c}'].$$

Such a problem might reasonably be called a standard minimum problem, and we adopt this terminology. These conventions are common, but not universal, in the literature. However, we see that if we set $\mathbb{A} = -\mathbb{A}'$, $\mathbf{b} = -\mathbf{b}'$, and $\mathbf{c} = -\mathbf{c}'$, then this problem is equivalent to a standard maximum problem (Exercise 1). Consequently, all results stated for the standard maximum problem have analogs for the standard minimum problem and vice versa. Since it is sufficient to state our results for only one form and since the geometry of the maximum problem is somewhat simpler, we state our results for problems in that form. However, we do not hesitate to apply them to minimum problems whenever it is convenient. We urge the reader to formulate the statement of each result for the standard minimum problem.

It is clear from these definitions that the situations described in Sec. 4.1 lead to standard problems. The details of this verification are left to the reader (Exercise 2).

We shall find some additional notation useful.

Definition: Let $[\mathbb{A}, \mathbf{b}, \mathbf{c}]$ define a standard maximum problem and let $\mathfrak{F}$ be its feasible set. A vector $\mathbf{x}^* \in \mathfrak{F}$ which satisfies Eq. (3.1) is said to be an *optimal* vector.

As an example illustrating these concepts, and one to which we shall return later, consider the following problem. Find $\mathbf{x} \in R^4$, $\mathbf{x} = (\xi_1, \xi_2, \xi_3$ $\xi_4)$, which maximizes

$$L(\mathbf{x}) = 2\xi_1 + 4\xi_2 + \xi_3 + \xi_4$$

over the set of all nonnegative vectors $\mathbf{x}$ which satisfy

$$\begin{aligned} \xi_1 + 3\xi_2 + 4\xi_4 &\leq 8, \\ 2\xi_1 + \xi_2 &\leq 6, \\ \xi_2 + 4\xi_3 + \xi_4 &\leq 6. \end{aligned}$$

Clearly, this is a standard problem with

$$\mathbb{A} = \begin{bmatrix} 1 & 3 & 0 & 4 \\ 2 & 1 & 0 & 0 \\ 0 & 1 & 4 & 1 \end{bmatrix}, \qquad \mathbf{b} = \begin{bmatrix} 8 \\ 6 \\ 6 \end{bmatrix}, \qquad \mathbf{c} = (2, 4, 1, 1).$$

A computation verifies that $\mathbf{x} = (2, 2, 1, 0)$ is a feasible vector and that $L(\mathbf{x}) = 13$. Actually $\mathbf{x}$ is optimal, although it is not easy to prove this without additional theory.

4.3.2 Dual Problems

One of the remarkable features of linear programming problems is the relation between the primal problem and an associated problem known as the dual problem. The dual of a given problem will be seen to be important from both a theoretical and a practical point of view. Indeed, it is only by considering the two problems simultaneously that one obtains the proper perspective of the subject as a whole.

In formulating the dual of the standard maximum problem, one again encounters the notion of a minimum problem.

Definition: The dual of the standard maximum problem $[\mathbb{A}, \mathbf{b}, \mathbf{c}]$ is the following standard minimum problem: Find nonnegative real numbers

$\eta_1, \ldots, \eta_m$ which satisfy

$$\sum_{i=1}^{m} \eta_i a_{ij} \geq \gamma_j, \qquad j = 1, 2, \ldots, n,$$

and which minimize the linear function

$$L'(\eta_1, \ldots, \eta_m) = \sum_{i=1}^{m} \beta_i \eta_i.$$

Again it is convenient to phrase our definition in vector notation.

Definition: Let $[\mathbb{A}, \mathbf{b}, \mathbf{c}]$ define a standard maximum problem. Set

$$\mathfrak{F}' = \{\mathbf{y} \in R^m : \mathbf{y} \geq \mathbf{0},\ \mathbb{A}^T\mathbf{y} \geq \mathbf{c}\}.$$

The dual of the above primal problem is that of finding $\mathbf{y}^* \in \mathfrak{F}'$ such that

$$\mathbf{y}^* \cdot \mathbf{b} = \min_{\mathbf{y} \in \mathfrak{F}'} [\mathbf{y} \cdot \mathbf{b}]. \tag{3.2}$$

In this definition the set $\mathfrak{F}'$ is known as the set of feasible vectors for the dual problem. In view of the comments made above concerning minimum problems, we could state the dual problem in terms of another maximum problem. In particular, the dual of the standard maximum problem defined by $[\mathbb{A}, \mathbf{b}, \mathbf{c}]$ is the standard maximum problem defined by $[-\mathbb{A}^T, -\mathbf{c}, -\mathbf{b}]$. We do not exploit this formulation of the dual problem for reasons which will become clear as we proceed. Recalling the definition of an optimal vector and our agreement concerning the statement of analogous results for minimum problems, we note that an optimal vector for the dual problem is a vector $\mathbf{y}^* \in \mathfrak{F}'$ which satisfies Eq. (3.2).

We shall illustrate these concepts by returning to the numerical example introduced above. If we introduce $\mathbf{y} = (\eta_1, \eta_2, \eta_3)$, then the dual of that problem is the one of minimizing the quantity

$$L'(\mathbf{y}) = 8\eta_1 + 6\eta_2 + 6\eta_3$$

over the set of all nonnegative vectors $\mathbf{y}$ which satisfy

$$\begin{aligned} \eta_1 + 2\eta_2 \qquad &\geq 2, \\ 3\eta_1 + \eta_2 + \eta_3 &\geq 4, \\ 4\eta_3 &\geq 1, \\ 4\eta_1 \qquad + \eta_3 &\geq 1. \end{aligned}$$

A computation shows that the vector $\mathbf{y} = (11/10, 9/20, 1/4)$ is feasible and that $L'(\mathbf{y}) = 13$. Combining this with the results obtained in the discussion of the original example, we have vectors $\mathbf{x}$ and $\mathbf{y}$, $\mathbf{x} \in \mathfrak{F}$ and $\mathbf{y} \in \mathfrak{F}'$ with $L(\mathbf{x}) = L'(\mathbf{y})$. It turns out that this is sufficient for both $\mathbf{x}$ and $\mathbf{y}$ to be optimal for the primal program and its dual, respectively. This fact is quite useful and it is contained in a more general result to which we now turn.

We proceed using an approach due to A. W. Tucker. The feasible set $\mathfrak{F}$ for a standard maximum problem consists of vectors $\mathbf{x} \geq \mathbf{0}$ for which $\mathbb{A}\mathbf{x} \leq \mathbf{b}$. Another way of looking at this is that the set $\mathfrak{F}$ consists of those nonnegative vectors $\mathbf{x} \in R^n$ for which there is a nonnegative vector $\mathbf{w} \in R^m$ satisfying $\mathbb{A}\mathbf{x} + \mathbf{w} = \mathbf{b}$. Likewise, the set $\mathfrak{F}'$ of feasible vectors for the dual problem consists of vectors $\mathbf{y} \geq \mathbf{0}$, $\mathbb{A}^T\mathbf{y} \geq \mathbf{c}$, or, equivalently, of those vectors $\mathbf{y} \in R^m$ for which there exists a nonnegative vector $\mathbf{v} \in R^n$ satisfying $\mathbb{A}^T\mathbf{y} - \mathbf{v} = \mathbf{c}$. Suppose for the moment that we consider the equations

$$\mathbb{A}\mathbf{x} + \mathbf{w} = \mathbf{b}, \qquad \mathbb{A}^T\mathbf{y} - \mathbf{v} = \mathbf{c}, \tag{3.3}$$

without any other constraints on the vectors $\mathbf{x}, \mathbf{y}, \mathbf{v}, \mathbf{w}$. It follows from these equations that

$$\mathbf{y} \cdot \mathbb{A}\mathbf{x} + \mathbf{y} \cdot \mathbf{w} = \mathbf{y} \cdot \mathbf{b}, \qquad \mathbf{x} \cdot \mathbb{A}^T\mathbf{y} - \mathbf{x} \cdot \mathbf{v} = \mathbf{x} \cdot \mathbf{c}.$$

Subtracting the second of these two equations from the first and using the fact that $\mathbf{y} \cdot \mathbb{A}\mathbf{x} = \mathbf{x} \cdot \mathbb{A}^T\mathbf{y}$, we obtain

$$\mathbf{y} \cdot \mathbf{b} - \mathbf{x} \cdot \mathbf{c} = \mathbf{y} \cdot \mathbf{w} + \mathbf{x} \cdot \mathbf{v}. \tag{3.4}$$

Equation (3.4), which holds for any vectors $\mathbf{x}, \mathbf{y}, \mathbf{v}, \mathbf{w}$ for which (3.3) holds, is known as *Tucker's duality relation*. In particular, it holds for vectors $\mathbf{x}$ and $\mathbf{y}$ which are feasible for the primal problem and its dual, respectively. In this case $\mathbf{v} \geq \mathbf{0}$ and $\mathbf{w} \geq \mathbf{0}$, and consequently $\mathbf{y} \cdot \mathbf{w} + \mathbf{x} \cdot \mathbf{v} \geq 0$. This and (3.4) yield the following result.

Lemma 1: Let $\mathfrak{F}$ be the set of feasible vectors for the standard maximum problem $[\mathbb{A}, \mathbf{b}, \mathbf{c}]$, and let $\mathfrak{F}'$ be the set of feasible vectors for the associated dual problem. For every $\mathbf{x} \in \mathfrak{F}$ and $\mathbf{y} \in \mathfrak{F}'$ we have

$$\mathbf{x} \cdot \mathbf{c} \leq \mathbf{y} \cdot \mathbf{b}. \tag{3.5}$$

We can now formulate a sufficient condition for optimality. It has to do with the special case of equality in (3.5).

Theorem 1: Let $\mathfrak{F}$ and $\mathfrak{F}'$ be as in Lemma 1, $\mathbf{x} \in \mathfrak{F}$, $\mathbf{y} \in \mathfrak{F}'$. If $\mathbf{x} \cdot \mathbf{c} = \mathbf{y} \cdot \mathbf{b}$, then $\mathbf{x}$ is optimal for the primal problem and $\mathbf{y}$ is optimal for its dual.

Proof: Let $\mathbf{x}'$ be any feasible vector for the primal problem. By Lemma 1 we have $\mathbf{x}' \cdot \mathbf{c} \leq \mathbf{y} \cdot \mathbf{b}$. But $\mathbf{y} \cdot \mathbf{b} = \mathbf{x} \cdot \mathbf{c}$ by hypothesis, and consequently

$L(\mathbf{x}') \leq L(\mathbf{x})$ for all feasible vectors $\mathbf{x}'$. It follows that $\mathbf{x}$ is optimal for the primal problem. A similar argument shows that $\mathbf{y}$ is optimal for the dual problem.

As the numerical example above indicates, this theorem provides a very useful test for optimality. In general, if one can conjecture on heuristic or other grounds what the optimal vectors might be, this result provides a partial test of the validity of the conjecture. Also, from a theoretical point of view, most proofs of the existence of optimal vectors are based on it.

There is still more to be obtained from Tucker's duality relation. If $\mathbf{x}$ and $\mathbf{y}$ are nonnegative vectors, then the nonnegativity of the vectors $\mathbf{v}$ and $\mathbf{w}$ implies the feasibility of $\mathbf{x}$ and $\mathbf{y}$ for the primal problem and its dual, respectively. Therefore the additional condition $\mathbf{x} \cdot \mathbf{v} + \mathbf{y} \cdot \mathbf{w} = 0$ implies that $\mathbf{y} \cdot \mathbf{b} = \mathbf{x} \cdot \mathbf{c}$ and thus the optimality of $\mathbf{x}$ and $\mathbf{y}$. Since $\mathbf{x}, \mathbf{y}, \mathbf{v}, \mathbf{w}$ are all nonnegative, $\mathbf{x} \cdot \mathbf{v} + \mathbf{y} \cdot \mathbf{w} = 0$ is equivalent to the two conditions $\mathbf{x} \cdot \mathbf{v} = 0$ and $\mathbf{y} \cdot \mathbf{w} = 0$. This relation between $\mathbf{x}$, $\mathbf{v}$ and $\mathbf{y}$, $\mathbf{w}$ is known as *complementary slackness*, and its meaning is as follows. The vector $\mathbf{w}$, a vector in R^m, is nonnegative and its jth coordinate can be different from zero only if the jth coordinate of the vector $\mathbf{y}$ is equal to zero. Also, the jth coordinate of $\mathbf{y}$ can be different from zero only if the jth coordinate of $\mathbf{w}$ is equal to zero. Similar assertions hold for the vectors $\mathbf{x}$ and $\mathbf{v}$. We state this conclusion formally.

Theorem 2: If $\mathbf{x}, \mathbf{y}, \mathbf{v}, \mathbf{w}$ are nonnegative vectors which satisfy (3.3) and the complementary slackness relation $\mathbf{x} \cdot \mathbf{v} = 0$, $\mathbf{y} \cdot \mathbf{w} = 0$, then $\mathbf{x}$ and $\mathbf{y}$ are optimal for the primal and dual problems, respectively.

Actually, Theorem 2 can be improved to an *if and only if* result. That is, if the vectors $\mathbf{x}, \mathbf{y}, \mathbf{v}, \mathbf{w}$ are nonnegative, then the complementary slackness relation $\mathbf{x} \cdot \mathbf{v} = 0$, $\mathbf{y} \cdot \mathbf{w} = 0$ is both necessary and sufficient for $\mathbf{x}$ and $\mathbf{y}$ to be optimal. To complete the proof of this assertion, we need to know that if $\mathbf{x}$ and $\mathbf{y}$ are optimal, then $\mathbf{x} \cdot \mathbf{c} = \mathbf{y} \cdot \mathbf{b}$. This is indeed true, and it is proved in Theorem 3. We shall return to the present discussion in Sec. 4.3.5.

The notion of complementary slackness is an important and useful one from both a theoretical and practical point of view. Some of its theoretical uses have been demonstrated in this section. We shall return to its practical implications in the exercises and in the projects.

At this point in our development, a number of questions should occur to the reader. First, do optimal vectors always exist, or are additional conditions necessary? Second, if an optimal vector exists, how does one find it? Third, can one attach any significance to the dual problem in the context of the primal problem?

The first of these questions will be answered in Sec. 4.3.5, where some general theorems on the existence of optimal vectors for linear programming problems are given. In Chap. 5 we shall give some constructive means of solving linear programming problems, including both geometrical methods and the well-known simplex algorithm. The next section of this chapter is devoted to a study of the dual problem in the particular setting of the two-by-three transportation problem of Sec. 2.2.

4.3.3 An Interpretation of the Dual of a Transportation Problem

The context for this discussion is that of Sec. 2.2, where the mobile home business of Mr. Wheeler was discussed. Briefly, Mr. Wheeler owns two factories where mobile homes are assembled and three distribution centers where they are sold. He assumes that in a specific month he will sell d_i mobile homes at distribution center i, $i = 1, 2, 3$. Accordingly, he plans to assemble $d_1 + d_2 + d_3$ homes, s_1 at the first factory and s_2 at the second. His concern is the transporting of the homes from the factories to the distribution centers.

Mr. Wheeler has estimated that there is a cost c_{ij} incurred when he moves one home from factory i to center j. This cost estimate is based, for example, on his evaluation of the costs of equipment, maintenance, insurance, labor, etc. In his evaluation of the situation, Mr. Wheeler discovers that there is another alternative available to him, namely, to have the homes transported by another party. In particular, Mr. Wheeler finds that a certain Mr. Fastbuck of the Firm Specializing in the Resale of Mobile Homes (FSRMH for short) is willing to transport the homes under an agreement where Mr. Wheeler may save on his moving costs. Certainly these costs cannot increase since he always has the option of moving the homes on his own. The proposal of Mr. Fastbuck is as follows: His company (FSRMH) will purchase all the homes assembled at the factories and sell them back to Mr. Wheeler at the distribution centers. Let ξ_1 be the price per home in dollars paid by FSRMH at factory number 1 and ξ_2 the corresponding price at the second factory. Likewise, let ξ_3 be the price at which Mr. Fastbuck will resell a mobile home to Mr. Wheeler at distribution center number 1, and let ξ_4 and ξ_5 be the corresponding prices at the second and third centers. From the point of view of Mr. Fastbuck there is a very natural question; namely, how should the prices $\{\xi_i\}_{i=1}^5$ be chosen so as to maximize the net income of FSRMH? Notice that there can be no unique solution to the problem. Indeed, Mr. Fastbuck can increase both the prices he pays and the prices at which he sells by a fixed amount and his net income remains the same.

We now consider the situation in more detail. Since Mr. Wheeler has the option of moving the homes himself, in order for FSRMH to get the business in the first place it is necessary that the price charged by Mr. Fastbuck at center j not exceed the price paid by him at factory i by more than c_{ij}. This can be expressed by the following inequalities on the prices $\{\xi_i\}_{i=1}^5$:

$$
\begin{aligned}
\xi_3 &\leq \xi_1 + c_{11} \\
\xi_3 &\leq \xi_2 + c_{21} \\
\xi_4 &\leq \xi_1 + c_{12} \\
\xi_4 &\leq \xi_2 + c_{22} \\
\xi_5 &\leq \xi_1 + c_{13} \\
\xi_5 &\leq \xi_2 + c_{23}
\end{aligned}
\quad \text{or} \quad
\begin{aligned}
\xi_3 - \xi_1 &\leq c_{11} \\
\xi_3 - \xi_2 &\leq c_{21} \\
\xi_4 - \xi_1 &\leq c_{12} \\
\xi_4 - \xi_2 &\leq c_{22} \\
\xi_5 - \xi_1 &\leq c_{13} \\
\xi_5 - \xi_2 &\leq c_{23}.
\end{aligned}
\tag{3.6}
$$

With regard to the comment on uniqueness made above, if $\{\xi_i\}_{i=1}^5$ satisfies (3.6), and if $\xi_i' = \xi_i + d$, $i = 1, 2, \ldots, 5$, then $\{\xi_i'\}_{i=1}^5$ also satisfies (3.6).

We assume that Mr. Wheeler sells and Mr. Fastbuck purchases s_i homes at factory i, $i = 1, 2$, and that Mr. Fastbuck sells and Mr. Wheeler buys d_j homes at center j, $j = 1, 2, 3$. Thus the net income from sales for FSRMH, excluding any expenses other than the purchase of homes, is

$$
I(\xi_1, \ldots, \xi_5) = \xi_3 d_1 + \xi_4 d_2 + \xi_5 d_3 - \xi_1 s_1 - \xi_2 s_2. \tag{3.7}
$$

The problem for Mr. Fastbuck is to select that vector $\mathbf{x} = (\xi_1, \xi_2, \xi_3, \xi_4, \xi_5)$ which satisfies (3.6) and which maximizes $I(\xi_1, \ldots, \xi_5)$ over the class of all such vectors. This problem is exactly the dual of the minimization of costs problem of Sec. 2.2, and we now turn toward making this duality evident.

We begin with the remark that just as the dual of the maximum problem determined by $[\mathbb{A}, \mathbf{b}, \mathbf{c}]$ is the minimum problem determined by $[\mathbb{A}^T, \mathbf{c}, \mathbf{b}]$, also the dual of the minimum problem defined by $[\mathbb{A}, \mathbf{b}, \mathbf{c}]$ is the maximum problem defined by $[\mathbb{A}^T, \mathbf{c}, \mathbf{b}]$. Next, returning to specifics, it is convenient to write the original problem, that is, the problem of Sec. 2.2, as a standard minimum problem. Introduce $\mathbf{y} = (\eta_1, \eta_2, \eta_3, \eta_4, \eta_5, \eta_6)$, where

$$
\begin{aligned}
\eta_1 &= f_{11}, & \eta_2 &= f_{12}, & \eta_3 &= f_{13}, \\
\eta_4 &= f_{21}, & \eta_5 &= f_{22}, & \eta_6 &= f_{23}.
\end{aligned}
$$

The original problem is that of finding a $\mathbf{y} \geq \mathbf{0}$ which minimizes

$$
\eta_1 c_{11} + \eta_2 c_{12} + \eta_3 c_{13} + \eta_4 c_{21} + \eta_5 c_{22} + \eta_6 c_{23}
$$

over the set of all solutions of the inequalities

$$
\begin{aligned}
-\eta_1 - \eta_2 - \eta_3 &\geq -s_1, & \eta_1 + \eta_4 &\geq d_1, \\
-\eta_4 - \eta_5 - \eta_6 &\geq -s_2, & \eta_2 + \eta_5 &\geq d_2, \\
& & \eta_3 + \eta_6 &\geq d_3.
\end{aligned}
$$

Using matrix notation, we have the formulation: Find $\mathbf{y}^*$ such that

$$\mathbf{y}^* \cdot \mathbf{c} = \min_{\mathbf{y} \in \mathfrak{F}} [\mathbf{y} \cdot \mathbf{c}],$$

where $\mathfrak{F} = \{\mathbf{y} : \mathbf{y} \geq \mathbf{0},\ \mathbb{A}\mathbf{y} \geq \mathbf{d}\}$, $\mathbf{c} = (c_{11}, c_{12}, c_{13}, c_{21}, c_{22}, c_{23})$, $\mathbf{d} = (-s_1, -s_2, d_1, d_2, d_3)$, and

$$\mathbb{A} = \begin{bmatrix} -1 & -1 & -1 & 0 & 0 & 0 \\ 0 & 0 & 0 & -1 & -1 & -1 \\ 1 & 0 & 0 & 1 & 0 & 0 \\ 0 & 1 & 0 & 0 & 1 & 0 \\ 0 & 0 & 1 & 0 & 0 & 1 \end{bmatrix}.$$

We mention that for the present discussion the requirement that the coordinates of $\mathbf{y}$ be integers has not been included.

This standard minimum problem has a formal dual, that is, a problem satisfying the definitions of Sec. 4.3.2. Since the original problem is a minimum problem, its dual is a maximum problem, namely that of maximizing $\mathbf{x} \cdot \mathbf{d}$. This linear function is to be maximized over the set of vectors $\mathbf{x}$ satisfying

$$\mathbf{x} \geq \mathbf{0},$$
$$\mathbb{A}^T\mathbf{x} \leq \mathbf{c}.$$

We specify the problem in more detail. We are to find vectors $\mathbf{x}$ satisfying $\mathbf{x} = (\xi_1, \xi_2, \xi_3, \xi_4, \xi_5) \geq \mathbf{0}$,

$$\begin{aligned} -\xi_1 + \xi_3 &\leq c_{11}, & -\xi_2 + \xi_3 &\leq c_{21}, \\ -\xi_1 + \xi_4 &\leq c_{12}, & -\xi_2 + \xi_4 &\leq c_{22}, \\ -\xi_1 + \xi_5 &\leq c_{13}, & -\xi_2 + \xi_5 &\leq c_{23}, \end{aligned}$$

and for which the quantity

$$-s_1\xi_1 - s_2\xi_2 + d_1\xi_3 + d_2\xi_4 + d_3\xi_5$$

is as large as possible. But it is now clear that the dual of the original problem, the problem of Mr. Wheeler in Sec. 2.2, is the same as the problem of maximizing (3.7) subject to (3.6)—the problem of maximizing the net income of FSRMH. Thus Mr. Fastbuck's net income maximization problem provides an interpretation of the dual of Mr. Wheeler's cost minimization problem.

4.3.4 Equivalent Problems

In this section we shall introduce some new ideas, and we shall reformulate some of the results obtained in the preceding sections.

This serves as preparation both for the next section, where we shall prove that under quite general assumptions optimal vectors always exist, and for the algorithm given in Chap. 5.

Recall that the standard maximum problem is one of maximizing $\mathbf{x} \cdot \mathbf{c}$ over the set of all vectors $\mathbf{x} \geq \mathbf{0}$ which satisfy $\mathbb{A}\mathbf{x} \leq \mathbf{b}$. It is occasionally convenient to consider problems in which the restrictions on $\mathbf{x}$ are somewhat different. For example, one could relax the requirement that $\mathbf{x}$ be nonnegative, or one could replace the inequality $\mathbb{A}\mathbf{x} \leq \mathbf{b}$ by the equality $\mathbb{A}\mathbf{x} = \mathbf{b}$. The latter situation is of sufficient importance to justify its own name.

Definition: Let $\mathbb{A}$ be an $m \times n$ matrix, $\mathbf{b} \in R^m$, $\mathbf{c} \in R^n$, and $\mathfrak{F} = \{\mathbf{x} \in R^n: \mathbf{x} \geq \mathbf{0}, \mathbb{A}\mathbf{x} = \mathbf{b}\}$. The *restricted maximum problem* is the problem of finding $\mathbf{x}^* \in R^n$ such that

$$\mathbf{x}^* \cdot \mathbf{c} = \max_{\mathbf{x} \in \mathfrak{F}} [\mathbf{x} \cdot \mathbf{c}].$$

There is a corresponding definition for the restricted minimum problem.

It is useful to observe that the distinction between the standard and restricted problems is only apparent and that they are actually equivalent. Here the term *equivalent* means that one can be written in the form of the other. To show this, we note first that the set of $\mathbf{x}$ for which $\mathbb{A}\mathbf{x} = \mathbf{b}$ is the same as the set of $\mathbf{x}$ for which $\mathbb{A}\mathbf{x} \leq \mathbf{b}$ and $\mathbb{A}\mathbf{x} \geq \mathbf{b}$. That is,

$$\{\mathbf{x} : \mathbb{A}\mathbf{x} = \mathbf{b}\} = \{\mathbf{x} : \mathbb{A}\mathbf{x} \leq \mathbf{b},\ -\mathbb{A}\mathbf{x} \leq -\mathbf{b}\}.$$

Next, the inequalities

$$\mathbb{A}\mathbf{x} \leq \mathbf{b},$$
$$-\mathbb{A}\mathbf{x} \leq -\mathbf{b}$$

can be written $\tilde{\mathbb{A}}\mathbf{x} \leq \tilde{\mathbf{b}}$, where the matrix $\tilde{\mathbb{A}}$ is $2m \times n$ and is given by

$$\tilde{\mathbb{A}} = \begin{bmatrix} \mathbb{A} \\ -\mathbb{A} \end{bmatrix},$$

and the vector $\tilde{\mathbf{b}}$ is the vector in R^{2m} with its first m coordinates equal to $\beta_1, \ldots, \beta_m$ and its second m coordinates equal to $-\beta_1, \ldots, -\beta_m$, that is,

$$\tilde{\mathbf{b}} = (\beta_1, \ldots, \beta_m, -\beta_1, \ldots, -\beta_m).$$

It follows that the restricted maximum problem can be expressed in the following way:

Let $\tilde{\mathbb{A}}$, $\tilde{\mathbf{b}}$, and $\mathbf{c}$ be as above and define

$$\tilde{\mathfrak{F}} = \{\mathbf{x} \in R^n : \mathbf{x} \geq \mathbf{0}, \tilde{\mathbb{A}}\mathbf{x} \leq \tilde{\mathbf{b}}\}.$$

Find $\mathbf{x}^* \in \tilde{\mathfrak{F}}$ such that

$$\mathbf{x}^* \cdot \mathbf{c} = \max_{\mathbf{x} \in \tilde{\mathfrak{F}}} [\mathbf{x} \cdot \mathbf{c}].$$

This last problem is clearly a standard maximum problem, and this proves half the result. The other half is contained in Exercise 5.

We have seen that the standard and restricted problems, although superficially different, are actually equivalent. This leads us to ask whether they are both special cases of a problem of a more comprehensive form. This is indeed the case, and it is to this topic that we now turn. It is our purpose to introduce a rather general linear programming problem in a way which lends itself best to the development of duality. To this end, let $\mathbb{A}$ be an $m \times n$ matrix with rows $\mathbf{a}_i$, $i = 1, \ldots, m$, $\mathbf{b} = (\beta_1, \ldots, \beta_m)$, and $\mathbf{c}$ be a vector in R^n.

Definition: Let $\mathbb{A}$ be an $m \times n$ matrix, $\mathbf{b} \in R^m$, and $\mathbf{c} \in R^n$. Let $\mathfrak{F}$ be the set of all vectors $\mathbf{x} \in R^n$ which satisfy

$$\begin{aligned} \xi_i &\geq 0 && \text{for } i \in Q, \\ \mathbf{x} \cdot \mathbf{a}_j &\leq \beta_j && \text{for } j \in P, \\ \mathbf{x} \cdot \mathbf{a}_j &= \beta_j && \text{for } j \in P^*, \end{aligned} \tag{3.8}$$

where Q is some subset, possibly empty and possibly the whole set, of the set of integers $\{1, 2, \ldots, n\}$; P is some subset of the set of integers $\{1, 2, \ldots, m\}$; and P^* is the complement of P with respect to the set $\{1, 2, \ldots, m\}$. The *general maximum problem* is that of finding $\mathbf{x}^* \in \mathfrak{F}$ such that

$$L(\mathbf{x}^*) = \mathbf{x}^* \cdot \mathbf{c} = \max_{\mathbf{x} \in \mathfrak{F}} L(\mathbf{x}).$$

As with standard problems, we use the adjective *primal* to refer to the problem which is initially posed. It is clear that the standard and restricted maximum problems are special cases: In the standard problem $Q = \{1, 2, \ldots, n\}$ and $P^* = \varnothing$, and in the restricted problem $Q = \{1, 2, \ldots, n\}$ and $P = \varnothing$. Since the general form includes the other forms as special cases, it is somewhat unexpected to find that one can also write the general problem in either standard or restricted form. We already know that a standard problem can be written as a restricted one, and therefore it is sufficient to show that the general problem can be written as a standard one. This fact is needed in the proof of the fundamental duality theorem of Sec. 4.3.5. For the proof of equivalence, it is convenient to rewrite the general problem slightly. Notice that inequalities occur in (3.8) in two places; namely as constraints on some of the coordinates of $\mathbf{x}$ and as $\mathbf{x} \cdot \mathbf{a}_j \leq \beta_j$. Also, the inequalities and equalities of the form $\mathbf{x} \cdot \mathbf{a}_j \leq \beta_j$ and $\mathbf{x} \cdot \mathbf{a}_j = \beta_j$ may not be grouped together. First, notice that each of the coordinate constraints

$\xi_i \geq 0$ can be written as $\mathbf{x} \cdot \mathbf{u}_i \geq 0$ or $\mathbf{x} \cdot (-\mathbf{u}_i) \leq 0$. This is the same form as the inequalities $\mathbf{x} \cdot \mathbf{a}_j \leq \beta_j$. Second, define $|Q| = q$ to be the number of elements in the set Q and $|P|$ to be the number of elements in the set P. Now define a matrix $\hat{\mathbb{A}}$ as follows: $\hat{\mathbb{A}}$ is an $(m + q) \times n$ matrix the first q rows of which are the vectors $-\mathbf{u}_i$ which occur in the coordinate constraints $\mathbf{x} \cdot (-\mathbf{u}_i) \leq 0$, the next $|P|$ rows are the row vectors of $\mathbb{A}$ which occur in the inequalities $\mathbf{x} \cdot \mathbf{a}_j \leq \beta_j$, and the final $m - |P|$ rows are the rows of $\mathbb{A}$ which occur in the equalities $\mathbf{x} \cdot \mathbf{a}_j = \beta_j$. Define $\hat{\mathbf{b}} \in R^{m+q}$ as the vector whose first q coordinates are zero and whose last m coordinates are those of the vector $\mathbf{b}$, $\hat{\mathbf{b}} = (\hat{\beta}_1, \ldots, \hat{\beta}_{m+q})$. Finally, set $m' = m + q$ and $r = q + |P|$. With these definitions, the general problem can be stated in the following form: Let $\hat{\mathbb{A}}$ be an $m' \times n$ matrix, $\hat{\mathbf{b}} \in R^{m'}$, and $\mathbf{c} \in R^n$. Let $\hat{\mathfrak{F}}$ be the set of all vectors $\mathbf{x} \in R^n$ which satisfy

$$\mathbf{x} \cdot \hat{\mathbf{a}}_j \leq \hat{\beta}_j, \qquad j = 1, 2, \ldots, r, \tag{3.9}$$

$$\mathbf{x} \cdot \hat{\mathbf{a}}_j = \hat{\beta}_j, \qquad j = r + 1, \ldots, m'. \tag{3.10}$$

Find $\mathbf{x}^*$ such that

$$L(\mathbf{x}^*) = \mathbf{x}^* \cdot \mathbf{c} = \max_{\mathbf{x} \in \hat{\mathfrak{F}}} L(\mathbf{x}).$$

The proof that the general problem, when written in this form, is equivalent to a standard problem is outlined in Exercise 6.

We now turn to the definition of the dual of a general problem. The original formulation of the general problem was selected because of the ease of expressing the dual. As previously, the dual is another optimization problem which is closely associated with the original one. Referring to (3.8) and the associated notation, the dual problem is to find $\mathbf{y}^* \in R^m$ such that

$$\mathbf{y}^* \cdot \mathbf{b} = \min_{\mathbf{y} \in \mathfrak{F}'} [\mathbf{y} \cdot \mathbf{b}],$$

where $\mathfrak{F}'$ is the set of all $\mathbf{y} = (\eta_1, \ldots, \eta_m)$ satisfying

$$\begin{aligned} \eta_i &\geq 0 && \text{for } i \in P, \\ \mathbf{y} \cdot \mathbf{a}^j &\geq \gamma_j && \text{for } j \in Q, \\ \mathbf{y} \cdot \mathbf{a}^j &= \gamma_j && \text{for } j \in Q^*. \end{aligned} \tag{3.11}$$

The sets P and Q are as in the definition of the general problem, and Q^* is the complement of Q with respect to $\{1, 2, \ldots, n\}$.

It is instructive to consider some special cases as examples. First, the dual of a general maximum problem which happens to be a standard maximum problem is just the dual of that problem, i.e., a standard minimum problem. Indeed, in this case $P^* = \varnothing$, so $P = \{1, 2, \ldots, m\}$; and $Q = \{1, 2, \ldots, n\}$, so $Q^* = \varnothing$. Next, the dual of the restricted maximum problem is the problem of finding the minimum of $\mathbf{y} \cdot \mathbf{b}$ subject to

$$\mathbf{y} \cdot \mathbf{a}^j \geq \gamma_j, \qquad j = 1, 2, \ldots, n.$$

That is, the dual of the restricted maximum problem is not the restricted minimum problem, but rather the problem of minimizing $\mathbf{y} \cdot \mathbf{b}$ over *all* (meaning unrestricted as to nonnegativity of coordinates) solutions of $\mathbb{A}^T\mathbf{y} \geq \mathbf{c}$.

The notions of feasible and optimal vectors, as introduced in Sec. 4.3.1 for standard problems, can be extended in the obvious way to general problems. A linear programming problem is said to be feasible if the set of feasible vectors for the problem is nonempty.

In view of the equivalences proved here, the reader may wonder why we chose to introduce a form as (apparently) complicated as that of the general problem. The justification is that this form is broad enough to include many of the linear problems encountered in practice, and thus it enables one to tell by inspection whether the theory applies.

4.3.5 Existence of Optimal Vectors

We return to one of the questions raised at the end of Sec. 4.3.2, namely, the question of whether a linear programming problem always has optimal vectors. To answer this question, we shall derive one of the fundamental results of the elementary theory of linear programming. In addition to being a useful theoretical tool, the following theorem is also basic to many of the applications of linear programming techniques. This theorem, as well as some of its relatives, is known as a *duality theorem*.

Theorem 3: Let $[\mathbb{A}, \mathbf{b}, \mathbf{c}]$ define a general maximum problem such that both the primal problem and its dual have feasible vectors. Then both the primal problem and its dual have optimal vectors, and the respective maximum and minimum values are equal.

The proof of this theorem requires a number of steps, and before we begin, it is appropriate to point out that the hypotheses are not idle ones. That is, there are linear programming problems, very simple ones in fact, for which either the primal problem or its dual or both have no feasible vectors. For example, consider the 2×2 standard maximum problem determined by $[\mathbb{A}, \mathbf{b}, \mathbf{c}]$ with

$$\mathbb{A} = \begin{bmatrix} -1 & 3 \\ 1 & -2 \end{bmatrix},$$

$$\mathbf{b} = (-1, -3),$$

$$\mathbf{c} = (1, 1).$$

The equations $\mathbb{A}\mathbf{x} \leq \mathbf{b}$ become

$$-\xi_1 + 3\xi_2 \leq -1,$$
$$\xi_1 - 2\xi_2 \leq -3.$$

Adding these two equations, we conclude that any vector $\mathbf{x}$ which satisfies $\mathbb{A}\mathbf{x} \leq \mathbf{b}$ must have a second coordinate ξ_2 which satisfies $\xi_2 \leq -4$. Thus the constraints have no nonnegative solutions, and therefore the standard maximum problem has no feasible vectors. On the other hand, the vector $\mathbf{y} = (3, 4)$ is feasible for the dual problem. In fact, $\mathbf{y}_t = (3t, 4t)$ is feasible for all $t \geq 1$, and since

$$\mathbf{y}_t \cdot \mathbf{b} = -15t,$$

it is obvious that the dual problem has no optimal vectors.

An example of a problem in which neither the primal standard problem nor its dual has feasible vectors is the standard maximum problem defined by $[\mathbb{A}, \mathbf{b}, \mathbf{c}]$ with

$$\mathbb{A} = \begin{bmatrix} 1 & -1 \\ -1 & 1 \end{bmatrix},$$

$$\mathbf{b} = (-2, 1),$$

and

$$\mathbf{c} = (2, -1).$$

We return to the proof of Theorem 3. To simplify our argument as much as possible, we make use of the equivalence of a general problem and a standard one, and we shall prove the theorem first for the case of a standard problem. More precisely, let our problem be defined by $[\mathbb{A}, \mathbf{b}, \mathbf{c}]$. We shall prove that if both the primal problem and its dual have feasible vectors, then there are vectors $\mathbf{x}^*$ and $\mathbf{y}^*$ which satisfy $\mathbf{x}^* \geq \mathbf{0}$, $\mathbf{y}^* \geq \mathbf{0}$, and

$$\mathbb{A}\mathbf{x}^* \leq \mathbf{b},$$
$$\mathbb{A}^T\mathbf{y}^* \geq \mathbf{c},$$
$$\mathbf{x}^* \cdot \mathbf{c} = \mathbf{y}^* \cdot \mathbf{b}.$$

Consequently, by Theorem 1, both $\mathbf{x}^*$ and $\mathbf{y}^*$ are optimal vectors.

Our proof is based on the use of Farkas' lemma and an appropriate matrix. As usual, we assume that $\mathbb{A}$ is an $m \times n$ matrix, $\mathbf{b} \in R^m$, and $\mathbf{c} \in R^n$. Introduce an $(m + n + 1) \times (2m + 2n + 1)$ matrix $\tilde{\mathbb{A}}$ defined by

$$\tilde{\mathbb{A}} = \begin{bmatrix} & \mathbb{A} & \mathbb{O} \\ \mathbb{I}_{m+n+1} & \mathbb{O} & -\mathbb{A}^T \\ & -\mathbf{c} & \mathbf{b} \end{bmatrix},$$

and the vector $\tilde{\mathbf{b}} = (\mathbf{b}, -\mathbf{c}, 0) \in R^{m+n+1}$.

Suppose for the moment that the following fact has been established:

Fact. For every $\mathbf{z} \in R^{m+n+1}$ such that $\tilde{\mathbb{A}}^T\mathbf{z} \geq \mathbf{0}$, we have $\tilde{\mathbf{b}} \cdot \mathbf{z} \geq 0$.

Farkas' lemma can now be applied. We conclude that there exist positive constants $\{\lambda_i\}_{i=1}^{2(m+n)+1}$ satisfying $\lambda_i \geq 0$, $i = 1, \ldots, 2(m+n)+1$, and

$$\tilde{\mathbf{b}} = \sum_{i=1}^{2(m+n)+1} \lambda_i \tilde{\mathbf{a}}^i. \tag{3.12}$$

Here $\tilde{\mathbf{a}}^i$, $i = 1, \ldots, 2(m+n)+1$, are the column vectors of the matrix $\tilde{\mathbb{A}}$. Define

$$\mathbf{w}^* = (\lambda_1, \ldots, \lambda_m), \qquad \mathbf{v}^* = (\lambda_{m+1}, \ldots, \lambda_{m+n}), \qquad \mu = \lambda_{m+n+1},$$
$$\mathbf{x}^* = (\lambda_{m+n+2}, \ldots, \lambda_{m+2n+1}), \qquad \mathbf{y}^* = (\lambda_{m+2n+2}, \ldots, \lambda_{2m+2n+1}).$$

Since $\lambda_i \geq 0$ for all i, it follows that the vectors $\mathbf{x}^*$, $\mathbf{y}^*$, $\mathbf{v}^*$, $\mathbf{w}^*$ are all non-negative. Equation (3.12) can be viewed as $m + n + 1$ scalar equations, the first m of which, written in matrix form, are

$$\mathbf{b} = \mathbf{w}^* + \mathbb{A}\mathbf{x}^*.$$

Thus, $\mathbf{x}^*$ is feasible for the primal problem. The $(m + 1)$th to the $(m + n)$th equations of (3.12), written in matrix form, are

$$-\mathbf{c} = \mathbf{v}^* - \mathbb{A}^T\mathbf{y}^* \qquad \text{or} \qquad \mathbb{A}^T\mathbf{y}^* - \mathbf{v}^* = \mathbf{c},$$

and consequently the vector $\mathbf{y}^*$ is feasible for the dual problem. The last equation of (3.12) is $\mu + (-\mathbf{c}) \cdot \mathbf{x}^* + \mathbf{b} \cdot \mathbf{y}^* = 0$ or $\mathbf{b} \cdot \mathbf{y}^* \leq \mathbf{c} \cdot \mathbf{x}^*$. But this implies, by Lemma 1 of Sec. 4.3.3, that $\mathbf{b} \cdot \mathbf{y}^* = \mathbf{c} \cdot \mathbf{x}^*$, and therefore, by Theorem 1, the vectors $\mathbf{x}^*$ and $\mathbf{y}^*$ are optimal for the primal problem and its dual, respectively. Since $\mathbf{x}^* \cdot \mathbf{c} = \mathbf{y}^* \cdot \mathbf{b}$ for one pair of optimal vectors, it is clear that this must hold for every pair of optimal vectors, and the conclusion of Theorem 3 is verified. Of course, all this is subject to the validity of the fact asserted above. A proof that the fact is indeed true is not difficult, and it is the topic of Exercise 7.

It is actually true that if either the primal problem or its dual or both have no feasible vectors, then neither problem has an optimal vector. Although this fact is not difficult to prove, we have no need for the result, and therefore we do not pursue the matter.

The above discussion establishes Theorem 3 for the special class of standard maximum problems. The theorem for general problems follows at once. Indeed, a general problem $\mathcal{G}$ is equivalent to a standard problem $\mathcal{S}$, and the dual of $\mathcal{G}$, which is another general problem, is equivalent to a standard problem $\mathcal{S}'$. The problem $\mathcal{S}'$ is actually the dual of $\mathcal{S}$ (Exercise 8), and the theorem follows.

There are many more refined results on the properties of optimal vectors for linear programming problems. It is not appropriate for us to enter into a detailed study of these results, and we content ourselves with a brief return to the concept of complementary slackness.

If $\mathbf{x}$ and $\mathbf{y}$ are optimal for the primal problem and its dual, respectively, then Theorem 3 assures us that $\mathbf{x} \cdot \mathbf{c} = \mathbf{y} \cdot \mathbf{b}$. Therefore, if $\mathbf{w}$ and $\mathbf{v}$ are defined by (3.3), then $\mathbf{w} \geq \mathbf{0}$, $\mathbf{v} \geq \mathbf{0}$, and $\mathbf{y} \cdot \mathbf{w} + \mathbf{x} \cdot \mathbf{v} = \mathbf{y} \cdot \mathbf{b} - \mathbf{x} \cdot \mathbf{c} = 0$. This implies that $\mathbf{y} \cdot \mathbf{w} = 0$ and that $\mathbf{x} \cdot \mathbf{v} = 0$. This is the other part of Theorem 2 promised earlier. We state the complete result as an expanded Theorem 2.

Theorem 2 (expanded version): Let $\mathbf{x}$, $\mathbf{y}$, $\mathbf{v}$, $\mathbf{w}$ be nonnegative vectors which satisfy (3.3). The vectors $\mathbf{x}$ and $\mathbf{y}$ are optimal for the primal problem and its dual, respectively, if and only if $\mathbf{x} \cdot \mathbf{v} = 0$ and $\mathbf{y} \cdot \mathbf{w} = 0$.

EXERCISES

1. Prove that the standard minimum problem defined by $[\mathbb{A}, \mathbf{b}, \mathbf{c}]$ is equivalent to the standard maximum problem defined by $[-\mathbb{A}, -\mathbf{b}, -\mathbf{c}]$. Here, *equivalent* means that the sets of feasible vectors are identical and that the sets of optimal vectors are identical.
2. Prove that the situations described in Sec. 4.1 lead to standard problems.
3. Interpret complementary slackness in the context of the 2×3 transportation problem discussed in Secs. 2.2 and 4.3.3.
4. Give an example of a 2×2 restricted maximum problem such that neither the primal problem nor its dual has a feasible vector.
5. Show that a standard maximum problem can be written in the form of a restricted maximum problem. *Hint:* Recall that a positive vector $\mathbf{x}$ satisfies $\mathbb{A}\mathbf{x} \leq \mathbf{b}$ if and only if there is a vector $\mathbf{w} \geq 0$ such that $\mathbb{A}\mathbf{x} + \mathbf{w} = \mathbf{b}$.
6. The purpose of this exercise is to show that each general problem can be written in standard form. The proof outlined here proceeds in a number of steps and provides an explicit method of translating a general problem into standard form.
   (a) For any real number λ define $\lambda^+ = (|\lambda| + \lambda)/2$ and $\lambda^- = (|\lambda| - \lambda)/2$. Using this notation, for each vector $\mathbf{x} \in R^n$ introduce the vectors
   $$\mathbf{x}^+ = (\xi_1^+, \ldots, \xi_n^+) \qquad \text{and} \qquad \mathbf{x}^- = (\xi_1^-, \ldots, \xi_n^-).$$
   Show that $\mathbf{x} \in R^n$ satisfies (3.9) and (3.10) if and only if $\mathbf{x}^+$ and $\mathbf{x}^-$ satisfy
   $$\mathbf{x}^+ \cdot \hat{\mathbf{a}}_i - \mathbf{x}^- \cdot \hat{\mathbf{a}}_i \leq \hat{\beta}_i, \qquad i = 1, 2, \ldots, m',$$
   $$\mathbf{x}^- \cdot \hat{\mathbf{a}}_i - \mathbf{x}^+ \cdot \hat{\mathbf{a}}_i \leq -\hat{\beta}_i, \qquad i = r+1, \ldots, m'.$$
   (b) Find a $(2m' - r) \times 2n$ matrix $\tilde{\mathbb{A}}$, and a vector $\tilde{\mathbf{b}} \in R^{2m'-r}$ such that if $\tilde{\mathbf{x}}$ is defined to be $(\xi_1^+, \ldots, \xi_n^+, \xi_1^-, \ldots, \xi_n^-)$, then the inequalities of part (a) can be written $\tilde{\mathbb{A}}\tilde{\mathbf{x}} \leq \tilde{\mathbf{b}}$.

(c) Show that there is a vector $\mathbf{x} \in R^n$ satisfying (3.9) and (3.10) if and only if there is a vector $\tilde{\mathbf{x}} \in R^{2n}$ satisfying $\tilde{\mathbf{x}} \geq 0$, $\tilde{\mathbb{A}}\tilde{\mathbf{x}} \leq \tilde{\mathbf{b}}$ [$\tilde{\mathbb{A}}$, $\tilde{\mathbf{b}}$ as in part (b)].

(d) Given a vector $\mathbf{c} \in R^n$, find a vector $\tilde{\mathbf{c}}$ such that maximizing $\mathbf{x} \cdot \mathbf{c}$ is equivalent to maximizing $\tilde{\mathbf{x}} \cdot \tilde{\mathbf{c}}$ [$\tilde{\mathbf{x}}$ as in part (b)].

(e) Complete the argument showing that each general problem can be written as a standard problem.

7. Let $\tilde{\mathbb{A}}$ and $\tilde{\mathbf{b}}$ be defined as in Sec. 4.3.5. Show that for every $\mathbf{z} \in R^{m+n+1}$ for which $\tilde{\mathbb{A}}^T\mathbf{z} \geq \mathbf{0}$ we have $\tilde{\mathbf{b}} \cdot \mathbf{z} \geq 0$.
   (a) Write $\mathbf{z} = (\mathbf{y}, \mathbf{x}, \tau)$, where $\mathbf{y} \in R^m$, $\mathbf{x} \in R^n$, and τ is a real number. Show that if $\tau > 0$, then $\tau^{-1}\mathbf{x}$ and $\tau^{-1}\mathbf{y}$ are feasible for the primal problem and its dual, respectively, and that $\tilde{\mathbf{b}} \cdot \mathbf{z} \geq 0$ follows from Lemma 1 of Sec. 4.3.
   (b) Show that if $\mathbf{z} = (\mathbf{y}, \mathbf{x}, 0)$ is such that $\tilde{\mathbb{A}}^T\mathbf{z} \geq \mathbf{0}$, then $\tilde{\mathbf{b}} \cdot \mathbf{z} \geq 0$. *Hint:* Use the fact that there are feasible vectors for the primal problem and its dual.

8. Let $\mathcal{G}$ be a general linear programming problem and suppose that $\mathcal{G}$ is equivalent to a standard problem $\mathcal{S}$ and that the dual of $\mathcal{G}$ is equivalent to another standard problem $\mathcal{S}'$. Prove that $\mathcal{S}'$ is the dual of $\mathcal{S}$.

9. Given a standard maximum problem determined by $\mathbb{A}$, $\mathbf{b}$, and $\mathbf{c}$, where $\mathbb{A}$ is a 2×2 matrix, suppose the set of vectors $\mathbf{x}$ satisfying $\mathbb{A}\mathbf{x} \leq \mathbf{b}$ is nonempty (the set of feasible vectors for the primal problem may or may not be empty). What can you say about the set of vectors $\mathbf{y}$ satisfying $\mathbb{A}^T\mathbf{y} \geq \mathbf{c}$? What about the set of feasible vectors for the dual problem?

4.4 LINEAR OPTIMIZATION WITH UNCERTAINTY

The models considered in this chapter have been, until now, deterministic ones. That is, we have assumed that the data needed in the model were known exactly. However, we have already pointed out that in many applications at least some of these quantities are not known with complete certainty. That this is the case does not invalidate the preceding discussion as regards real applications. Indeed, it is remarkable that these deterministic techniques have been so useful when applied to situations in which uncertainty abounds. There have been a number of attempts to systematically extend linear optimization methods to situations involving random variations, but there is no general theory corresponding to that developed in Secs. 4.2 and 4.3. One of the fundamental difficulties is that the problem can be stated in many forms and that no single method has been discovered which is useful in all problems. This section gives some illustrations of the sort of techniques which can be used in an attempt to take uncertainty into account. The examples were chosen to give an idea of the techniques and no one of them is pursued in great detail.

4.4.1 Estimated Parameters

The basic problem considered here is a standard minimum problem: Minimize $\mathbf{x} \cdot \mathbf{c}$ over all $\mathbf{x} \geq \mathbf{0}$ which satisfy $\mathbb{A}\mathbf{x} \geq \mathbf{b}$. The most primitive assumption which can be made concerning the data other than pure determinism is that there are known error bounds. That is, the quantities $\mathbb{A}$, $\mathbf{b}$, and $\mathbf{c}$ are not known exactly, but their values are known to within specific errors. Consider again the health care example of Sec. 4.1. In this example, the matrix $\mathbb{A} = (a_{ij})$ is the matrix of nutrient concentrations, and a_{ij} is the concentration of the ith nutrient in the jth food. It is extremely unlikely that this concentration is known exactly, but it is also quite likely that one can obtain reliable estimates. For example, a certain concentration may be known to be greater than $a - \epsilon$ but less than $a + \epsilon$. It is conventional to say that this concentration is $a \pm \epsilon$. Likewise, γ_i is the concentration of the contaminant in the ith food, and it is likely to be subject to similar uncertainties. It seems reasonable to suppose that if one has data which are not known precisely but which are known to lie within certain bounds, then one should be able to obtain estimates on the minimum value of $\mathbf{x} \cdot \mathbf{c}$. This is, in fact, the case, and we shall show how upper and lower bounds on minimum $\mathbf{x} \cdot \mathbf{c}$ can be obtained from the solutions of certain deterministic problems.

Let the meaning of the symbol $\leq$ when used between two $m \times n$ matrices mean entrywise inequality. Suppose that the only knowledge available concerning the data of the system is contained in the estimates

$$\begin{aligned} \mathbb{A}_L &\leq \mathbb{A} \leq \mathbb{A}_U, \\ \mathbf{b}_L &\leq \mathbf{b} \leq \mathbf{b}_U, \\ \mathbf{c}_L &\leq \mathbf{c} \leq \mathbf{c}_U, \end{aligned} \tag{4.1}$$

where $\mathbb{A}_L$ and $\mathbb{A}_U$ are $m \times n$ matrices, $\mathbf{b}_L$ and $\mathbf{b}_U$ are vectors in R^m, and $\mathbf{c}_L$ and $\mathbf{c}_U$ are vectors in R^n. Everything except $\mathbb{A}$, $\mathbf{b}$, $\mathbf{c}$ is assumed to be known exactly. Phrased differently, although we do not know $\mathbb{A}$, $\mathbf{b}$, and $\mathbf{c}$, we do know approximate values and error bounds for each of them. It is natural to expect, and it is true, that more precise estimates give more accurate final results.

The precise quantitative answer to the estimation question is the following. There are numbers e_L and e_U depending only on $\mathbb{A}_L$, $\mathbb{A}_U$, $\mathbf{b}_L$, $\mathbf{b}_U$, $\mathbf{c}_L$, and $\mathbf{c}_U$ such that for *any* $\mathbb{A}$, $\mathbf{b}$, $\mathbf{c}$ satisfying (4.1) we have

$$e_L \leq \min\,[\mathbf{x} \cdot \mathbf{c}] \leq e_U,$$

where the minimum is taken over all $\mathbf{x} \geq \mathbf{0}$ which satisfy $\mathbb{A}\mathbf{x} \geq \mathbf{b}$.

It is not difficult to verify this assertion and in the process to give an explicit expression for e_L and e_U. In what follows, it is to be understood that each time $\mathbb{A}$, $\mathbf{b}$, and $\mathbf{c}$ appear they are subject to (4.1). We begin by considering an auxiliary problem in which the data are completely known. The appropriate problem is the following: Find the minimum of $\mathbf{x} \cdot \mathbf{c}_U$, for $\mathbf{x} \in \mathfrak{A}$, where the feasible set $\mathcal{A}$ is defined as

$$\mathcal{A} = \{\mathbf{x} \in R^n : \mathbf{x} \geq \mathbf{0}, \mathbb{A}_L\mathbf{x} \geq \mathbf{b}_U\}.$$

Let $\mathfrak{F}$ be the feasible set for the original problem, i.e.,

$$\mathfrak{F} = \{\mathbf{x} \in R^n : \mathbf{x} \geq \mathbf{0}, \mathbb{A}\mathbf{x} \geq \mathbf{b}\}.$$

Then $\mathcal{A} \subset \mathfrak{F}$. Indeed, since $\mathbb{A} - \mathbb{A}_L \geq \mathbb{O}$ and $\mathbf{b}_U - \mathbf{b} \geq \mathbf{0}$, we have

$$\mathbb{A}\mathbf{x} - \mathbf{b} = (\mathbb{A} - \mathbb{A}_L)\mathbf{x} + (\mathbf{b}_U - \mathbf{b}) + (\mathbb{A}_L\mathbf{x} - \mathbf{b}_U) \geq \mathbf{0}$$

for all $\mathbf{x} \in \mathcal{A}$. Also, $\mathbf{c}_U - \mathbf{c} \geq 0$ and $\mathbf{x} \geq \mathbf{0}$, so that

$$\mathbf{c}_U \cdot \mathbf{x} = (\mathbf{c}_U - \mathbf{c}) \cdot \mathbf{x} + \mathbf{c} \cdot \mathbf{x} \geq \mathbf{c} \cdot \mathbf{x}.$$

Combining these results, we conclude that

$$\min_{\mathbf{x} \in \mathfrak{F}} [\mathbf{x} \cdot \mathbf{c}] \leq \min_{\mathbf{x} \in \mathcal{A}} [\mathbf{x} \cdot \mathbf{c}] \leq \min_{\mathbf{x} \in \mathcal{A}} [\mathbf{x} \cdot \mathbf{c}_U].$$

Defining e_U by

$$e_U = \min_{\mathbf{x} \in \mathcal{A}} [\mathbf{x} \cdot \mathbf{c}_U],$$

we have half the desired result. Note that e_U is obtained as the solution to a deterministic linear programming problem. A lower bound can be obtained by entirely similar methods (Exercise 1).

It is clear that the technique described in this section is quite crude. Unless one has fairly precise estimates on the data, the error bounds on the optimal value may be quite large. Also, if one knows, for example, that the value of a parameter is in the interval $(a - \epsilon, a + \epsilon)$ but is twice as likely to be in $(a, a + \epsilon)$ as in $(a - \epsilon, a)$, then there is no way to take advantage of this additional information. It is worthwhile to consider a method in which this information can be used.

4.4.2 Parameters Given as Random Variables

In many situations one knows more about the parameters than simply their ranges. It is not uncommon to know, or to assume known, the values which can be assumed together with the respective probabilities that these values will occur. More precisely, in many cases it is possible to interpret the data as random variables whose associated probability distributions are known.

To illustrate this technique, we return again to the health care example of Sec. 4.1. We assume that the nutrient concentrations and the minimum daily requirements are known exactly; that is, the random variables corresponding to these quantities assume a specific value with probability 1. The contaminant concentrations are subject to uncertainty in the following precise sense. Although the exact value of the contaminant concentration in each food is unknown, the range of values of the concentration in each food as well as the associated probabilities are known. Thus one is interested in studying $\mathbf{x} \cdot \mathbf{c}$, where the coordinates of $\mathbf{c}$ are random variables and $\mathbf{x}$ is subject to $\mathbf{x} \geq \mathbf{0}$, $\mathbb{A}\mathbf{x} \geq \mathbf{b}$, $\mathbb{A}$ and $\mathbf{b}$ known quantities.

The selection of an optimal diet must be made in the light of this uncertainty, and one must decide what the optimization problem means in this case. One possible approach is to minimize the expected consumption of the contaminant. Thus, if, as usual, E denotes expected value, the problem is: Find the minimum of $E[\mathbf{x} \cdot \mathbf{c}]$ for $\mathbf{x} \in \mathfrak{F}$, where $\mathfrak{F} = \{\mathbf{x} \in R^n : \mathbf{x} \geq \mathbf{0}, \mathbb{A}\mathbf{x} \geq \mathbf{b}\}$. The diet or consumption vector $\mathbf{x}$ is to be chosen subject to this uncertainty. Thus, since taking expected values is a linear operation,

$$E[\mathbf{x} \cdot \mathbf{c}] = \mathbf{x} \cdot (E[\gamma_1], \ldots, E[\gamma_n]),$$

where $E[\gamma_i]$ is the expected value of the random variable γ_i. If we denote the vector $(E[\gamma_1], \ldots, E[\gamma_n])$ by $\bar{\mathbf{c}}$, then the stochastic optimization problem introduced above reduces to the deterministic problem: Find the minimum of $\mathbf{x} \cdot \bar{\mathbf{c}}$ for $\mathbf{x} \in \mathfrak{F}$. In this case, therefore, the stochastic problem was replaced by a deterministic one in which the values of the deterministic parameters were taken to be the expected values of the stochastic ones.

It should be emphasized that in more complicated situations this technique must be used with care. In general it is necessary to verify that one has a right to replace a random variable by its expected value before proceeding. In fact, this is not usually the case. To illustrate the complications which arise even in very simple examples, we consider the following.

Suppose that Mr. O. V. Ripe has 100 units of a perishable commodity which is to be sold and that he is willing to sell any number of units or fractions thereof to a specific buyer. There are two markets, one free and the other controlled. In the controlled market he can sell as much as he wishes at \$3.00 per unit. He is uncertain how much he can sell in the free market, but he estimates that the demand is greater than 50 and less than 100 units. For simplicity he assumes that the demand is uniformly distributed on that interval. Thus, if demand, i.e., the maximum number that can be sold on the free market, is denoted by u, then

$$\text{Prob}\,[u > t] = \begin{cases} 1, & 0 \leq t \leq 50, \\ \dfrac{100 - t}{50}, & 50 \leq t \leq 100. \end{cases}$$

In real terms this means that if he ships 50 or fewer units to the free market, then he is certain to sell them all, but if he ships x units, $x > 50$, to the free market, then the probability of his selling them is $(100 - x)/50$. The selling price on the free market is \$10.00 per unit, and those units shipped to the free market but not sold must be disposed of at no gain. Thus, if a unit is shipped to the free market but not sold, Mr. Ripe receives nothing and must pay for shipping. Finally, suppose that shipping costs are \$1.00 and \$5.00 per unit to the controlled and free markets, respectively.

Let x and y denote the number of units of the commodity shipped by Mr. Ripe to the free and controlled markets, respectively. Clearly, once he decides how much to ship to the free market, it is in his best interest to ship the remainder to the controlled market. Thus $x + y = 100$. Assume that Mr. Ripe is concerned with making his "profit" as large as possible. In this case, because the demand at the free market is uncertain, the profit itself is uncertain. To make the problem precise, we agree that it is the expected value of the profit which is to be maximized. For a fixed demand u at the free market, the profit depends on the amounts shipped to the two markets. That is, it depends on x and y, or since $y = 100 - x$, it can be written as a function of x alone:

$$\text{profit}(x) = \begin{cases} 2y + 5x = 200 + 3x, & 0 \leq x \leq u \leq 100, \\ 2y + 5u - 5(x - u) = 200 + 10u - 7x, & 0 \leq u \leq x \leq 100. \end{cases}$$

The distribution function for demand u evaluated at t is $\text{Prob}[u \leq t] = 1 - \text{Prob}[u > t]$, and therefore using the expression for $\text{Prob}[u > t]$ and the standard formula for expected value,

$$E[\text{profit}](x) = \begin{cases} 200 + 3x, & 0 \leq x \leq 50, \\ \displaystyle\int_{50}^{x} (200 + 10u - 7x)\frac{du}{50} + \int_{x}^{100} (200 + 3x)\frac{du}{50}, & 50 \leq x \leq 100, \end{cases}$$

$$= \begin{cases} 200 + 3x, & 0 \leq x \leq 50, \\ \dfrac{-500 + 130x - x^2}{10}, & 50 \leq x \leq 100. \end{cases}$$

We conclude that the maximum value of $E[\text{profit}]$ is \$372.50 and is assumed when $x = 65$, i.e., when 65 units are shipped to the free market. The graph of $E[\text{profit}]$ as a function of x is given in Fig. 4–2.

In light of the above comments concerning the replacement of a random variable by its expected value, it is interesting to compare this result with that obtained by computing the profit with demand on the free market

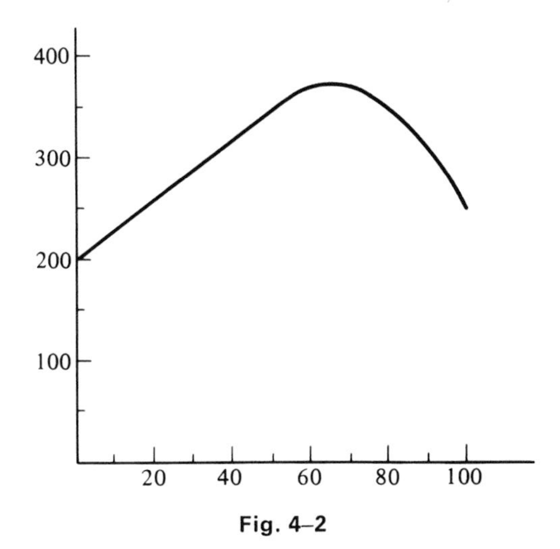

Fig. 4–2

equal to its expected value, namely 75 units. In this case

$$\text{profit}(x) = \begin{cases} 200 + 3x, & 0 \leq x \leq 75, \\ 950 - 7x, & 75 \leq x \leq 100, \end{cases}$$

and the maximum value of the profit is \$425.00, which is assumed when $x = 75$. Obviously here is a case where the uncritical replacement of a random variable by its expected value severely distorts the result.

Finally, we give a heuristic argument which supports the conclusion that the optimal allocation is 65 units to the free market. To this end, we observe that each unit which is shipped to the free market and not sold there actually means a loss of \$7.00 to Mr. Ripe. Indeed, he loses the \$2.00 profit from a sale on the controlled market plus the \$5.00 transportation charge for shipping the unit to the free market. On the other hand, each unit shipped to the free market and sold there means a gain of \$3.00 over the amount for which it could have been sold on the controlled market. That is, the free market offers a \$3.00 gain or a \$7.00 loss as compared to the controlled market depending on whether the unit is sold or not. But the demand is uniformly distributed on [50, 100], and consequently the optimal schedule for Mr. Ripe is $3/(7 + 3)$ of the way along this interval; that is, $x = 50 + \frac{3}{10}(50) = 65$ units.

EXERCISES

1. Use the estimates (4.1) to obtain a lower bound for $\min_{\mathbf{x} \in \mathfrak{F}} [\mathbf{x} \cdot \mathbf{c}]$.
2. Estimate the minimal cost of shipping goods in a 2×3 transportation problem with

$d_1 = 6 \pm 1$, $d_2 = 9 \pm 2$, $d_3 = 10 \pm 2$, $s_1 = 15 \pm 3$,
$s_2 = 10 \pm 2$, $c_{11} = 70$, $c_{12} = 70$, $c_{13} = 60$, $c_{21} = 80$,
$c_{22} = 60$, $c_{23} = 80$,

and each of the shipping costs subject to a 10% error. (Note that our methods only provide an estimate in one direction.)

3. Compute the maximum of the expected value of the profit for Mr. O. V. Ripe when the probability distribution for demand is

$$\text{(a) Prob[demand} > t] = \begin{cases} 1, & 0 \le t \le 50 \\ \left(\dfrac{100 - t}{50}\right)^2, & 50 \le t \le 100, \end{cases}$$

$$\text{(b) Prob[demand} > t] = \begin{cases} 1, & 0 \le t \le 50 \\ 1 - \left(\dfrac{t - 50}{50}\right)^2, & 50 \le t \le 100. \end{cases}$$

Graph E[profit] as a function of the amount shipped to the free market in each of these cases. Find the expected value of the demand in each case and compute the profit with these demands.

4. Construct a model for the allocation problem of Mr. Ripe if the controlled market also has a demand uniformly distributed on [50, 100]. The other conditions of Sec. 4.4.2 remain the same. Note that in this case one must consider the possibility that $x + y < 100$.

5. Consider the situation described in Sec. 4.4.2 if only an integer number of units can be sold in each market. Assume that each demand between 50 and 100 units is equiprobable in the free market, and suppose that the controlled market can absorb as much as desired. Find the optimal allocation of the commodity to maximize profit.

4.5 PROJECTS

4.5.1 Pollution on the Peacock

Peacock Paper Products, Inc. is a thriving producer of bulk paper products located on what used to be the scenic Peacock River. Last year, while producing at an optimal rate, the firm had a net profit of 1.7 million dollars. The manufacturing facilities are equipped to produce two types of bulk paper. The first, a high-quality bleached paper known as type A paper, yields the firm a net profit of \$10 per ton, and the second, a rough unbleached kraft paper called type B paper, yields a net profit of \$7 per ton. Facilities limit production to 200,000 tons per year of which at most 100,000 tons can be of type A.

This year the People for the Preservation of Peacock River, a conservation group, brought to the attention of the state legislature the high level of water pollution resulting from the paper manufacturing process. They pointed out that water from the Peacock River was used as a washing agent and then returned to the river. Officials of the paper company conceded that quite a large amount of water was used in their manufacturing processes. In fact, 2000 gallons of water was used in the production of each ton of type A paper and 1000 gallons for each ton of type B. Measurement of water flowing from the production facilities indicated that water used in the production of paper of type A was being returned to the river with a residue of 20 pounds of noxious chemicals per 1000 gallons of water. Water used in the production of type B paper carried 25 pounds of these chemicals per 1000 gallons of water. The legislature established a criterion of 3 pounds per 1000 gallons as "safe" and imposed a fine for amounts in excess of that quantity. If $x(\geq 3)$ is the number of pounds of chemicals of this type being returned to the river per 1000 gallons of water, then the fine is to be $(x - 3)/10$ dollars per 1000 gallons of water used. In addition, they determined that a total of no more than 2.5×10^8 gallons of water could be used by the company in any year.

The management realizes the significance of this act for their business and immediately begins a study of means to reduce the chemical content of the water being returned to the river. Investigation indicated that there were several options available to the company. Mechanical filters or chemical precipitators, either regular or high efficiency, could be installed to treat the water after use. These devices could be installed on the outflow lines from the production facilities of each type of paper independently. Alternatively, the company could build a sedimentation lagoon which would treat the outflow from both facilities simultaneously. Data regarding these options are given in Table 4–1. The column labeled Content gives the chemical content of the

Table 4–1

Type of treatment	Content (lb) (per 1000 gal) Type A	Content (lb) (per 1000 gal) Type B	Cost ($) (per 1000 gal) Type A	Cost ($) (per 1000 gal) Type B
Mechanical filter	11	12	0.60	0.50
Chemical precipitator	5	8	1.00	0.60
High-efficiency chemical precipitator	4	7	1.10	0.90
Sedimentation lagoon	3	3	*	*

treated outflow water for each production unit. The Cost column gives the cost of installing and operating the various devices for production of each type of paper. In each case the type of treatment is specified in the column to the far left and the figures are given for each 1000 gallons of water used. The sedimentation lagoon requires additional comment. Consulting engineers estimate the cost of building a sedimentation lagoon at 1.5 million dollars and that it will be adequate for treating all water used by the plant for five years, after which time it will be filled and therefore useless. Management therefore agrees that if a lagoon is to be built it must be amortized at a rate of $300,000 per year regardless of how much it is used.

Problem 1. Determine which treatment devices (if any) should be installed and how production should be allocated between paper of types A and B so as to maximize net profit.

Problem 2. Formulate and interpret the dual of the maximization problem posed in Problem 1.

Problem 3. Discuss what modifications in the model would make it correspond more closely to reality.

4.5.2 Quadratic Objective Functions

The theory of a linear model is simply inadequate for many situations arising in practice, and more elaborate models must be constructed. We consider here one of the most obvious generalizations, that of a quadratic function to be minimized. Let $\mathbb{A}$ be an $m \times n$ matrix, $\mathbf{b} \in R^m$, and $\mathbb{C}$ a symmetric $n \times n$ matrix for which $\mathbf{x} \cdot (\mathbb{C}\mathbf{x}) \geq 0$ for all $\mathbf{x} \in R^n$. Let

$$\mathfrak{F} = \{\mathbf{x} \in R^n : \mathbf{x} \geq \mathbf{0}, \mathbb{A}\mathbf{x} \geq \mathbf{b}\},$$

and consider the problem of finding $\tilde{\mathbf{x}} \in \mathfrak{F}$ for which

$$\tilde{\mathbf{x}} \cdot (\mathbb{C}\tilde{\mathbf{x}}) = \min_{\mathbf{x} \in \mathfrak{F}} [\mathbf{x} \cdot (\mathbb{C}\mathbf{x})].$$

Such a vector $\tilde{\mathbf{x}}$ is said to be *optimal* just as in the linear case.

Problem 1. Find an optimal vector $\tilde{\mathbf{x}}$ in the special case

$$\mathbb{C} = \begin{bmatrix} 1 & 0 \\ 0 & 1 \end{bmatrix}, \quad \mathbb{A} = \begin{bmatrix} 1 & 3 \\ 2 & 4 \end{bmatrix}, \quad \text{and} \quad \mathbf{b} = \begin{bmatrix} 5 \\ 8 \end{bmatrix}.$$

Problem 2. Let $\tilde{\mathbf{x}}$ be an optimal vector for the quadratic problem and consider the standard minimum problem defined by $[\mathbb{A}, \mathbf{b}, \mathbf{c}]$, where $\mathbf{c} = \mathbb{C}\tilde{\mathbf{x}}$. Show that $\tilde{\mathbf{x}}$ is also optimal for this linear problem. *Hint:* Show that $\tilde{\mathbf{x}}$ is feasible for the linear problem and then compare $\tilde{\mathbf{x}} \cdot \mathbf{c}$ with $\mathbf{x} \cdot \mathbf{c}$ for any other feasible $\mathbf{x}$.

Problem 3. The dual of the standard minimum problem introduced in Problem 2 is a standard maximum problem defined by $[\mathbb{A}^T, \mathbf{c}, \mathbf{b}]$. If the original quadratic problem has an optimal vector, then so does this standard maximum problem. Why?

Problem 4. Define the dual of the quadratic problem as follows: Let

$$\mathfrak{G} = \{(\mathbf{y}, \mathbf{z}), \mathbf{y} \in R^m, \mathbf{z} \in R^n : \mathbf{y} \geq \mathbf{0}, \mathbb{A}^T\mathbf{y} - \mathbb{C}^T\mathbf{z} \leq \mathbf{0}\}.$$

Find $(\tilde{\mathbf{y}}, \tilde{\mathbf{z}}) \in \mathcal{G}$ such that

$$-\tilde{\mathbf{z}}\cdot(\mathbb{C}\tilde{\mathbf{z}}) + 2\tilde{\mathbf{y}}\cdot\mathbf{b} = \max_{(\mathbf{y},\mathbf{z})\in\mathcal{G}} [-\mathbf{z}\cdot(\mathbb{C}\mathbf{z}) + 2\mathbf{y}\cdot\mathbf{b}].$$

Prove that if $\tilde{\mathbf{x}}$ is optimal for the primal quadratic minimum problem, then there is an optimal vector $(\tilde{\mathbf{y}}, \tilde{\mathbf{z}})$ for the dual problem and

$$\tilde{\mathbf{x}}\cdot(\mathbb{C}\,\tilde{\mathbf{x}}) = -\tilde{\mathbf{z}}\cdot(\mathbb{C}\tilde{\mathbf{z}}) + 2\tilde{\mathbf{y}}\cdot\mathbf{b}.$$

Equivalently, the last assertion can be written as

$$\min_{\mathbf{x}\in\mathcal{F}} [\mathbf{x}\cdot(\mathbb{C}\mathbf{x})] = \max_{(\mathbf{y},\mathbf{z})\in\mathcal{G}} [-\mathbf{z}\cdot(\mathbb{C}\mathbf{z}) + 2\mathbf{y}\cdot\mathbf{b}].$$

Hint. Show that if $\tilde{\mathbf{x}}$ is optimal for the primal quadratic minimum problem and $\tilde{\mathbf{y}}$ is optimal for the standard maximum problem introduced in Problem 2, then the vector $(\tilde{\mathbf{y}}, \tilde{\mathbf{x}})$ is contained in $\mathcal{G}$. Compare the values of the objective function for $(\tilde{\mathbf{y}}, \tilde{\mathbf{x}})$ and for any other $(\mathbf{y}, \mathbf{z}) \in \mathcal{G}$.

4.5.3 Registration at Big State

The registrar at Big State University is responsible for designing a registration procedure for the 3000 unlucky students who did not complete registration by mail. The procedure is to set up stations in a large gymnasium and schedule students to arrive at intervals so that the process proceeds as smoothly and as rapidly as possible. The registrar decides that there are three essentially different stations to be set up. The first handles initial processing, the second handles class assignment, and the third the payment of fees. Students must progress through these stations in order. Also, the student population consists of three dissimilar groups, freshmen, other undergraduates, and graduate and professional students. The estimated time required at each station for a single student of each type is given in Table 4–2. Also, it is anticipated that of the 3000 students to be processed

Table 4–2

	Initial processing (minutes)	Class assignment (minutes)	Payment of fees (minutes)
Freshman	10	15	5
Other undergraduates	8	10	3
Graduate and professional studies	7	5	3

there will be 1500 freshmen, 1000 other undergraduates, and 500 graduate and professional students. There are 100 workers available to handle registration, and each is capable of serving at any station. However, once a worker has been assigned to a station, he must remain at that station for the entire period of registration.

Problem 1. Identify the factors which must be considered in determining the minimum time necessary for registration.

Problem 2. Discuss precisely how the factors identified in Problem 1 influence the registration time.

Problem 3. Specify a scheme which allows for registration to be completed in 20 hours.

REFERENCES

Textbooks and monographs on the theory and computational aspects of linear programming began appearing in significant numbers in the late 1950s. These books are now available at all levels, and, in a very elementary version, the subject forms a part of what is commonly known as *finite mathematics*.

The book [KF] consists of 18 chapters each of which is relatively independent of the others and considers a particular example in detail. Several of these examples lead to models involving optimization questions, and Chap. 13 in particular is closely related to the work of this chapter. In [Da], [CC], [Wa], and Part 2 of [DaV] there are discussions of several situations in economic planning and business which lead to linear programming problems and their generalizations. The article [Ko] gives a linear optimization model for the study of air pollution, and [Kr] a similar model for the allocation of resources in a political campaign.

For a more detailed mathematical presentation, we refer the reader to the textbooks [Da], [Ga], and [V]. Although the emphasis is primarily on methods, [Si] contains much which is relevant to this chapter. The fundamental papers [Go], [GoT], and [T] are accessible to readers with a knowledge of linear algebra.

Chapters on linear programming with uncertainty are included in [Da], [V], and [Wa]. The paper [FeDa] is an application of stochastic linear programming to a real problem.

[CC] Charnes, A., and W. W. Cooper, *Management Models and Industrial Applications of Linear Programming*, Vols. I and II. New York: Wiley, 1961.

[Da] Dantzig, G. B., *Linear Programming and Extensions*. Princeton, N.J.: Princeton University Press, 1963.

[DaV] Dantzig, G. B., and A. F. Veinott, eds., *Mathematics of the Decision Sciences*, Parts 1 and 2 (Lectures in Applied Mathematics, Vols. XI and XII). American Mathematical Society, Providence, R.I., 1968.

[F] Farkas, J., "Theorie der einfachen Ungleichungen," *Journal für die Reine und Angewandte Mathematik*, **124** (1902), 1–27.

[FeDa] FERGUSON, A. R., and G. B. DANTZIG, "The Allocation of Aircraft to Routes—An Example of Linear Programming Under Uncertain Demand," *Management Science*, **3** (1956), 45–73.

[Ga] GALE, D., *The Theory of Linear Economic Models*. New York: McGraw-Hill, 1960.

[Go] GOLDMAN, A. J., "Resolution and Separation Theorems for Polyhedral Convex Sets," *Annals of Mathematics Studies*, No. 38, Princeton University Press, Princeton, N.J. (1956), 41–52.

[GoT] GOLDMAN, A. J., and A. W. TUCKER, "Theory of Linear Programming," *Annals of Mathematics Studies*, No. 38, Princeton University Press, Princeton, N.J. (1956), 53–98.

[KF] KAUFMANN, A., and R. FAURE, *Introduction to Operations Research*. New York: Academic Press, 1968.

[Ko] KOHN, R. E., "A Mathematical Programming Model for Air Pollution Control," *School Science and Mathematics*, **69** (June 1969), 487–494.

[Kr] KRAMER, G. H., "A Decision-Theoretic Analysis of a Problem in Political Campaigning," *Mathematical Applications in Political Science*, Vol. II (J. L. Bernd, ed.). Dallas: Southern Methodist University Press, 1966.

[Si] SIMONNARD, M., *Linear Programming*, tr. W. S. Jewell. Englewood Cliffs, N.J.: Prentice-Hall, 1966.

[T] TUCKER, A. W., "Dual Systems of Homogeneous Linear Relations," *Annals of Mathematics Studies*, No. 38, Princeton University Press, Princeton, N.J. (1956), 3–18.

[V] VAJDA, S., *Mathematical Programming*. Reading, Mass.: Addison-Wesley, 1961.

[Wa] WAGNER, H. M., *Principles of Management Science*. Englewood Cliffs, N.J.: Prentice-Hall, 1970.

5 Geometric and Computational Aspects of Linear Optimization

5.0 INTRODUCTION

In this chapter we shall turn from the algebraic theory of linear optimization to the geometry of the theory and to the practical computational problem of actually finding explicit solutions. Our aims are modest in that we make no attempt to give an in-depth study of either the geometry or all the relevant computational techniques. Indeed, we are content to give only one well-known algorithm, together with the geometry which is useful in studying this algorithm. Moreover, even though in practice this algorithm is usually carried out on a computer, we shall not develop computer programs in this discussion. We shall simply describe the algorithm and, to some extent, show when and why it works.

The main topic of this chapter is the algorithm known as the simplex method. This algorithm was introduced in the late 1940s by George Dantzig, and it has been the basis for most computational methods for linear programming problems. We begin our treatment of the simplex method with a section on the geometric aspects of linear programming problems. We concentrate on those aspects which are important for the subsequent work with the algorithm. In the second section we give the algorithm of the simplex method and conditions under which the method is successful.

5.1 GEOMETRIC ASPECTS OF LINEAR OPTIMIZATION

We begin this section by considering a simple example of a linear optimization problem. Thus, let

$$\mathbb{A} = \begin{bmatrix} 1 & 2 \\ 3 & 4 \end{bmatrix}, \qquad \mathbf{b} = \begin{pmatrix} 5 \\ 12 \end{pmatrix}, \qquad \mathbf{c} = \begin{pmatrix} 5 \\ 8 \end{pmatrix}.$$

Consider the problem of maximizing the quantity $L(\mathbf{x}) = \mathbf{x} \cdot \mathbf{c}$, where

$$\mathbf{x} = \begin{pmatrix} \xi_1 \\ \xi_2 \end{pmatrix}$$

is subject to the constraints $\mathbf{x} \geq \mathbf{0}$ and $\mathbb{A}\mathbf{x} \leq \mathbf{b}$. Explicitly, these constraints are given by the four inequalities

$$\begin{aligned} \xi_1 + 2\xi_2 &\leq 5, \\ 3\xi_1 + 4\xi_2 &\leq 12, \\ \xi_1 &\geq 0, \\ \xi_2 &\geq 0. \end{aligned} \tag{1.1}$$

We want to examine the set of solutions of system (1.1) from a geometrical point of view. This set of solutions is the set of feasible vectors for the linear optimization problem posed above. We begin by considering each of the inequalities in (1.1) to be the determining equation for a set of points in the ξ_1–ξ_2 plane. Namely, we associate with each inequality the set of points in R^2 which satisfy that inequality. For example, the first inequality, $\xi_1 + 2\xi_2 \leq 5$, determines the set $\mathfrak{F}_1 = \{\mathbf{x}: \xi_1 + 2\xi_2 \leq 5\}$. If we graph the four sets which are obtained in this manner, we obtain the graphs shown in Fig. 5–1.

A vector $\mathbf{x}$ is feasible if it satisfies all four of the inequalities of (1.1). In other words, it must be in each of the sets $\mathfrak{F}_1$, $\mathfrak{F}_2$, $\mathfrak{F}_3$ and $\mathfrak{F}_4$. Hence the set of feasible vector $\mathfrak{F}$ is given by $\mathfrak{F} = \mathfrak{F}_1 \cap \mathfrak{F}_2 \cap \mathfrak{F}_3 \cap \mathfrak{F}_4$. This set is shown in Fig. 5–2.

The set $\mathfrak{F}$ is bounded by four lines, and these lines determine four *corner* points of $\mathfrak{F}$: (0, 0), (0, 5/2), (2, 3/2), and (4, 0). If we evaluate the function $L(\mathbf{x}) = 5\xi_1 + 8\xi_2$ at these corner points, then we obtain $L(0, 0) = 0$, $L(0, 5/2) = 20$, $L(2, 3/2) = 22$, and $L(4, 0) = 20$, Thus, at least among

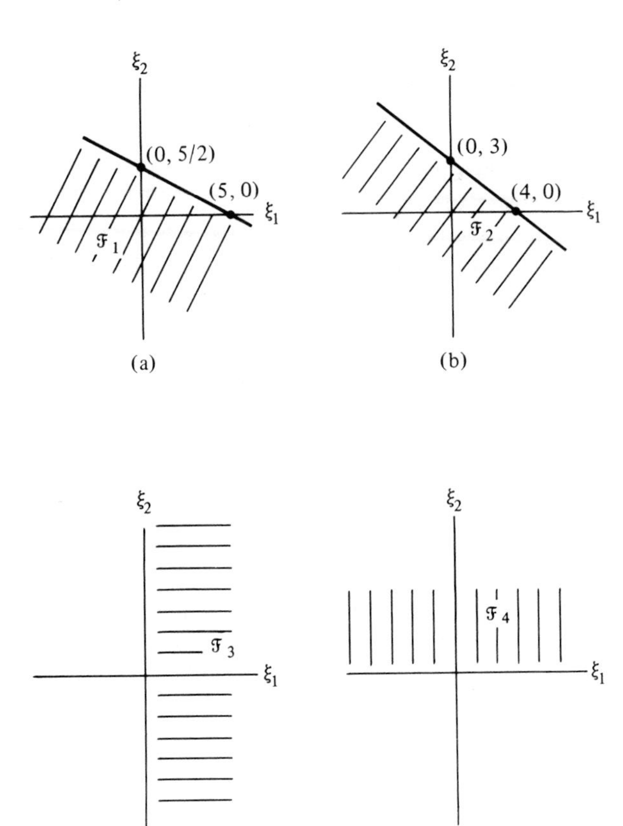
ξ_2
(0, 5/2)
(5, 0)
ξ_1
$\mathfrak{F}_1$
(a)
ξ_2
(0, 3)
(4, 0)
ξ_1
$\mathfrak{F}_2$
(b)
ξ_2
$\mathfrak{F}_3$
ξ_1
(c)
ξ_2
$\mathfrak{F}_4$
ξ_1
(d)

Fig. 5–1

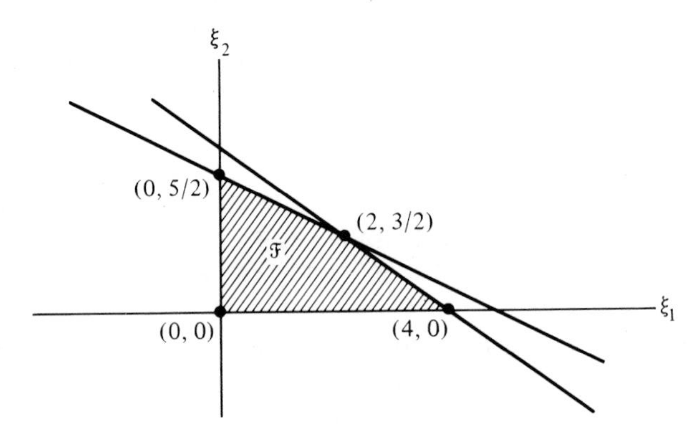
ξ_2
(0, 5/2)
(2, 3/2)
$\mathfrak{F}$
ξ_1
(0, 0)
(4, 0)

Fig. 5–2

the corner points of $\mathfrak{F}$, the maximum of L is 22. Since there are infinitely many points of $\mathfrak{F}$ which are not corner points, it is not obvious that the maximum of L over the entire set $\mathfrak{F}$ is also 22. We shall prove this result in a more general setting later in the section. We prove it in this special case by considering the dual problem.

The dual problem in this case is the problem of minimizing $\mathbf{y} \cdot \mathbf{b}$, where

$$\mathbf{y} = \begin{pmatrix} \eta_1 \\ \eta_2 \end{pmatrix}$$

is subject to the constraints $\mathbb{A}^T\mathbf{y} \geq \mathbf{c}$ and $\mathbf{y} \geq \mathbf{0}$. These constraints are given by the following inequalities:

$$\begin{aligned} \eta_1 + 3\eta_2 &\geq 5, \\ 2\eta_1 + 4\eta_2 &\geq 8, \\ \eta_1 &\geq 0, \\ \eta_2 &\geq 0. \end{aligned} \tag{1.2}$$

As in the primal problem, each inequality determines a set of points in the plane. We take the intersection of these four sets to obtain the set of feasible vectors for the dual problem. This set is shown in Fig. 5–3. The corner points

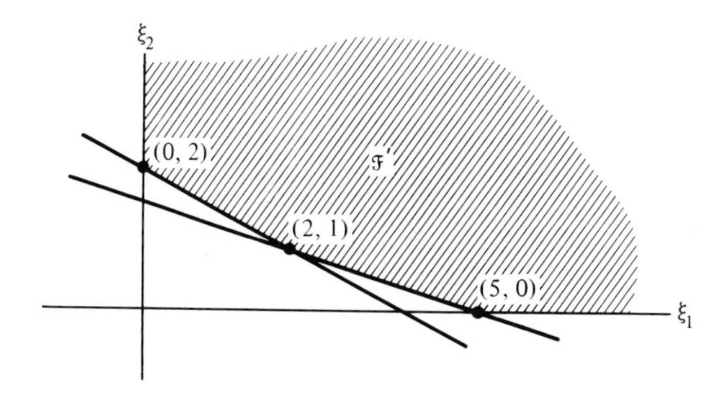

Fig. 5–3

of the feasible set for the dual problem are the points (0, 2), (2, 1), and (5, 0). If we evaluate the function $L'(\mathbf{y}) = \mathbf{y} \cdot \mathbf{b} = 5\eta_1 + 12\eta_2$ at these corner points, we obtain the values $L'(0, 2) = 24$, $L'(2, 1) = 22$, and $L'(5, 0) = 25$. Therefore, on the corner points, the function L' has a minimum of 22 at the point (2, 1). However, it now follows that this is actually a minimum of L' for all feasible points in $\mathfrak{F}'$ because of our information about the primal problem. Indeed, we know that $L'(2, 1) = 22 = L(2, 3/2)$, and according to Lemma 1 of Sec. 4.3.2, $L'(\mathbf{y}) \geq L(\mathbf{x})$ for all feasible vectors **x** and **y**. Therefore, the vectors

$$\mathbf{x} = \begin{pmatrix} 2 \\ 3/2 \end{pmatrix} \quad \text{and} \quad \mathbf{y} = \begin{pmatrix} 2 \\ 1 \end{pmatrix}$$

are optimal vectors for the primal and dual problems, respectively. Thus we have shown that, at least in this special case, it is possible to find the optimal vectors by checking only a finite set of special vectors. These special vectors correspond to the corner points of the set of feasible vectors. Our aim is to show that this situation is not unusual and that, in fact, similar statements can be made about more general problems where the vectors are not in R^2. As a first step in carrying out this aim, we must consider the geometry of some particular figures in the space R^n, $n \geq 2$.

5.1.1 Half-Spaces and Hyperplanes

In the example considered above the set of feasible vectors is determined by the inequalities of system (1.1). Each of these inequalities is an equality for a set of points in R^2 which forms a line. This line divides the space R^2 into two half-spaces. One of these half-spaces contains all the points which satisfy the inequality, and the other half-space contains all those points of R^2 which do not satisfy the inequality. Thus the feasible set $\mathfrak{F}$ is the intersection of four half-spaces, each one determined by one of the constraints of the problem.

Suppose that we consider a standard maximum problem where the matrix $\mathbb{A}$ has dimension 3×3 instead of 2×2. Then the feasible vectors **x** are elements of R^3, and the constraints of the problem are inequalities which are satisfied by subsets of R^3. If the matrix $\mathbb{A} = (a_{ij})$ has rows $\mathbf{a}_1, \mathbf{a}_2, \mathbf{a}_3$ and if **b** is the vector $(\beta_1, \beta_2, \beta_3)$ and **x** is the vector (ξ_1, ξ_2, ξ_3), then the constraints are given by the following system:

$$\begin{aligned} \mathbf{a}_1 \cdot \mathbf{x} &\leq \beta_1, \\ \mathbf{a}_2 \cdot \mathbf{x} &\leq \beta_2, \\ \mathbf{a}_3 \cdot \mathbf{x} &\leq \beta_3, \\ \xi_1 &\geq 0, \\ \xi_2 &\geq 0, \\ \xi_3 &\geq 0. \end{aligned} \tag{1.3}$$

In this system the first inequality becomes an equality for the set given by $\{\mathbf{x}: a_{11}\xi_1 + a_{12}\xi_2 + a_{13}\xi_3 = \beta_1\}$. We recognize that this is the equation of a plane in R^3. This plane divides the set R^3 into halves. One half consists of all points which satisfy the inequality $\mathbf{a}_1 \cdot \mathbf{x} \leq \beta_1$, and the other half consists of points which do not satisfy the inequality. Similarly, each of the other five inequalities of (1.3) has as solutions of the corresponding equality a set of points which constitutes a plane in R^3. Each of these planes separates R^3 into two half-spaces. In one half-space the inequality is satisfied and in the other half-space it is not satisfied. A vector $\mathbf{x}$ is feasible for the problem if it satisfies all six inequalities, i.e., if it lies in each of the six half-spaces defined by the inequalities. Therefore the set of feasible vectors is the intersection of six half-spaces, each one determined by a plane.

We now turn to the general case and assume that $\mathbb{A}$ is an $m \times n$ matrix. Then, $\mathbf{b} \in R^m$ and the feasible vectors $\mathbf{x}$ are in R^n. The problem now has $m + n$ constraints, and they are given by inequalities of the form

$$\begin{aligned} \mathbf{a}_1 \cdot \mathbf{x} &\leq \beta_1 \\ &\vdots \\ \mathbf{a}_m \cdot \mathbf{x} &\leq \beta_m \\ \xi_1 &\geq 0 \\ &\vdots \\ \xi_n &\geq 0. \end{aligned} \tag{1.4}$$

The inequality $\mathbf{a}_i \cdot \mathbf{x} \leq \beta_i$ is satisfied by some subset of vectors in R^n, and this inequality is an equality for the set $\mathcal{H}_i = \{\mathbf{x}: a_{i1}\xi_1 + a_{i2}\xi_2 + \cdots + a_{in}\xi_n = \beta_i\}$. Motivated by the terminology in R^3, this set $\mathcal{H}_i$ is called a hyperplane in R^n. We make this notion precise.

Definition: A subset $\mathcal{H}$ of R^n is said to be a *hyperplane* if there exists a constant k and a vector $\mathbf{w} \in R^n$ such that $\mathcal{H} = \{\mathbf{x} \in R^n: \mathbf{w} \cdot \mathbf{x} = k\}$.

To check his understanding of this definition, we suggest that the reader verify that lines are hyperplanes in R^2 and that planes are hyperplanes in R^3.

Every hyperplane in R^n is the dividing set for two half-spaces. The first space consists of all points $\mathbf{x}$ in R^n which satisfy $\mathbf{w} \cdot \mathbf{x} \leq k$ and the second consists of all the other points of R^n. Since each of the inequalities of (1.4) is an equality for a set of points forming a hyperplane (Exercise 1), we see that the set of feasible vectors is once again the intersection of certain half-spaces. In this general case, there are $m + n$ half-spaces: one half-space for each of the inequalities $\mathbf{a}_i \cdot \mathbf{x} \leq \beta_i$, $i = 1, \ldots, m$, and one half-space for each of the inequalities $\xi_i \geq 0$, $i = 1, \ldots, n$.

To provide illustrations of these ideas, we consider an example of the situation when $n = m = 3$.

Example. Let

$$\mathbb{A} = \begin{bmatrix} 1 & 1 & 0 \\ 0 & 1 & 1 \\ 1 & 0 & 1 \end{bmatrix}, \qquad \mathbf{b} = \begin{pmatrix} 1 \\ 1 \\ 1 \end{pmatrix}, \qquad \text{and} \qquad \mathbf{c} = \begin{pmatrix} 2 \\ 2 \\ 2 \end{pmatrix}.$$

Find $\mathbf{x}$ such that $\mathbb{A}\mathbf{x} \leq \mathbf{b}$, $\mathbf{x} \geq \mathbf{0}$, and $\mathbf{x} \cdot \mathbf{c}$ is a maximum for all such $\mathbf{x}$.

In this example there are six inequalities which provide the constraints for the optimization. These inequalities determine six hyperplanes (ordinary planes in this case) and six half-spaces in R^3. The six inequalities form the following system:

$$\begin{aligned} \xi_1 + \xi_2 &\leq 1, \\ \xi_2 + \xi_3 &\leq 1, \\ \xi_1 + \xi_3 &\leq 1, \\ \xi_1 &\geq 0, \\ \xi_2 &\geq 0, \\ \xi_3 &\geq 0. \end{aligned} \tag{1.5}$$

The first three inequalities of (1.5) determine the half-spaces shown in Fig. 5–4(a), (b), and (c), respectively. The arrows are directed into the half-space of points which satisfy the appropriate inequality.

Each of the last three inequalities of (1.5) is an equality for one of the coordinate planes in R^3. Hence the set of feasible vectors for this problem lies in the first octant, and *under* each of the planes shown in Fig. 5–4. By

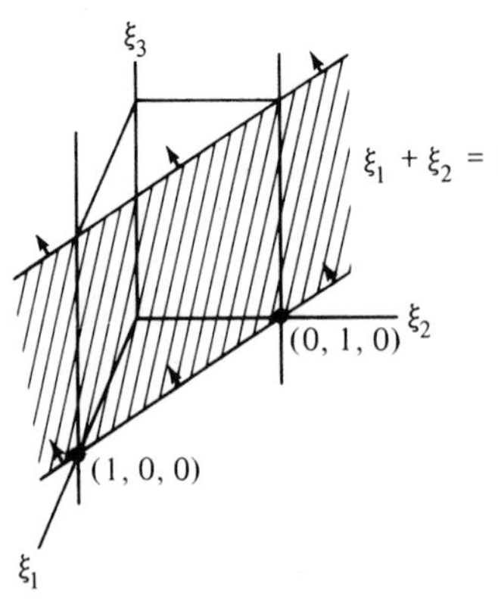

Fig. 5–4(a)

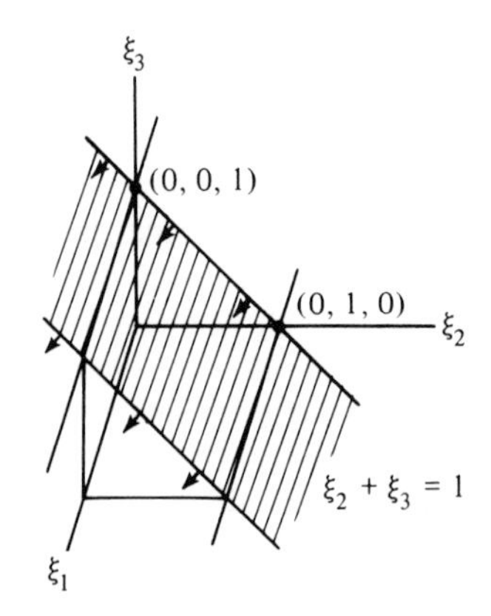

Fig. 5–4(b)

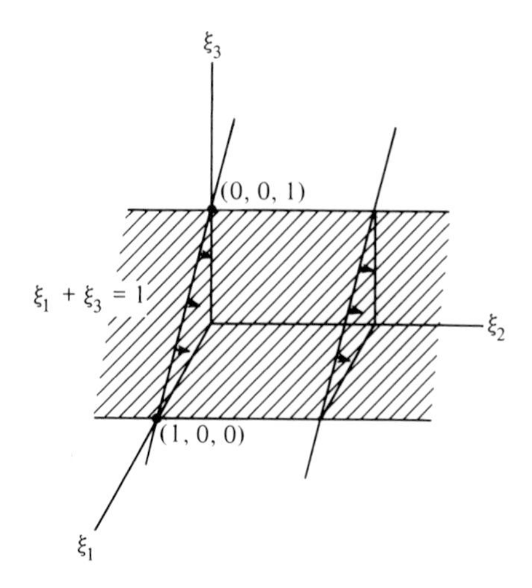

Fig. 5–4(c)

under we mean here that half-space which is determined by the plane and which contains the origin. This set is shown in Fig. 5–5.

The set $\mathfrak{F}$ of feasible vectors, which is shown in Fig. 5–5, has corner points given by (0, 0, 0), (1, 0, 0), (0, 1, 0), (0, 0, 1), and $(\frac{1}{2}, \frac{1}{2}, \frac{1}{2})$. If we evaluate the function $L(\mathbf{x}) = \mathbf{x} \cdot \mathbf{c}$ at these points, we find that the maximum is taken at $(\frac{1}{2}, \frac{1}{2}, \frac{1}{2})$, where $L(\frac{1}{2}, \frac{1}{2}, \frac{1}{2}) = 3$. Now, in the earlier example in this section the maximum at the corner points was the maximum over the whole feasible set. This was shown by considering the dual problem and by using results from Chap. 4. We could use the same method in this case; however, we prefer to adopt a different approach.

Let k be a constant, and consider the equation $L(\mathbf{x}) = \mathbf{x} \cdot \mathbf{c} = k$. Recall that this is the equation of a plane in R^3. We denote this plane by π_k (i.e., $\pi_k = \{\mathbf{x}: \mathbf{x} \cdot \mathbf{c} = k\}$), and we consider the intersection of this plane and the set $\mathfrak{F}$. This intersection may consist of a single point or many points or may be empty, depending on the choice of k. Now suppose for a moment that the intersection is not empty and that $\mathbf{z} \in \mathfrak{F} \cap \pi_k$. Then $\mathbf{z}$ is feasible for the optimization problem under consideration and, moreover, $L(\mathbf{z}) = k$. Since the problem seeks to maximize the function L over the set $\mathfrak{F}$, $\mathbf{z}$ will be an optimal vector provided that k is the maximum value taken by L on the set $\mathfrak{F}$. Thus this optimization problem can be restated as the problem of finding the maximum value of k for which $\pi_k \cap \mathfrak{F}$ is not empty. Let us make this observation more precise. Suppose that $k_0 = \max\{k: \pi_k \cap \mathfrak{F} \neq \varnothing\}$ and that $\mathbf{z} \in \pi_{k_0} \cap \mathfrak{F}$ (the maximum exists because $\pi_k \cap \mathfrak{F}$ is closed and bounded; see Appendix A). If there is a vector $\mathbf{x} \in \mathfrak{F}$ such that $L(\mathbf{x}) > L(\mathbf{z})$, then one can choose $k' = L(\mathbf{x})$, and this gives $\mathbf{x} \in \pi_{k'} \cap \mathfrak{F}$, $k' > k_0$. But this contra-

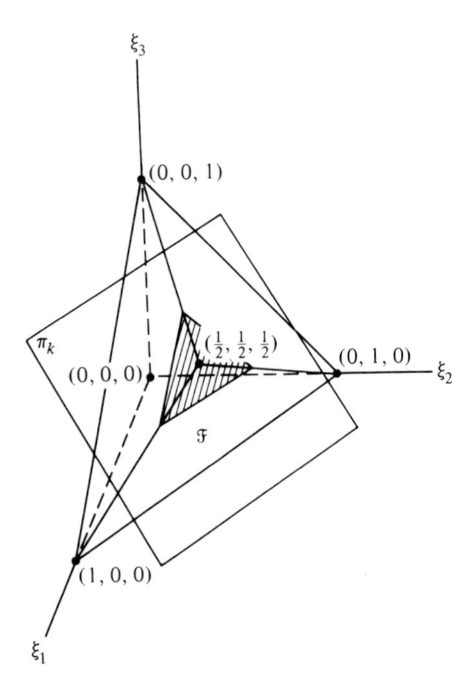

Fig. 5–5

dicts the definition of k_0, and hence we must have $L(\mathbf{x}) \leq L(\mathbf{z})$ for all $\mathbf{x} \in \mathfrak{F}$ and $\mathbf{z} \in \pi_{k_0} \cap \mathfrak{F}$. Therefore $\mathbf{z}$ is optimal and $L(\mathbf{z}) = k_0$ is the maximum of L over $\mathfrak{F}$.

We now consider this new interpretation of the primal optimization problem from a geometric point of view. Using elementary analytic geometry, we know that the constant k is essentially a measure of the perpendicular distance from the origin to the plane π_k. We say *essentially* because the actual distance is always a multiple of $|k|$. In this example, the distance is $|k|/(2\sqrt{3})$. Thus an increase in k corresponds to a movement of the plane π_k away from the origin. If we examine the set $\mathfrak{F}$ of Fig. 5–5, we see that as π_k moves away from the origin, the set $\pi_k \cap \mathfrak{F}$ (the shaded triangle of Fig. 5–5) becomes smaller. Finally, we reach a point where any increase in k will mean that $\pi_k \cap \mathfrak{F}$ is empty. This is the point $(\frac{1}{2}, \frac{1}{2}, \frac{1}{2})$, where $k = 3$. If $k > 3$, then $\pi_k \cap \mathfrak{F} = \varnothing$, and if $0 \leq k < 3$, then there are numbers k', $k' > k$ for which the sets $\pi_{k'} \cap \mathfrak{F}$ are nonempty. Thus the point $(\frac{1}{2}, \frac{1}{2}, \frac{1}{2})$ is the optimal vector for the original optimization problem, and the maximum value of L is 3.

We have now given two examples in which the standard primal problem can be solved by simply examining certain special points called corner points. We next make this notion of corner point more precise, and then we shall obtain the corresponding result for problems in R^n. To obtain this general result, we introduce a new class of sets which includes the half-spaces of R^n as a particular subclass. These sets are called convex sets.

5.1.2 Convex Sets

If we look at a plane in R^3 and distinguish two points on the same side of the plane, then the line between these points lies completely on one side of the plane. It is this idea which forms the basis of the definition of a convex set.

Definition: Let $\mathbf{x}, \mathbf{y} \in R^n$. Then the *line segment between* $\mathbf{x}$ and $\mathbf{y}$ is the set defined by

$$[\mathbf{x}, \mathbf{y}] = \{\mathbf{z}_\alpha : \mathbf{z}_\alpha = \alpha\mathbf{x} + (1 - \alpha)\mathbf{y}, 0 \leq \alpha \leq 1\}.$$

We recommend that the reader choose some points $\mathbf{x}, \mathbf{y}$ in R^3 and verify that, indeed, in the usual geometric sense of the word line, $[\mathbf{x}, \mathbf{y}]$ is the line segment between these points.

Definition: A set $\mathcal{K}$ in R^n is said to be *convex* if $\mathbf{x}, \mathbf{y} \in \mathcal{K} \Rightarrow [\mathbf{x}, \mathbf{y}] \subset \mathcal{K}$. Thus a set is convex if for every two points in the set the line segment between them is also contained in the set.

In Fig. 5–6, the sets $\mathcal{J}$, $\mathcal{G}$, and $\mathcal{M}$ are convex, while the set $\mathcal{L}$ is not convex.

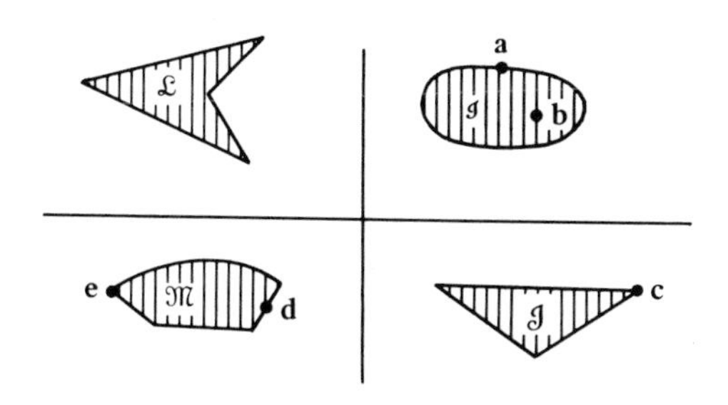

Fig. 5–6

In any convex set there are certain special points which correspond to the corner points discussed in Sec. 5.1.1. These are called extreme points and they are defined as follows:

Definition: A point $\mathbf{e}$ in a convex set $\mathcal{K}$ is said to be an *extreme point* of $\mathcal{K}$ if there do not exist points $\mathbf{x}, \mathbf{y} \in \mathcal{K}$, $\mathbf{x} \neq \mathbf{y}$, such that $\mathbf{e} \in [\mathbf{x}, \mathbf{y}]$, $\mathbf{x} \neq \mathbf{e} \neq \mathbf{y}$.

In Fig. 5–6 the points **a**, **c**, and **e** are extreme, while **b** and **d** are not.

There is an extensive theory of convex sets in R^n and also in more general spaces. We shall develop only a small part of this theory; in particular, we shall develop that part of the theory which we need to continue our discussion of linear optimization.

Lemma: If $\mathbf{a} \in R^n$ and k is any real constant, then each of the sets $\mathcal{S}_1 = \{\mathbf{x} \in R^n: \mathbf{a} \cdot \mathbf{x} \leq k\}$, $\mathcal{S}_2 = \{\mathbf{x} \in R^n: \mathbf{a} \cdot \mathbf{x} < k\}$, and $\mathcal{S}_3 = \{\mathbf{x} \in R^n: \mathbf{a} \cdot \mathbf{x} = k\}$ is a convex set.

Proof: Exercise 3.

Theorem 1: If $\mathcal{C}$ is a collection of convex subsets of R^n, then the set

$$\mathcal{G} = \bigcap_{\mathcal{S} \in \mathcal{C}} \mathcal{S}$$

is also a convex subset of R^n.

Proof: Let $\mathbf{x}, \mathbf{y}$ be elements of $\mathcal{G}$. Then for each $\mathcal{S} \in \mathcal{C}$, $\mathbf{x}, \mathbf{y} \in \mathcal{S}$. Since $\mathcal{S}$ is convex, this implies that $[\mathbf{x}, \mathbf{y}] \subset \mathcal{S}$. But since $\mathcal{S}$ is an arbitrary set in $\mathcal{C}$, this gives $[\mathbf{x}, \mathbf{y}] \subset \mathcal{G}$, and hence $\mathcal{G}$ is convex. Q.E.D.

We can combine the lemma above and Theorem 1 to show that the set of feasible vectors for the primal problem is a convex set. We state this result as a theorem.

Theorem 2: Let $\mathbb{A}$ be an $m \times n$ matrix, $\mathbf{b} \in R^m$, and $\mathbf{c} \in R^n$. The set $\mathfrak{F}$ of feasible vectors for the problem of maximizing $\mathbf{x} \cdot \mathbf{c}$ subject to $\mathbb{A}\mathbf{x} = \mathbf{b}$ and $\mathbf{x} \geq \mathbf{0}$ is a convex set.

Proof: Let $\mathcal{H}_i = \{\mathbf{x}: \mathbf{a}_i \cdot \mathbf{x} = \beta_i\}$, $i = 1, 2, \ldots, m$, and $\mathcal{H}_i = \{\mathbf{x}: \xi_{i-m} \geq 0\}$, $i = m + 1, m + 2, \ldots, m + n$. Then, by the lemma above, $\mathcal{H}_i$ is a convex set for $i = 1, \ldots, m$. Actually the same lemma also shows that $\mathcal{H}_i$ is convex for $i = m + 1, \ldots, m + n$. To verify this, we take $k = 0$ and $\mathbf{a} = (0, \ldots, \overset{i\text{th}}{-1}, 0, \ldots, 0)$. Then $\mathbf{a} \cdot \mathbf{x} \leq k$ is the same as $-\xi_i \leq 0$, or equivalently $\xi_i \geq 0$, and this is the desired inequality. Finally, $\mathfrak{F} = \bigcap_{i=1}^{m+n} \mathcal{H}_i$, and hence by Theorem 1 the set $\mathfrak{F}$ is convex.

We remark that Theorem 2 also holds if we replace $\mathbb{A}\mathbf{x} = \mathbf{b}$ by $\mathbb{A}\mathbf{x} \leq \mathbf{b}$. The proof is identical and we do not state this as a separate result. In fact, there are many results of this section which are not stated in all possible forms. Our aims are limited, and, in general, we attempt to give only those forms which we intend to use in the future.

Our next result deals with the extreme points of the set $\mathfrak{F}$ of feasible vectors. This will be important for the theoretical discussion of the simplex method which follows Sec. 5.2. It will help to show how and why the simplex method works.

Theorem 3: Let $\mathbb{A}$ be an $m \times n$ matrix, $\mathbf{b} \in R^m$, and $\mathfrak{F}$ be the set of solutions of the system $\mathbb{A}\mathbf{x} = \mathbf{b}$, $\mathbf{x} \geq \mathbf{0}$. Then,

1. $\mathbf{b} = \mathbf{0} \Rightarrow \mathbf{x} = \mathbf{0}$ is the only extreme point of $\mathfrak{F}$.
2. $\mathbf{b} \neq \mathbf{0} \Rightarrow \mathbf{0} \notin \mathfrak{F}$. Moreover, if $\mathbf{x} = (\xi_1, \ldots, \xi_n) \in \mathfrak{F}$ and $\xi_i > 0$ for $i \in \{i_1, \ldots, i_r\}$ and $\xi_i = 0$ for $i \notin \{i_1, \ldots, i_r\}$, then $\mathbf{x}$ is extreme for $\mathfrak{F}$ if and only if the columns $\mathbf{a}^{i_1}, \mathbf{a}^{i_2}, \ldots, \mathbf{a}^{i_r}$ are independent in R^m (see Appendix C).
3. If $\mathbf{x}$ is an extreme point of $\mathfrak{F}$, then $\mathbf{x}$ has at most m positive coordinates.
4. $\mathfrak{F}$ has only a finite number of extreme points.

Proof

1. Exercise 4.
2. Assume that $\mathbf{x} \in \mathfrak{F}$, $\xi_i > 0$ for $i \in \{i_1, \ldots, i_r\}$, and $\xi_1 = 0$ for $i \notin \{i_1, \ldots, i_r\}$.

We first assume that $\mathbf{a}^{i_1}, \ldots, \mathbf{a}^{i_r}$ are independent and we shall show that $\mathbf{x}$ is extreme in $\mathfrak{F}$. Suppose that $\mathbf{x}$ is not extreme in $\mathfrak{F}$. Then by the definition there exists $\mathbf{y}, \mathbf{z} \in \mathfrak{F}$ and α, $0 < \alpha < 1$, such that $\mathbf{x} = \alpha\mathbf{y} + (1 - \alpha)\mathbf{z}$, $\mathbf{y} \neq \mathbf{z}$. Since $\mathbf{b} \neq \mathbf{0}$, it follows that neither of the nonnegative vectors $\mathbf{y}$ and $\mathbf{z}$ can be the zero vector. Hence each of them must have at least one positive coordinate. Now, $\mathbf{x} = \alpha\mathbf{y} + (1 - \alpha)\mathbf{z}$, so if $\mathbf{y} = (\eta_1, \ldots, \eta_n)$ and $\mathbf{z} = (\zeta_1, \ldots, \zeta_n)$, then $0 = \xi_i = \alpha\eta_i + (1 - \alpha)\zeta_i$ for $i \notin \{i_1, \ldots, i_r\}$. Since $\alpha > 0$, it follows that $\eta_i = \zeta_i = 0$ for $i \notin \{i_1, \ldots, i_r\}$. Therefore the nonnegative components of $\mathbf{y}$ and $\mathbf{z}$ must be among the coordinates $\{\eta_{i_1}, \ldots, \eta_{i_r}\}$ and $\{\zeta_{i_1}, \ldots, \zeta_{i_r}\}$, respectively. Next, $\mathbf{y}$ and $\mathbf{z}$ satisfy $\mathbb{A}\mathbf{y} = \mathbf{b}$ and $\mathbb{A}\mathbf{z} = \mathbf{b}$, and hence, in view of the above information on the coordinates of $\mathbf{y}$ and $\mathbf{z}$, we have

$$\eta_{i_1}\mathbf{a}^{i_1} + \cdots + \eta_{i_r}\mathbf{a}^{i_r} = \mathbf{b},$$
$$\zeta_{i_1}\mathbf{a}^{i_1} + \cdots + \zeta_{i_r}\mathbf{a}^{i_r} = \mathbf{b}.$$

Equivalently, $(\eta_{i_1} - \zeta_{i_1})\mathbf{a}^{i_1} + \cdots + (\eta_{i_r} - \zeta_{i_r})\mathbf{a}^{i_r} = \mathbf{0}$. But the vectors $\mathbf{a}^{i_1}, \ldots, \mathbf{a}^{i_r}$ are independent, and therefore $\eta_{i_1} = \zeta_{i_1}, \eta_{i_2} = \zeta_{i_2}, \ldots, \eta_{i_r} = \zeta_{i_r}$. This implies that $\mathbf{y} = \mathbf{z}$, which contradicts the original assumption about $\mathbf{y}$ and $\mathbf{z}$. Hence we have shown that if $\mathbf{a}^{i_1}, \ldots, \mathbf{a}^{i_r}$ are independent, then $\mathbf{x}$ is extreme.

We next turn to the converse, and we assume that $\mathbf{x}$ is extreme, $\xi_i > 0$

for $i \in \{i_1, \ldots, i_r\}$, and $\xi_1 = 0$ for $i \notin \{i_1, \ldots, i_r\}$. We seek to show that $\mathbf{a}^{i_1}, \ldots, \mathbf{a}^{i_r}$ are independent. Suppose that these columns are not independent. Then there must exist constants $\gamma_1, \gamma_2, \ldots, \gamma_r$, not all zero, such that

$$\gamma_1 \mathbf{a}^{i_1} + \cdots + \gamma_r \mathbf{a}^{i_r} = \mathbf{0}. \tag{1.6}$$

Also, $\mathbf{x} \in \mathfrak{F}$ so that $\mathbb{A}\mathbf{x} = \mathbf{b}$, or, equivalently,

$$\xi_{i_1} \mathbf{a}^{i_1} + \cdots + \xi_{i_r} \mathbf{a}^{i_r} = \mathbf{b}. \tag{1.7}$$

Therefore we can multiply both sides of (1.6) by an arbitrary constant $\pm\epsilon$, and then add the result to (1.7) to obtain

$$(\xi_{i_1} \pm \epsilon\gamma_1)\mathbf{a}^{i_1} + \cdots + (\xi_{i_r} \pm \epsilon\gamma_r)\mathbf{a}^{i_r} = \mathbf{b}. \tag{1.8}$$

Since the coordinates ξ_{i_k}, $k = 1, \ldots, r$, are all positive numbers, we can choose ϵ small enough so that $\xi_{i_k} + \epsilon\gamma_k > 0$ and $\xi_{i_k} - \epsilon\gamma_k > 0$, $k = 1, 2, \ldots, r$. Now define vectors $\mathbf{y}$ and $\mathbf{z}$ whose coordinates are η_k and ζ_k, respectively: $\eta_k = \xi_k - \epsilon\gamma_k$ for $k \in \{i_i, \ldots, i_r\}$, $\eta_k = 0$ for $k \notin \{i_1, \ldots, i_r\}$, and $\zeta_k = \xi_k + \epsilon\gamma_k$ for $k \in \{i_1, \ldots, i_r\}$, $\zeta_k = 0$ for $k \notin \{i_1, \ldots, i_r\}$. From Eq. (1.8) we see that $\mathbf{y}$ and $\mathbf{z}$ are in $\mathfrak{F}$. Also, not all the constants γ_i are zero, and so $\mathbf{y} \neq \mathbf{z}$. But, $\frac{1}{2}\mathbf{y} + \frac{1}{2}\mathbf{z} = \mathbf{x}$, because $\frac{1}{2}(\xi_k - \epsilon\gamma_k + \xi_k + \epsilon\gamma_k) = \xi_k$. Hence $\mathbf{x}$ lies on the line between $\mathbf{y}$ and $\mathbf{z}$. This contradicts our original assumption that $\mathbf{x}$ was extreme. Hence our second assumption, namely that the columns $\mathbf{a}^{i_1}, \ldots, \mathbf{a}^{i_r}$ are dependent, has lead to a contradiction. We conclude that these columns are independent, and the proof of part 2 is complete.

3. Exercise 5.
4. Our goal is to show that $\mathfrak{F}$ has only a finite number of extreme points.

We do this by showing that the set of extreme points is not larger than another set which is known to be finite. We have already shown that if $\mathbf{x}$ is an extreme point, then the columns of $\mathbb{A}$ which correspond to the positive coordinates of $\mathbf{x}$ form an independent set. We now show that no two distinct extreme points can have exactly the same coordinates positive. Thus suppose that $\mathbf{x}$ and $\mathbf{y}$ are extreme points and that $\mathbf{x}$ has positive coordinates ξ_{i_k}, $k = 1, \ldots, r$, while $\mathbf{y}$ has positive coordinates η_{i_k}, $k = 1, \ldots, r$. Since $\mathbf{x}$ and $\mathbf{y}$ are feasible, we have

$$\xi_{i_1} \mathbf{a}^{i_1} + \cdots + \xi_{i_r} \mathbf{a}^{i_r} = \mathbf{b},$$
$$\eta_{i_1} \mathbf{a}^{i_1} + \cdots + \eta_{i_r} \mathbf{a}^{i_r} = \mathbf{b},$$

and subtracting, we obtain

$$(\xi_{i_1} - \eta_{i_1})\mathbf{a}^{i_1} + \cdots + (\xi_{i_r} - \eta_{i_r})\mathbf{a}^{i_r} = \mathbf{0}.$$

But the independence of the columns $\mathbf{a}^{i_1}, \ldots, \mathbf{a}^{i_r}$ implies that $\xi_{i_k} = \eta_{i_k}$, $k = 1, \ldots, r$, and that $\mathbf{x} = \mathbf{y}$. Thus we have shown that at most one extreme point can have a certain subset of the coordinates positive. This concludes the proof because there are only finitely many ways of choosing subsets of the n coordinates.
Q.E.D.

We are now ready to discuss the problem of finding maximums and minimums of certains functions over convex sets. The functions of interest to us here are linear functions.

5.1.3 Linear Functions on Convex Sets

Recall that a function f is said to be *linear* if $f(\alpha\mathbf{x} + \beta\mathbf{y}) = \alpha f(\mathbf{x}) + \beta f(\mathbf{y})$ for all constants α, β and vectors $\mathbf{x}$, $\mathbf{y}$ in the domain of f. In this subsection, we are interested in the maximum and minimum values of a linear function which is defined over a certain convex set. The result we seek is that the function always assumes its maximum and minimum values at extreme points of the convex set. We choose to work in the special setting which seems best for our needs, but we hasten to add that the result is also true in a more general context. The interested reader should consult [BeGH] in this regard.

At one point in the proof of the desired result we shall find it necessary to use some facts from analysis. Forms of these results which are appropriate for our use are discussed in the appendix on calculus (Appendix A).

The main result of this section is the following.

Theorem 4: Let f be a linear function defined on a closed, bounded, convex subset $\mathcal{K}$ of R^n. Then f assumes its maximum and minimum values at extreme points of $\mathcal{K}$.

Remark. In the special setting used in Theorem 4 (closed, bounded sets), the theorem is not directly applicable to most linear programming problems because the set of feasible vectors for either the primal problem or dual problem is unbounded. However, it is a relatively simple matter to modify the set of feasible vectors to obtain a bounded set, and it can be shown that the original linear programming problem can be solved by using the modified set of feasible vectors. We shall return to this topic in Sec. 5.2.3 (the comments following Theorem 3).

Proof: Every linear function on R^n is continuous, and every continuous function defined on a closed and bounded set assumes a maximum and minimum value there (see Appendix A). Thus we need only show that there are extreme points where the maximum and minimum values are assumed.

Let $\mathbf{x}_0$ be any point in $\mathcal{K}$ where f assumes a maximum. We first suppose that $\mathbf{x}_0$ is unique, so that at every other point $\mathbf{x}$ of K we have $f(\mathbf{x}) < f(\mathbf{x}_0)$. Then we claim that $\mathbf{x}_0$ is an extreme point of K. Indeed, if $\mathbf{x}_0$ is not an extreme point, then there exist points $\mathbf{y}$ and $\mathbf{z}$ in K such that $\mathbf{x}_0 = \alpha\mathbf{y} + (1 - \alpha)\mathbf{z}$, where $0 < \alpha < 1$. Now, by the linearity of f, we have

$$f(\mathbf{x}_0) = f(\alpha\mathbf{y} + (1 - \alpha)\mathbf{z}) = \alpha f(\mathbf{y}) + (1 - \alpha)f(\mathbf{z}).$$

But $f(\mathbf{y}) < f(\mathbf{x}_0)$ and $f(\mathbf{z}) < f(\mathbf{x}_0)$; hence we obtain $f(\mathbf{x}_0) < \alpha f(\mathbf{x}_0) + (1 - \alpha)f(\mathbf{x}_0) = f(\mathbf{x}_0)$. This contradiction shows that if $\mathbf{x}_0$ is unique, then it is an extreme point, Next, suppose that $\mathbf{x}_0$ is not unique, and let $\mathcal{M} = \{\mathbf{x}: f(\mathbf{x}) = f(\mathbf{x}_0)\}$. Then the set $\mathcal{M}$ is closed and bounded and there is a point $\mathbf{x}_1$ which is at a maximum distance from $\mathbf{x}_0$ in the set $\mathcal{M}$ (this is a consequence of Theorem A.1, Sec. A.3, Appendix A). We now claim that $\mathbf{x}_1$ is an extreme point of $\mathcal{K}$. As above, if $\mathbf{x}_1$ is not an extreme point of $\mathcal{K}$, then there exists α, $0 < \alpha < 1$, and vectors $\mathbf{y}$, $\mathbf{z}$ such that $\mathbf{x}_1 = \alpha\mathbf{y} + (1 - \alpha)\mathbf{z}$. Now $\mathbf{y}$ and $\mathbf{z}$ must both be in the set $\mathcal{M}$, because otherwise we obtain the contradiction

$$\begin{aligned} f(\mathbf{x}_1) = f(\alpha\mathbf{y} + (1 - \alpha)\mathbf{z}) &= \alpha f(\mathbf{y}) + (1 - \alpha)f(\mathbf{z}) \\ &< \alpha f(\mathbf{x}_1) + (1 - \alpha)f(\mathbf{x}_1) = f(\mathbf{x}_1). \end{aligned}$$

Thus, $\mathbf{y}$ and $\mathbf{z}$ are in $\mathcal{M}$ and $\mathbf{x}_1$ lies on the line between $\mathbf{y}$ and $\mathbf{z}$. Actually, $\mathbf{x}_0$, $\mathbf{x}_1$, $\mathbf{y}$, and $\mathbf{z}$ all lie in the same plane π, and the situation is illustrated in Fig. 5–7, where the plane of the figure is taken to be the plane π. But this

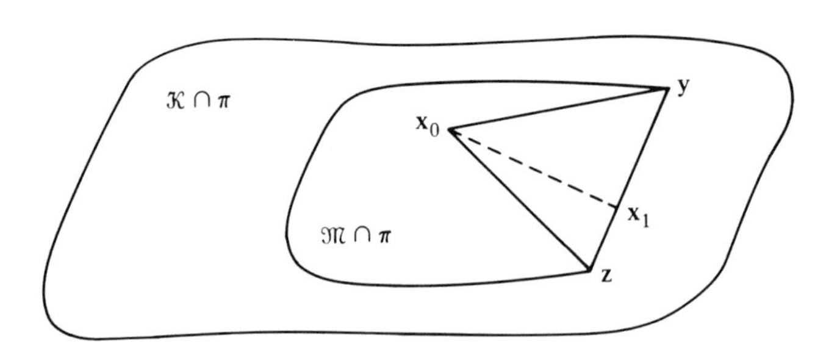

Fig. 5–7

figure indicates that we have a contradiction. Indeed, $\mathbf{x}_1$ was chosen as one of the points in $\mathcal{M}$ farthest from $\mathbf{x}_0$, and yet a simple geometric argument (Exercise 6) shows that one of the points $\mathbf{y}$ or $\mathbf{z}$ must be farther from $\mathbf{x}_0$ than $\mathbf{x}_1$. Hence we conclude that $\mathbf{x}_1$ must be an extreme point, and the proof that the maximum value is assumed at an extreme point is complete. The proof for minimum values follows by looking at the function $g = -f$. The details are left to the reader (Exercise 7). Q.E.D.

Since the function $L(\mathbf{x}) = \mathbf{x} \cdot \mathbf{c}$ is linear and since the set of feasible vectors for a linear optimization problem is a convex set, we have as a direct corollary the following result.

Corollary: Let $\mathbb{A}$ be an $m \times n$ matrix, $\mathbf{b} \in R^m$, and $\mathbf{c} \in R^n$. If the set $\mathfrak{F} = \{\mathbf{x}: \mathbb{A}\mathbf{x} = \mathbf{b}, \mathbf{x} \geq \mathbf{0}\}$ is bounded and nonempty, then the function $L(\mathbf{x}) = \mathbf{x} \cdot \mathbf{c}$ assumes its maximum value at an extreme point of $\mathfrak{F}$.

This concludes our short discussion of linear functions on convex sets. We now return to the geometry of linear optimization where we apply these results.

5.1.4 Linear Programming and Convex Sets

We have derived a number of useful facts about the geometry of linear optimization. First, we know that the set $\mathfrak{F}$ of feasible vectors is a convex set. This is true for both the restricted problem and the standard problem. This convex set is the intersection of either a finite number of hyperplanes or a finite number of half-spaces, depending on the problem being considered. Second, we know that this convex set has a finite number of special points called extreme points and that for $\mathfrak{F}$ bounded and nonempty, the maximum value of $L(\mathbf{x}) = \mathbf{x} \cdot \mathbf{c}$ is always assumed at one of these extreme points. From the definition of extreme point is it intuitively clear that the extreme points lie on the *boundary* of the convex set $\mathfrak{F}$. We could give a rigorous proof of this fact, but first it would be necessary to make the concept of boundary precise by introducing some definitions from topology. A digression into topology would take us rather far afield, and we choose not to proceed in this direction. Our goal here is to convey a feeling for the geometry of linear optimization, and this can be accomplished without being absolutely precise. In the next section, we shall give a rigorous treatment of the algebraic aspects of the process of optimization.

We can now utilize the geometry developed above to give a geometric introduction to the algorithm known as the simplex method. This algorithm is a systematic method for finding an optimal vector for the problem of maximizing $\mathbf{x} \cdot \mathbf{c}$ subject to $\mathbb{A}\mathbf{x} = \mathbf{b}$ and $\mathbf{x} \geq \mathbf{0}$. Naturally, it can find an optimal vector only if one exists, and as a first step the algorithm either determines a feasible vector or else shows that one does not exist. Assuming that the set of feasible vectors is nonempty and that it is bounded, the simplex method determines an optimal vector by systematically checking the extreme points to see if they are optimal. The algorithm proceeds in such a manner that it is not necessary to check every extreme point. In fact, it is designed so that only "better" vectors are checked at successive stages. That is, if $\mathbf{e}_1$ is the first extreme point checked, then a second extreme point $\mathbf{e}_2$ will be checked if and only if $L(\mathbf{e}_2) > L(\mathbf{e}_1)$.

In geometric terms, the algorithm begins at some corner point $\mathbf{e}_1$ of the set $\mathfrak{F}$. It checks whether this point is optimal, or, equivalently, if $L(\mathbf{e}_1) = k$ is the largest choice of k for which $\pi_k \cap \mathfrak{F}$ is not empty. If $\mathbf{e}_1$ is not optimal, then there must exist a corner point $\mathbf{e}_2$, which is farther from the origin, in the sense that it lies on a parallel hyperplane $\pi_{k'}$ for $k' > k$. The algorithm then moves to this corner point $\mathbf{e}_2$ and checks whether it is optimal. If it also is not optimal, the process continues by moving to another "better" extreme point. Since there are only finitely many extreme points, the process must reach an optimal vector in a finite number of steps. In the next section we shall show how this process is carried out in practice. The actual algorithm can be presented independently of any geometry; however, we think it is useful for the reader to keep the geometry in mind.

EXERCISES

1. Show that the inequality $\xi_i \geq 0$ is an equality for a set of points constituting a hyperplane in R^n.
2. Show that hyperplanes of R^n are translates of subspaces of R^n which have dimension $n - 1$; i.e., if $\mathcal{H}$ is a hyperplane, then there exists a vector $\mathbf{t}$ such that $\mathcal{H}_t = \{\mathbf{y}: \mathbf{y} = \mathbf{x} - \mathbf{t}, \mathbf{x} \in \mathcal{H}\}$ is an $n - 1$ dimensional subspace of R^n.
3. Prove the lemma of Sec. 5.1.2.
4. Prove part 1 of Theorem 3.
5. Prove part 3 of Theorem 3. You may use part 2.
6. In the proof of Theorem 4, show that one of the points $\mathbf{y}$ or $\mathbf{z}$ is farther from $\mathbf{x}_0$ than the point $\mathbf{x}_1$. *Hint:* Use the fact that $\mathbf{x}_0$, $\mathbf{x}_1$, $\mathbf{y}$, and $\mathbf{z}$ all lie in one plane, together with some elementary plane geometry.
7. Prove Theorem 4 for minimums by considering the function $g = -f$.
8. Show that the point $(\frac{1}{2}, \frac{1}{2}, \frac{1}{2})$ is an optimal vector for the example in Sec. 5.1.1 by using the dual problem.
9. Solve the following problems by the geometrical method used on the example of Sec. 5.1.1:

   (a) Maximize $\xi_1 + \xi_2$ subject to
$$4\xi_1 + 2\xi_2 \leq 4,$$
$$2\xi_1 + 4\xi_2 \leq 4,$$
$$\xi_1 \geq 0,$$
$$\xi_2 \geq 0.$$

   (b) Maximize $\xi_1 + \xi_2 + \xi_3$ subject to
$$2\xi_1 + 4\xi_2 + 8\xi_3 \leq 8,$$
$$8\xi_1 + 2\xi_2 + 4\xi_3 \leq 8,$$
$$4\xi_1 + 8\xi_2 + 2\xi_3 \leq 8,$$

$$\xi_1 \geq 0,$$
$$\xi_2 \geq 0,$$
$$\xi_3 \geq 0.$$

10. Solve the following standard maximum problems:

(a) Maximize $\mathbf{x} \cdot \mathbf{c}$ subject to $\mathbb{A}\mathbf{x} \leq \mathbf{b}$, $\mathbf{x} \geq \mathbf{0}$, where

$$\mathbb{A} = \begin{bmatrix} 1 & 4 \\ 4 & 1 \\ 1 & 1 \end{bmatrix}, \qquad \mathbf{b} = \begin{pmatrix} 4 \\ 4 \\ 3/2 \end{pmatrix}, \qquad \mathbf{c} = (1, 8).$$

(b) Maximize $\mathbf{x} \cdot \mathbf{c}$ subject to $\mathbb{A}\mathbf{x} \leq \mathbf{b}$, $\mathbf{x} \geq \mathbf{0}$, where

$$\mathbb{A} = \begin{bmatrix} 1 & 4 \\ 4 & 1 \\ 1 & 1 \\ 2/3 & 2/5 \end{bmatrix}, \qquad \mathbf{b} = \begin{pmatrix} 4 \\ 4 \\ 2 \\ 1 \end{pmatrix}, \qquad \mathbf{c} = (1, 1).$$

5.2 COMPUTATIONAL ASPECTS OF LINEAR OPTIMIZATION

In this section we digress somewhat from the main theme of the book which is model building and model development. We consider here the purely computational problem of efficiently finding feasible and optimal vectors for linear programming problems. A similar digression would be possible in connection with almost every model in the text. Indeed, the computational aspects are of prime importance in the use and testing of a model, and the reader should be aware of this aspect of the study. However, in general such considerations would add substantially to the length and depth of the book and are omitted. We make an exception in this case for a number of reasons. First, linear programming models are very widely used, and much of their popularity stems from the nature of the algorithms available for use with these models. These algorithms are relatively simple to understand and they are very adaptable to machine computation. Second, the algorithms for linear programming are also useful in the study of certain models involving competition. These models fall into the subject known as game theory and they will be treated in Chap. 6. Finally, and perhaps most basic, we present this material as an illustration of the type of consideration that could be given to related problems for other models.

In light of the above remarks it is sufficient for our purposes to limit our study of the computational aspects of linear optimization to the well-known simplex method. We begin the discussion of this algorithm by once again considering the transportation problem which was discussed in both Chaps. 2 and 4. In Sec. 2.2, we derived a method for finding an optimal shipping schedule for the transportation problem with two supply areas and two sale areas in the special case where supply and demand are equal. In the derivation a process of elimination is used to reduce the number of variables to 1. Then this variable is determined by examining the quantity which is to be minimized. Thus the solution is carried out in steps. First one finds all shipping schedules which fill the demand and are consistent with the supply available. Next, a "best" or optimal shipping schedule is determined. We seek a method for carrying out a similar process for more general problems. As a first step in obtaining this method we discuss another specific example, this time using the oil refinery setting of Chap. 4.

Suppose for this illustration that there are two oil refineries denoted by R_1 and R_2 and two terminals denoted by T_1 and T_2. We suppose that the production at refineries R_1 and R_2 is 1700 and 800 barrels per day, respectively, and that the demand at terminals T_1 and T_2 is 1000 and 1500 barrels per day, respectively. In addition, if the transportation cost to move 100 barrels of oil from refinery R_i to terminal T_j is denoted by c_{ij}, then we assume that $c_{11} =$ \$8, $c_{12} =$ \$12, $c_{21} =$ \$10, and $c_{22} =$ \$15. Also, as in our previous discussions of transportation problems, if it costs D dollars to move b barrels of oil, then we assume that it costs kD dollars to move kb barrels, for any $k > 0$.

We first phrase this problem as a linear programming problem. To this end, we let $\mathbf{y} = (\eta_1, \eta_2, \eta_3, \eta_4)$ be the unknown shipping schedule; i.e., η_1 is the number of hundreds of barrels of oil to be shipped from refinery R_1 to terminal T_1, η_2 is the number from refinery R_1 to terminal T_2, η_3 is the number from refinery R_2 to terminal T_1, and η_4 is the number from refinery R_2 to terminal T_2. We also let $\mathbf{f} = (8, 12, 10, 15)$ and $\mathbf{d} = (10, 15, 17, 8)$. Then the problem is to find a $\mathbf{y}$ such that $\mathbf{y} \geq \mathbf{0}$, $\mathbb{B}\mathbf{y} = \mathbf{d}$, and $L(\mathbf{y}) = \mathbf{y} \cdot \mathbf{f}$ is a minimum for such $\mathbf{y}$. Here $\mathbb{B}$ is the matrix given by

$$\mathbb{B} = \begin{bmatrix} 1 & 0 & 1 & 0 \\ 0 & 1 & 0 & 1 \\ 1 & 1 & 0 & 0 \\ 0 & 0 & 1 & 1 \end{bmatrix}.$$

We are seeking a systematic approach to this problem and, subsequently, to more complicated problems. It is useful to subdivide the problem into a graduated set of subproblems in the following manner:

1. Develop a method which can be used to find a vector $\mathbf{y}$ such that $\mathbb{B}\mathbf{y} = \mathbf{d}$.
2. Develop a method which can be used to find a nonnegative vector $\mathbf{y}$ such that $\mathbb{B}\mathbf{y} = \mathbf{d}$.
3. Develop a method which can be used to find a nonnegative vector $\mathbf{y}$ such that $\mathbb{B}\mathbf{y} = \mathbf{d}$ and in addition $L(\mathbf{y}) = \mathbf{y} \cdot \mathbf{f}$ is a minimum for such $\mathbf{y}$.

We first give our attention to Problem 1. This problem, which the reader is likely to have encountered previously, is of great independent interest. Indeed, it is the question of finding a method of solving systems of linear equations. The method which is usually used in elementary treatments of such problems is known as the Gauss elimination method. It consists of systematically modifying the system of equations by successively eliminating variables from some of the equations. This procedure is carried out so that the solution set can ultimately be obtained by inspection of the final modified form of the equations. We now consider this elimination process from a different point of view. Our aim is to develop a modified process which will yield optimal vectors for linear programming problems.

5.2.1 Systems of Equations and the Replacement Method

Let $\mathbf{b}^1$, $\mathbf{b}^2$, $\mathbf{b}^3$, and $\mathbf{b}^4$ be the four columns of the matrix $\mathbb{B}$. To solve the problem $\mathbb{B}\mathbf{y} = \mathbf{d}$, we need a $\mathbf{y} = (\eta_1, \eta_2, \eta_3, \eta_4)$ such that $\mathbf{d} = \eta_1\mathbf{b}^1 + \eta_2\mathbf{b}^2 + \eta_3\mathbf{b}^3 + \eta_4\mathbf{b}^4$. In other words, we must write $\mathbf{d}$ as a linear combination of the column vectors of $\mathbb{B}$. One method of obtaining such a linear combination is first to write $\mathbf{d}$ as a linear combination of some other set of vectors and then express this second set of vectors in terms of the column vectors. This is the procedure we shall follow. However, before we proceed, it is useful to introduce a new definition and some additional notation.

Definition: Let $V = \{\mathbf{v}_1, \mathbf{v}_2, \ldots, \mathbf{v}_p\}$ be an independent ordered set of vectors in R^n, and let $W = \{\mathbf{w}_1, \mathbf{w}_2, \ldots, \mathbf{w}_q\}$ be another ordered set of vectors, each of which is a linear combination of the vectors of V. Then the *tableau* for the set W with respect to the set V is the $p \times q$ array $T = (t_{ij})$, where t_{ij} is defined by the expression

$$\mathbf{w}_j = \sum_{i=1}^{p} t_{ij}\mathbf{v}_i, \qquad j = 1, 2, \ldots, q.$$

Pictorially, we represent the tableau as follows:

$$T = \begin{array}{c|cccc|} & \mathbf{w}_1 & \mathbf{w}_2 & \cdots & \mathbf{w}_q \\ \hline \mathbf{v}_1 & t_{11} & t_{12} & \cdots & t_{1q} \\ \mathbf{v}_2 & t_{21} & t_{22} & \cdots & t_{2q} \\ \cdot & \cdot & \cdot & & \cdot \\ \cdot & \cdot & \cdot & & \cdot \\ \cdot & \cdot & \cdot & & \cdot \\ \mathbf{v}_p & t_{p1} & t_{p2} & & t_{pq} \\ \hline \end{array}$$

Returning to our discussion of the problem $\mathbb{B}\mathbf{y} = \mathbf{d}$, there is a natural choice for the original vectors to use in expressing $\mathbf{d}$, namely, the unit column vectors given by

$$\mathbf{u}^1 = \begin{pmatrix} 1 \\ 0 \\ 0 \\ 0 \end{pmatrix}, \quad \mathbf{u}^2 = \begin{pmatrix} 0 \\ 1 \\ 0 \\ 0 \end{pmatrix}, \quad \mathbf{u}^3 = \begin{pmatrix} 0 \\ 0 \\ 1 \\ 0 \end{pmatrix}, \quad \text{and} \quad \mathbf{u}^4 = \begin{pmatrix} 0 \\ 0 \\ 0 \\ 1 \end{pmatrix}.$$

Then $\mathbf{d} = 10\mathbf{u}^1 + 15\mathbf{u}^2 + 17\mathbf{u}^3 + 8\mathbf{u}^4$.

The next step is to replace the vectors $\mathbf{u}^i$ by the column vectors $\mathbf{b}^j$. Naturally, as vectors $\mathbf{u}^i$ are replaced by vectors $\mathbf{b}^j$, the coefficients in the linear expression for $\mathbf{d}$ will change. Also, in general, we can only make a change of $\mathbf{b}^j$ for $\mathbf{u}^i$ if the resulting set of vectors is still an independent set. Otherwise, it may not be possible to write $\mathbf{d}$ in terms of the new set of vectors. Phrased in terms of the two sets V and W of the above definition, we have the following result about such replacements.

Theorem 1 (The Replacement Theorem): If $t_{lm} \neq 0$ in the tableau T, then the ordered set $V' = \{\mathbf{v}_1, \mathbf{v}_2, \ldots, \mathbf{v}_{l-1}, \mathbf{w}_m, \mathbf{v}_{l+1}, \ldots, \mathbf{v}_p\}$ is an independent set of vectors. Moreover, the entries in the tableau T', for the set W with respect to the set V', are given by

$$t'_{ij} = \begin{cases} t_{ij} - \dfrac{t_{im}}{t_{lm}} t_{lj}, & i \neq l, \\[2ex] \dfrac{t_{lj}}{t_{lm}}, & i = l. \end{cases}$$

Proof: First we show that the vectors of the set V' form an independent set. Thus, suppose that there exist real numbers $\{\lambda_i\}_{i=1}^p$ such that $\sum_{i \neq l} \lambda_i \mathbf{v}_i + \lambda_l \mathbf{w}_m = \mathbf{0}$. Then, since $\mathbf{w}_m = \sum_{i=1}^p t_{im}\mathbf{v}_i$, we have

$$\sum_{i \neq l} \lambda_i \mathbf{v}_i + \lambda_l \sum_{i=1}^p t_{im} \mathbf{v}_i = \mathbf{0}.$$

Equivalently,

$$\sum_{i \neq l} (\lambda_i + \lambda_l t_{im}) \mathbf{v}_i + \lambda_l t_{lm} \mathbf{v}_l = \mathbf{0}.$$

Now the set V is independent, so each of the coefficients $(\lambda_i + \lambda_l t_{im})$, $i \neq l$, and $\lambda_l t_{lm}$ is zero. Since $t_{lm} \neq 0$, we have $\lambda_l = 0$, and consequently $\lambda_i = 0$, $i = 1, \ldots, p$. Therefore the set V' is an independent set of vectors.

Next, we must show that when $\mathbf{w}_j$ is written in terms of the vectors of the set V' the coefficients are given by the t'_{ij}s of the theorem. We show this by taking the sum $\sum_{i \neq l} t'_{ij}\mathbf{v}_i + t'_{lj}\mathbf{w}_m$ and showing that it is, in fact, $\mathbf{w}_j$. The steps are as follows:

$$\begin{aligned}
\sum_{i \neq l} t'_{ij}\mathbf{v}_i + t'_{lj}\mathbf{w}_m &= \sum_{i \neq l}\left[t_{ij} - \frac{t_{im}}{t_{lm}}t_{lj}\right]\mathbf{v}_i + \frac{t_{lj}}{t_{lm}}\mathbf{w}_m \\
&= \sum_{\substack{i=1 \\ i \neq l}}^{p}\left[t_{ij} - \frac{t_{im}}{t_{lm}}t_{lj}\right]\mathbf{v}_i + \frac{t_{lj}}{t_{lm}}\left[\sum_{i=1}^{p} t_{im}\mathbf{v}_i\right] \\
&= \sum_{i=1}^{p} t_{ij}\mathbf{v}_i + \sum_{\substack{i=1 \\ i \neq l}}^{p}\left[\frac{t_{im}}{t_{lm}}t_{lj} - \frac{t_{im}}{t_{lm}}t_{lj}\right]\mathbf{v}_i \\
&= \mathbf{w}_j.
\end{aligned}$$

Remarks. The replacement operation of Theorem 1 is exactly the same as the elimination operation traditionally used in solving systems of equations. For example, suppose that $\mathbb{A} = (a_{ij})$ is the matrix of coefficients for a system of equations $\mathbb{A}\mathbf{x} = \mathbf{b}$. The first step in the elimination process is to choose an equation and a variable which has a nonzero coefficient in that equation. Then, using this coefficient, one eliminates this variable from all other equations of the system. If $a_{lm} \neq 0$ is the nonzero coefficient which is chosen, then to eliminate the mth variable from equation i, one multiplies equation l by $[-(a_{im}/a_{lm})]$ and adds this equation to equation i. In the new ith equation, the coefficient of the jth variable is given by $a_{ij} - (a_{im}/a_{lm})a_{lj}$. Thus we see that this corresponds exactly to obtaining the entries in the tableau T' as given by Theorem 1.

We now return to the transportation problem and use the replacement technique to solve the problem $\mathbb{B}\mathbf{y} = \mathbf{d}$. We take the unit column vectors $\{\mathbf{u}^i\}$ to be the set V, and the column vectors of $\mathbb{B}$ are the first $q - 1$ vectors of the set W. The last vector, $\mathbf{w}_q$, is chosen to be the vector $\mathbf{d}$. Thus, $p = 4$, $q = 5$, and the initial tableau is given by

$T^{(0)} =$

	$\mathbf{b}^1$	$\mathbf{b}^2$	$\mathbf{b}^3$	$\mathbf{b}^4$	$\mathbf{d}$
$\mathbf{u}^1$	1	0	1	0	10
$\mathbf{u}^2$	0	1	0	1	15
$\mathbf{u}^3$	1	1	0	0	17
$\mathbf{u}^4$	0	0	1	1	8

$= (t_{ij}^{(0)}).$

Here the superscript 0 is used to indicate that this is the initial tableau. As we proceed through the process, we shall use the superscript to indicate the stage currently being considered. Thus after one step we shall have a new matrix T, which will be denoted by $T^{(1)}$.

We distinguish the vector $\mathbf{d}$ from the vectors $\mathbf{b}^i$ because it plays a different role in the use of the tableau. We are interested in replacing the vectors $\mathbf{u}^i$ by vectors $\mathbf{b}^j$. The vector $\mathbf{d}$ cannot take part in the replacement operation. Instead, we are interested in the coefficients for the expansion of $\mathbf{d}$ in terms of the vectors at the left of the tableau. These coefficients occur in the column under $\mathbf{d}$, and when we have replaced each of the vectors $\mathbf{u}^i$ by vectors $\mathbf{b}^j$, then these coefficients will give the solution to the problem $\mathbb{B}\mathbf{y} = \mathbf{d}$.

Since $t_{11}^{(0)} = 1 \neq 0$, we first replace $\mathbf{u}^1$ by $\mathbf{b}^1$. We call the new tableau $T^{(1)}$, and by Theorem 1 it is given by

$T^{(1)} = (t_{ij}^{(1)}) =$

	$\mathbf{b}^1$	$\mathbf{b}^2$	$\mathbf{b}^3$	$\mathbf{b}^4$	$\mathbf{d}$
$\mathbf{b}^1$	1	0	1	0	10
$\mathbf{u}^2$	0	1	0	1	15
$\mathbf{u}^3$	0	1	−1	0	7
$\mathbf{u}^4$	0	0	1	1	8

Next, $t_{22}^{(1)} = 1 \neq 0$, and so we replace $\mathbf{u}^2$ by $\mathbf{b}^2$. The resulting tableau, $T^{(2)}$, is given by

$T^{(2)} = (t_{ij}^{(2)}) =$

	$\mathbf{b}^1$	$\mathbf{b}^2$	$\mathbf{b}^3$	$\mathbf{b}^4$	$\mathbf{d}$
$\mathbf{b}^1$	1	0	1	0	10
$\mathbf{b}^2$	0	1	0	1	15
$\mathbf{u}^3$	0	0	−1	−1	−8
$\mathbf{u}^4$	0	0	1	1	8

We pause at this point to remind the reader of the information carried in the last column of the tableau. For example, this column in tableau $T^{(2)}$ tells us that

$$\mathbf{d} = 10\mathbf{b}^1 + 15\mathbf{b}^2 - 8\mathbf{u}^3 + 8\mathbf{u}^4.$$

Thus we do not yet have $\mathbf{d}$ as a linear combination of the vectors $\mathbf{b}^i$, and hence our replacement process is not complete.

Continuing the process, since $t_{43}^{(2)} = 1 \neq 0$, we replace $\mathbf{u}^4$ by $\mathbf{b}^3$. $T^{(3)}$ is then given by

$T^{(3)} = (t_{ij}^{(3)}) =$

	$\mathbf{b}^1$	$\mathbf{b}^2$	$\mathbf{b}^3$	$\mathbf{b}^4$	$\mathbf{d}$
$\mathbf{b}^1$	1	0	0	−1	2
$\mathbf{b}^2$	0	1	0	1	15
$\mathbf{u}^3$	0	0	0	0	0
$\mathbf{b}^3$	0	0	1	1	8

At this point we stop. After three applications of our procedure we have a solution of the problem $\mathbb{B}\mathbf{y} = \mathbf{d}$. This is true, even though we have not replaced $\mathbf{u}^3$, because the coefficient of $\mathbf{u}^3$ is zero in the expansion of $\mathbf{d}$. In fact, we have $\mathbf{d} = 2\mathbf{b}^1 + 15\mathbf{b}^2 + 0\mathbf{u}^3 + 8\mathbf{b}^3$. Our conclusion is that one solution of the system of equations $\mathbb{B}\mathbf{y} = \mathbf{d}$ is given by $\mathbf{y} = (2, 15, 8, 0)$. Note that the solution vector is *not* $(2, 15, 0, 8)$ because the entries at the left of tableau $T^{(3)}$ are not in their natural order. The solution vector must be formed by taking into account the order of the vectors at the left of the tableau.

Soon we shall refine this replacement technique so that it can be applied to optimization problems. However, we first give another example of using the method to solve a system of equations. Our example from the transportation problem was atypical because the entries in the coefficient matrix were all 0s and 1s. Our second example will be more typical in this respect. Also, it will naturally lead into the refinements which are needed. The problem to be considered is that of finding a column vector $\mathbf{y}$ which minimizes $\mathbf{y} \cdot \mathbf{b}$ subject to the constraints $\mathbf{y} \geq \mathbf{0}$, $\mathbb{A}\mathbf{y} = \mathbf{c}$, where $\mathbf{b} = (3, -2, 5, 2)$,

$$\mathbf{c} = \begin{pmatrix} 10 \\ 5 \\ 17 \end{pmatrix}, \quad \text{and} \quad \mathbb{A} = \begin{bmatrix} 1 & 2 & 0 & 5 \\ 0 & 1 & 1 & 2 \\ 1 & 3 & 2 & 8 \end{bmatrix}.$$

We begin by finding a column vector $\mathbf{y}$ which satisfies $\mathbb{A}\mathbf{y} = \mathbf{c}$. We use the replacement method, consecutively replacing $\mathbf{u}^1$ by $\mathbf{a}^1$, $\mathbf{u}^2$ by $\mathbf{a}^2$, and $\mathbf{u}^3$ by $\mathbf{a}^3$. The tableaus are the following:

$$T^{(0)} = \begin{array}{c|cccc|c|} & \mathbf{a}^1 & \mathbf{a}^2 & \mathbf{a}^3 & \mathbf{a}^4 & \mathbf{c} \\ \hline \mathbf{u}^1 & 1 & 2 & 0 & 5 & 10 \\ \mathbf{u}^2 & 0 & 1 & 1 & 2 & 5 \\ \mathbf{u}^3 & 1 & 3 & 2 & 8 & 17 \\ \hline \end{array}, \qquad T^{(1)} = \begin{array}{c|cccc|c|} & \mathbf{a}^1 & \mathbf{a}^2 & \mathbf{a}^3 & \mathbf{a}^4 & \mathbf{c} \\ \hline \mathbf{a}^1 & 1 & 2 & 0 & 5 & 10 \\ \mathbf{u}^2 & 0 & 1 & 1 & 2 & 5 \\ \mathbf{u}^3 & 0 & 1 & 2 & 3 & 7 \\ \hline \end{array},$$

$$T^{(2)} = \begin{array}{c|cccc|c|} & \mathbf{a}^1 & \mathbf{a}^2 & \mathbf{a}^3 & \mathbf{a}^4 & \mathbf{c} \\ \hline \mathbf{a}^1 & 1 & 0 & -2 & 1 & 0 \\ \mathbf{a}^2 & 0 & 1 & 1 & 2 & 5 \\ \mathbf{u}^3 & 0 & 0 & 1 & 1 & 2 \\ \hline \end{array}, \qquad T^{(3)} = \begin{array}{c|cccc|c|} & \mathbf{a}^1 & \mathbf{a}^2 & \mathbf{a}^3 & \mathbf{a}^4 & \mathbf{c} \\ \hline \mathbf{a}^1 & 1 & 0 & 0 & 3 & 4 \\ \mathbf{a}^2 & 0 & 1 & 0 & 1 & 3 \\ \mathbf{a}^3 & 0 & 0 & 1 & 1 & 2 \\ \hline \end{array}.$$

Therefore, $\mathbf{y} = (4, 3, 2, 0)$ is a solution of the system $\mathbb{A}\mathbf{y} = \mathbf{c}$.

This process of replacement raises many questions in regard to its applicability for linear programming problems. First, it is highly nonunique, and by making different replacements, other solutions will be obtained. Thus it seems very likely that the solution may, in fact, have some coordinates which are negative. This did not happen in the examples given here; however, it could have happened. In fact, if the replacements $\mathbf{a}^1 \longrightarrow \mathbf{u}^3$, $\mathbf{a}^2 \longrightarrow \mathbf{u}^2$, and $\mathbf{a}^4 \longrightarrow \mathbf{u}^1$ are made in our second example, then the solution obtained is $\mathbf{y} = (-2, 1, 0, 2)$. Since linear programming problems which we consider usually require that the feasible vectors be nonnegative, we shall have to refine the replacement method so that it will yield nonnegative solutions of systems of equations when such solutions exist. Also, even assuming that we have a method for obtaining feasible vectors, we still need a method for obtaining optimal vectors. It is this last problem to which we turn first. Therefore, assuming for the moment that we are able to obtain feasible vectors for linear programming problems, we shall show how to obtain optimal vectors by means of the replacement method. As is our custom, we first consider an example and then proceed to generalize our results.

5.2.2 Recognizing and Obtaining Optimal Vectors

Returning to the above example, we recall that the use of the replacement method has led to the vector $\mathbf{y}_0 = (4, 3, 2, 0)$ as a solution to the system $\mathbb{A}\mathbf{y} = \mathbf{c}$. We are interested in the problem of minimizing $\mathbf{y} \cdot \mathbf{b}$ subject to $\mathbf{y} \geq \mathbf{0}$ and $\mathbb{A}\mathbf{y} = \mathbf{c}$. Hence this particular $\mathbf{y}$ is a feasible vector for the minimization problem. *Is it an optimal vector*? To check whether this vector is optimal, we must compare the value of $\mathbf{y}_0 \cdot \mathbf{b}$ with $\mathbf{y} \cdot \mathbf{b}$ for other feasible vectors $\mathbf{y}$. For us to make such a comparison, we must know what the other feasible vectors look like. Since the last coordinate in the vector $\mathbf{y}_0$ is 0, it follows that the expansion of $\mathbf{c}$ in terms of the columns of the matrix $\mathbb{A}$ which is defined by $\mathbf{y}_0$ does not actually contain $\mathbf{a}^4$. In fact, this expansion is

$$\mathbf{c} = 4\mathbf{a}^1 + 3\mathbf{a}^2 + 2\mathbf{a}^3 + 0\mathbf{a}^4.$$

Inspection of $\mathbf{a}^1$, $\mathbf{a}^2$, and $\mathbf{a}^3$ shows that these vectors are independent and hence that this is the only expansion of $\mathbf{c}$ which uses only $\mathbf{a}^1$, $\mathbf{a}^2$, and $\mathbf{a}^3$. Therefore all other feasible vectors, if any exist, yield expansions in which $\mathbf{a}^4$ has a nonzero coefficient. We introduce the vector $\mathbf{a}^4$ formally into the above expansion for $\mathbf{c}$ by adding and subtracting the quantity $\eta_4\mathbf{a}^4$. Here the quantity η_4 is an arbitrary nonnegative real number. We have

$$\mathbf{c} = (4\mathbf{a}^1 + 3\mathbf{a}^2 + 2\mathbf{a}^3 - \eta_4\mathbf{a}^4) + \eta_4\mathbf{a}^4.$$

It is useful at this point to write the vector $\mathbf{a}^4$ occurring in the parentheses in an equivalent form using $\mathbf{a}^1$, $\mathbf{a}^2$, and $\mathbf{a}^3$. Recalling the discussion of the preceding section, we know that the appropriate form is given by the column under $\mathbf{a}^4$ in the tableau which yielded $\mathbf{y}_0$. Thus, $\mathbf{a}^4 = 3\mathbf{a}^1 + 1\mathbf{a}^2 + 1\mathbf{a}^3$, and

$$\begin{aligned} \mathbf{c} &= [4\mathbf{a}^1 + 3\mathbf{a}^2 + 2\mathbf{a}^3 - \eta_4(3\mathbf{a}^1 + 1\mathbf{a}^2 + 1\mathbf{a}^3)] + \eta_4\mathbf{a}^4 \\ &= (4 - 3\eta_4)\mathbf{a}^1 + (3 - \eta_4)\mathbf{a}^2 + (2 - \eta_4)\mathbf{a}^3 + \eta_4\mathbf{a}^4. \end{aligned} \tag{2.1}$$

The coefficient η_4 is arbitrary at this point, and we would like to choose it in such a way that it yields a feasible vector $\mathbf{y}$ for which $\mathbf{y} \cdot \mathbf{b} < \mathbf{y}_0 \cdot \mathbf{b}$. For $\mathbf{y}$ to be feasible, the coefficients of each column $\mathbf{a}^1$, $\mathbf{a}^2$, $\mathbf{a}^3$, and $\mathbf{a}^4$ must be nonnegative: i.e., $\mathbf{y} = (4 - 3\eta_4, 3 - \eta_4, 2 - \eta_4, \eta_4) \geq \mathbf{0}$. Next, we compute $\mathbf{y} \cdot \mathbf{b}$ using $\mathbf{y}$ as written above. We have

$$\begin{aligned} \mathbf{y} \cdot \mathbf{b} &= 3(4 - 3\eta_4) - 2(3 - \eta_4) + 5(2 - \eta_4) + 2\eta_4 \\ &= 12 - 6 + 10 - (3\cdot 3 - 2\cdot 1 + 5\cdot 1 - 2)\eta_4 \\ &= 16 - [(3\cdot 3 - 2\cdot 1 + 5\cdot 1) - 2]\eta_4. \end{aligned} \tag{2.2}$$

We note that actually $\mathbf{y} \cdot \mathbf{b} = 16 - 10\eta_4$, although the form (2.2) will be more useful later.

At this point we seek to compare $\mathbf{y} \cdot \mathbf{b}$ with $\mathbf{y}_0 \cdot \mathbf{b}$. It is useful to introduce some new terminology for use in carrying out this comparison. Although we introduce these terms in the setting currently under consideration, they will be of value in discussing the general situation as well.

Definition: In the problem of minimizing $\mathbf{y} \cdot \mathbf{b}$ subject to $\mathbb{A}\mathbf{y} = \mathbf{c}$ and $\mathbf{y} \geq \mathbf{0}$, we say that the *actual cost* of column $\mathbf{a}^j$ of $\mathbb{A}$ is β_j, the jth coordinate of $\mathbf{b}$.

Before stating our next definition, it is useful to establish a convention. Let I be a subset of the set of integers $\{1, 2, \ldots, q\}$. By $\{\mathbf{a}^i\}_{i \in I}$ we mean the subset of $\{\mathbf{a}^1, \ldots, \mathbf{a}^q\}$ consisting of those elements whose superscript is contained in I. Summation is treated similarly.

Definition: If column $\mathbf{a}^j$ can be written in terms of an independent set $\{\mathbf{a}^i\}_{i \in I}$ of columns of $\mathbb{A}$,

$$\mathbf{a}^j = \sum_{i \in I} t_{ij} \mathbf{a}^i,$$

then the *equivalent cost of* $\mathbf{a}^j$, *in terms of* $\{\mathbf{a}^i\}_{i \in I}$, is

$$\nu_j = \sum_{i \in I} t_{ij} \beta_i.$$

We now make use of these definitions in the problem under consideration. The actual cost of column $\mathbf{a}^4$ is $\beta_4 = 2$. The equivalent cost of $\mathbf{a}^4$, in terms of $\mathbf{a}^1$, $\mathbf{a}^2$, and $\mathbf{a}^3$, is $\nu_4 = 3\cdot 3 + 1\cdot(-2) + 1\cdot 5 = 12$. This follows at once from the fact that $\mathbf{a}^4 = 3\mathbf{a}^1 + 1\mathbf{a}^2 + 1\cdot\mathbf{a}^3$. Therefore we can write the expression $\mathbf{y} \cdot \mathbf{b}$ in the form

$$\mathbf{y} \cdot \mathbf{b} = 16 - (\nu_4 - \beta_4)\eta_4.$$

We are attempting to minimize the quantity $\mathbf{y} \cdot \mathbf{b}$, and we immediately see that if $\nu_4 > \beta_4$, then it is to our advantage to choose $\eta_4 > 0$. Also, if

$\nu_4 < \beta_4$, then in order to have $\mathbf{y} \cdot \mathbf{b}$ as small as possible, η_4 must be zero, since it cannot be negative. In the case under consideration $\nu_4 = 12 > 2 = \beta_4$, and thus we should use a vector $\mathbf{y}$ for which $\eta_4 > 0$.

The above discussion has given us a criterion for the introduction of column $\mathbf{a}^4$ into the expansion of $\mathbf{c}$. The test involves a comparison of the actual cost of $\mathbf{a}^4$ and the equivalent cost of $\mathbf{a}^4$ in terms of the columns currently used in the expansion of $\mathbf{c}$. If the actual cost is less than the equivalent cost, then $\mathbf{a}^4$ should be introduced (i.e., η_4 should be taken to be positive). If the actual cost is not less than the equivalent cost, then $\mathbf{a}^4$ should not be introduced (i.e., η_4 should be taken equal to 0).

We have now shown that the vector $\mathbf{y}$ should have a fourth coordinate, η_4, which is positive. A natural question to ask at this point is, How should η_4 be chosen? To answer this question, we must consider the manner in which η_4 affects the coefficients of $\mathbf{a}^1$, $\mathbf{a}^2$, and $\mathbf{a}^3$. Recall Eq. (2.1), and note that as η_4 increases, the coefficients of $\mathbf{a}^1$, $\mathbf{a}^2$, and $\mathbf{a}^3$ decrease. We also have $\mathbf{y} \cdot \mathbf{b} = 16 - (12 - 2)\eta_4 = 16 - 10\eta_4$. Thus $\mathbf{y} \cdot \mathbf{b}$ continues to decrease as long as η_4 increases. However, we are constrained in choosing η_4 by the requirement that the coefficients of $\mathbf{a}^1$, $\mathbf{a}^2$, and $\mathbf{a}^3$, i.e., η_1, η_2, and η_3, must be nonnegative. Since these coefficients decrease as η_4 increases, we are led to increase η_4 until we reach a value of η_4 for which one of the coefficients η_1, η_2, or η_3 has become 0 and any further increase of η_4 will make it negative. Inspection of η_1, η_2, and η_3 shows that we should choose $\eta_4 = 4/3$. This gives $\eta_1 = 0$, $\eta_2 = 5/3$, and $\eta_3 = 2/3$, or, equivalently, $\mathbf{y}_1 = (0, 5/3, 2/3, 4/3)$ is a feasible vector for which $\mathbf{y}_1 \cdot \mathbf{b} < \mathbf{y}_0 \cdot \mathbf{b}$.

The vector $\mathbf{y}_1$ can also be obtained by our replacement method. It corresponds to replacing $\mathbf{a}^1$ by $\mathbf{a}^4$ in the tableau for $\mathbf{y}_0$. The two tableaus are as follows:

$$T^{(3)} = \begin{array}{c} \\ \mathbf{a}^1 \\ \mathbf{a}^2 \\ \mathbf{a}^3 \end{array} \begin{array}{cccc|c} \mathbf{a}^1 & \mathbf{a}^2 & \mathbf{a}^3 & \mathbf{a}^4 & \mathbf{c} \\ \hline 1 & 0 & 0 & 3 & 4 \\ 0 & 1 & 0 & 1 & 3 \\ 0 & 0 & 1 & 1 & 2 \end{array}, \qquad T^{(4)} = \begin{array}{c} \\ \mathbf{a}^4 \\ \mathbf{a}^2 \\ \mathbf{a}^3 \end{array} \begin{array}{cccc|c} \mathbf{a}^1 & \mathbf{a}^2 & \mathbf{a}^3 & \mathbf{a}^4 & \mathbf{c} \\ \hline 1/3 & 0 & 0 & 1 & 4/3 \\ -1/3 & 1 & 0 & 0 & 5/3 \\ -1/3 & 0 & 1 & 0 & 2/3 \end{array}.$$

We note that $\mathbf{y}_1 \cdot \mathbf{b} = \frac{8}{3} < 16 = \mathbf{y}_0 \cdot \mathbf{b}$, so that $\mathbf{y}_1$ is actually better than $\mathbf{y}_0$ in terms of the linear programming problem.

We now want to generalize the process which was used above to obtain $\mathbf{y}_1$ from $\mathbf{y}_0$. In general, we consider the problem of minimizing $\mathbf{y} \cdot \mathbf{b}$ subject to $\mathbf{y} \geq \mathbf{0}$ and $\mathbb{A}\mathbf{y} = \mathbf{c}$. Here we assume that $\mathbb{A}$ is $p \times q$, $\mathbf{c}$ is $p \times 1$, and both $\mathbf{y}$ and $\mathbf{b}$ are $1 \times q$. We also assume that a feasible vector $\mathbf{y}_0$ is given. Finally, we assume that this feasible vector gives a coefficient of 0 to column $\mathbf{a}^m$ in the expansion of $\mathbf{c}$ defined by $\mathbb{A}\mathbf{y}_0 = \mathbf{c}$ and $t_{km} > 0$ for some k. Then, to decide if the column $\mathbf{a}^m$ should be in the expansion for $\mathbf{c}$, we compare the actual cost β_m with the equivalent cost ν_m. The tableau looks as follows:

$T =$

	$\mathbf{a}^1$	$\mathbf{a}^2$	$\cdots$	$\mathbf{a}^m$	$\cdots$	$\mathbf{a}^q$	$\mathbf{c}$
$\mathbf{a}^{i_1}$	t_{11}	t_{12}	$\cdots$	t_{1m}	$\cdots$	t_{1q}	η_{i_1}
$\mathbf{a}^{i_2}$	t_{21}	t_{22}	$\cdots$	t_{2m}	$\cdots$	t_{2q}	η_{i_2}
$\vdots$	$\vdots$					$\vdots$	$\vdots$
$\mathbf{a}^{i_p}$	t_{p1}	t_{p2}	$\cdots$	t_{pm}	$\cdots$	t_{pq}	η_{i_p}

where the index set of the definition of equivalent cost is given by $I = \{i_1, i_2, \ldots, i_p\}$. Thus the equivalent cost of $\mathbf{a}^m$ is $\nu_m = \sum_{k=1}^{p} t_{km}\beta_{i_k}$. If $\nu_m \leq \beta_m$, then the vector $\mathbf{a}^m$ should not be added to the solution. On the other hand, if $\nu_m > \beta_m$, then it is advantageous to bring the vector $\mathbf{a}^m$ into the solution, and this can be done by replacing one of the columns at the left of the tableau by $\mathbf{a}^m$. Actually, however, there is another condition which also must be satisifed before we should carry out such a replacement. *There must be a positive entry in the position* t_{lm} *before we can replace* $\mathbf{a}^{i_l}$ *by* $\mathbf{a}^m$. This is necessary because of the assumptions used in the development of the replacement method and in its use in obtaining optimal vectors. We shall proceed with the case in which a positive entry exists in the column under $\mathbf{a}^m$. The alternative case is considered in Theorem 3 and Exercise 2.

If $\nu_m > \beta_m$ and there are positive entries in the column under $\mathbf{a}^m$, then which column vector at the left of the tableau should $\mathbf{a}^m$ replace? The answer to this question is as follows:

> *Examine all positive entries* t_{km}, *find the minimum of the corresponding ratios* η_{i_k}/t_{km}, *and replace one such column which gives this minimum by* $\mathbf{a}^m$.

The reasoning which leads to this answer is the same as that which leads to replacing $\mathbf{a}^1$ by $\mathbf{a}^4$ in the above example. Namely, if it is advantageous to bring $\mathbf{a}^m$ into the expansion of $\mathbf{c}$, then it should be brought in with as large a coefficient as is possible. We have

$$\begin{aligned}\mathbf{c} &= \eta_{i_1}\mathbf{a}^{i_1} + \eta_{i_2}\mathbf{a}^{i_2} + \cdots + \eta_{i_p}\mathbf{a}^{i_p} - \eta_m\mathbf{a}^m + \eta_m\mathbf{a}^m \\ &= (\eta_{i_1} - \eta_m t_{1m})\mathbf{a}^{i_1} + (\eta_{i_2} - \eta_m t_{2m})\mathbf{a}^{i_2} + \cdots + (\eta_{i_p} - \eta_m t_{pm})\mathbf{a}^{i_p} + \eta_m\mathbf{a}^m.\end{aligned}$$

Thus, we have a situation where if $t_{km} > 0$, then as η_m increases, the coefficient of $\mathbf{a}^{i_k}$ decreases. Hence, we increase η_m until one of these coefficients is 0 or, equivalently, until $\eta_m = \min\{\eta_{i_k}/t_{km} : k \in \{1, \ldots, p\} \text{ and } t_{km} > 0\}$. Inspection of the tableau and the replacement method shows that this is exactly what happens when $\mathbf{a}^m$ replaces a column on the left for which the ratio is a minimum.

We have now given a method for taking a feasible vector and, subject to certain hypotheses, obtaining a better feasible vector. We shall go into the question of convergence of this process later. Now we turn to the question of how we know when we have an optimal vector. We need a test which will tell us that no further improvement is possible. Recalling our criterion for replacement, we are led to conjecture that such a test can be phrased in terms of actual and equivalent cost. This is indeed the case, and we now prove the following theorem on optimality.

Theorem 2 (Checking Optimality): Let $\mathbf{y}$ be a feasible vector for the restricted minimum problem defined by $[\mathbb{A}, \mathbf{c}, \mathbf{b}]$. Let $I \subset \{1, \ldots, q\}$ be an index set, and suppose that $\{\mathbf{a}^i\}_{i \in I}$ is an independent set such that $S[\{\mathbf{a}^i\}_{i \in I}] = S[\{\mathbf{a}^i\}_{i=1, \ldots, q}]$. Moreover, assume that $\mathbf{y} = (\eta_1, \ldots, \eta_q)$ satisfies $\eta_i = 0$, $i \notin I$. Then the condition $\nu_m \leq \beta_m$ for all $m \in [\{1, \ldots, q\} \setminus I]$ implies that $\mathbf{y}$ is optimal.

Note. If $X = \{\mathbf{x}_1, \ldots, \mathbf{x}_q\}$ is a set of vectors, all in the same space, then the notation $S[X]$ refers to the *span* of the set X. The reader is referred to Appendix C for the definition and further discussion.

Proof: We base our proof of optimality on Theorem 2 of Chap. 4. The dual of this minimization problem is to maximize $\mathbf{x} \cdot \mathbf{c}$ subject to $\mathbb{A}^T\mathbf{x} \leq \mathbf{b}$ (Exercise 1). Now, from the corollary to Theorem C.5 of Appendix C, we know that there exists a vector $\mathbf{x} \in R^p$ such that $\mathbf{a}^i \cdot \mathbf{x} = \beta_i$, $i \in I$. Also, each column $\mathbf{a}^m$ is contained in $S[\{\mathbf{a}^i\}_{i \in I}]$. If $\mathbf{a}^m = \sum_{i \in I} t_{im}\mathbf{a}^i$, then

$$\begin{aligned} \mathbf{a}^m \cdot \mathbf{x} &= \sum_{i \in I} t_{im}\mathbf{a}^i \cdot \mathbf{x} \\ &= \sum_{i \in I} t_{im}\beta_i \\ &\leq \beta_m. \end{aligned}$$

But stated differently, this means that $\mathbb{A}^T\mathbf{x} \leq \mathbf{b}$ so that the vector $\mathbf{x}$ is feasible for the dual problem. Next, if $\mathbf{y} = (\eta_1, \ldots, \eta_q)$, $\mathbf{x} = (\xi_1, \ldots, \xi_p)$, and $\mathbb{A}\mathbf{y} = \mathbf{c}$, then it follows that

$$\begin{aligned} \mathbf{x} \cdot \mathbf{c} &= \sum_{i=1}^{p} \xi_i(\mathbf{a}_i \cdot \mathbf{y}) \\ &= \sum_{j=1}^{q} (\mathbf{a}^j \cdot \mathbf{x})\eta_j \\ &= \sum_{j=1}^{q} \beta_j\eta_j \\ &= \mathbf{y} \cdot \mathbf{b}. \end{aligned}$$

In this string of equalities all but the one $\sum_{j=1}^{q} (\mathbf{a}^j \cdot \mathbf{x})\eta_j = \sum_{j=1}^{q} \beta_j\eta_j$ are obvious. This crucial equality follows because in all cases $\mathbf{a}^j \cdot \mathbf{x} \leq \beta_j$, and whenever $\eta_j \neq 0$, we have $\mathbf{a}^j \cdot \mathbf{x} = \beta_j$. Therefore $\mathbf{x} \cdot \mathbf{c} = \mathbf{y} \cdot \mathbf{b}$, and by the results of Chap. 4 both $\mathbf{x}$ and $\mathbf{y}$ are optimal. Q.E.D.

The reader should compare the final steps of the proof of Theorem 2 with Tucker's duality relation and the discussion of complementary slackness in Sec. 4.3.2.

Theorem 2 gives a criterion for optimality. To check this criterion, it is useful to add another row to the tableau containing the feasible vectors. We again consider a restricted minimum problem defined by $[\mathbb{A}, \mathbf{c}, \mathbf{b}]$. Suppose that we have a feasible vector $\mathbf{y}$, which is expressed in terms of the columns of the set $\{\mathbf{a}^i\}_{i \in I} = \{\mathbf{a}^{i_1}, \mathbf{a}^{i_2}, \ldots, \mathbf{a}^{i_p}\}$ and for which the tableau has the form

$$T = \begin{array}{c|cccc|c|} & \mathbf{a}^1 & \mathbf{a}^2 & \cdots & \mathbf{a}^q & \mathbf{c} \\ \hline \mathbf{a}^{i_1} & t_{11} & t_{12} & \cdots & t_{1q} & \eta_{i_1} \\ \mathbf{a}^{i_2} & t_{21} & t_{22} & \cdots & t_{2q} & \eta_{i_2} \\ \vdots & \vdots & \vdots & & \vdots & \vdots \\ \mathbf{a}^{i_p} & t_{p1} & t_{p2} & \cdots & t_{pq} & \eta_{i_p} \\ \hline \end{array}\,.$$

We form a new vector $\mathbf{v}$ according to the definition $\mathbf{v} = (v_1, v_2, \ldots, v_q; v_0)$, where $v_0 = \sum_{i=1}^{p} \eta_{k_i}\beta_{k_i}$; $v_j = \sum_{i=1}^{p} t_{ij}\beta_{k_i}, j = 1, 2, \ldots, q$; and $\mathbf{b} = (\beta_1, \ldots, \beta_q)$. Finally, we define the $(p+1)$th row of the tableau T to be the coordinates of the vector $\mathbf{v} - \tilde{\mathbf{b}}$, where $\tilde{\mathbf{b}} = (\beta_1, \ldots, \beta_q, 0)$. In this way the $(p+1)$th row of T will immediately show whether the feasible vector $\mathbf{y}$ is actually an optimal vector. From Theorem 2, this vector is optimal if $v_j \leq \beta_j, j = 1, 2, \ldots, q$, i.e., if the first q entries of this last row are all nonpositive. The $(q+1)$th entry is just the value of $\mathbf{y} \cdot \mathbf{b}$, the quantity which is being minimized.

The real usefulness of this $(p+1)$th row is that as new tableaus are formed for new feasible vectors, this last row does not have to be computed from its basic definition. Instead, it can be obtained by using the same algorithm which is used to obtain the other rows of the new tableau. To show that this is true, we recall that the entries in T', the new tableau, are given by the formula

$$t'_{ij} = \begin{cases} t_{ij} - \dfrac{t_{im}}{t_{lm}} t_{lj}, & i \neq l, \\ \dfrac{t_{lj}}{t_{lm}}, & i = l. \end{cases} \tag{2.3}$$

This formula assumes that the pivot element is t_{lm}; i.e., $\mathbf{a}^m$ is replacing $\mathbf{a}^{i_l}$. Now let $\mathbf{v}'$ be the vector defined for T' in the same way that $\mathbf{v}$ was defined for T. Then the last row of T' is $\mathbf{v}' - \tilde{\mathbf{b}}$, where the jth entry, $j = 1, \ldots, q$, is given by

$$\begin{aligned} v'_j - \beta_j &= \sum_{\substack{i=1 \\ i \neq l}}^{p} t'_{ij}\beta_{k_i} + t'_{lj}\beta_m - \beta_j \\ &= \sum_{\substack{i=1 \\ i \neq l}}^{p} \left[t_{ij} - \frac{t_{im}}{t_{lm}} t_{lj} \right] \beta_{k_i} + \frac{t_{lj}}{t_{lm}} \beta_m - \beta_j \\ &= \sum_{i=1}^{p} t_{ij}\beta_{k_i} - t_{lj}\beta_{k_l} - \left[\sum_{i=1}^{p} t_{im}\beta_{k_i} \right] \frac{t_{lj}}{t_{lm}} \\ &\quad + \frac{t_{lm}}{t_{lm}} t_{lj}\beta_{k_l} + \frac{t_{lj}}{t_{lm}} \beta_m - \beta_j \\ &= (v_j - \beta_j) - \frac{(v_m - \beta_m)}{t_{lm}} t_{lj}. \end{aligned}$$

Similarly the $(q + 1)$th entry is given by

$$\begin{aligned} v'_0 &= \sum_{\substack{i=1 \\ i \neq l}}^{p} \eta'_{k_i}\beta_{k_i} + \eta_m\beta_m \\ &= \sum_{\substack{i=1 \\ i \neq l}}^{p} \left[\eta_{k_i} - \frac{t_{im}}{t_{lm}} \eta_{k_l} \right] \beta_{k_i} + \frac{\eta_{k_l}}{t_{lm}} \beta_m \\ &= \sum_{i=1}^{p} \eta_{k_i}\beta_{k_i} - \eta_{k_l}\beta_{k_l} - \left[\sum_{i=1}^{p} t_{im}\beta_{k_i} \right] \frac{\eta_{k_l}}{t_{lm}} \\ &\quad + \frac{t_{lm}}{t_{lm}} \beta_{k_l}\eta_{k_l} + \frac{\eta_{k_l}}{t_{lm}} \beta_m \\ &= v_0 - \frac{(v_m - \beta_m)}{t_{lm}} \eta_{k_l}. \end{aligned}$$

But these results are the same as the results obtained by applying Eq. (2.3) to the last row of the tableau T. Thus, as new feasible vectors are obtained by the optimization technique, one can check if the new vector is optimal by examining the $(p + 1)$th row of the new tableau. If the first q entries are nonpositive and the other hypotheses of Theorem 2 are satisfied, then the vector is optimal.

We illustrate this idea of a checking row by returning to our earlier example. The tableau corresponding to the feasible vector $\mathbf{y} = (4, 3, 2, 0)$ has the form

$T =$

	$\mathbf{a}^1$	$\mathbf{a}^2$	$\mathbf{a}^3$	$\mathbf{a}^4$	$\mathbf{c}$
$\mathbf{a}^1$	1	0	0	3	4
$\mathbf{a}^2$	0	1	0	1	3
$\mathbf{a}^3$	0	0	1	1	2
$\mathbf{v} - \tilde{\mathbf{b}}$	0	0	0	10	16

The last row of tableau T (the checking row) is obtained as follows: The entry under $\mathbf{a}^1$ is the difference between the equivalent cost v_1 of $\mathbf{a}^1$ and the actual cost β_1 of $\mathbf{a}^1$. Here the equivalent cost is in terms of $\mathbf{a}^1$, $\mathbf{a}^2$, and $\mathbf{a}^3$ since they are the entries at the left of the tableau. Writing $\mathbf{a}^1$ in terms of $\mathbf{a}^1$, $\mathbf{a}^2$, and $\mathbf{a}^3$, we have $\mathbf{a}^1 = 1 \cdot \mathbf{a}^1 + 0 \cdot \mathbf{a}^2 + 0 \cdot \mathbf{a}^3$ (i.e., the entries under $\mathbf{a}^1$ are 1, 0, 0). Hence the equivalent cost of $\mathbf{a}^1$ is $\nu_1 = 1 \cdot \beta_1 + 0 \cdot \beta_2 + 0 \cdot \beta_3 = \beta_1$. Therefore the entry in the checking row under $\mathbf{a}^1$ is $\nu_1 - \beta_1 = \beta_1 - \beta_1 = 0$. Similarly, the entries under $\mathbf{a}^2$ and $\mathbf{a}^3$ are 0 since these vectors also appear at the left of the tableau. In fact, it is true in general that if any vector $\mathbf{a}^m$ appears at the left of the tableau, then the entry in the checking row under $\mathbf{a}^m$ is $\nu_m - \beta_m = \beta_m - \beta_m = 0$.

Column $\mathbf{a}^4$ does not appear at the left of tableau T. The entries under $\mathbf{a}^4$ show that $\mathbf{a}^4 = 3 \cdot \mathbf{a}^1 + 1 \cdot \mathbf{a}^2 + 1 \cdot \mathbf{a}^3$. Hence the equivalent cost of $\mathbf{a}^4$ in terms of $\mathbf{a}^1, \mathbf{a}^2$, and $\mathbf{a}^3$ is $\nu_4 = 3 \cdot \beta_1 + 1 \cdot \beta_2 + 1 \cdot \beta_3 = 3 \cdot 3 + 1 \cdot (-2) + 1 \cdot 5 = 12$. The actual cost of $\mathbf{a}^4$ is $\beta_4 = 2$, and hence in the checking row the entry under $\mathbf{a}^4$ is $\nu_4 - \beta_4 = 12 - 2 = 10$. The entry under $\mathbf{c}$ is simply the equivalent cost of $\mathbf{c}$, $\nu_0 = 4 \cdot 3 + 3 \cdot (-2) + 2(5) = 16$. The first q $(= 4)$ entries in the last row contain the positive number 10. Hence this feasible vector is not optimal. However, after replacing $\mathbf{a}^1$ by $\mathbf{a}^4$ the tableau is

$T' =$

	$\mathbf{a}^1$	$\mathbf{a}^2$	$\mathbf{a}^3$	$\mathbf{a}^4$	$\mathbf{c}$
$\mathbf{a}^4$	1/3	0	0	1	4/3
$\mathbf{a}^2$	−1/3	1	0	0	5/3
$\mathbf{a}^3$	−1/3	0	1	0	2/3
$\mathbf{v}' - \tilde{\mathbf{b}}$	−10/3	0	0	0	8/3

In this tableau the first q coordinates of the last row are all nonpositive. Moreover, $\mathbf{a}^2$, $\mathbf{a}^3$, and $\mathbf{a}^4$ are independent and span the column space. Hence the vector $\mathbf{y} = (0, 5/3, 2/3, 4/3)$ is optimal, and the minimum of $\mathbf{y} \cdot \mathbf{b}$ is 8/3.

5.2.3 Finding Feasible Vectors

In Secs. 5.2.1 and 5.2.2, we have described a method by which one feasible vector can be modified to obtain another feasible vector which is better for the optimization problem. We have not yet shown how to obtain an initial feasible vector to use at the start of this optimization process. We now turn to this question, and in particular we seek a method for obtaining a feasible vector for the problem of minimizing $\mathbf{y} \cdot \mathbf{b}$ subject to $\mathbf{y} \geq \mathbf{0}$ and $\mathbb{A}\mathbf{y} = \mathbf{c}$. For the benefit of readers who may use a standard computer code, we note that this process of finding a feasible vector is called *Phase I* of the simplex method.

The first step in finding a feasible vector is to rewrite the problem so that $\mathbf{c} \geq \mathbf{0}$. This is easily done by multiplying certain of the equations in the system $\mathbb{A}\mathbf{y} = \mathbf{c}$ by -1. Henceforth, we shall assume that this was done initially and that therefore $\mathbf{c} \geq \mathbf{0}$. Next, we introduce a new problem which is formed from this original one. This new problem is to find nonnegative vectors $\mathbf{z}$ and $\mathbf{y}$ which satisfy $\mathbb{A}\mathbf{y} + \mathbf{z} = \mathbf{c}$ and for which $\mathbf{z} \cdot \mathbf{u}$ is a minimum. Here $\mathbf{u} = (1, 1, \ldots, 1) \in R^p$. Clearly, if the problem of minimizing $\mathbf{y} \cdot \mathbf{b}$ subject to $\mathbf{y} \geq \mathbf{0}$ and $\mathbb{A}\mathbf{y} = \mathbf{c}$ has a feasible vector $\mathbf{y}_0$, then our new problem has an optimal vector given by $\mathbf{z} = \mathbf{0}$ and $\mathbf{y} = \mathbf{y}_0$. The minimum of $\mathbf{z} \cdot \mathbf{u}$ is thus 0. Conversely, if the second problem has optimal vectors $\mathbf{z}$ and $\mathbf{y}$ and if the minimum of $\mathbf{z} \cdot \mathbf{u}$ is 0, then $\mathbf{z} = \mathbf{0}$ and $\mathbb{A}\mathbf{y} = \mathbf{c}$. Hence a vector $\mathbf{y}$ which is optimal for the second problem is feasible for the first problem if and only if the associated vector $\mathbf{z} = \mathbf{0}$. In this second problem we can think of the unknown vectors as being combined into a single vector of dimension $p + q$ (we are assuming the matrix $\mathbb{A}$ has dimension $p \times q$). In particular, this unknown vector can be written as $\mathbf{w} = (\mathbf{y}, \mathbf{z}) = (\eta_1, \eta_2, \ldots, \eta_q, \zeta_1, \zeta_2, \ldots, \zeta_p)$. With this selection of $\mathbf{w}$, the new coefficient matrix is $[\mathbb{A}, \mathbb{I}]$, where $\mathbb{I}$ is the $p \times p$ identity matrix. Also, since we are minimizing $\mathbf{z} \cdot \mathbf{u} = \zeta_1 + \cdots + \zeta_p$, the new cost vector is $\mathbf{h} = (0, 0, \ldots, 0, 1, 1, \ldots, 1) \in \mathbb{R}^{q+p}$, where there are q coordinates of 0 and p coordinates of 1. The second problem can now be stated as the problem of minimizing $\mathbf{w} \cdot \mathbf{h}$ subject to $\mathbf{w} \geq \mathbf{0}$ and $[\mathbb{A}, \mathbb{I}]\mathbf{w} = \mathbf{c}$. What have we gained by creating this second problem? It would seem that we have gained very little, because the second problem is the same type as the original problem—and, moreover, it is bigger. However, the second problem has a distinct advantage: It has a ready-made feasible vector. Thus, for this problem, optimization can begin immediately. The feasible vector is obtained by taking $\mathbf{w}_0 = (\mathbf{0}, \mathbf{c}) \in \mathbb{R}^{q+p}$; i.e., $\mathbf{w}_0 = (0, \ldots, 0, \gamma_1, \gamma_2, \ldots, \gamma_p)$. This is feasible because $\mathbf{c} \geq \mathbf{0}$ and $[\mathbb{A}, \mathbb{I}]\mathbf{w}_0 = \mathbf{0} + \mathbb{I}\mathbf{c} = \mathbf{c}$. Therefore, using $\mathbf{w}_0$ as an initial feasible vector, we can apply the method of optimization of Sec. 5.2.2. Moreover, this second problem always has optimal vectors (see Exercise 6). There are two cases to consider at this point. First, suppose that we obtain an optimal vector $\mathbf{w}$ for which the last p coordinates are 0. Then the original problem has a feasible vector given by the first q coordinates. Next, if the optimal vector $\mathbf{w}$ does not have the last p coordinates equal to 0 then the first problem does not have a feasible vector.

There is an important question which has been ignored in the discussion of this section and preceding sections. Namely, does the process of optimization always yield an optimal vector in a finite number of steps? We already know that certain problems do not have optimal vectors; however, our question also applies to problems that do possess an optimal vector. It seems possible that even if a problem has an optimal vector, the optimization method may not yield it. Instead, the process continues to yield better and better feasible vectors without obtaining a best vector. If this were the case, then clearly the method given here for obtaining feasible vectors is useless because it requires an optimal vector for a second problem. Fortunately, this awkward situation does not occur; the process does yield an optimal vector when one exists. We shall prove this fact in the next section. The remainder of this section is devoted to working out an example of the process of finding a feasible vector for one problem by finding an optimal vector for an auxiliary problem.

Original Problem. Let

$$\mathbb{A} = \begin{bmatrix} 1 & 2 & 0 & 5 \\ 0 & 1 & 1 & 2 \\ 1 & 3 & 2 & 8 \end{bmatrix}, \qquad \mathbf{b} = (3, -2, 5, 2), \qquad \text{and} \qquad \mathbf{c} = \begin{pmatrix} 10 \\ 5 \\ 17 \end{pmatrix}.$$

Find $\mathbf{y}$ such that $\mathbf{y} \cdot \mathbf{b}$ is a minimum subject to $\mathbf{y} \geq \mathbf{0}$ and $\mathbb{A}\mathbf{y} = \mathbf{c}$.

Auxiliary Problem. Find a vector $\mathbf{w} = (\mathbf{y}, \mathbf{z})$ such that $\mathbf{w} \cdot \mathbf{h}$ is a minimum subject to $\mathbf{w} \geq \mathbf{0}$ and $[\mathbb{A}, \mathbb{I}]\mathbf{w} = \mathbf{c}$, where $\mathbf{h} = (0, 0, 0, 0, 1, 1, 1)$ and $\mathbf{w} = (\eta_1, \eta_2, \eta_3, \eta_4, \zeta_1, \zeta_2, \zeta_3)$. From our discussion above, we know this second problem has a feasible vector which is given by $\mathbf{w}_0 = (0, 0, 0, 0, 10, 5, 17)$. This vector, $\mathbf{w}_0$, corresponds to an expansion of $\mathbf{c}$ in terms of the last three columns of the matrix $[\mathbb{A}, \mathbb{I}]$. These columns are the unit vectors, and hence the tableau for $\mathbf{w}_0$ is the following:

$T^{(0)} =$

	$\mathbf{a}^1$	$\mathbf{a}^2$	$\mathbf{a}^3$	$\mathbf{a}^4$	$\mathbf{u}^1$	$\mathbf{u}^2$	$\mathbf{u}^3$	$\mathbf{c}$
$\mathbf{u}^1$	1	2	0	5	1	0	0	10
$\mathbf{u}^2$	0	1	1	2	0	1	0	5
$\mathbf{u}^3$	1	3	2	8	0	0	1	17
$\mathbf{v}^{(0)} - \tilde{\mathbf{h}}$	2	6	3	15	0	0	0	32

We use the optimization method, and successively replace $\mathbf{u}^1$ by $\mathbf{a}^1$, $\mathbf{u}^2$ by $\mathbf{a}^2$, and $\mathbf{u}^3$ by $\mathbf{a}^3$. The tableaus are

$T^{(1)} =$

	$\mathbf{a}^1$	$\mathbf{a}^2$	$\mathbf{a}^3$	$\mathbf{a}^4$	$\mathbf{u}^1$	$\mathbf{u}^2$	$\mathbf{u}^3$	$\mathbf{c}$
$\mathbf{a}^1$	1	2	0	5	1	0	0	10
$\mathbf{u}^2$	0	1	1	2	0	1	0	5
$\mathbf{u}^3$	0	1	2	3	−1	0	1	7
$\mathbf{v}^{(1)} - \tilde{\mathbf{h}}$	0	2	3	5	−2	0	0	12

$T^{(2)} =$

	$\mathbf{a}^1$	$\mathbf{a}^2$	$\mathbf{a}^3$	$\mathbf{a}^4$	$\mathbf{u}^1$	$\mathbf{u}^2$	$\mathbf{u}^3$	$\mathbf{c}$
$\mathbf{a}^1$	1	0	−2	1	1	−2	0	0
$\mathbf{a}^2$	0	1	1	2	0	1	0	5
$\mathbf{u}^3$	0	0	1	1	−1	−1	1	2
$\mathbf{v}^{(2)} - \tilde{\mathbf{h}}$	0	0	1	1	−2	−2	0	2

and

$T^{(3)} =$

	$\mathbf{a}^1$	$\mathbf{a}^2$	$\mathbf{a}^3$	$\mathbf{a}^4$	$\mathbf{u}^1$	$\mathbf{u}^2$	$\mathbf{u}^3$	$\mathbf{c}$
$\mathbf{a}^1$	1	0	0	3	−1	−4	2	4
$\mathbf{a}^2$	0	1	0	1	1	2	−1	3
$\mathbf{a}^3$	0	0	1	1	−1	−1	1	2
$\mathbf{v}^{(3)} - \tilde{\mathbf{h}}$	0	0	0	0	−1	−1	−1	0

The last row of the tableau $T^{(3)}$ is nonpositive. Hence we have obtained an optimal vector for the auxiliary problem. This optimal vector is $\mathbf{w} = (4, 3, 2, 0, 0, 0, 0) = (\mathbf{y}_0, \mathbf{z})$. Thus, $\mathbf{z} = \mathbf{0}$ and $\mathbf{y}_0 = (4, 3, 2, 0)$ is a feasible vector for the original problem. This is the same feasible vector that we used earlier, and hence there is no need to continue and again obtain an optimal vector for the original problem. In general, of course, the next step is to use this feasible vector for the original problem and begin the process of optimization. We note that it is possible to continue via the last tableau used in the second problem. Naturally, only the columns of the matrix $\mathbb{A}$ are retained and the last row must be recomputed to reflect the fact that the cost vector is now $\mathbf{b}$ and not $\mathbf{h}$. The first tableau for the original problem would be given by

$T^{(4)} =$

	$\mathbf{a}^1$	$\mathbf{a}^2$	$\mathbf{a}^3$	$\mathbf{a}^4$	$\mathbf{c}$
$\mathbf{a}^1$	1	0	0	3	4
$\mathbf{a}^2$	0	1	0	1	3
$\mathbf{a}^3$	0	0	1	1	2
$\mathbf{v}^{(4)} - \tilde{\mathbf{b}}$	0	0	0	12	16

5.2.4 Convergence and Degeneracy

We now turn to the question which we posed and left open in Sec. 5.2.3. This question can be phrased in the following manner. Given that a linear programming problem has a feasible vector and assuming that the method of optimization can be applied to this vector and subsequent feasible vectors, does the method of optimization always yield an optimal vector in a finite number of steps? Subject to certain assumptions, we shall show that the answer to this question is yes. In practice, some of these assumptions are not needed, and we shall comment on these and other practical aspects of the result after we state and prove it.

Theorem 3 (A Convergence Theorem): Let $\mathbb{A}$ be a $p \times q$ matrix, $\mathbf{c} \in R^p$, and $\mathbf{b} \in R^q$. Assume that the vector $\mathbf{y}_0$ is a feasible vector for the problem of minimizing $\mathbf{y} \cdot \mathbf{b}$ subject to the constraints $\mathbf{y} \geq \mathbf{0}$ and $\mathbb{A}\mathbf{y} = \mathbf{c}$. Also assume that the $p \times (q + 1)$ matrix $[\mathbb{A}, \mathbf{c}]$ has the property that every set of p columns is an independent set. Then, either the method of optimization will reach a state where it cannot be applied, in which case there does not exist an optimal vector, or, in a finite number of steps, an optimal vector will be obtained.

Note that the statement of this theorem implicitly assumes that the $p \times (q + 1)$ matrix $[\mathbb{A}, \mathbf{c}]$ has at least p columns. See Exercise 13 for a justification of this assumption.

Proof: One part of the proof is considered in the Exercises. This is the case where the process of optimization reaches a state where a replacement is called for but the replacement cannot be carried out. This would happen if $\nu_m - \beta_m > 0$ for some m, but $t_{lm} < 0$ for each l, $l = 1, \ldots, p$ (see Exercise 2).

The alternative to the situation mentioned in the above paragraph is the situation where the optimization process can always be carried out. We seek to show that we obtain an optimal vector in a finite number of steps. To show this, it is sufficient to show that each step of the process yields a new feasible vector which is better than the previous vector. This will give the desired result because there are only finitely many ways of choosing p columns from a total of q columns and hence only finitely many different feasible vectors which use only p columns. In fact, there are

$$\binom{q}{p} = \frac{q!}{p!(q-p)!}$$

different sets of p columns, each is an independent set, and from the hypothesis of the theorem, $\mathbf{c}$ is never a linear combination of less than p columns. Hence, there are at most $\binom{q}{p}$ feasible vectors which use exactly p columns for the expansion of $\mathbf{c}$. Now suppose that the feasible vector $\mathbf{y}_0$ uses the columns $\mathbf{a}^{i_1}, \mathbf{a}^{i_2}, \ldots, \mathbf{a}^{i_p}$ in the expansion of $\mathbf{c}$. Also, let the coefficients of these be $\eta_{i_1}, \eta_{i_2}, \ldots, \eta_{i_p}$, i.e., $\mathbf{c} = \eta_{i_1}\mathbf{a}^{i_1} + \cdots + \eta_{i_p}\mathbf{a}^{i_p}$. Suppose that the method of optimization calls for replacing column $\mathbf{a}^{i_l}$ by $\mathbf{a}^m$ and that the replacement can be carried out; i.e., $t_{lm} > 0$. Carrying out the replacement and calling the new feasible vector $\mathbf{y}_1$, we have

$$\mathbf{c} = \eta'_{i_1}\mathbf{a}^{i_1} + \cdots + \eta'_{i_{l-1}}\mathbf{a}^{i_{l-1}} + \eta_m\mathbf{a}^m + \eta'_{i_{l+1}}\mathbf{a}^{i_{l+1}} + \cdots + \eta'_{i_p}\mathbf{a}^{i_p},$$

where the coefficients are given by the formula

$$\begin{aligned} \eta'_{i_k} &= \eta_{i_k} - \frac{\eta_{i_l} t_{km}}{t_{lm}}, \qquad k \neq l, \\ \eta_m &= \frac{\eta_{i_l}}{t_{lm}}. \end{aligned} \tag{2.4}$$

The cost of the vector $\mathbf{y}_1$ is given by

$$\mathbf{y}_1 \cdot \mathbf{b} = v_0' = \sum_{\substack{k=1 \\ k \neq l}}^{p} \eta_{i_k}' \beta_{i_k} + \eta_m \beta_m,$$

while the cost of $\mathbf{y}_0$ is

$$\mathbf{y}_0 \cdot \mathbf{b} = v_0 = \sum_{k=1}^{p} \eta_{i_k} \beta_{i_k}.$$

Therefore

$$v_0 - v_0' = \sum_{\substack{k=1 \\ k \neq l}}^{p} (\eta_{i_k} - \eta_{i_k}') \beta_{i_k} + \eta_{i_l} \beta_{i_l} - \eta_m \beta_m.$$

Using the equivalent form of η_{i_k}' and η_m given by (2.4), we obtain

$$\begin{aligned} v_0 - v_0' &= \sum_{\substack{k=1 \\ k \neq l}}^{p} \frac{\eta_{i_l} t_{km}}{t_{lm}} \beta_{i_k} + \eta_{i_l} \beta_{i_l} - \frac{\eta_{i_l} \beta_m}{t_{lm}} \\ &= \frac{\eta_{i_l}}{t_{lm}} \left\{ \left[\sum_{k=1}^{p} t_{km} \beta_{i_k} \right] - \beta_m \right\} \\ &= \frac{\eta_{i_l}}{t_{lm}} [v_m - \beta_m]. \end{aligned}$$

The quantity on the last line in this series of equalities is positive as a result of our hypothesis. To be more explicit, we have assumed that

1. A replacement was called for; thus $v_m > \beta_m$.
2. The replacement was possible; thus $t_{lm} > 0$.
3. $\mathbf{c}$ and the columns $\mathbf{a}^{i_1}, \mathbf{a}^{i_2}, \ldots, \mathbf{a}^{i_p}$ have the property that every subset of p vectors is independent; thus $\eta_{i_l} > 0$.

Therefore, $v_0 - v_0' > 0$, and the new feasible vector has a smaller cost than the old feasible vector. Continuing this process, if another replacement can be made, then the new feasible vector $\mathbf{y}_2$ will again have a cost smaller than that of $\mathbf{y}_1$ or $\mathbf{y}_0$. Therefore each new feasible vector is different from the earlier feasible vectors. Since each is expressed in terms of a different set of p columns there are only finitely many replacements which can be made before we must reach an optimal vector. Q.E.D.

Comments on Theorem 3

1. Theorem 3 has an intimate connection with the geometrical results of Sec. 5.1. The feasible vectors which are obtained under the hypotheses of the theorem are extreme points of the set of feasible vectors. This follows

from Theorem 3 of Sec. 5.1 and the fact that these feasible vectors yield an expansion of $\mathbf{c}$ in terms of independent columns of $\mathbb{A}$. Since there are only a finite number of extreme points and since each such succeeding vector is better than the previous ones, the simplex method will reach the best extreme point in a finite number of steps. However, by the corollary to Theorem 4 of Sec. 5.1, if the set of feasible vectors is bounded, then the best extreme vector is also a best vector for the whole set. The condition of boundedness corresponds to the possibility of continually carrying out the replacement operation whenever it is called for by the comparison of actual and equivalent cost. Thus, as the theorem concludes, the simplex method then yields an optimal vector. Conversely, if a replacement is called for but cannot be made, then it can be shown that the set of feasible vectors is either empty or is not bounded. These topics are considered in more detail in the Exercises.

2. The assumptions of Theorem 3 certainly do not always hold for linear programming problems. If the assumption about the independence of the columns of each $p \times p$ submatrix in $[\mathbb{A}, \mathbf{c}]$ is violated, we say that the problem is *degenerate*. Example 2 of Sec. 4.2.2 can easily be modified to give a problem which is degenerate. In that example $p = 3, q = 4$, and $[\mathbb{A}, \mathbf{c}]$ is the 3×5 matrix given by

$$[\mathbb{A}, \mathbf{c}] = \begin{bmatrix} 1 & 2 & 0 & 5 & 10 \\ 0 & 1 & 1 & 2 & 5 \\ 1 & 3 & 2 & 8 & 17 \end{bmatrix}.$$

We modify $\mathbb{A}$ by adding the column

$$\mathbf{u}^3 = \begin{pmatrix} 0 \\ 0 \\ 1 \end{pmatrix}$$

and denote the new coefficient matrix by $\mathbb{A}_1$. Then the matrix $[\mathbb{A}_1, \mathbf{c}]$ is the 3×6 matrix

$$\begin{bmatrix} 1 & 2 & 0 & 5 & 0 & 10 \\ 0 & 1 & 1 & 2 & 0 & 5 \\ 1 & 3 & 2 & 8 & 1 & 17 \end{bmatrix}.$$

In this matrix, the submatrix

$$\begin{bmatrix} 2 & 0 & 10 \\ 1 & 0 & 5 \\ 3 & 1 & 17 \end{bmatrix}$$

has dependent columns. Thus, for example, if we let $\mathbf{b}_1 = (\mathbf{b}, 0) = (3, -2,$

5, 2, 0), then the problem of minimizing $\mathbf{y} \cdot \mathbf{b}_1$ subject to $\mathbf{y} \geq \mathbf{0}$ and $\mathbb{A}_1\mathbf{y} = \mathbf{c}$ is a degenerate problem. In this problem it is possible to use the optimization method to obtain vectors which are written in terms of different sets of p columns and for which the cost is the same (see Exercise 7). The details of the proof of Theorem 3 show that this situation could not occur without degeneracy. It follows that in a problem involving degeneracy there is a danger of getting into a loop of feasible vectors where successive replacements make no improvement. In this situation the sequence of feasible vectors consists of repeating blocks of vectors, and an optimal vector is never attained. Fortunately, this looping is very rare. Because of round-off error in computation, a degenerate problem usually acts like a nondegenerate problem, and the optimization method yields a solution. Moreover, there is actually a more complicated version of the simplex method which will handle degenerate problems. We do not discuss this other version, but the reader can find it presented in [Ga] and other books which treat computational methods in linear programming.

3. Theorem 3 of Sec. 5.1 and Theorem 3 here are theoretical results about the simplex method. They both involve the concept of an independent set of vectors. It is often difficult to decide if a given set of vectors is independent or not, and hence for computational purposes these theorems are not considered to be practical results. Therefore, if one is presented with a linear programming problem, it is not usual to begin by trying to show that the hypothesis of independence is satisfied. Instead, it is generally assumed that the simplex method will work and it is immediately used. Modifications can be made if it turns out that this assumption is false.

4. We remarked in regard to Theorem 4 of Sec. 5.1 that the set of feasible vectors for a standard linear programming problem (and other such problems) may well be unbounded. However, it is always a convex set in the positive octant of $R^n(\mathbf{x} \geq \mathbf{0})$, and it is known that such a set can be made artifically bounded by adding the constraint $\sum \xi_i \leq M$ for a large M (this is called the Charnes-Cooper regularization method). Then, if the new problem has a feasible vector, it has an optimal vector, and the original problem has no optimal vector if and only if the regularized problem has all optimal vectors satisfying $\sum \xi_i = M$.

We next turn to the application of the simplex method to problems which are not in restricted form.

5.2.5 Inequalities and Slack Variables

Our discussion of the simplex method has been limited to restricted linear programming problems. In Chap. 4 it was shown that all standard and general linear programming problems can be written as restricted problems, and hence the simplex method can also be applied to such problems. It is useful to give some examples of such applications. We concentrate on the standard problem which asks for the maximization of $\mathbf{x} \cdot \mathbf{c}$ subject to $\mathbf{x} \geq \mathbf{0}$ and $\mathbb{A}\mathbf{x} \leq \mathbf{b}$. In particular, let

$$\mathbb{A} = \begin{bmatrix} 1 & 4 \\ 3 & 1 \\ 1 & 1 \end{bmatrix}, \qquad \mathbf{b} = \begin{pmatrix} 24 \\ 21 \\ 9 \end{pmatrix}, \qquad \text{and} \qquad \mathbf{c} = \begin{pmatrix} 2 \\ 5 \end{pmatrix}.$$

The first step in the solution of this problem is to convert the system of inequalities $\mathbb{A}\mathbf{x} \leq \mathbf{b}$ into a system of equalities. As noted in Sec. 4.3.2. this can be accomplished by writing $\mathbb{A}\mathbf{x} + \mathbf{w} = \mathbf{b}$, where $\mathbf{w}$ is an unknown nonnegative vector. Thus we add ω_i to the ith equation, we obtain the system

$$\begin{aligned} \xi_1 + 4\xi_2 + \omega_1 &= 24, \\ 3\xi_1 + \xi_2 + \omega_2 &= 21, \\ \xi_1 + \xi_2 + \omega_3 &= 9, \end{aligned} \tag{2.5}$$

and we set

$$\mathbf{w} = \begin{pmatrix} \omega_1 \\ \omega_2 \\ \omega_3 \end{pmatrix}.$$

This new system has the form

$$\begin{bmatrix} 1 & 4 & 1 & 0 & 0 \\ 3 & 1 & 0 & 1 & 0 \\ 1 & 1 & 0 & 0 & 1 \end{bmatrix} \begin{pmatrix} \xi_1 \\ \xi_2 \\ \omega_1 \\ \omega_2 \\ \omega_3 \end{pmatrix} = \mathbf{b}, \qquad \text{or} \qquad [\mathbb{A}, \mathbb{I}] \begin{pmatrix} \mathbf{x} \\ \mathbf{w} \end{pmatrix} = \mathbf{b}.$$

The vector $\mathbf{w}$ is called the *slack* vector since it picks up the slack in the inequalities to make them equalities. If we were working with the dual problem, then we would have a system of inequalities of the form $\mathbb{A}^T\mathbf{y} \geq \mathbf{c}$. In this case we would subtract a nonnegative number from the left-hand side of each inequality so as to form a vector equality.

System (2.5) is in a form where we can apply the simplex method as it was developed in Secs. 5.2.2 and 5.3.2. The vector $\mathbf{b}$ is already positive, and so it is not necessary to multiply any of the equations by -1. In general, such multiplications would be the first step. The second step would be to find a feasible vector or show that none exists. In this special case we can obtain a feasible vector by inspection. The system (2.5) has the form $\mathbb{A}\mathbf{x} + \mathbf{w} = \mathbf{b}$. Since $\mathbf{b} > \mathbf{0}$, we can obtain a feasible vector by setting $\mathbf{x} = \mathbf{0}$ and

$\mathbf{w} = \mathbf{b}$. In general, feasible vectors are not so easily obtained and it will be necessary to use the technique developed in Sec. 5.2.3. In system (2.5) the coefficient matrix is $[\mathbb{A}, \mathbb{I}]$ and the unknown vector is $\hat{\mathbf{x}} = (\mathbf{x}, \mathbf{w})$. Before we can begin the simplex method, we must decide on the *value* vector. The term *value* is used here instead of *cost* because the problem is one of maximization (see Exercise 8). Since the original problem is to maximize $\mathbf{x} \cdot \mathbf{c}$, we give a value of zero to each of the columns in the matrix $\mathbb{I}$. Hence the value vector is given by $\hat{\mathbf{c}} = (\mathbf{c}, \mathbf{0})$. Therefore the restricted problem which is obtained is the problem of maximizing $\hat{\mathbf{x}} \cdot \hat{\mathbf{c}} = \mathbf{x} \cdot \mathbf{c} + \mathbf{w} \cdot \mathbf{0} = \mathbf{x} \cdot \mathbf{c}$ subject to $\hat{\mathbf{x}} \geq \mathbf{0}$ and $[\mathbb{A}, \mathbb{I}]\hat{\mathbf{x}} = \mathbf{b}$. Using the feasible vector $(\mathbf{0}, \mathbf{b})$, the first tableau is the following:

$T^{(0)} =$

	$\mathbf{a}^1$	$\mathbf{a}^2$	$\mathbf{u}^1$	$\mathbf{u}^2$	$\mathbf{u}^3$	$\mathbf{b}$
$\mathbf{u}^1$	1	4	1	0	0	24
$\mathbf{u}^2$	3	1	0	1	0	21
$\mathbf{u}^3$	1	1	0	0	1	9
$\mathbf{v} - \hat{\mathbf{c}}$	-2	-5	0	0	0	0

.

We are now maximizing, and in view of the relation between standard maximum and minimum problems (Chapter 4), it is now necessary to remove all negative numbers from the last row of the tableau. We use our criteria for making replacements and successively replace $\mathbf{u}^2$ by $\mathbf{a}^1$, $\mathbf{u}^3$ by $\mathbf{a}^2$, and then $\mathbf{u}^1$ by $\mathbf{u}^2$. The successive tableaus are as follows:

$T^{(1)} =$

	$\mathbf{a}^1$	$\mathbf{a}^2$	$\mathbf{u}^1$	$\mathbf{u}^2$	$\mathbf{u}^3$	$\mathbf{b}$
$\mathbf{u}^1$	0	11/3	1	$-1/3$	0	17
$\mathbf{a}^1$	1	1/3	0	1/3	0	7
$\mathbf{u}^3$	0	2/3	0	$-1/3$	1	2
$\mathbf{v}^{(1)} - \hat{\mathbf{c}}$	0	$-13/3$	0	2/3	0	14

,

$T^{(2)} =$

	$\mathbf{a}^1$	$\mathbf{a}^2$	$\mathbf{u}^1$	$\mathbf{u}^2$	$\mathbf{u}^3$	$\mathbf{b}$
$\mathbf{u}^1$	0	0	1	3/2	$-11/2$	6
$\mathbf{a}^1$	1	0	0	1/2	$-1/2$	6
$\mathbf{a}^2$	0	1	0	$-1/2$	3/2	3
$\mathbf{v}^{(2)} - \hat{\mathbf{c}}$	0	0	0	$-3/2$	13/2	27

,

$$T^{(3)} = \begin{array}{r|cccccc|c|} & \mathbf{a}^1 & \mathbf{a}^2 & \mathbf{u}^1 & \mathbf{u}^2 & \mathbf{u}^3 & & \mathbf{b} \\ \hline \mathbf{u}^2 & 0 & 0 & 2/3 & 1 & -11/3 & & 4 \\ \mathbf{a}^1 & 1 & 0 & -1/3 & 0 & 4/3 & & 4 \\ \mathbf{a}^2 & 0 & 1 & 1/3 & 0 & -1/3 & & 5 \\ \hline \mathbf{v}^{(3)} - \hat{\mathbf{c}} & 0 & 0 & 1 & 0 & 1 & & 33 \\ \hline \end{array}.$$

The bottom row of $T^{(3)}$ consists entirely of nonnegative numbers. Hence, we have obtained an optimal vector. It is given by $\hat{\mathbf{x}} = (\mathbf{x}, \mathbf{w}) = (4, 5, 0, 4, 0)$. Therefore the solution to the original problem is given by $\mathbf{x} = (4, 5)$. The slack vector $\mathbf{w}$ is given by $\mathbf{w} = (0, 4, 0)$. This vector can be used to draw important conclusions about the dual problem. Since the first and third coordinates of $\mathbf{w}$ are zero, we know that the first and third equations of $\mathbb{A}\mathbf{x} \leq \mathbf{b}$ are equalities when $\mathbf{x} = (4, 5)$. Also, the second equation is a strict inequality because the second slack variable is a 4. Now, recall that the dual problem is the problem of minimizing $\mathbf{y} \cdot \mathbf{b}$ subject to $\mathbf{y} \geq \mathbf{0}$ and $\mathbb{A}^T\mathbf{y} \geq \mathbf{c}$. Here, $\mathbf{y} = (\eta_1, \eta_2, \eta_3)$ has three coordinates; and, from Chap. 4, the expanded version of Theorem 2, we know that these coordinates are positive only if the corresponding equations in $\mathbb{A}\mathbf{x} \leq \mathbf{b}$ are equalities. Therefore, for an optimal $\mathbf{y}$, we must have $\eta_1 \geq 0, \eta_2 = 0, \eta_3 \geq 0$. In fact, tableau $T^{(3)}$ tells us more than just the form of an optimal vector for the dual problem. It actually yields an optimal vector. This optimal vector is obtained from the last row of the tableau, and it consists of the entries under the slack columns $\mathbf{u}^1, \mathbf{u}^2, \mathbf{u}^3$. We define $\mathbf{y} = (\eta_1, \eta_2, \eta_3)$ by taking η_i to be the entry under $\mathbf{u}^i$, $i = 1, 2, 3$. Thus, $\mathbf{y} = (1, 0, 1)$. If we check this vector in the dual problem, we have

$$\mathbb{A}^T\mathbf{y} = \begin{bmatrix} 1 & 3 & 1 \\ 4 & 1 & 1 \end{bmatrix} \begin{pmatrix} 1 \\ 0 \\ 1 \end{pmatrix} = \begin{pmatrix} 2 \\ 5 \end{pmatrix} = \mathbf{c},$$

and $\mathbf{y} \geq \mathbf{0}$, and so $\mathbf{y}$ is feasible for the dual problem. Moreover, $\mathbf{y} \cdot \mathbf{b} = 24 \cdot 1 + 0 \cdot 21 + 9 \cdot 1 = 33 = \mathbf{x} \cdot \mathbf{c}$, and consequently $\mathbf{y}$ is optimal for the dual problem. Naturally, it is no accident that tableau $T^{(3)}$ yields a solution for both the primal problem and the dual problem. This is true in general, and it is one reason that the simplex method is so useful and easy to check. The reasons behind this simultaneous solution of the primal problem and dual problem are explored in the problem set which follows this section (in particular Exercise 12).

The work above is for a special problem. In general the method is exactly the same for any standard problem. Thus the first step in maximizing $\mathbf{x} \cdot \mathbf{c}$ subject to $\mathbf{x} \geq \mathbf{0}$ and $\mathbb{A}\mathbf{x} \leq \mathbf{b}$ is to add a slack vector $\mathbf{w}$ to form a restricted problem. The restricted problem is that of finding a vector $\hat{\mathbf{x}} = (\mathbf{x}, \mathbf{w})$ such that $[\mathbb{A}, \mathbb{I}]\hat{\mathbf{x}} = \mathbf{b}$, $\hat{\mathbf{x}} \geq 0$, and $\hat{\mathbf{x}} \cdot (\mathbf{c}, \mathbf{0}) = \mathbf{x} \cdot \mathbf{c}$ is a maximum for such vectors. If $\mathbf{b} \geq \mathbf{0}$, then a feasible vector is always given by $(\mathbf{0}, \mathbf{b})$. If some components of $\mathbf{b}$ are negative, then the method of Sec. 5.2.3 can be used to obtain a feasible vector. If the problem has an optimal vector then the final tableau will also yield a solution of the dual problem.

In a manner similar to that given for solving a primal problem in standard form, one can directly solve the dual standard problem and the general linear programming problem and its dual. In each case, the problem is converted to a restricted problem and the simplex method is used. These ideas are also explored in the Exercises which follow this section.

5.2.6 Summary of the Simplex Method

In this section we summarize the details of the simplex method in a flow chart (Fig. 5–8). This chart outlines the computations and decisions which go into the solution of a linear programming problem by this method. Accompanying the chart there are notes which discuss the specific rules needed to carry out the steps indicated in the chart. These notes coordinate the flow chart and the techniques developed earlier in this section.

Notes

1. In Sec. 4.3.4 it was shown that every general problem can be written as a standard problem and that every standard problem can be written as a restricted problem. It is easily shown that every restricted maximum problem can be written as a restricted minimum problem. Thus every linear programming problem given in one of these forms can be converted into a restricted minimum problem.

2. The vector equation $\mathbb{A}\mathbf{y} = \mathbf{c}$ contains p scalar equations (we assume here that $\mathbb{A}$ is a $p \times q$ matrix). For each negative coordinate of $\mathbf{c}$ we multiply the corresponding scalar equation by -1. This gives a system of equations in which the right-hand side consists of nonnegative constants. We retain the notation $\mathbb{A}\mathbf{y} = \mathbf{c}$ for the new system of equations.

3. It may happen that a feasible vector is known from the structure of the original problem. For example, consider the problem of maximizing $\mathbf{x} \cdot \mathbf{c}$ subject to $\mathbf{x} \geq \mathbf{0}$ and $\mathbb{A}\mathbf{x} \leq \mathbf{b}$. If this problem is converted into a restricted minimization problem, then we obtain the problem of minimizing $-(\mathbf{x}, \mathbf{w}) \cdot (\mathbf{c}, \mathbf{0}) = -\mathbf{x} \cdot \mathbf{c}$, subject to $(\mathbf{x}, \mathbf{w}) \geq 0$ and

$$[\mathbb{A}, \mathbb{I}]\begin{pmatrix} \mathbf{x} \\ \mathbf{w} \end{pmatrix} = \mathbf{b}.$$

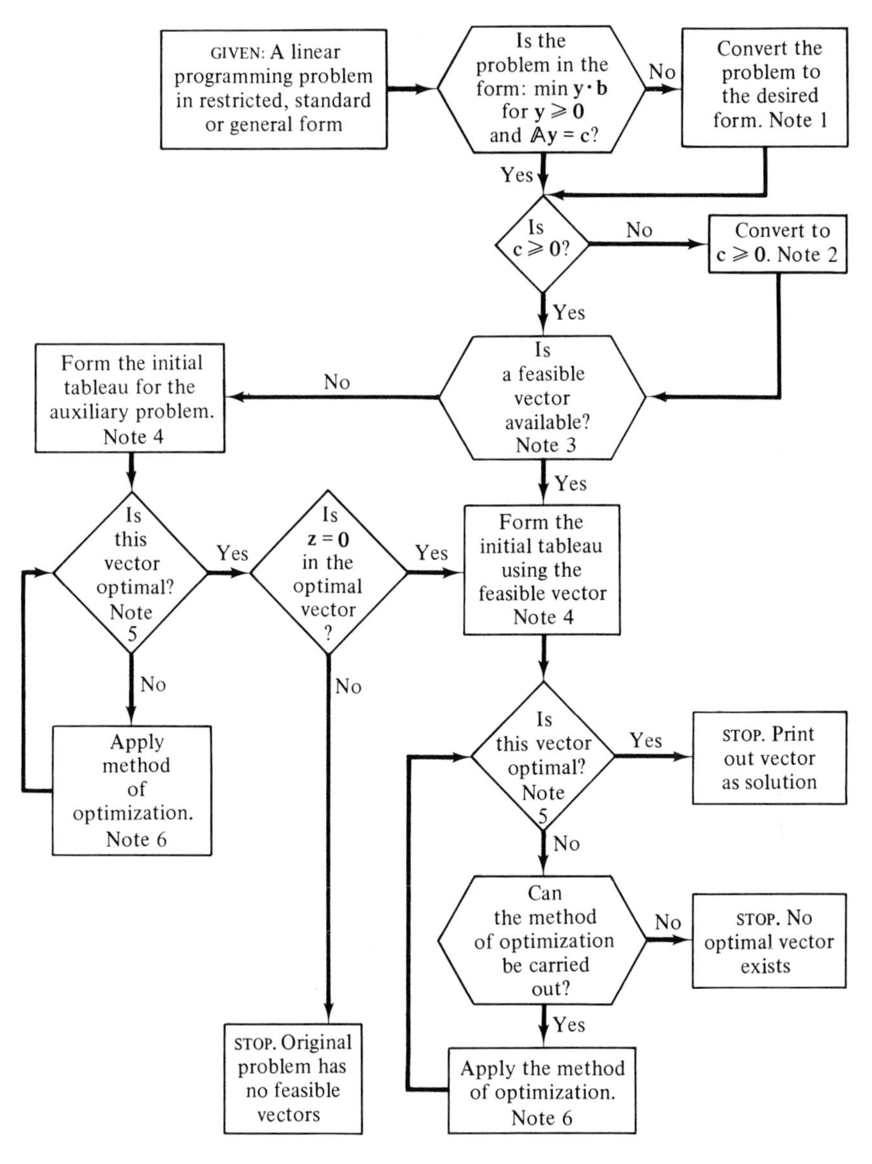

Fig. 5–8

If $\mathbf{b} \geq \mathbf{0}$, then a feasible vector is always given by $\mathbf{x} = \mathbf{0}$ and $\mathbf{w} = \mathbf{b}$. This situation is discussed in detail in Sec. 5.2.5, where an example is worked out.

4. Section 5.2.3 describes a method for obtaining a feasible vector (if one exists) for the problem of minimizing $\mathbf{y} \cdot \mathbf{b}$ subject to $\mathbf{y} \geq \mathbf{0}$ and $\mathbb{A}\mathbf{y} = \mathbf{c}$. The method involves the use of a solution of the problem of minimizing

$(\mathbf{y}, \mathbf{z}) \cdot (\mathbf{0}, \mathbf{u}) = \mathbf{z} \cdot \mathbf{u}$ subject to $(\mathbf{y}, \mathbf{z}) \geq \mathbf{0}$ and

$$[\mathbb{A}, \mathbb{I}]\begin{pmatrix} \mathbf{y} \\ \mathbf{z} \end{pmatrix} = \mathbf{c}.$$

$\mathbf{u}$ is a vector in the same space as $\mathbf{z}$ with all coordinates equal to 1. This auxiliary problem has the natural feasible vector given by $\mathbf{y} = \mathbf{0}$ and $\mathbf{z} = \mathbf{c}$. The method of Sec. 5.2.2 can then be applied.

5. An optimality criterion is discussed in Sec. 5.2.2. The criterion is checked by means of a technique involving the addition of a row to the tableau. A solution is optimal if the last row is nonpositive. See Sec. 5.2.2 for an example.

6. A method of optimization is discussed in Sec. 5.2.2. This method is based on the replacement technique, and a replacement is called for if there are positive entries in the last row of the tableau (i.e., actual cost is less than equivalent cost). The replacement can be made if there is at least one positive entry in the column above the positive entry in the last row. For example, suppose that a feasible vector is given in terms of the independent set of columns $\{\mathbf{a}^i\}_{i \in I}$. Also, suppose that $m \notin I$ and that $v_m - \beta_m > 0$. Under these circumstances, the column $\mathbf{a}^m$ should replace one of the columns $\{\mathbf{a}^i\}_{i \in I}$. Such a replacement can be made if there exists a tableau entry t_{lm} which is positive. The column $\mathbf{a}^m$ then replaces any column $\mathbf{a}^k$ such that the ratio β_k/t_{km} satisfies $\beta_k/t_{km} = \min\{\beta_i/t_{im}: i \in I, t_{im} > 0\}$.

EXERCISES

1. Find the dual of the restricted maximum problem. Use the definition of dual as given for general linear programming problems in Chap. 4.

2. Establish the assertion in the proof of Theorem 3 in Sec. 5.2 that if a replacement is called for but it cannot be made, then no optimal vector exists. What does this assertion say about the boundedness of the set of feasible vectors?

3. Use the replacement method of Sec. 5.2.1 to solve the system of equations $\mathbb{A}\mathbf{x} = \mathbf{b}$ for the following choices of $\mathbb{A}$ and $\mathbf{b}$:

   (a)
$$\mathbb{A} = \begin{bmatrix} -1 & 2 & -1 & 3 & 2 \\ 0 & -2 & 0 & -2 & 1 \\ 1 & 3 & 1 & 0 & 1 \end{bmatrix}, \quad \mathbf{b} = \begin{pmatrix} 14 \\ -10 \\ 5 \end{pmatrix},$$

   (b)
$$\mathbb{A} = \begin{bmatrix} 1 & 2 & 3 \\ 3 & 2 & 1 \\ 1 & -2 & 1 \\ 10 & -5 & 10 \end{bmatrix}, \quad \mathbf{b} = \begin{pmatrix} 14 \\ 10 \\ 0 \\ 30 \end{pmatrix}.$$

4. Use the simplex method to solve the linear programming problems in Exercise 9, Sec. 5.1.

5. Use the simplex method to solve the following linear programming problems:
(a) Minimize $\eta_1 - \eta_2 + \eta_3 - \eta_4$ subject to $\mathbf{y} = (\eta_1, \eta_2, \eta_3, \eta_4) \geq 0$ and

$$\begin{aligned} \eta_1 + \eta_2 - \eta_3 + \eta_4 &= 0, \\ \eta_1 + 2\eta_2 - 3\eta_3 + 4\eta_4 &= -2, \\ 4\eta_1 + 3\eta_2 - 2\eta_3 + \eta_4 &= 2. \end{aligned}$$

(b) Minimize $\eta_1 - \eta_2 + \eta_3 - \eta_4$ subject to the same conditions as (a) plus $\eta_1 - 2\eta_2 + 3\eta_3 - 4\eta_4 = 10$.

6. In Sec. 5.2.3, a method is given for finding feasible vectors for restricted minimum problems. This method depends on finding an optimal vector for an auxiliary problem. Show that this auxiliary problem always has an optimal vector by using results from Chap. 4.

7. Let

$$\mathbb{A}_1 = \begin{bmatrix} 1 & 2 & 0 & 5 & 0 \\ 0 & 1 & 1 & 2 & 0 \\ 1 & 3 & 2 & 0 & 1 \end{bmatrix}, \qquad \mathbf{c} = \begin{pmatrix} 10 \\ 5 \\ 17 \end{pmatrix},$$

and $\mathbf{b}_1 = (3, -2, 5, 2, 0)$. Consider the problem of minimizing $\mathbf{y} \cdot \mathbf{b}_1$ subject to the conditions $\mathbb{A}_1\mathbf{y} = \mathbf{c}$, $\mathbf{y} \geq 0$. Show that there are two feasible vectors for this problem, $\mathbf{y}_1$ and $\mathbf{y}_2$, such that $\mathbf{y}_1 \neq \mathbf{y}_2$ and $\mathbf{y}_1 \cdot \mathbf{b}_1 = \mathbf{y}_2 \cdot \mathbf{b}_1$. Discuss this situation from a geometric point of view.

8. In Sec. 5.2, the simplex method is developed for the problem of minimizing $\mathbf{y} \cdot \mathbf{b}$ subject to $\mathbf{y} \geq 0$ and $\mathbb{A}\mathbf{y} = \mathbf{c}$. Show how the simplex method can be directly used on the similar problem of maximizing $\mathbf{x} \cdot \mathbf{c}$ subject to $\mathbf{x} \geq 0$ and $\mathbb{A}\mathbf{x} = \mathbf{b}$. In this case use the term *value* in the same way that we previously used *cost* with regard to the columns of $\mathbb{A}$ and feasible vectors.

9. Consider the problem of maximizing $\mathbf{x} \cdot \mathbf{u}^1$ subject to $\mathbf{x} \geq 0$ and $\mathbb{A}\mathbf{x} = \mathbf{b}$, where

$$\mathbb{A} = \begin{bmatrix} 1 & 1 & 0 \\ 0 & 1 & 1 \end{bmatrix}, \qquad \mathbf{b} = \begin{pmatrix} 1 \\ 0 \end{pmatrix}, \qquad \text{and} \qquad \mathbf{u}^1 = \begin{pmatrix} 1 \\ 0 \\ 0 \end{pmatrix}.$$

Construct the tableau for this problem using columns $\mathbf{a}^1$ and $\mathbf{a}^2$ as the independent set yielding a feasible vector. Show that this feasible vector is optimal but that the criterion for optimality is not satisfied. Finally, find a basis for the solution (i.e., independent vectors at the left of the tableau) for which the optimality criterion is satisfied.

10. Consider the phenomenon of Exercise 9 in the general case. Show that it cannot happen unless $\mathbf{b}$ is a linear combination of less than p columns of $\mathbb{A}$. That is, show that if $\mathbf{b}$ cannot be written as a linear combination of less than p columns of $\mathbb{A}$ and if $\mathbf{x}$ is an optimal vector with at most p nonzero coordinates, then $\mathbf{x}$ satisfies the optimality criterion.

11. Let $\mathbb{A}$ be an $n \times n$ matrix. We say that $\mathbb{A}$ is invertible if there exists an $n \times n$ matrix $\mathbb{B}$ such that $\mathbb{A}\mathbb{B} = \mathbb{B}\mathbb{A} = \mathbb{I}$ (Appendix C). If a square matrix $\mathbb{A}$ has an inverse, then the replacement method of Sec. 5.2.1 can be used to obtain this inverse. Using the terminology of Sec. 5.2.1, the inverse is obtained in the following manner:

(a) Let W and V be the sets

$$W = \{\mathbf{a}^1, \ldots, \mathbf{a}^n, \mathbf{u}^1, \ldots, \mathbf{u}^n\},$$
$$V = \{\mathbf{u}^1, \ldots, \mathbf{u}^n\}.$$

(b) Form the tableau of W with respect to V. This tableau has the form

$$T = \begin{array}{c|c|c|} & \mathbf{a}^1 \cdots \mathbf{a}^n & \mathbf{u}^1 \cdots \mathbf{u}^n \\ \hline \mathbf{u}^1 \\ \vdots & \mathbb{A} & \mathbb{I} \\ \mathbf{u}^n \\ \hline \end{array}.$$

(c) Replace as many $\mathbf{u}^i$s by $\mathbf{a}^i$s as possible and arrange the left-hand side of the tableau so that the $\mathbf{a}^i$s are first and in order. Thus, if all $\mathbf{u}^i$s are replaced, then the new tableau has the form

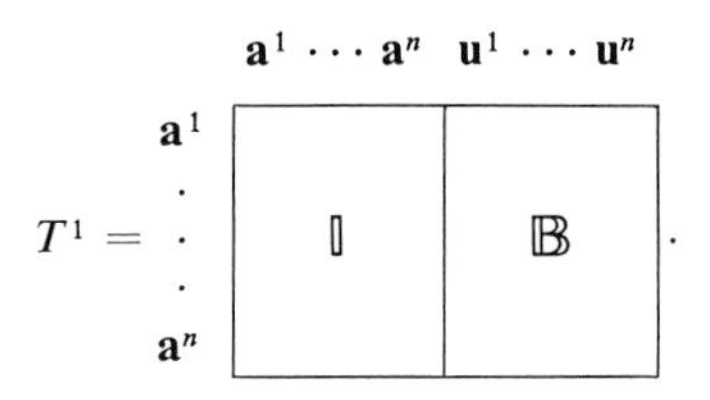

(d) If all the $\mathbf{u}^i$s are replaced by $\mathbf{a}^i$s, then $\mathbb{A}$ has an inverse and this inverse is the matrix $\mathbb{B}$ in tableau T^1 of part c. $\mathbb{B}$ is usually denoted by $\mathbb{A}^{-1}$.

Establish the assertion of part d.

12. In Sec. 5.2.5 it was asserted that the simplex method simultaneously solved the primal standard maximum problem and the dual standard minimum problem. In this exercise the reader is asked to establish this assertion under certain conditions on the matrix $\mathbb{A}$. Specifically, let $\mathbb{A}$ be a $p \times p$ matrix which is invertible, and let $\mathbf{b}$ be a nonnegative vector in R^p. Consider the problem of maximizing $\mathbf{x} \cdot \mathbf{c}$ subject to $\mathbb{A}\mathbf{x} \leq \mathbf{b}$ and $\mathbf{x} \geq 0$. We let $\mathbf{w}$ be the slack vector, and we form the initial tableau as follows:

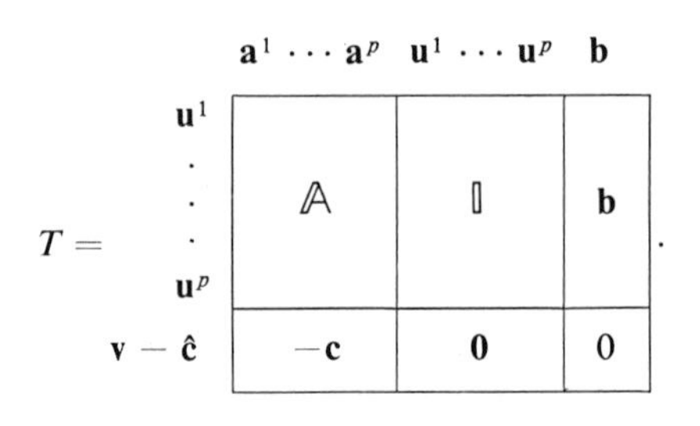

Next, we replace all the $\mathbf{u}^i$s by $\mathbf{a}^i$s and obtain the tableau

$T' =$

	$\mathbf{a}^1 \ \mathbf{a}^2 \cdots \mathbf{a}^p$	$\mathbf{u}^1 \ \mathbf{u}^2 \cdots \mathbf{u}^p$	$\mathbf{b}$
$\mathbf{a}^1$ $\vdots$ $\mathbf{a}^p$	$\mathbb{I}$	$\mathbb{A}^{-1}$	$\mathbf{x}$
$\mathbf{v}' - \hat{\mathbf{c}}$	$\mathbf{z}$	$\mathbf{y}$	$\mathbf{x} \cdot \mathbf{c}$

Show that the vector $\mathbf{y}$ of tableau T' is an optimal vector for the problem of minimizing $\mathbf{y} \cdot \mathbf{b}$ subject to $\mathbf{y} \geq 0$ and $\mathbb{A}^T\mathbf{y} \geq \mathbf{c}$. *Hint:* Use the facts that $\mathbf{y} = \mathbf{c}\mathbb{A}^{-1}$ if and only if $\mathbf{y}\mathbb{A} = \mathbf{c}$ and that $\mathbf{y}\mathbb{A} = \mathbf{c}$ if and only if $\mathbb{A}^T\mathbf{y} = \mathbf{c}$.

13. Let $\mathbb{A}$ be a $p \times q$ matrix, $p > q$ and $\mathbf{c} \in R^p$. Then there is a matrix $\hat{\mathbb{A}}$ of dimension $p' \times q$, $p' \leq q$, $\hat{\mathbf{c}} \in R^{p'}$ such that

$$\{\mathbf{x} \in R^q : \mathbb{A}\mathbf{x} = \mathbf{c}\} = \{\mathbf{x} \in R^q : \hat{\mathbb{A}}\mathbf{x} = \hat{\mathbf{c}}\}.$$

5.3 PROJECTS

5.3.1 For Computer Buffs

Write a computer program for the simplex method as applied to a restricted minimum problem where the matrix $\mathbb{A}$ has four rows and five columns.

5.3.2 Quadratic Geometry

In Project 4.5.2 quadratic objective functions were introduced as a generalization of linear objective functions. The geometry discussed in Sec. 5.1 is based in part on a linear objective function. Consider the geometry which results from a quadratic function. First consider a two-by-two example and study this in detail. Then consider the general case.

5.3.3 The Efficient Use of Rats

A psychologist wishes to conduct two types of studies which he designates types I and II. Each experiment requires white rates, gray rats, and black rats in the following numbers. One experiment of type I requires 5 white rats, 1 gray rat, and 2 black rats, while one experiment of type II requires 2 white rats, 3 gray rats, and 2 black rats. The psychologist has available for use in his study 100 white rats, 60 gray rats, and 50 black rats. Also, the psychologist has decided that each experiment of type II has twice the "value" of an experiment of type I.

Problem 1. How many experiments of each type should be conducted in order to maximize the value of the study?

Problem 2. Same as Problem 1 with the restriction that at least ten experiments of type I must be conducted.

Remark. Note that the number of experiments of each type must be an integer.

Problem 3. Consider the development of an algorithm to solve problems such as 1 and 2 above. The simplex method is not appropriate because the answer may not be an integer. Can you think of another approach to such problems?

REFERENCES

The references of Chap. 4 are also appropriate for this chapter. In particular, we again cite references [Da], [Ga], and [Si] since they are especially relevant to the geometric and computational aspects of linear optimization.

The geometry associated with problems of linear programming in R^2 is covered in [Gl] and [Sp]. The same topic in higher dimensions is treated in detail in both [BeGH] and [DPZ]. The basic concepts of convex sets and functions are given in [BeGH], and there is an in-depth treatment in [Val].

Section 5.2 on the simplex method is modeled after the treatment in [Ga], although the setting of that discussion is different. The spirit and level of [Ga] are somewhat similar to those of this text, and we recommend it for additional study of linear programming and related topics. The simplex method is also treated in almost every other book on linear programming listed below. Specifically, we note that it is covered in [Da], [Gl], and [Sp]. Generalizations of the topics considered here can be found in [BeGH], [DPZ], and [Kr].

[BeGH] BERGE, C., and A. GHOUILA-HOURI, *Programming, Games and Transportation Networks*. London: Methuen, 1965.

[Da] DANTZIG, G. B., *Linear Programming and Extensions*. Princeton, N.J.: Princeton University Press, 1963.

[DPZ] DUFFIN, R. J., E. L. PETERSON, and C. ZENER, *Geometric Programming*. New York: Wiley, 1967.

[Ga] GALE, D., *The Theory of Linear Economic Models*. New York: McGraw-Hill, 1960.

[Gl] GLICKSMAN, A. M., *An Introduction to Linear Programming and the Theory of Games*. New York: Wiley, 1963.

[Kr] KREKÓ, B., *Linear Programming*, tr. J. H. L. Ahrens and C. M. Safe. London: Sir Isaac Pitman & Sons, Ltd., 1968.

[Si] SIMONNARD, M., *Linear Programming*, tr. W. S. Jewell. Englewood Cliffs, N.J.: Prentice-Hall, 1966.

[Sp] SPIVEY, W. Allen, *Linear Programming: An Introduction*. New York: Macmillan, 1963.

[Val] VALENTINE, F. A., *Convex Sets*. New York: McGraw-Hill, 1964.

6 Models Involving Chance, Choice, and Competition

6.0 INTRODUCTION

In Chaps. 4 and 5 we applied linear methods to problems of optimization, allocation, and assignment. In certain cases the same or related methods can be applied to problems which involve decision making in a competitive environment. However, these linear methods do not in general suffice for the solution of such problems, and thus new methods and techniques must be developed. The purpose of this chapter is to introduce the reader to a selection of the many problems which may be treated by means of a model—or a class of models—involving randomness (chance) and strategy (educated choice). Problems with these basic characteristics are quite common, and they represent a setting which we shall study in detail. We devote most of our attention to two particular types of models. For one of these there exists a well-developed mathematical theory which is closely related to the theory of linear optimization. The other type of model is quite different in nature, and our discussion of it will be much less complete.

We have already encountered one example of a problem involving chance and choice, and we briefly discussed a model for it. We refer to the political party strategy problem of Sec. 2.5. The situation there can be summarized as follows: Each party must make a decision (when to broadcast), and the consequences of that decision depend on chance (the probability of being effective) and strategy (the opposing party's decision). We may think of the two parties here as players in a game, and it is in this general setting (of players in games) where we shall study all similar problems. Of course, in the general competitive situation (called a *game* henceforth) the relationship between the participants (now called *players*) may be very complex. First, the number of players usually varies with the game; sometimes there are many players and sometimes as few as two. Second, although the rules which determine the gains (or losses) of the players are determined in advance, they may vary with the game. Several players may gain from the loss of a single player; or, perhaps, one player may gain while many players lose; or it may even be that all players lose while a nonplayer gains (for example, a neutral country during a war). As an illustration of some of these possibilities we shall begin our study by presenting two examples. Each is a special case of a problem which will subsequently be examined in detail.

Problem A: A Price War. The well-known village, Smalltown, U.S.A., contains exactly two service stations which sell gasoline to the local citizens. These stations compete for the same fixed number of customers, and essentially the only variable subject to their control is the price that they charge for gasoline. Thus they both clean the windows, check the oil, and provide reasonably courteous service. Each station determines for itself the price it will charge, but this price must reflect the prevailing facts of business life. At the time under consideration the going rate for regular gasoline (the only item we shall consider) is 40 cents a gallon. The local operators pay a total of 32 cents a gallon to the oil companies which provide gasoline and to state and federal governments in taxes; they keep the remaining 8 cents a gallon as their gross profit. However, the oil companies encourage price competition by giving the local dealers the option of lowering their price to 35 cents a gallon. If the retail price is 35 cents a gallon, then the oil company and taxes take 30 cents and the local dealer keeps only 5 cents. It is reasonable to assume that at this lower rate of return the local dealer needs a substantial increase in sales to make a profit since most of his expenses are fixed and independent of the price of gasoline. For example, rent, labor, and maintenance costs do not decrease when the price of gas is lowered. Thus, if the dealer's volume of sales does not increase, then he loses money, or at best makes less profit at the lower price. However, there is another aspect to be considered. Each dealer also knows that if one station is selling gas at 5 cents a gallon less than the other, then the station with the lower price will indeed experience a significant increase in sales. Simultaneously the station with the higher price suffers a corresponding loss of business and an associated loss of income. To be more specific, suppose that both dealers know that their net income is approximately $100 per day when they both sell gasoline at a price of 40 cents a gallon. If both set their price at 35 cents a gallon, then they lose \$50 a day; and if one has a price of 40 cents and the other has a price of 35 cents, then the station with the higher price loses \$100 per day and the station with the lower price makes about \$200 per day.

The Problem. Formulate a day-to-day schedule of price selection for each local dealer.

Problem B: The Border Battle. The alpine village of Osohi is having trouble with a gang of smugglers. They enter the village at night and remove priceless art treasures. The village is defended by the Captain of the Border and his force of four patrolmen. The smugglers are directed by the crafty crook known as Major Nuisance, and he always has three smugglers in his gang. The village can be entered only at two spots: the mountain pass and the river outlet of Lake Osohi. The four patrolmen must defend these entry points each evening as the three smugglers always try to enter somewhere. If either side has a superior number of men at an entry point, then these men capture the opposing force, and they control the entry point for the night. If the forces are equal in strength, then the match is a draw, the smugglers do not enter, and neither side captures anyone. If the smugglers capture control of an entry point, then they receive the benefits of a night of smuggling, and also they have some patrolmen to hold for ransom or exchange purposes. Suppose that the capture of a single individual by his opponents is given the value $+1$, and suppose that each group considers a night of smuggling equal in value to the capture of a patrolman. Hence when the smugglers control an entry point, their reward is $k =$ (number of patrolmen captured) $+ 1$. If the patrolmen capture some smugglers, then they count a reward of $+1$ for each man captured. The assignment of values might be thought of as reflecting the assumption that these captured men can be used to exchange for captured patrolmen in a one-to-one manner.

The Problem. How should the Captain of the Border (C.B. from now on) and Major Nuisance (M.N. from now on) deploy their men?

Problems A and B are illustrations of the sort of topics discussed in the mathematical discipline known as game theory. This theory has received considerable attention during the last 25 or 30 years, and it and its applications are the subject matter of this chapter. We shall begin with the simplest case: the case of two contestants in a competition where the gain of one is the loss of the other.

6.1 TWO-PERSON ZERO-SUM GAMES

A basic characteristic of problems A and B above is that of competitive choice. Each individual has certain choices available to him, and he knows the consequences of these choices for each choice of his opponent. However, and this is an essential feature, he must make his choice independent of any knowledge of his opponent's decision. Since this situation is similar in nature to many parlor games, it is traditional to refer to any such problem as a *game*, and to call the individuals involved *players*. We now proceed with the formulation of a mathematical model, restricting ourselves for the time being to games with only two players.

Let S be the set consisting of the choices of action for one of the players, say player Red, and let T be the set of choices of action for the second player, say player Cream. Then each element of the cartesian product $S \times T$ will correspond to a pair of choices, one made by player Red and one by player Cream. The consequence to each player for each pair of choices is assumed known and is usually called the *payoff* for that player. Thus the payoff for a player is a function defined on $S \times T$. Let us look at Problem A from this point of view. We let c denote the *cooperative* choice of maintaining the price of gasoline at 40 cents per gallon and n the *noncooperative* alternative. We have $S = \{c, n\} = T$ and $S \times T = \{(c, c), (c, n), (n, c), (n, n)\}$. Select one of the dealers (arbitrarily) as Red and let $p_R(s, t)$ denote the payoff to Red when he elects choice $s \in S$ and his opponent uses choice $t \in T$. The information provided with the example gives $p_R(c, c) = \$100.00$, $p_R(c, n) = -\$100.00$, $p_R(n, c) = \$200.00$, and $p_R(n, n) = -\$50.00$. Notice that the function p_R has domain $S \times T$ and that its range is the set of possible payoffs to player Red. There is a corresponding function p_C which associates with each element of $S \times T$ the appropriate payoff for player Cream, the other dealer. The definitions can easily be extended to the case where there are more than two sets of opposing interests.

In certain types of games the payoff functions exhibit special features. A particularly interesting case, and one for which the theory is well developed, is that of games where p_C is equal to the negative of p_R. This corresponds to a situation where the gain of one player is matched by the loss of the second player. It is clear that under such conditions we need to be concerned only with a single payoff function. An example of a game of this sort is provided by Problem B. We proceed now to give formal definitions which cover games of the latter type. We shall return to Problem A and certain related problems later (Sec. 6.2).

6.1.1 Basic Characteristics

A *two-person zero-sum* game Γ consists of a pair of sets S and T and a function p defined on $S \times T$. The elements of S and T are called the *pure strategies for the players Red and Cream*, respectively. The function p is the *payoff function* to Red. If $s \in S$ and $t \in T$, then $p(s, t)$ is called the gain for Red when he uses strategy s and Cream uses strategy t. If $p(s, t) < 0$, then this gain is negative (i.e., Red suffers a loss). The corresponding gain (or loss) to Cream is $-p(s, t)$. We denote the game Γ by $\Gamma = (S, T, p)$.

We now show that problem B can be viewed as a *two-person zero-sum* game. We do this by finding appropriate sets S and T and a function p. We let C.B. be the Red player. Since he must decide how to deploy his four men in defense of the two entry points, each of his strategies involves two numbers, the number of men at the first entry and the number of men at the second entry. We denote such a strategy by an ordered pair (a, b), where a is the number of men at the first entry point and b is the number at the second entry point. Since $a + b = 4$, we obtain the strategy set $S = \{(4, 0), (3, 1), (2, 2), (1, 3), (0, 4)\}$. Likewise, the strategy set for M.N. is $T = \{(3, 0), (2, 1), (1, 2), (0, 3)\}$. It is convenient to define and exhibit the payoff function p by means of a matrix, and for this reason games such as this one are often called *matrix games*. We associate the elements of S with rows, the elements of T with columns, and the quantity $p(s, t)$ with the matrix element in the row associated with s and the column associated with t. It is for this reason that we have chosen to call the players Red (row) and Cream (column). We obtain the following matrix:

		M.N. or Cream Player			
		(3, 0)	(2, 1)	(1, 2)	(0, 3)
C.B. or Red Player	(4, 0)	+3	+1	0	−1
	(3, 1)	0	+2	−1	−2
	(2, 2)	−3	+1	+1	−3
	(1, 3)	−2	−1	+2	0
	(0, 4)	−1	0	+1	+3

For example, $p((2, 2), (1, 2)) = +1$ because, if C.B. defends each entry point with 2 men and if M.N. sends one man to entry 1 and two men to entry 2, then C.B. captures the man sent to entry 1, and the action at entry 2 is a draw. Thus the net gain for C.B. is $+1$.

As we shall see later, it is useful to take the point of view that each game will be played a number of times. Adopting this perspective, we can obtain some information about the border battle by simply inspecting the payoff matrix. For example, from C.B.'s point of view it does not seem wise for him to use strategy (2, 2) very often. Indeed, this strategy will sometimes cost him three men while it can never win him more than one, and no other strategy can cost him more than two men. In addition, it appears that strategies (0, 4) or (4, 0) are each better than (3, 1) or (1, 3). Does this mean that C.B. should always use strategy (4, 0)? If he did so, then soon M.N. would catch on to his plan and M.N. would then use (0, 3) and take an entry point. It would seem, therefore, that it may not be desirable for a player to consistently use the same strategy no matter what its merits may be. Indeed, a possible tactic for C.B. would be to use (4, 0) half of the time and (0, 4) the other half of the time. That is, he might decide to mix his pure strategies. This idea of mixing pure strategies is an important and useful idea in game theory.

6.1.2 Mixed Strategies and Optimal Strategies

Definition: A *mixed strategy* for the Red player in the game $\Gamma = (S, T, p)$ is a real-valued function σ defined on S such that $\sigma(s) \geq 0$ for each $s \in S$ and $\sum_{s \in S} \sigma(s) = 1$.

This definition of a mixed strategy σ associates a number between 0 and 1 with each pure strategy. The real number $\sigma(s)$ can be interpreted as the probability of using pure strategy s on a specific play of the game when mixed strategy σ is adopted. From this point of view, a mixed strategy is simply a formula for deciding how often each pure strategy should be used. We denote a mixed strategy for Cream by τ and note that τ is a function defined on T. The function τ has properties like those of σ. Namely,

1. τ is a nonnegative function defined on T.
2. $\sum_{t \in T} \tau(t) = 1$.

We refer to σ and τ as *strategy functions* as well as mixed strategies.

It is useful at this point to observe that the set of pure strategies can be considered to be a subset of the set of mixed strategies. Indeed, each pure strategy can be associated with a mixed strategy of a special type, namely a mixed strategy whose strategy function is 0 for all but one pure strategy. For example, if $s_1 \in S$, then the mixed strategy associated with s_1 is the function σ_1 defined by

$$\sigma_1(s) = \begin{cases} 0, & s \neq s_1, \\ 1, & s = s_1. \end{cases}$$

Thus the mixed strategy σ_1 corresponds to playing the pure strategy s_1 with probability 1 and all other pure strategies with probabilities 0.

In this section we are concerned exclusively with games in which the sets of pure strategies are finite sets. In such cases it is very convenient to give a second (and equivalent) formulation of the definition of a mixed strategy. First, suppose that each of the strategy sets is ordered, say $S = \{s_1, s_2, \ldots, s_m\}$ and $T = \{t_1, t_2, \ldots, t_n\}$. Note that this is an abuse of the set notation since we now suppose that the elements of S and T have a specific order. We prefer this abuse to the introduction of a new symbol. A mixed strategy σ for Red is completely defined by specifying the m values $\sigma(s_1), \sigma(s_2), \ldots, \sigma(s_m)$ or, if we agree to maintain the ordering for S, by specifying an m-tuple of positive real numbers whose sum is 1. Thus mixed strategies for Red can be thought of as special vectors in R^m, and the definition given above is equivalent to the following.

Definition: A *mixed strategy* for the Red player in the game $\Gamma = (S, T, p)$, where S consists of m pure strategies, is a probability vector $\boldsymbol{\sigma} \in R^m$. The coordinates of $\boldsymbol{\sigma}$ are denoted by $\sigma(s_1), \ldots, \sigma(s_m)$, or generically by $\sigma(s)$.

Clearly, mixed strategies for the player Cream can be defined to be probability vectors in the same way as has been done for Red.

The association between pure strategies and mixed strategies which we discussed above is easily represented with the definition of mixed strategies as probability vectors. Pure strategies are associated with mixed strategies which are unit vectors.

Next, it is necessary for us to consider the meaning of the term payoff when it is used in connection with two mixed strategies $\boldsymbol{\sigma}$ and $\boldsymbol{\tau}$. When the players are using mixed strategies, it is not possible to give a meaning to the notion of a payoff for a single play. Indeed, the payoff is defined for a pair of pure strategies, and we do not know which pure strategy each player will be using in any specific play of the game. However, we do know how often each will use a given pure strategy "in the long run," since this is precisely the information furnished by the strategy function. Thus, if Red uses mixed strategy $\boldsymbol{\sigma}$ and Cream uses mixed strategy $\boldsymbol{\tau}$ and if they play a large number of games, then we can compute the expected gain per game for each player. We denote this expected gain per game for Red by $p(\boldsymbol{\sigma}, \boldsymbol{\tau})$. Using elementary probability theory, we have

$$p(\boldsymbol{\sigma}, \boldsymbol{\tau}) = \sum_{s \in S} \sum_{t \in T} \sigma(s)\tau(t)p(s, t). \tag{1.1}$$

This quantity is known as the *expected payoff per game* or simply the payoff for Red when he uses mixed strategy $\boldsymbol{\sigma}$ and Cream uses mixed strategy $\boldsymbol{\tau}$. If the sets S and T are finite and if we let $\mathbb{G} = (g_{ij})$ denote the payoff or gain matrix, i.e., $g_{ij} = p(s_i, t_j)$, then the expected payoff can be written in vector form as $p(\boldsymbol{\sigma}, \boldsymbol{\tau}) = \boldsymbol{\sigma}\mathbb{G}\boldsymbol{\tau}$. Since we are assuming that the strategy sets are finite, we shall use this matrix notation whenever it is convenient. In each case, it is understood that the strategy sets are ordered and the matrix $\mathbb{G}$ and the mixed strategies $\boldsymbol{\sigma}$ and $\boldsymbol{\tau}$ reflect this ordering.

Let $\mathfrak{S}$ denote the set of mixed strategies for Red, and let $\mathfrak{T}$ denote the set of mixed strategies for Cream. As noted above, we can consider S to be a subset of $\mathfrak{S}$ and T to be a subset of $\mathfrak{T}$. Also, there is an expected payoff function p defined on $\mathfrak{S} \times \mathfrak{T}$ by Eq. (1.1). Note that when this is restricted to $S \times T$, it agrees with the original payoff function defined earlier. This justifies the dual use of the symbol p since the new function is simply an extension of the original function.

We return to Problem B to illustrate these ideas. Suppose, for example, that Red decides to use only strategies (4, 0) and (0, 4), each one half the time. That is, he adopts the mixed strategy σ, where σ is defined by

$$\sigma(s) = \begin{cases} 0, & s = (3, 1), (2, 2), (1, 3), \\ \frac{1}{2}, & s = (4, 0), (0, 4). \end{cases}$$

Using vector notation, we have $\boldsymbol{\sigma} = (1/2, 0, 0, 0, 1/2)$. Let $\boldsymbol{\tau} = (\alpha, \beta, \gamma, \delta)$ be any strategy vector for Cream. Then $0 \le \alpha, \beta, \gamma, \delta \le 1$ and $\alpha + \beta + \gamma + \delta = 1$. Using the definition of the payoff function, Eq. (1.1), we have

$$\begin{aligned} p(\boldsymbol{\sigma}, \boldsymbol{\tau}) = \boldsymbol{\sigma}\mathfrak{G}\boldsymbol{\tau} &= \tfrac{1}{2}(3\alpha + \beta - \delta - \alpha + \gamma + 3\delta) \\ &= \tfrac{1}{2}(2\alpha + 2\delta + \beta + \gamma) \\ &= 1 - \frac{\beta + \gamma}{2}. \end{aligned}$$

Recall that β and γ are nonnegative and that their sum is always ≤ 1. Consequently, $p(\boldsymbol{\sigma}, \boldsymbol{\tau}) = 1 - [(\beta + \gamma)/2] \ge \frac{1}{2}$. We conclude from this calculation that Red can make his expected gain at least 1/2 by playing strategies (4, 0) and (0, 4) only. His gain will actually be $>$ 1/2 if Cream uses (3, 0) or (0, 3) at all. This leads us to consider the notion of a solution of a game.

Definition: A game $\Gamma = (S, T, p)$ is said to have *value* v if there exist mixed strategies $\boldsymbol{\sigma}_0$ and $\boldsymbol{\tau}_0$ for Red and Cream, respectively, such that $v = p(\boldsymbol{\sigma}_0, \boldsymbol{\tau}_0)$ and $p(\boldsymbol{\sigma}_0, \boldsymbol{\tau}) \ge p(\boldsymbol{\sigma}_0, \boldsymbol{\tau}_0) \ge p(\boldsymbol{\sigma}, \boldsymbol{\tau}_0)$ for all $\boldsymbol{\sigma} \in \mathfrak{S}$ and $\boldsymbol{\tau} \in \mathfrak{T}$. The mixed strategies $\boldsymbol{\sigma}_0$ and $\boldsymbol{\tau}_0$ are called *optimal strategies* and the triple $(\boldsymbol{\sigma}_0, \boldsymbol{\tau}_0, v)$ is called a *solution* of the game.

As will be shown below, optimal strategies need not be unique. The value of the game is uniquely defined (Exercise 1).

The discussion above has shown that if the border battle game has a solution, then the value of the game must be at least 1/2. We now prove that the game actually has a value of 5/9. Later we shall show how this value can be obtained from the information given in the statement of Problem B. Let $\boldsymbol{\sigma}_0 = (4/9, 0, 1/9, 0, 4/9)$ be a strategy for the Red player, C.B., and let $\boldsymbol{\tau} = (\alpha, \beta, \gamma, 1 - \alpha - \beta - \gamma)$ be any strategy for the Cream player, M.N. We have

$$\begin{aligned} p(\boldsymbol{\sigma}_0, \boldsymbol{\tau}) = \boldsymbol{\sigma}_0\mathfrak{G}\boldsymbol{\tau} &= \tfrac{4}{9}(3\alpha + \beta - 1 + \alpha + \beta + \gamma) \\ &\quad + \tfrac{1}{9}(-3\alpha + \beta + \gamma - 3 + 3\alpha + 3\beta + 3\gamma) \\ &\quad + \tfrac{4}{9}(-\alpha + \gamma + 3 - 3\alpha - 3\beta - 3\gamma) \\ &= \tfrac{5}{9}. \end{aligned}$$

Thus, if C.B. uses $\boldsymbol{\sigma}_0$, then his expected gain is always 5/9, independent of the strategy chosen by M.N.

Next, let $\boldsymbol{\tau}_0 = (1/18, 4/9, 4/9, 1/18)$ be a strategy for M.N., and let $\boldsymbol{\sigma} = (\alpha, \beta, \gamma, \delta, 1 - \alpha - \beta - \gamma - \delta)$ be any strategy for C.B. Then,

$$\begin{aligned} p(\boldsymbol{\sigma}, \boldsymbol{\tau}_0) = \boldsymbol{\sigma}\mathfrak{G}\boldsymbol{\tau}_0 &= \tfrac{1}{18}(3\alpha - 3\gamma - 2\delta - 1 + \alpha + \beta + \gamma + \delta) \\ &\quad + \tfrac{8}{18}(\alpha + 2\beta + \gamma - \delta) \\ &\quad + \tfrac{8}{18}(-\beta + \gamma + 2\delta + 1 - \alpha - \beta - \gamma - \delta) \\ &\quad + \tfrac{1}{18}(-\alpha - 2\beta - 3\gamma + 3 - 3\alpha - 3\beta - 3\gamma - 3\delta) \\ &= \tfrac{5}{9} - \tfrac{2}{9}(\beta + \delta). \end{aligned}$$

Since β and δ are nonnegative, we see that $p(\boldsymbol{\sigma}, \boldsymbol{\tau}_0) \leq \frac{5}{9}$ for all $\boldsymbol{\sigma}$. Therefore, $\boldsymbol{\sigma}_0$ and $\boldsymbol{\tau}_0$ are optimal strategies and the value of this game is $v = \frac{5}{9}$.

Let us now return to the general situation and consider the question of determining the most desirable plan of action for the players of a game. We look at the outcome from the point of view of Red, and we make a fundamental assumption regarding his behavior. We suppose that his goal is to *maximize his long-run return or payoff.* We emphasize that this assumption is crucial to what follows, and we add that it may well not be true for certain players of two-person zero-sum games. A player might instead seek an occasional large payoff, or he may adopt any of a number of goals. However, for the time being we assume that he seeks *the largest possible average payoff.* We make a corresponding assumption for Cream, i.e., we assume that Cream seeks to maximize his average gain—or, in view of the zero-sum assumption, this is equivalent to Cream seeking to minimize the average gain of Red. Now, for any pure strategy s chosen by Red, he can be certain of a payoff of at least $\min_{t \in T} p(s, t)$, where the minimum is taken over all pure strategies for Cream. But, in addition, Red is at liberty to choose which pure strategy he will use, and consequently, he can make his choice in such a way that his payoff will be at least $\max_{s \in S} \min_{t \in T} p(s, t)$.

We now make use of our assumption that the game is zero-sum, that is, the payoff to Cream is the negative of the payoff to Red. Then for any pure strategy t which Cream chooses, he can be certain of a payoff of at least

$$\min_{s \in S} (-p(s, t)) = -\max_{s \in S} p(s, t).$$

Another way of saying this is that no matter which pure strategy t is played by Cream, Red will obtain no more than $\max_{s \in S} p(s, t)$. The pure strategy t to be used is still at Cream's disposal, and he will select t so that Red receives no more than $\min_{t \in T} \max_{s \in S} p(s, t)$.

In summary, there is a way for Red to play, that is, there is a pure strategy for Red, for which he obtains a payoff of at least $\max_{s \in S} \min_{t \in T} p(s, t)$, and there is a way for Cream to play, that is, a pure strategy for Cream, for which Red obtains no more than $\min_{t \in T} \max_{s \in T} p(s, t)$.

From this discussion one expects that in general

$$\max_{s \in S} \min_{t \in T} p(s, t) \leq \min_{t \in T} \max_{s \in S} p(s, t), \tag{1.2}$$

and, indeed, this is the case. To verify this inequality in a rigorous manner, we set

$$\max_{s \in S} \min_{t \in T} p(s, t) = p(\bar{s}, \bar{t}) \qquad \text{and} \qquad \min_{t \in T} \max_{s \in S} p(s, t) = p(\tilde{s}, \tilde{t}).$$

In terms of the payoff matrix, this means that $p(\bar{s}, \bar{t})$ is the minimum of the $\bar{s}$th row. Thus, in particular, $p(\bar{s}, \bar{t}) \leq p(\bar{s}, \tilde{t})$. Similarly, $p(\tilde{s}, \tilde{t})$ is the maximum of $\tilde{t}$th column and in particular $p(\bar{s}, \tilde{t}) \leq p(\tilde{s}, \tilde{t})$. Taking these together, we have $p(\bar{s}, \bar{t}) \leq p(\bar{s}, \tilde{t}) \leq p(\tilde{s}, \tilde{t})$, and the proof of the inequality is complete. These inequalities and the reasoning behind them are basic to that aspect of the study of games known as min–max theory.

Inequality (1.2) may be an equality or a strict inequality depending on the payoff function p. If p is given by the matrix

$$\begin{bmatrix} 2 & 1 \\ 1 & 2 \end{bmatrix},$$

then $\max_s \min_t p(s, t) = 1$, while $\min_t \max_s p(s, t) = 2$. An easy example in which equality holds is given by a matrix all of whose entries are the same.

The case where the inequality (1.2) becomes an equality is of particular interest. Suppose that the common value of the right-hand side and the left-hand side of (1.2) is v. Then Red can select a pure strategy $\bar{s}$ so that his payoff is at least v and Cream can select a pure strategy $\bar{t}$ so that the payoff to Red is at most v. That is, $p(s, \bar{t}) \leq p(\bar{s}, \bar{t}) \leq p(\bar{s}, t)$ and $p(\bar{s}, \bar{t}) = v$. Since Red cannot do better than to choose $\bar{s}$ and Cream cannot do better than to choose $\bar{t}$, these pure strategies are *optimal strategies*. Notice that in addition to the properties indicated above, the optimal strategies have the further property that if Red were to announce in advance that he intended to use pure strategy $\bar{s}$, then Cream cannot use this information to reduce the payoff to Red. Likewise, if Cream announced in advance that he intended to use $\bar{t}$, then Red cannot use this information to increase his payoff.

If equality holds in (1.2), then the payoff matrix (and the game) is said to have a *saddle point* at $(\bar{s}, \bar{t})$, and the number $v = p(\bar{s}, \bar{t})$ is the *value of the game*. Note that a game may have several saddle points, but the payoff must be the same at each of them. In such cases there are several distinct optimal strategies, one pair for each saddle point. Also, there may be an entry in the payoff matrix which is equal to the value of the game but which is not a saddle point. For example, consider the game with payoff matrix

$$\begin{bmatrix} 1 & -3 & -2 \\ 2 & 5 & 4 \\ 2 & 3 & 2 \end{bmatrix}.$$

The value of the game is 2 $[\max_s \min_t p(s, t) = \min_t \max_s p(s, t) = 2]$, but the entry in row 3, column 3 is not a saddle point.

It is very important to note that not all games have a saddle point. In particular, the border battle game does not have one. For such games, we have

$$\max_{s \in S} \min_{t \in T} p(s, t) < \min_{t \in T} \max_{s \in S} p(s, t).$$

The left-hand side of the above inequality represents the greatest payoff that Red can be assured of obtaining, and the right-hand side represents the least payoff that Cream must make.

A simple example will be useful here. Suppose that a two-person game has payoff matrix (for Red) as follows:

$$\begin{bmatrix} 4 & 0 \\ 2 & 3 \end{bmatrix}.$$

Here, $\max_{s \in S} \min_{t \in T} p(s, t) = 2$ and $\min_{t \in T} \max_{s \in S} p(s, t) = 3$, and thus there is no saddle point. Red can play so as to guarantee himself a payoff of 2 units and Cream can play so that he will have to pay no more than 3 units. Since Red will try to win more than 2 and Cream will try to pay less than 3, Red expects to win between 2 and 3 per play. Now, if Cream knows in advance which strategy Red is going to use, then he can be certain of a loss of 0 or 2, in any case actually less than 3. Thus Red is at a disadvantage if Cream knows his strategy and it is to his advantage to prevent Cream from learning his intentions. One method for Red to guarantee that Cream will not be able to predict his strategy is to choose a strategy at random. For similar reasons Cream should choose his strategy at random.

Thus in nonsaddle-point games we are led naturally to the idea of a mixed strategy. Instead of choosing a pure strategy, the player selects a probability distribution over his set of strategies, and then an appropriate random device selects the particular strategy to be used in a given play of the game. A mixed strategy is simply another name for this probability distribution. The outcome of the game is measured in terms of expected gain.

We illustrate the above ideas by considering the two-by-two game with payoff matrix $\begin{bmatrix} 4 & 0 \\ 2 & 3 \end{bmatrix}$. Suppose that the row player in this game uses the mixed strategy $\boldsymbol{\sigma}^r = (r, 1 - r)$, where $0 \leq r \leq 1$, and that the column player uses the mixed strategy $\boldsymbol{\tau}^t = (t, 1 - t)$, $0 \leq t \leq 1$. Then the expected payoff to the row player is

$$\begin{aligned} p(\boldsymbol{\sigma}^r, \boldsymbol{\tau}^t) = \boldsymbol{\sigma}^r \mathbb{G} \boldsymbol{\tau}^t &= 4 \cdot r \cdot t + 0 \cdot r \cdot (1 - t) + 2 \cdot (1 - r) \cdot t + 3 \cdot (1 - r) \cdot (1 - t) \\ &= (3 - t) + (5t - 3)r. \end{aligned}$$

We see from this formula that if the column player chooses $t = 0$ (i.e., he uses his second pure strategy every time), then the row player can penalize the column player for this choice by taking $r = 0$. In this case the payoff to the row player is $p(\boldsymbol{\sigma}^0, \boldsymbol{\tau}^0) = 3$. Similarly, if the column player uses $\boldsymbol{\tau}^1$, then the row player should use $\boldsymbol{\sigma}^1$ so that the return to the row player is $p(\boldsymbol{\sigma}^1, \boldsymbol{\tau}^1) = 4$. However, if the column player chooses $t = 3/5$, then no matter what strategy is chosen by the row player the payoff to the row player will not exceed 12/5. This is the smallest total to which the column player can restrict the long-term gains of the row player and in this sense the strategy (3/5, 2/5) is *optional* for the column player. Similarly, we can define an optional strategy for the row player and this is found to be (1/5, 4/5). We now turn to the general case.

Recall that by (1.1) if R (Red) selects mixed strategy $\boldsymbol{\sigma}$ and C (Cream) selects mixed strategy $\boldsymbol{\tau}$, then the expected gain of R is given by

$$p(\boldsymbol{\sigma}, \boldsymbol{\tau}) = \boldsymbol{\sigma} \mathbb{G} \boldsymbol{\tau} = \sum_{\substack{s \in S \\ t \in T}} \sigma(s) \tau(t) p(s, t).$$

Player R can therefore be certain of an expected gain of at least $\min_{\boldsymbol{\tau} \in \mathfrak{J}} p(\boldsymbol{\sigma}, \boldsymbol{\tau})$ where the minimum is taken over all mixed strategies available to C. However, as in the discussion of payoffs with pure strategies, R is at liberty to select his mixed strategy, and as a result he can make his choice in such a way as to guarantee an expected gain of at least

$$\max_{\boldsymbol{\sigma} \in \mathfrak{S}} \min_{\boldsymbol{\tau} \in \mathfrak{J}} p(\boldsymbol{\sigma}, \boldsymbol{\tau}).$$

On the other hand, for each mixed strategy $\boldsymbol{\tau}$ selected by C, the maximum expected gain for R is $\max_{\boldsymbol{\sigma} \in \mathfrak{S}} p(\boldsymbol{\sigma}, \boldsymbol{\tau})$, and C can select his mixed strategy in such a way that this expected gain is no more than

$$\min_{\boldsymbol{\tau} \in \mathfrak{J}} \max_{\boldsymbol{\sigma} \in \mathfrak{S}} p(\boldsymbol{\sigma}, \boldsymbol{\tau}).$$

This discussion provides heuristic support for the inequality

$$\max_{\boldsymbol{\sigma} \in \mathfrak{S}} \min_{\boldsymbol{\tau} \in \mathfrak{J}} p(\boldsymbol{\sigma}, \boldsymbol{\tau}) \leq \min_{\boldsymbol{\tau} \in \mathfrak{J}} \max_{\boldsymbol{\sigma} \in \mathfrak{S}} p(\boldsymbol{\sigma}, \boldsymbol{\tau}). \tag{1.3}$$

An argument similar to that given for the payoff of a game played with pure strategies establishes the result rigorously.

Although inequalities (1.2) for pure strategies and (1.3) for mixed strategies are formally similar, the following remarkable theorem shows that there is a fundamental difference between them. This difference enables one to solve all two-person zero-sum games with the use of mixed strategies.

Fundamental Theorem of Game Theory: Let $\Gamma = (S, T, p)$ be any two-person zero-sum game. Then

$$\max_{\sigma \in \mathfrak{S}} \min_{\tau \in \mathfrak{T}} p(\boldsymbol{\sigma}, \boldsymbol{\tau}) = \min_{\tau \in \mathfrak{T}} \max_{\sigma \in \mathfrak{S}} p(\boldsymbol{\sigma}, \boldsymbol{\tau}). \tag{1.4}$$

The proof of this theorem will be given later. We proceed here with a discussion of the implications of the result.

The common value of the two sides of equality (1.4) is necessarily the value of the game. We denote this common value by v and consider any pair of strategies $(\bar{\boldsymbol{\sigma}}, \bar{\boldsymbol{\tau}})$ such that $p(\bar{\boldsymbol{\sigma}}, \bar{\boldsymbol{\tau}}) = v$. Then

$$\begin{aligned} &p(\bar{\boldsymbol{\sigma}}, \bar{\boldsymbol{\tau}}) = v \\ &p(\bar{\boldsymbol{\sigma}}, \boldsymbol{\tau}) \geq v, \qquad \text{for all } \boldsymbol{\tau} \in \mathfrak{T}, \\ &p(\boldsymbol{\sigma}, \bar{\boldsymbol{\tau}}) \leq v, \qquad \text{for all } \boldsymbol{\sigma} \in \mathfrak{S}. \end{aligned}$$

This confirms that the triple $(\bar{\boldsymbol{\sigma}}, \bar{\boldsymbol{\tau}}, v)$ is a solution of the game Γ.

The game Γ is said to have a *pure solution* if there exists $\bar{s} \in S$ and $\bar{t} \in T$ such that

$$p(\bar{s}, \bar{t}) = v = \max_{\sigma \in \mathfrak{S}} \min_{\tau \in \mathfrak{T}} p(\boldsymbol{\sigma}, \boldsymbol{\tau}) = \min_{\tau \in \mathfrak{T}} \max_{\sigma \in \mathfrak{S}} p(\boldsymbol{\sigma}, \boldsymbol{\tau}).$$

Suppose that both of the players know a solution pair $(\bar{\boldsymbol{\sigma}}, \bar{\boldsymbol{\tau}})$ of the game Γ. In particular, R knows that if he uses mixed strategy $\bar{\boldsymbol{\sigma}}$, then his expected gain per game is at least v no matter what C does. Similarly, C knows that he can use strategy $\bar{\boldsymbol{\tau}}$ and be assured that the expected gain for R will never exceed v, no matter what choice R makes. It is in this sense that the strategies $\bar{\boldsymbol{\sigma}}$ and $\bar{\boldsymbol{\tau}}$ deserve to be called optimal or best strategies. Also, if both R and C use an optimal strategy, then the expected gain to R (and loss to C) is v, the value of the game.

We turn next to the practical problem of finding solutions to two-person zero-sum games. We first discuss a geometrical method which can be used when at least one of the two players has only two pure strategies. Then we consider the solution of a general two-person zero-sum game by means of linear programming. The basic connection between linear programming and two-person zero-sum games will lead naturally to a proof of the fundamental theorem stated above.

6.1.3 Geometrical Methods

Let $\Gamma = (S, T, p)$ be a two-person zero-sum game and let $\mathbb{G}$ be its payoff matrix. Thus we are assuming that S and T are finite and $\mathbb{G} = (g_{ij})$, where $g_{ij} = p(s_i, t_j)$. If S has m elements and T has n elements, then $\mathbb{G}$ is an $m \times n$ matrix. Thus, if $\boldsymbol{\sigma}$ and $\boldsymbol{\tau}$ are mixed strategies for Red and Cream, respectively, we have

$$p(\boldsymbol{\sigma}, \boldsymbol{\tau}) = \boldsymbol{\sigma}\mathbb{G}\boldsymbol{\tau}.$$

In this section we concentrate on the case $m = 2$. Thus $S = \{s_1, s_2\}$ and the row player has only two pure strategies. Therefore a mixed strategy for the row player is a vector $\boldsymbol{\sigma}^r = (r, 1 - r)$, where $0 \leq r \leq 1$. We are interested in finding r_0, $0 \leq r_0 \leq 1$, such that $\boldsymbol{\sigma}^{r_0}$ is an optimal strategy. That is,

$$p(\boldsymbol{\sigma}^{r_0}, \boldsymbol{\tau}) = \max_{\boldsymbol{\sigma} \in \mathbb{S}} \min_{\boldsymbol{\tau} \in \mathfrak{I}} \boldsymbol{\sigma}\mathbb{G}\boldsymbol{\tau} = \max_{0 \leq r \leq 1} \min_{\boldsymbol{\tau} \in \mathfrak{I}} \boldsymbol{\sigma}^r\mathbb{G}\boldsymbol{\tau} \qquad \text{for all } \boldsymbol{\tau}.$$

We begin our search for r_0 by considering the n choices of the vector $\boldsymbol{\tau}$ which correspond to the pure strategies for the column player. For each pure strategy $\boldsymbol{\tau}_i$, $i = 1, \ldots, n$, the expression $p(\boldsymbol{\sigma}^r, \boldsymbol{\tau}_i)$ is a linear function of the real variable r. The graph of each such function is a straight line, and we can plot these n straight lines and compare them. We illustrate the process with a specific example. Let

$$\mathbb{G} = \begin{bmatrix} 1 & -1 & 1 \\ -2 & 2 & 0 \end{bmatrix}.$$

Then

$$\begin{aligned} p(\boldsymbol{\sigma}^r, \boldsymbol{\tau}_1) &= r - 2(1 - r) = 3r - 2 = p_1(r), \\ p(\boldsymbol{\sigma}^r, \boldsymbol{\tau}_2) &= -r + 2(1 - r) = 2 - 3r = p_2(r), \\ p(\boldsymbol{\sigma}^r, \boldsymbol{\tau}_3) &= r + 0(1 - r) = r = p_3(r), \end{aligned}$$

and the graphs of these functions are shown in Fig. 6–1. The vertical axis in this figure is measured in units of payoff to the row player. We see that if the column player is restricted to the three pure strategies, then the worst thing that can happen to the row player is that his payoff is on the heavy polygonal line which bounds the graph from below.

Since the row player wants to maximize his payoff, he should choose r so that the associated payoff is the highest point on this polygonal line. In the example this point is at the intersection of the graphs of p_1 and p_2, and setting $p_1(r) = p_2(r)$ we find that $r = 2/3$. Hence, if the row player uses strategy $\boldsymbol{\sigma}^{2/3} = (2/3, 1/3)$ and if the column player uses any pure strategy, then the row player will obtain a payoff of at least $p_1(2/3) = p_2(2/3) = 0$. But this is not a complete analysis because the column player may elect to use a mixed strategy. What can be said in this case? Suppose that the column player uses strategy $\boldsymbol{\tau} = (\alpha, \beta, \gamma)$, where $0 \leq \alpha, \beta, \gamma \leq 1$ and $\alpha + \beta + \gamma = 1$. Then the payoff when the row player uses $\boldsymbol{\sigma}^r = (r, 1 - r)$ is

$$p(\boldsymbol{\sigma}^r, \boldsymbol{\tau}) = (r, 1 - r)\mathbb{G}\begin{pmatrix} \alpha \\ \beta \\ \gamma \end{pmatrix} = \alpha p_1(r) + \beta p_2(r) + \gamma p_3(r).$$

For each choice of r, $0 \leq r \leq 1$, there are three possible cases which can arise in comparing $p_1(r)$, $p_2(r)$, and $p_3(r)$. They are

Case 1: $p_1(r) \leq p_2(r)$ and $p_1(r) \leq p_3(r)$,

Case 2: $p_2(r) \leq p_3(r)$ and $p_2(r) \leq p_1(r)$, or

Case 3: $p_3(r) \leq p_1(r)$ and $p_3(r) \leq p_2(r)$.

In case 1 we obtain $\boldsymbol{\sigma}^r\mathbb{G}\boldsymbol{\tau} = \alpha p_1(r) + \beta p_2(r) + \gamma p_3(r) \geq \alpha p_1(r) + \beta p_1(r) + \gamma p_1(r) = p_1(r)$. Similarly, for cases 2 and 3 we obtain $\boldsymbol{\sigma}^r\mathbb{G}\boldsymbol{\tau} \geq p_2(r)$ and $\boldsymbol{\sigma}^r\mathbb{G}\boldsymbol{\tau} \geq p_3(r)$, respectively. Thus in all cases the expected return to the row player is bounded below by the return which corresponds to the column player using a pure strategy. Hence the heavy line in Fig. 6–1 is indeed a lower bound for the expected return to the row player. Therefore, if the row player maximizes his payoff against each of the pure strategies of his opponent, then he will also have maximized it against every mixed strategy. We can now conclude that an optimal strategy for the row player is (2/3, 1/3).

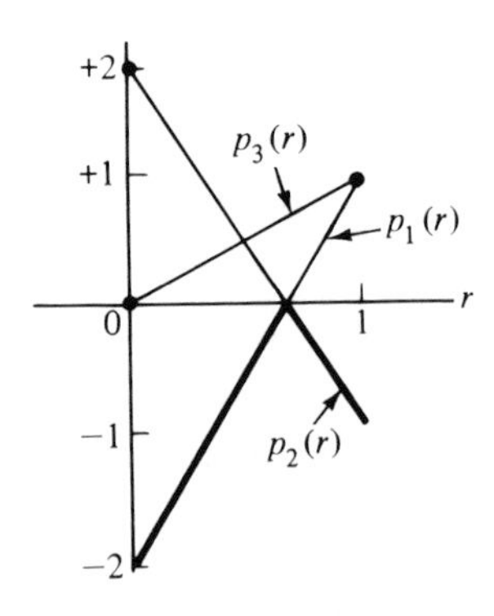

Fig. 6–1

What is an optimal strategy for the column player C? We have already seen that for each strategy of R, C can do best by using a pure strategy. However, as noted earlier, if he always uses the same pure strategy, then his opponent could take advantage of this fact. Hence he should somehow randomly mix at least two strategies. But which strategies and in what proportions? This question can be answered by again examining Fig. 6–1. We identify those pure strategies for C for which the associated graphs $p_i(r)$ have the lowest point of intersection in the figure. They are strategies 1 and 2. This lowest point indicates the best payoff that R can obtain given that C uses these two strategies in the proper mix. We illustrate the idea by letting C use the mixed strategy $\boldsymbol{\tau}^{r\prime} = (r, 1 - r, 0)$. Then the payoff to R is a function of r, and whenever R uses a pure strategy, the graph of this payoff function is a straight line. In fact, if we define the functions q_1 and q_2 by

$$\boldsymbol{\sigma}_1 \mathbb{G} \boldsymbol{\tau}^r = \quad 2r - 1 = q_1(r),$$
$$\boldsymbol{\sigma}_2 \mathbb{G} \boldsymbol{\tau}^r = -4r + 2 = q_2(r),$$

then the graphs of q_1 and q_2 are as in Fig. 6–2. The vertical axis in Fig. 6–2 gives the payoff to R when he uses a pure strategy and C uses the strategy $(r, 1 - r, 0)$. Thus the heavy polygonal line which bounds the graph from above is the best payoff that R can obtain using a pure strategy. However, just as before this is also the best that R can do using mixed strategies. Hence C should choose r to obtain an expected payoff which is the lowest point on the polygonal line. In this example the desired point is the intersection of the graphs of q_1 and q_2. Setting $q_1(r) = q_2(r)$, we obtain $r = 1/2$, and hence an optimal strateby for C is given by $\bar{\boldsymbol{\tau}} = (1/2, 1/2, 0)$. The value of this game is

$$q_1(1/2) = q_2(1/2) = p_1(2/3) = p_2(2/3) = 0.$$

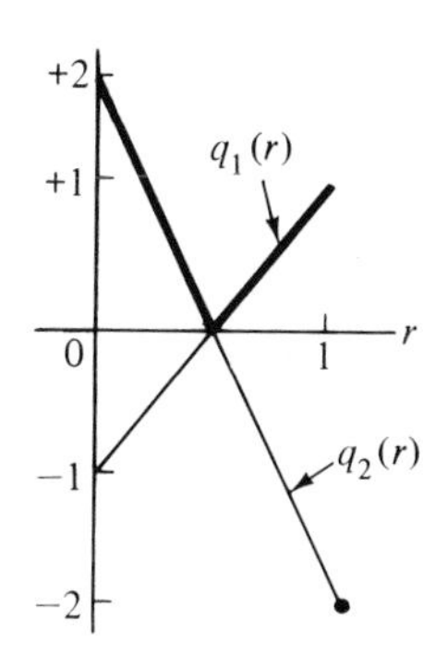

Fig. 6–2

We could have deduced immediately from Fig. 6–1 that C should not use strategy t_3. This follows since the payoff to R is always less for strategy t_1 than for t_3, and hence it is never to C's advantage to use t_3. In general, one can shows that if R has only two pure strategies, then C has an optimal mixed strategy which uses only two pure strategies (Exercise 11).

The method illustrated above is easily extended to payoff matrices with more than three columns. A summary of the solution of a $2 \times n$ game is given below. Of course, an $n \times 2$ game can be solved in the same manner.

Summary of Geometrical Method

Let $\mathbb{G}$ $(2 \times n)$ be the payoff matrix for the game $\Gamma = (S, T, p)$.

1. Graph the functions p_i, $i = 1, 2, \ldots, n$, defined by

$$p_i(r) = \boldsymbol{\sigma}^r \mathbb{G} \boldsymbol{\tau}_i, \qquad i = 1, 2, \ldots, n, \qquad \boldsymbol{\sigma}^r = (r, 1 - r).$$

2. Find the maximum point on the polygonal line bounding the graphs of $p_1, \ldots, p_n$ from below. Let k and l be defined by the property that the graphs of $p_k(r)$ and $p_l(r)$ lie on this polygon and intersect at this highest point, $k < l$.
3. Solve for the value r_0 for which $p_k(r_0) = p_l(r_0)$.
4. An optimal strategy for R is given by $\boldsymbol{\sigma}^{r_0} = (r_0, 1 - r_0)$.
5. Let $\boldsymbol{\tau}^r = (0, \ldots, 0, r, 0, \ldots, 0, 1 - r, 0, \ldots, 0)$, where r is the kth coordinate and $1 - r$ is the lth coordinate. Graph the functions $q_1(r) = \boldsymbol{\sigma}_1 \mathbb{G} \boldsymbol{\tau}^r$, $q_2(r) = \boldsymbol{\sigma}_2 \mathbb{G} \boldsymbol{\tau}^r$.
6. Solve for the value r_1 for which $q_1(r_1) = q_2(r_1)$.
7. An optimal strategy for C is $\boldsymbol{\tau}^{r_1}$.
8. The value of the game is $v = p_1(r_0) = r_2(r_0) = q_1(r_1) = q_2(r_1)$.

This method can be extended to games in which at least one player has at most three pure strategies. However, the details are quite complicated, and except in special situations other methods are preferable.

6.1.4 Linear Programming and a Proof of the Fundamental Theorem

We are now ready to prove that every two-person zero-sum game has a solution using mixed strategies. The proof will be constructive and will show that we can obtain a solution to every such game by means of the methods of Chap. 5. We proceed by first proving three lemmas which serve both to establish some background for the proof of the fundamental theorem and to indicate the connection between linear programming and two-person zero-sum games.

Lemma 1: If $(\boldsymbol{\sigma}_0, \boldsymbol{\tau}_0, v)$ is a solution of the game $\Gamma = (S, T, p)$ and if $a > 0$ and b are constants, then $(\boldsymbol{\sigma}_0, \boldsymbol{\tau}_0, av + b)$ is a solution of the game $\Gamma_{a,b} = (S, T, ap + b)$.

Proof: Since $(\boldsymbol{\sigma}_0, \boldsymbol{\tau}_0, v)$ is a solution of $\boldsymbol{\Gamma} = (S, T, p)$, we know that for all $\boldsymbol{\sigma} \in \mathcal{S}$ and $\boldsymbol{\tau} \in \mathfrak{T}$ the following inequality holds:

$$\sum_{s\in S}\sum_{t\in T}\sigma_0(s)\tau(t)p(s, t) \geq v \geq \sum_{s\in S}\sum_{t\in T}\sigma(s)\tau_0(t)p(s, t). \tag{1.5}$$

Next, recall that for every $\boldsymbol{\sigma} \in \mathcal{S}$ and $\boldsymbol{\tau} \in \mathfrak{T}$ we have $\sum_{s\in S}\sigma(s) = 1$ and $\sum_{t\in T}\tau(t) = 1$. Therefore we obtain the new inequality

$$\begin{aligned}\sum_{s\in S}\sum_{t\in T}\sigma_0(s)\tau(t)[ap(s, t) + b] &= a\sum_{s\in S}\sum_{t\in T}\sigma_0(s)\tau(t)p(s, t) + b\sum_{s\in S}\sum_{t\in T}\sigma_0(s)\tau(t)\\ &= a\sum_{s\in S}\sum_{t\in T}\sigma_0(s)\tau(t)p(s, t) + b\\ &\geq av + b \qquad \text{for } \tau\in\mathfrak{T}. \end{aligned}\tag{1.6}$$

Similarly, we have

$$\sum_{s\in S}\sum_{t\in T}\sigma(s)\tau_0(t)[ap(s, t) + b] \leq av + b \qquad \text{for } \sigma \in \mathcal{S}.$$

Combining these inequalities we have the desired result, i.e., $(\boldsymbol{\sigma}_0, \boldsymbol{\tau}_0, av + b)$ is a solution for the game $\boldsymbol{\Gamma}_{a,b} = (S, T, ap + b)$. Q.E.D.

Our next lemma gives an equivalent formulation for the definition of the solution of a game. We adopt the convention that the vectors $\boldsymbol{\sigma}$ are row vectors and that the vectors $\boldsymbol{\tau}$ are column vectors.

Lemma 2: The triple $(\boldsymbol{\sigma}_0, \boldsymbol{\tau}_0, v)$ is a solution of the game $\boldsymbol{\Gamma} = (S, T, p)$ with payoff matrix $\mathbb{G}$ if and only if $\boldsymbol{\sigma}_0\mathbb{G} \geq (v, \ldots, v)$ and

$$\mathbb{G}\boldsymbol{\tau}_0 \leq \begin{pmatrix} v\\ \cdot\\ \cdot\\ \cdot\\ v\end{pmatrix}.$$

Proof: For $\boldsymbol{\sigma} \in \mathcal{S}$ and $\boldsymbol{\tau} \in \mathfrak{T}$ we have $p(\boldsymbol{\sigma}, \boldsymbol{\tau}) = \boldsymbol{\sigma}\mathbb{G}\boldsymbol{\tau}$. Hence $(\boldsymbol{\sigma}_0, \boldsymbol{\tau}_0, v)$ is a solution of the game with payoff function p and matrix $\mathbb{G}$ if and only if

$$\boldsymbol{\sigma}_0\mathbb{G}\boldsymbol{\tau} \geq v > \boldsymbol{\sigma}\mathbb{G}\boldsymbol{\tau}_0 \tag{1.7}$$

for all $\boldsymbol{\sigma} \in \mathcal{S}$, $\boldsymbol{\tau} \in \mathfrak{T}$. In particular, (1.7) must hold when $\boldsymbol{\sigma}$ and $\boldsymbol{\tau}$ are chosen to be the unit basis vectors. If, as is our custom, S and T contain m and n strategies, respectively, then $\mathbb{G}$ is an $m \times n$ matrix, and $\boldsymbol{\sigma} \in R^m$, $\boldsymbol{\tau} \in R^n$. Hence (1.7) implies the following:

$$\begin{aligned}\boldsymbol{\sigma}_0\mathbb{G}\mathbf{u}^i &\geq v, \qquad i = 1, 2, \ldots, n,\\ \mathbf{u}_i\mathbb{G}\boldsymbol{\tau}_0 &\leq v, \qquad i = 1, 2, \ldots, m.\end{aligned}\tag{1.8}$$

This is the same as the statements $\boldsymbol{\sigma}_0\mathbb{G} \geq (v, \ldots, v)$ and

$$\mathbb{G}\boldsymbol{\tau}_0 \leq \begin{pmatrix} v \\ \cdot \\ \cdot \\ \cdot \\ v \end{pmatrix},$$

and hence half of the lemma is established.

We turn to the converse; that is, we prove that if $(\boldsymbol{\sigma}_0, \boldsymbol{\tau}_0, v)$ satisfies (1.8), then it also satisfies (1.7). This is easily seen by writing the probability vectors $\boldsymbol{\sigma}$ and $\boldsymbol{\tau}$ as linear combinations of the unit basis vectors. For example, if

$$\boldsymbol{\sigma} = (\sigma(s_1), \ldots, \sigma(s_m)) = \sum_{i=1}^{m} \sigma(s_i)\mathbf{u}_i,$$

then

$$\boldsymbol{\sigma}\mathbb{G}\boldsymbol{\tau}_0 = \sum_{i=1}^{m} \sigma(s_i)\mathbf{u}_i\mathbb{G}\boldsymbol{\tau}_0 \leq \sum_{i=1}^{m} \sigma(s_i)v = v\left[\sum_{i=1}^{m} \sigma(s_i)\right] = v.$$

Thus (1.8) is equivalent to the condition that $(\boldsymbol{\sigma}_0, \boldsymbol{\tau}_0, v)$ is a solution of the game. We complete the proof by noting that $\boldsymbol{\sigma}_0\mathbb{G}\mathbf{u}^i$ is the ith coordinate of the vector $\boldsymbol{\sigma}_0\mathbb{G}$ and, similarly, that $\mathbf{u}_i\mathbb{G}\boldsymbol{\tau}_0$ is the ith coordinate of the vector $\mathbb{G}\boldsymbol{\tau}_0$. Therefore (1.8) can be rewritten in the form $\boldsymbol{\sigma}_0\mathbb{G} \geq (v, \ldots, v) \in R^n$ and

$$\mathbb{G}\boldsymbol{\tau}_0 \leq \begin{pmatrix} v \\ \cdot \\ \cdot \\ \cdot \\ v \end{pmatrix} \in R^m.$$

Q.E.D.

The next lemma introduces linear programming into our study of game theory.

Lemma 3: Let $\mathbb{G}$ be the $m \times n$ payoff matrix for a two-person zero-sum game, and assume that $\mathbb{G}$ is positive ($\mathbb{G} > \mathbb{O}$). Then the following linear programming problem has an optimal vector: Find the minimum of $\mathbf{x}\cdot(1, 1, \ldots, 1)$, for $\mathbf{x} \in \mathfrak{F}$ where

$$\mathfrak{F} = \{\mathbf{x} \in R^m : \mathbf{x} \geq \mathbf{0}, \mathbf{x}\mathbb{G} \geq (1, 1, \ldots, 1) \in R^n\}.$$

Proof: The dual problem of the given one is to find $\mathbf{y} \in R^n$ which maximizes $\mathbf{y}\cdot(1, \ldots, 1)$ subject to $\mathbf{y} \geq \mathbf{0}$ and

$$\mathbb{G}\mathbf{y} \leq \begin{pmatrix} 1 \\ \cdot \\ \cdot \\ \cdot \\ 1 \end{pmatrix} \in R^m.$$

This dual problem has a feasible vector $\mathbf{y} = \mathbf{0}$. The primal problem also has feasible vectors; indeed, $\mathbb{G} > \mathbb{O}$, and hence we need only choose an $\mathbf{x}$ with sufficiently large positive coordinates to satisfy $\mathbf{x} \geq \mathbf{0}$ and $\mathbf{x}\mathbb{G} \geq (1, \ldots, 1)$. Thus both the primal and dual problems have feasible vectors, and it follows from Theorem 3 of Sec. 4.3 that they both have optimal vectors. Q.E.D.

We are now ready for our main result on two-person zero-sum games.

Fundamental Theorem of Game Theory: Every two-person zero-sum game has a solution.

Proof: Let $\mathbf{\Gamma} = (S, T, p)$ be any two-person zero-sum game, and let $\mathbb{G}$ be the associated payoff matrix. Choose k so that $\mathbf{\Gamma}_k = (S, T, p + k)$ has a positive matrix. Denote this matrix by $\mathbb{G}_k$. Next, let $\mathbf{x}_0$ and $\mathbf{y}_0$ be optimal vectors for the primal and dual problems introduced in Lemma 3, where the matrix $\mathbb{G}_k$ plays the role of the $\mathbb{G}$ used there. Then $\mathbf{x}_0$ and $\mathbf{y}_0$ satisfy $\mathbf{x}_0 \geq \mathbf{0}$, $\mathbf{y}_0 \geq \mathbf{0}$, and

$$\mathbf{x}_0\mathbb{G}_k \geq (1, 1, \ldots, 1), \qquad \mathbb{G}_k\mathbf{y}_0 \leq \begin{pmatrix} 1 \\ 1 \\ \cdot \\ \cdot \\ \cdot \\ 1 \end{pmatrix}.$$

Moreover,

$$\mathbf{x}_0 \cdot (1, 1, \ldots, 1) = \mathbf{y}_0 \cdot (1, \ldots, 1), \tag{1.9}$$

and we denote this common value by M. Define $\boldsymbol{\sigma}_0$ and $\boldsymbol{\tau}_0$ by

$$\boldsymbol{\sigma}_0 = \frac{\mathbf{x}_0}{M} \qquad \text{and} \qquad \boldsymbol{\tau}_0 = \frac{\mathbf{y}_0}{M}.$$

The nonnegativity of $\mathbf{x}_0$ and $\mathbf{y}_0$, together with condition (1.9), guarantees that $\boldsymbol{\sigma}_0$ and $\boldsymbol{\tau}_0$ are probability vectors. Also, the conditions on $\mathbf{x}_0\mathbb{G}_k$ and $\mathbb{G}_k\mathbf{y}_0$ imply that

$$\boldsymbol{\sigma}_0\mathbb{G}_k \geq \left(\frac{1}{M}, \ldots, \frac{1}{M}\right) \qquad \text{and} \qquad \mathbb{G}_k\boldsymbol{\tau}_0 \leq \begin{pmatrix} \frac{1}{M} \\ \cdot \\ \cdot \\ \cdot \\ \frac{1}{M} \end{pmatrix}.$$

Finally, by Lemma 2 we conclude that $(\boldsymbol{\sigma}_0, \boldsymbol{\tau}_0, 1/M)$ is a solution of

$\Gamma = (S, T, p + k)$, and by Lemma 1 we have that $(\boldsymbol{\sigma}_0, \boldsymbol{\tau}_0, v)$, $v = (1/M) - k$, is a solution of the game $\Gamma = (S, T, p)$. This completes the proof.

6.1.5 The Simplex Method and Games

The result of the preceding subsection provides a method for solving any two-person zero-sum game. We simply form the associated linear programming problem, and then use the simplex method to solve the latter problem. We illustrate this method with two examples.

Example 1. Let

$$\mathbb{G} = \begin{pmatrix} -2 & 1 & 4 \\ 1 & 4 & -2 \end{pmatrix}$$

be the payoff matrix of a two-person zero-sum game. The first step in solving the game is to form a new game for which the associated matrix is positive. In this example we can use Lemma 1 with $k = 3$. We obtain

$$\mathbb{G}_3 = \begin{pmatrix} 1 & 4 & 7 \\ 4 & 7 & 1 \end{pmatrix}.$$

The associated linear programming problem is to find

$$\mathbf{y} = \begin{pmatrix} \eta_1 \\ \eta_2 \\ \eta_3 \end{pmatrix}$$

such that $\eta_1 + \eta_2 + \eta_3$ is a maximum subject to $\mathbf{y} \geq \mathbf{0}$ and

$$\mathbb{G}_3\mathbf{y} \leq \begin{pmatrix} 1 \\ 1 \end{pmatrix}.$$

We use the simplex method, and in particular the techniques of Sec. 5.2. The resulting tableaus are as follows (the circled elements indicate the row and column used in replacement):

	$\mathbf{g}^1$	$\mathbf{g}^2$	$\mathbf{g}^3$	$\mathbf{u}^1$	$\mathbf{u}^2$	$\mathbf{b}$
$\mathbf{u}^1$	1	4	7	1	0	1
$\mathbf{u}^2$	(4)	7	1	0	1	1
$\mathbf{v} - \tilde{\mathbf{c}}$	−1	−1	−1	0	0	0

	$\mathbf{g}^1$	$\mathbf{g}^2$	$\mathbf{g}^3$	$\mathbf{u}^1$	$\mathbf{u}^2$	$\mathbf{b}$
$\mathbf{u}^1$	0	$\frac{9}{4}$	$\boxed{\frac{27}{4}}$	1	$-\frac{1}{4}$	$\frac{3}{4}$
$\mathbf{g}^1$	1	$\frac{7}{4}$	$\frac{1}{4}$	0	$\frac{1}{4}$	$\frac{1}{4}$
$\mathbf{v}^{(1)} - \tilde{\mathbf{c}}$	0	$\frac{3}{4}$	$-\frac{3}{4}$	0	$\frac{1}{4}$	$\frac{1}{4}$

	$\mathbf{g}^1$	$\mathbf{g}^2$	$\mathbf{g}^3$	$\mathbf{u}^1$	$\mathbf{u}^2$	$\mathbf{b}$
$\mathbf{g}^3$	0	$\frac{1}{3}$	1	$\frac{4}{27}$	$-\frac{1}{27}$	$\frac{1}{9}$
$\mathbf{g}^1$	1	$\frac{5}{3}$	0	$-\frac{1}{27}$	$\frac{7}{27}$	$\frac{2}{9}$
$\mathbf{v}^{(2)} - \tilde{\mathbf{c}}$	0	1	0	$\frac{1}{9}$	$\frac{2}{9}$	$\frac{1}{3}$

Therefore this computation gives

$$\mathbf{y}_0 = \begin{pmatrix} 2/9 \\ 0 \\ 1/9 \end{pmatrix}, \qquad \mathbf{x}_0 = (1/9, 2/9),$$

and $M = 1/3$. Hence, following the reasoning of the proof of the fundamental theorem, we have

$$\boldsymbol{\sigma}_0 = (1/3, 2/3), \qquad \boldsymbol{\tau}_0 = \begin{pmatrix} 2/3 \\ 0 \\ 1/3 \end{pmatrix},$$

and $v = 1/(1/3) - 3 = 0$.

Example 2. The Border Battle Problem. Recall that the payoff matrix for this problem is

$$\mathbb{G} = \begin{bmatrix} 3 & 1 & 0 & -1 \\ 0 & 2 & -1 & -2 \\ -3 & 1 & 1 & -3 \\ -2 & -1 & 2 & 0 \\ -1 & 0 & 1 & 3 \end{bmatrix}.$$

We add $+4$ to each entry in $\mathbb{G}$ to obtain a matrix with only positive entries, and we call the new matrix $\mathbb{G}_4$:

$$\mathbb{G}_4 = \begin{bmatrix} 7 & 5 & 4 & 3 \\ 4 & 6 & 3 & 2 \\ 1 & 5 & 5 & 1 \\ 2 & 3 & 6 & 4 \\ 3 & 4 & 5 & 7 \end{bmatrix}.$$

The associated linear programming problems are the following:

Find $\mathbf{x} = (\xi_1, \xi_2, \xi_3, \xi_4, \xi_5)$, which minimizes $\xi_1 + \xi_2 + \xi_3 + \xi_4 + \xi_5$, subject to the constraints $\mathbf{x} \geq \mathbf{0}$, $\mathbf{x}\mathbb{G}_4 \geq (1, 1, 1, 1)$.

Find $\mathbf{y} = \begin{pmatrix} \eta_1 \\ \eta_2 \\ \eta_3 \\ \eta_4 \end{pmatrix}$ which maximizes $\eta_1 + \eta_2 + \eta_3 + \eta_4$ subject to the constraints $\mathbf{y} \geq \mathbf{0}$, $\mathbb{G}_4\mathbf{y} \leq \begin{pmatrix} 1 \\ 1 \\ 1 \\ 1 \\ 1 \end{pmatrix}$.

We solve both of these problems by applying the techniques of Sec. 5.2. The sequence of tableaus for the maximization problem is as follows:

	$\mathbf{g}^1$	$\mathbf{g}^2$	$\mathbf{g}^3$	$\mathbf{g}^4$	$\mathbf{u}^1$	$\mathbf{u}^2$	$\mathbf{u}^3$	$\mathbf{u}^4$	$\mathbf{u}^5$	$\mathbf{b}$
$\mathbf{u}^1$	(7)	5	4	3	1	0	0	0	0	1
$\mathbf{u}^2$	4	6	3	2	0	1	0	0	0	1
$\mathbf{u}^3$	1	5	5	1	0	0	1	0	0	1
$\mathbf{u}^4$	2	3	6	4	0	0	0	1	0	1
$\mathbf{u}^5$	3	4	5	7	0	0	0	0	1	1
$\mathbf{v} - \tilde{\mathbf{c}}$	-1	-1	-1	-1	0	0	0	0	0	0

	$\mathbf{g}^1$	$\mathbf{g}^2$	$\mathbf{g}^3$	$\mathbf{g}^4$	$\mathbf{u}^1$	$\mathbf{u}^2$	$\mathbf{u}^3$	$\mathbf{u}^4$	$\mathbf{u}^5$	$\mathbf{b}$
$\mathbf{g}^1$	1	$\frac{5}{7}$	$\frac{4}{7}$	$\frac{3}{7}$	$\frac{1}{7}$	0	0	0	0	$\frac{1}{7}$
$\mathbf{u}^2$	0	($\frac{22}{7}$)	$\frac{5}{7}$	$\frac{2}{7}$	$-\frac{4}{7}$	1	0	0	0	$\frac{3}{7}$
$\mathbf{u}^3$	0	$\frac{30}{7}$	$\frac{31}{7}$	$\frac{4}{7}$	$-\frac{1}{7}$	0	1	0	0	$\frac{6}{7}$
$\mathbf{u}^4$	0	$\frac{11}{7}$	$\frac{34}{7}$	$\frac{22}{7}$	$-\frac{2}{7}$	0	0	1	0	$\frac{5}{7}$
$\mathbf{u}^5$	0	$\frac{13}{7}$	$\frac{23}{7}$	$\frac{40}{7}$	$-\frac{3}{7}$	0	0	0	1	$\frac{4}{7}$
$\mathbf{v}^{(1)} - \tilde{\mathbf{c}}$	0	$-\frac{2}{7}$	$-\frac{3}{7}$	$-\frac{4}{7}$	$\frac{1}{7}$	0	0	0	0	$\frac{1}{7}$

	$\mathbf{g}^1$	$\mathbf{g}^2$	$\mathbf{g}^3$	$\mathbf{g}^4$	$\mathbf{u}^1$	$\mathbf{u}^2$	$\mathbf{u}^3$	$\mathbf{u}^4$	$\mathbf{u}^5$	$\mathbf{b}$
$\mathbf{g}^1$	1	0	$\frac{9}{22}$	$\frac{4}{11}$	$\frac{3}{11}$	$-\frac{5}{22}$	0	0	0	$\frac{1}{22}$
$\mathbf{g}^2$	0	1	$\frac{5}{22}$	$\frac{1}{11}$	$-\frac{2}{11}$	$\frac{7}{22}$	0	0	0	$\frac{3}{22}$
$\mathbf{u}^3$	0	0	$\frac{38}{11}$	$\frac{2}{11}$	$\frac{7}{11}$	$-\frac{15}{11}$	1	0	0	$\frac{3}{11}$
$\mathbf{u}^4$	0	0	$\frac{9}{2}$	3	0	$-\frac{1}{2}$	0	1	0	$\frac{1}{2}$
$\mathbf{u}^5$	0	0	$\frac{63}{22}$	$\frac{61}{11}$	$-\frac{1}{11}$	$-\frac{13}{22}$	0	0	1	$\frac{7}{22}$
$\mathbf{v}^{(2)} - \tilde{\mathbf{c}}$	0	0	$-\frac{4}{11}$	$-\frac{6}{11}$	$\frac{1}{11}$	$\frac{1}{11}$	0	0	0	$\frac{2}{11}$

	$\mathbf{g}^1$	$\mathbf{g}^2$	$\mathbf{g}^3$	$\mathbf{g}^4$	$\mathbf{u}^1$	$\mathbf{u}^2$	$\mathbf{u}^3$	$\mathbf{u}^4$	$\mathbf{u}^5$	$\mathbf{b}$
$\mathbf{g}^1$	1	0	0	$\frac{13}{38}$	$\frac{15}{76}$	$-\frac{5}{76}$	$-\frac{9}{76}$	0	0	$\frac{1}{76}$
$\mathbf{g}^2$	0	1	0	$\frac{3}{38}$	$-\frac{17}{76}$	$\frac{31}{76}$	$-\frac{5}{76}$	0	0	$\frac{9}{76}$
$\mathbf{g}^3$	0	0	1	$\frac{1}{19}$	$\frac{7}{38}$	$-\frac{15}{38}$	$\frac{11}{38}$	0	0	$\frac{3}{38}$
$\mathbf{u}^4$	0	0	0	$\frac{105}{38}$	$-\frac{63}{76}$	$\frac{97}{76}$	$-\frac{99}{76}$	1	0	$\frac{11}{76}$
$\mathbf{u}^5$	0	0	0	$\frac{205}{38}$	$-\frac{47}{76}$	$\frac{41}{76}$	$-\frac{63}{76}$	0	1	$\frac{7}{76}$
$\mathbf{v}^{(3)} - \tilde{\mathbf{c}}$	0	0	0	$-\frac{10}{19}$	$\frac{3}{19}$	$-\frac{1}{19}$	$\frac{2}{19}$	0	0	$\frac{4}{19}$

	$\mathbf{g}^1$	$\mathbf{g}^2$	$\mathbf{g}^3$	$\mathbf{g}^4$	$\mathbf{u}^1$	$\mathbf{u}^2$	$\mathbf{u}^3$	$\mathbf{u}^4$	$\mathbf{u}^5$	$\mathbf{b}$
$\mathbf{g}^1$	1	0	0	0	*	*	*	0	*	$\frac{3}{410}$
$\mathbf{g}^2$	0	1	0	0	*	*	*	0	*	$\frac{24}{205}$
$\mathbf{g}^3$	0	0	1	0	*	*	*	0	*	$\frac{16}{205}$
$\mathbf{u}^4$	0	0	0	0	*	*	*	1	*	$\frac{4}{41}$
$\mathbf{g}^4$	0	0	0	1	*	*	*	0	*	$\frac{7}{410}$
$\mathbf{v}^{(4)} - \tilde{\mathbf{c}}$	0	0	0	0	$\frac{4}{41}$	0	$\frac{1}{41}$	0	$\frac{4}{41}$	$\frac{9}{41}$

In the last tableau the $*$s represent fractions whose exact values need not be computed for the solution of the problem. The bottom row of this last tableau has only positive elements, and consequently we have obtained optimal vectors for the linear programming problems. These optimal vectors are

$$\mathbf{x}_0 = (4/41, 0, 1/41, 0, 4/41) \quad \text{and} \quad \mathbf{y}_0 = \begin{pmatrix} 3/410 \\ 24/205 \\ 16/205 \\ 7/410 \end{pmatrix}.$$

Using the notation of the proof of the fundamental theorem, $M = \sum \xi_i = \sum \eta_i = 9/41$. Thus optimal strategies are given by

$$\boldsymbol{\sigma}_0 = (4/9, 0, 1/9, 0, 4/9) \qquad \text{and} \qquad \boldsymbol{\tau}_0 = \begin{pmatrix} 1/30 \\ 24/45 \\ 16/45 \\ 7/90 \end{pmatrix}.$$

The value of the border battle game is $v = (1/M) - k = [1/(9/41)] - 4 = 5/9$. Note that the simplex method has given us the same $\boldsymbol{\sigma}_0$ and v as we had earlier but that the strategy $\boldsymbol{\tau}_0$ is different. This points out the fact that optimal strategies are not necessarily unique. As a check on this particular $\boldsymbol{\tau}_0$, we let $\boldsymbol{\sigma} = (\alpha, \beta, \gamma, \delta, 1 - \alpha - \beta - \gamma - \delta)$ and then compute $p(\boldsymbol{\sigma}, \boldsymbol{\tau}_0)$. We have

$$p(\boldsymbol{\sigma}, \boldsymbol{\tau}_0) = (\alpha, \beta, \gamma, \delta, 1 - \alpha - \beta - \gamma - \delta)\mathbb{G}\begin{pmatrix} 1/30 \\ 24/45 \\ 16/45 \\ 7/90 \end{pmatrix}$$

$$= (\alpha, \beta, \gamma, \delta, 1 - \alpha - \beta - \gamma - \delta)\begin{pmatrix} 3 & 1 & 0 & -1 \\ 0 & 2 & -1 & -2 \\ -3 & 1 & 1 & -3 \\ -2 & -1 & 2 & 0 \\ -1 & 0 & 1 & 3 \end{pmatrix}\begin{pmatrix} 1/30 \\ 24/45 \\ 16/45 \\ 7/90 \end{pmatrix}$$

$$= (\alpha, \beta, \gamma, \delta, 1 - \alpha - \beta - \gamma - \delta)\begin{pmatrix} 5/9 \\ 5/9 \\ 5/9 \\ 1/9 \\ 5/9 \end{pmatrix}$$

$$= \tfrac{5}{9} - \tfrac{4}{9}\delta.$$

Since $\delta \geq 0$, we see that $p(\boldsymbol{\sigma}, \boldsymbol{\tau}_0) \leq 5/9$ for every choice of $\boldsymbol{\sigma}$. Therefore $\boldsymbol{\tau}_0$ is indeed an optimal strategy.

EXERCISES

1. Show that the value of a two-person zero-sum game is uniquely defined.
2. Show that a game with the payoff matrix

$$\mathbb{G} = \begin{bmatrix} a & a \\ b & c \end{bmatrix}$$

always has a saddle point.

3. Show that a necessary and sufficient condition for a two-person zero-sum game (or its payoff matrix) to have a saddle point is that there is an element in the payoff matrix which is simultaneously the minimum of its row and the maximum of its column.

4. Assume that the game with the payoff matrix

$$\mathbb{G} = \begin{bmatrix} a & b \\ c & d \end{bmatrix}$$

does not have a saddle point. Solve the game in terms of the parameters a, b, c, and d.

5. Use the geometrical methods of Sec. 6.1.3 to solve the games with the following payoff matrices:

(a) $\begin{bmatrix} 1 & 2 \\ 3 & 0 \end{bmatrix}$. (b) $\begin{bmatrix} 1 & 2 \\ 3 & 4 \end{bmatrix}$. (c) $\begin{bmatrix} 1 & 2 & 3 \\ 3 & 2 & 1 \end{bmatrix}$.

(d) $\begin{bmatrix} 1 & 2 & 3 \\ -3 & -2 & -1 \end{bmatrix}$. (e) $\begin{bmatrix} -1 & 5 & 1 & -2 \\ 1 & -3 & -2 & 5 \end{bmatrix}$.

(f) $\begin{bmatrix} -1 & 2 \\ 3 & 0 \\ 0 & 2 \end{bmatrix}$. (g) $\begin{bmatrix} 1 & 4 \\ 5 & -1 \\ 2 & 0 \\ 7 & -3 \end{bmatrix}$.

6. Consider the following statement: There exists a two-person zero-sum game Γ for which the row player R has exactly two distinct optimal strategies. Prove or disprove this statement either by giving an example which shows it is true or by providing a proof that it cannot be true.

7. Solve the following games by using the methods of Sec. 6.1.5:

(a) $\begin{bmatrix} 4 & 0 \\ 2 & 3 \end{bmatrix}$. (b) $\begin{bmatrix} 1 & 2 & 3 \\ -3 & -2 & -1 \end{bmatrix}$.

(c) $\begin{bmatrix} 0 & 2 & 1 & 1 \\ 1 & 0 & 2 & 0 \\ 2 & 1 & 0 & 2 \end{bmatrix}$. (d) $\begin{bmatrix} 1 & 4 \\ 5 & -1 \\ 2 & 0 \\ 7 & -3 \end{bmatrix}$.

8. Mr. Honorable (Mr. H) and Mr. Rogue (Mr. R) are engaged in a four-step duel. They have the option of firing or not firing after each step and only at these times is it legal to fire. After each step they must simultaneously decide whether or not to fire. That is, after a step, one player cannot wait until the other fires before he decides whether or not he will fire. Neither knows if or when his opponent fires, with the obvious exception of when he is hit. Each has only one bullet. Mr. H places \$2.00 and Mr. R places \$1.00 in a pot before starting the duel. If there is a winner, then he gets the pot. If the duel ends in a draw, then the pot is split by the contestants or their heirs. The probability that a given player will hit his opponent is dependent on the number of steps taken. After moving k steps, this probability is $k/4$ for Mr. H and $k^2/16$ for Mr. R.

(a) Show that this situation can be formulated as a two-person zero-sum game, and find the payoff matrix for this game. *Hint:* See Sec. 2.5; use expected returns.

(b) Solve the two-person zero-sum game which describes this duel.

9. Let the $m \times n$ matrix $\mathbb{G}$ be the payoff matrix for a two-person zero-sum game $\mathbf{\Gamma}$. Suppose that $(\boldsymbol{\sigma}_0, \boldsymbol{\tau}_0, -v)$ is a solution for the game with payoff matrix $-\mathbb{G}^T$. Find a solution for $\mathbf{\Gamma}$.

10. Show that the optimal strategy $\boldsymbol{\sigma}_0 = (4/9, 0, 1/9, 0, 4/9)$ is unique for the border battle game.

11. Show that in any two-person zero-sum game in which the row player has only two pure strategies the column player has an optimal mixed strategy which uses only two pure strategies. *Hint:* Begin by considering the case in which the column player has exactly three pure strategies, and write the payoff to the row player in the form $p(\boldsymbol{\sigma}^{r_0}, \boldsymbol{\tau}) = \alpha p_1(r_0) + \beta p_2(r_0) + \gamma p_3(r_0)$. Here $\boldsymbol{\sigma}^{r_0}$ is an optimal strategy for the row player and $\boldsymbol{\tau} = (\alpha, \beta, \gamma)$, $\alpha + \beta + \gamma = 1$, is any strategy for the column player. Show that at least one of the quantities α, β, γ can be chosen to be 0 by comparing the sizes of $p_1(r_0)$, $p_2(r_0)$, and $p_3(r_0)$.

6.2 TWO-PERSON NONZERO-SUM GAMES

Our development of the theory of two-person zero-sum games depends heavily on the assumption that whenever the payoff to one player is an amount a, then the payoff to the other player is the amount $-a$. This assumption provides the name *zero-sum*, and because of it only one payoff function need be considered. However, this assumption is false for many (perhaps most) of the situations which have models fitting our definition of a game. As one example of this we note that the first problem discussed in this chapter (the gas war) does not have a zero-sum game as an appropriate model. This problem is so stated that both players can gain or lose at the same time. The general theory of nonzero-sum games is not nearly as well developed as that of the games studied in the preceding section. However, so that our study will present a balanced picture of the subject, this section is devoted to a discussion of certain aspects of such games. We first make a number of remarks about nonzero-sum games in general (these are continued in the Exercises), and then we concentrate on those games which are similar in nature to a gas war problem. Although this provides a somewhat limited perspective for the reader, our aims are better served by a reasonably detailed study of the theory and applications of a particular nonzero-sum game than by a more mathematical study of general games. Also, the game which we shall discuss at length (the Prisoner's Dilemma) is of interest to investigators in many different fields, and it is likely that the reader will encounter some version of it elsewhere.

6.2.1 General Remarks

In any two-person game there are two payoff functions to be considered, one for each of the players. If the players have only a finite number of pure strategies, then these payoff functions can be represented by two payoff matrices. For example, consider a game in which the row player has the pure strategy set $S = \{1, 2\}$, the column player has the pure strategy set $T = \{1, 2, 3\}$, and the payoff functions are given by $p_R(i, j) = |i - j|$ and $p_C(i, j) = i^2 - j$. These payoff functions are represented by the matrices

$$\mathbb{G}(R) = \begin{array}{c} \\ 1 \\ 2 \end{array}\!\!\begin{array}{c} \begin{array}{ccc} 1 & 2 & 3 \end{array} \\ \begin{bmatrix} 0 & 1 & 2 \\ 1 & 0 & 1 \end{bmatrix} \end{array} \quad \text{and} \quad \mathbb{G}(C) = \begin{array}{c} \\ 1 \\ 2 \end{array}\!\!\begin{array}{c} \begin{array}{ccc} 1 & 2 & 3 \end{array} \\ \begin{bmatrix} 0 & -1 & -2 \\ 3 & 2 & 1 \end{bmatrix} \end{array}.$$

Here, the entry in the ith row and jth column of $\mathbb{G}(R)$ is the gain to the row player when he uses strategy i and the column player uses strategy j. The payoff matrix $\mathbb{G}(C)$, which gives the payoff to the column player, is defined similarly.

The matrices $\mathbb{G}(R)$ and $\mathbb{G}(C)$ can be combined into a single matrix which contains all the information defining p_R and p_C. This is done by letting the entry in the ith row and jth column of the new matrix be the ordered pair (a, b), where a is the i–j entry of $\mathbb{G}(R)$ and b is the i–j entry of $\mathbb{G}(C)$. In the above example this matrix of ordered pairs is

$$\mathbb{G} = \begin{array}{c} \\ 1 \\ 2 \end{array}\!\!\begin{array}{c} \begin{array}{ccc} 1 & 2 & 3 \end{array} \\ \begin{bmatrix} (0, 0) & (1, -1) & (2, -2) \\ (1, 3) & (0, 2) & (1, 1) \end{bmatrix} \end{array}.$$

In this way we can consider every (finite) two-person game to be determined by a matrix $\mathbb{G} = (g_{ij})$, where the matrix entries g_{ij} are ordered pairs of real numbers. Conversely, any matrix of this form can be considered to be the payoff matrix for a two-person game.

Consider a two-person game with the payoff functions p_R and p_C for the row and column players, respectively. If $p_R + p_C$ is a constant function [i.e., $p_R(i, j) + p_C(i, j)$ is independent of the strategies i and j], then the game is said to be a *constant-sum* game. It is a simple matter to convert any constant-sum game to a zero-sum game by adding a fixed amount to the payoff functions. For example, if $p_R + p_C = K$, then by defining $\tilde{p}_R = p_R + [(-k)/2]$ and $\tilde{p}_C = p_C + [(-K)/2]$, we have $\tilde{p}_R + \tilde{p}_C = 0$. Thus the game with the payoff functions $\tilde{p}_R$ and $\tilde{p}_C$ is a zero-sum game. Also, it is easy to show that every solution of this zero-sum game yields a solution of the original constant-sum game (Exercise 7). We conclude that although the results of Sec. 6.1 are stated for zero-sum games, they also apply to all (finite) two-person constant-sum games. Thus, to study something new, we consider *two-person nonconstant-sum* games, i.e., those two-person games for which $p_R + p_C$ is not a constant function. We refer to these games by the shorter title of nonzero-sum games, keeping in mind that all constant-sum games are equivalent to zero-sum games.

It is possible to apply the mathematical techniques of constant-sum games to nonconstant-sum games; however, the results are not nearly as useful in this case. As an indication of the complications which may arise, we consider the problem of defining the notion of an *optimal strategy*. Here we encounter one of the more significant complications arising in nonzero-sum games. In zero-sum games an optimal strategy is defined by considering expected payoffs, and then in some sense each player has an optimal strategy when his strategy maximizes his expected payoff. However, this approach does not work with general (nonzero-sum) games. In fact, for these games it may be difficult to establish that any sort of optimal strategy is really optimal in the usual sense. An example will help to illustrate the point.

Consider the two-person game with the following payoff matrix:

$$\mathbb{G} = \begin{array}{c} \\ 1 \\ 2 \end{array} \begin{array}{c} \begin{array}{cc} 1 & 2 \end{array} \\ \begin{bmatrix} (2, 1) & (0, 0) \\ (0, 0) & (1, 2) \end{bmatrix} \end{array}.$$

If the row player plays strategy 1, then his payoff is either 0 or 2, and if he plays strategy 2 his payoff is 0 or 1. Thus it can be argued that he would do best by playing strategy 1, for if he consistently plays strategy 1, then his opponent does best (for himself) by also playing strategy 1. Consequently the "expected" payoff to the row player is 2. However, the reasoning is the same for the column player and he might well conclude that he should play strategy 2 and hope that his opponent diagnoses his intentions and also plays 2 to maximize his gain. In this way both players would obtain a return of 0. Of course, if the players are able to communicate with one another, then they might well agree on a method of play whereby 50% of the time they would both play strategy 1 and the other 50% of the time they would both play strategy 2. This would give a maximum total gain to both players as a pair, and since it is equally divided it would seem to fit the concept of an optimal strategy. But what if no communication is possible? What sequence of strategies should each player adopt so as to maximize his individual return? This is a difficult question and there is no easy answer. The question is an important one since many nonzero-sum games have much in common with this example. One must distinguish between the cases where communication is and is not legal. In either case it may be difficult to define the concept optimal strategy in a satisfactory manner. One possibility is to consider *equilibrium pairs* instead of optimal strategies. This is discussed in Exercises 8 and 9, and bargaining and threats are discussed briefly in Exercise 10. Another possibility in situations like this is to first determine how people actually play such a game. Then after this information is available, one can try to build a theory which accounts for the observed behavior. This has been done for a number of nonzero-sum games, and we shall report on the game for which the most work has been done.

6.2.2 The Prisoner's Dilemma

In the gas war problem of Sec. 6.0 each station manager must choose between two possible prices for his regular gasoline. Each knows the rewards and penalties for the various combinations of choices available to the other manager and himself. However, he does not know the other manager's actual decision. In this situation each manager is faced with an instance of the Prisoner's Dilemma.

The Prisoner's Dilemma is the name given to a certain two-person nonzero-sum game which has been extensively studied in recent years. The results of these studies are reported in books and journal articles, some devoted just to the game itself and others to a discussion of various applications of the game in fields such as economics, psychology, and law. Several of these studies are cited in the References at the end of this chapter.

The name *Prisoner's Dilemma* designates (see [LuR]) a hypothetical situation involving two suspected criminals. The suspects have been arrested on suspicion of robbery and they have been separated since their arrest. The district attorney interviews each suspect and makes the following statement to each man: "We would like you to confess and save the state the costs of prosecuting your case. If both of you do not confess, then we have enough evidence to convict each of you on a misdemeanor charge, and it is certain that you will receive the maximum penalty of one year in jail. However, if you confess and your partner does not, then we will arrange for you to receive a suspended sentence, and he will receive 10 years in prison. Naturally, if you do not confess and he does, then the sentences are reversed. Finally, if you both confess, then you will each receive a sentence of 5 years in prison."

What should the suspects do? They are not able to communicate with each other, and hence they are unable to form a binding agreement not to confess. Each prisoner must make his own decision independent of any knowledge of his partner's choice. Let us consider the situation from the point of view of one of the suspects. His fate depends on both the unknown actions of his partner and on his own decision. There are two cases to consider. First, if his partner does not confess, then assuming that he seeks to minimize his stay in jail he should confess and go free rather than spend a year in jail. Second, if his partner does confess, then he should also confess so that he spends 5 years in prison instead of 10 years. Thus *in all cases* the suspect should confess. But the situation is completely symmetrical, and hence the other suspect should also confess. Thus the logical outcome of this reasoning is that both suspects confess, and consequently both receive sentences of five years in prison. Now the dilemma is clear. If each suspect does what he apparently should do (confess), then each suspect ends up rather badly off, while, on the other hand, if they do what they apparently should not do (remain silent), then they end up with a significantly better return (sentence). In such a situation there does not seem to be a rational solution. Both choices can be supported by logical reasoning.

We shall examine the Prisoner's Dilemma from both mathematical and nonmathematical points of view. Our first observation is that, as in Sec. 6.1, it is useful to consider a sequence of such situations, as opposed to the study of a single play of a "game" of this type. Thus the players may compare the prospects for long-term gains and/or losses with those for short-term gains and/or losses. In this regard the gas war problem is a better illustration of the situation than is the dilemma faced by the prisoners, the difference being the long-term nature of the setting. The station managers must choose a price every few days for perhaps years to come. However, the typical thief does not face a Prisoner's Dilemma every day. The nature of the payoffs limits the number of times that any two partners in crime will play this game to a reasonably small number.

We have not yet shown that the Prisoner's Dilemma is really a game in our technical sense of the term. With this in mind our next step is to introduce some terminology and to present an abstract game which displays the essential features of the Prisoner's Dilemma. In other words, we form a mathematical model for such situations.

Let the two players be denoted by the letters A and B, and suppose that each player has two pure strategies, which are denoted by C_A and D_A for A and by C_B and D_B for B. The letters C and D are used here to suggest the actions of cooperate (with each other) and defect (from the coalition). There are four possible outcomes for a play of this game, and these are denoted by CC, CD, DC, and DD. CD represents that outcome which occurs when player A uses strategy C and player B uses strategy D, and likewise for the other outcomes. Note that CD is an outcome, an ordered pair of payoffs, and not simply a single payoff. We assume no communication between players, and so each must make his choice independent of a knowledge of the other's choice. We also assume that each player has a certain preference ranking of the outcomes and that he plays so as to achieve the outcome with the highest preference ranking regardless of the payoff to his opponent. It is this assumption which characterizes the Prisoner's Dilemma among a whole class of two-person nonzero-sum games. In particular, we assume that player A prefers outcome DC to CC, CC to DD, and DD to CD. Player B prefers CD to CC, CC to DD, and DD to DC. Moreover, these preferences are transitive in every case. With this notation we may represent the outcome of an abstract Prisoner's Dilemma game (APDG) by the matrix below.

		Strategies for player B	
		C_B	D_B
Strategies for player A	C_A	(R_A, R_B)	(S_A, T_B)
	D_A	(T_A, S_B)	(P_A, P_B)

For example, R_A is the payoff to A for outcome CC, and S_B is the payoff to B for outcome DC. This terminology is due to Anatol Rapoport, who gives the following justification: R is the reward (for cooperating) payoff, T is the temptation (to take advantage of your partner) payoff, S is the sucker (for believing in your partner) payoff, and P is the punishment (for not cooperating) payoff. The preferences of the players imply that $T_A > R_A > P_A > S_A$, and similarly for B. This yields a problem with eight preference parameters. If we make the additional assumption that payoffs are invariant with respect to permutation of the players—that is, $R_A = R_B$, $S_A = S_B$, $T_A = T_B$, and $P_A = P_B$—then the number of parameters in the above matrix is reduced to four. The number of parameters can be decreased further with additional assumptions. A common one which is appropriate for some empirical situations and which makes the model much more tractable is $T_A > R_A > 0$, $P_A = -R_A$, and $S_A = -T_A$. There are only two parameters in this case.

We conclude this discussion of the definitions by remarking that the usual Prisoner's Dilemma is indeed an example of an APDG. To verify this, we note that the payoff matrix for the dilemma faced by the criminal suspects is given by

	C	D
C	$(-1, -1)$	$(-10, 0)$
D	$(0, -10)$	$(-5, -5)$

Here the units are years of confinement, and we have $T = 0$, $R = -1$, $P = -5$, and $S = -10$. Thus, $T > R > P > S$, and our assertion is confirmed.

6.2.3 Nonmathematical Aspects of the Prisoner's Dilemma

Much of the work which has been done with the Prisoner's Dilemma is nonmathematical in nature. It is descriptive or experimental, and it consists of an effort to determine and describe how people actually react when they are faced with a decision similar in nature to that of a player in an APDG. Since numerous journal articles and even complete books have been devoted to these studies, it is presumptuous to expect that we can give an adequate picture of this work in a few paragraphs. At best we can report on only a few of these studies in an effort to indicate the nature of the questions asked by the investigator and what sort of answers were obtained. One can argue that such discussions are not properly mathematics and hence the title of this subsection. However, for investigators using game-theoretic models, such topics are highly nontrivial, and we feel that it is appropriate to introduce them at this time.

In one study of the Prisoner's Dilemma (see [L] for details) the intention was to investigate certain factors which were believed to be important in determining a player's tendency toward adopting the strategy C (cooperate). The factors studied were the following:

1. The number of trials of the game.
2. The relative sizes of the payoffs.
3. The nature of the opposing player as indicated by his play of the game.

The subjects were given a fixed amount of money before play began and their payoffs were added to or subtracted from that amount. Some of the results obtained can be summarized as follows.

1. It was conjectured that a tendency toward cooperation was fostered by having many plays of the game. This conjecture was generally supported by the experimental evidence; however, other factors could offset the effect of having a game involving many plays. In particular, if the subjects knew that they were near the end of play, then there was an increased tendency toward defection and the associated double-crossing of the opposing player. Also, the size of the payoffs must be considered in drawing conclusions based on length of play. For example, there is some experimental support for the assertion that the expected value of the number of attempts at cooperation exceeds that of the number of attempts at defection, provided that the number of trials n satifies

$$n > k\left[\frac{P - S}{R - P}\right].$$

Here the parameter k depends on the subject population. In one particular study the population consisted of Reed college undergraduates and k was estimated to be approximately equal to 3.

2. It was conjectured that by varying the size of the payoffs cooperation could be made easier to achieve and to maintain. In particular it was felt that if T were large with respect to R, then cooperation would be difficult to achieve because the temptation to defect would be too large in comparison to the reward for cooperation. On the other hand, it was believed that cooperation would be enhanced by making $R - P$ large, since this is the incentive to change from defection to cooperation. Finally it was conjectured that cooperation would be encouraged by making S larger. In this way the prospects of receiving the sucker payment are not so unattractive and the subject is willing to take the risk. There is experimental work which tends to support these conjectures. In [L] the last conjecture had the strongest support. There was a marked increase in attempts at cooperation as S was significantly increased.

3. It is a common technique of experimental design in such work for the investigator to adopt the tactic of using a stooge as one of the players. This device permits the testing of responses of many subjects to certain patterns of play by their opponent. This design was utilized in [L], where the different play patterns of the stooges were colorfully and suggestively called Stalin, Khrushchev, Coolidge, and Gandhi. Stalin always played D, Khrushchev almost always played D but occasionally played C, Coolidge played C only if his opponent played C four times in a row, and Gandhi always played C.

Against Stalin and Coolidge (which were analyzed together) the subjects would make a small number of tries at cooperation (about 9 in 100 or more trials). This number was relatively independent of the size of the payoff.

Against Khrushchev the subjects reacted very favorably to the early hints at cooperation. However, by the fourth hint the subjects tended to refuse to react since their early responses were not rewarded.

Against Gandhi the subjects were quick to exploit an apparent weakness in their opponent. In fact every player that initially double-crossed Gandhi after having once cooperated eventually ended up by playing only D.

Results of experimental work with this game have rarely been conclusive. It is an area of active concern to researchers in psychology and sociology, and the methods and questions are becoming increasingly complex.

6.2.4 A Stochastic Model for the Prisoner's Dilemma

Several broadly based studies of the Prisoner's Dilemma have been conducted by Anatol Rapoport and associates at the University of Michigan. They have concentrated on accumulating data for the purpose of building and testing models, and we shall present here one of the models which resulted from their work. See [RC] for more complete discussions of this study.

Adopting the point of view that the game is to be played a number of times in succession, it is reasonable to formulate some type of stochastic model. A Markov chain model will be developed here. To proceed with such a model, it is necessary to specify the states of the process and the transition probabilities. Since each play of the game results in a unique outcome, it is natural to take the outcomes as states of the Markov process. It follows from the definitions that there are four possible outcomes for each play of an APDG, and we have denoted them by CC, CD, DC, and DD. Thus we are led to consider a Markov process with four states, which for convenience we identify with the same symbols as the outcomes, namely CC, CD, DC, and DD. A sequence of plays of the game corresponds to a passage of the process through the set of states or a subset of this set. The specific way in which the process passes from state to state depends on the way in which the players play the game, and different assumptions regarding the play will lead to different results. To form a definite model, it is necessary to make certain explicit assumptions. The model proposed here is based on the following assumptions regarding the players:

Assumptions

1. The players are psychologically identical. With respect to the model, we interpret this to mean that the players are equally likely to respond to a given situation in a certain way.
2. There is no communication between the players. Each player knows the results of all previous plays of the game, but he does not know in advance the intentions of his opponent on the current play.
3. The following probabilities are constant over repeated plays of the game:

$x = \Pr[\text{Play } C \mid \text{strategies on preceding play were } C \text{ and } C]$,

$y = \Pr[\text{Play } C \mid \text{strategies on preceding play were } C \text{ and } D,\ D \text{ by opponent}]$,

$z = \Pr[\text{Play } C \mid \text{strategies on preceding play were } D \text{ and } C,\ C \text{ by opponent}]$,

$w = \Pr[\text{Play } C \mid \text{strategies on preceding play were } D \text{ and } D]$.

Notice that in view of assumption 1 these probabilities are the same for both players.

Example. One of the simplest possible cases is that in which the players select strategies according to the following elementary decision rule: If a play results in an R or T (reward or temptation) payoff, then continue it, and if a play results in a P or S (punishment or sucker) payoff, then switch. This decision rule leads to an absorbing Markov chain model for the game, and the associated process is absorbed into state CC after at most three plays. This rule gives values $x = 1$, $y = 0$, $z = 0$, and $w = 1$ to the above probabilities.

Suppose that on the initial play of the game the players both select strategy C. Thus we take CC as the initial state of the process. This assumption is not crucial to what follows and is made only to provide a well-defined Markov chain. If we introduce the notation $\tilde{x} = 1 - x$, $\tilde{y} = 1 - y$, $\tilde{z} = 1 - z$, and $\tilde{w} = 1 - w$ and if we agree to order the states as CC, CD, DC, DD, then the transition matrix for the process is

$$\mathbb{P} = \begin{bmatrix} x^2 & x\tilde{x} & \tilde{x}x & \tilde{x}^2 \\ yz & y\tilde{z} & \tilde{y}z & \tilde{y}\tilde{z} \\ zy & z\tilde{y} & \tilde{z}y & \tilde{z}\tilde{y} \\ w^2 & w\tilde{w} & \tilde{w}w & \tilde{w}^2 \end{bmatrix}.$$

Thus the transition matrix for the above example is

$$\mathbb{P} = \begin{bmatrix} 1 & 0 & 0 & 0 \\ 0 & 0 & 0 & 1 \\ 0 & 0 & 0 & 1 \\ 1 & 0 & 0 & 0 \end{bmatrix}.$$

This is obviously the matrix of an absorbing process.

Let $p_{CC}^{(i)}$, $p_{CD}^{(i)}$, $p_{DC}^{(i)}$, and $p_{DD}^{(i)}$ denote the probabilities that the process is in the states CC, CD, DC, and DD, respectively, on the ith play of the game. We have the following relations:

$$\begin{aligned}
p_{CC}^{(i+1)} &= x^2 p_{CC}^{(i)} + yz p_{CD}^{(i)} + zy p_{DC}^{(i)} + w^2 p_{DD}^{(i)}, \\
p_{CD}^{(i+1)} &= x(1-x)p_{CC}^{(i)} + y(1-z)p_{CD}^{(i)} + (1-y)zp_{DC}^{(i)} + w(1-w)p_{DD}^{(i)}, \\
p_{DC}^{(i+1)} &= x(1-x)p_{CC}^{(i)} + (1-y)zp_{CD}^{(i)} + y(1-z)p_{DC}^{(i)} + w(1-w)p_{DD}^{(i)}, \\
p_{DD}^{(i+1)} &= (1-x)^2 p_{CC}^{(i)} + (1-y)(1-z)p_{CD}^{(i)} + (1-y)(1-z)p_{DC}^{(i)} + (1-w)^2 p_{DD}^{(i)}.
\end{aligned}$$

We shall investigate the asymptotic behavior of $p_{CC}^{(i)}$ and $p_{CD}^{(i)}$ as i becomes large. To this end, it is convenient to introduce a final assumption concerning the nature of the players.

Assumption 4: The behavior of the players is such that in the long run there is a chance they will select any pair of strategies.

In mathematical terms we assume that the Markov chain is a regular one. Thus there is a unique stationary vector $\mathbf{s}$ whose coordinates are stable probabilities for the process, i.e., $\mathbf{s}\mathbb{P} = \mathbf{s}$. Recall (Chap. 3) that the coordinates of $\mathbf{s}$ are all positive. If $\mathbf{s} = (\sigma_1, \sigma_2, \sigma_3, \sigma_4)$, then $\mathbf{s}$ can be determined from the system of equations

$$\mathbf{s}(\mathbb{P} - \mathbb{I}) = \mathbf{0},$$
$$\sigma_1 + \sigma_2 + \sigma_3 + \sigma_4 = 1.$$

The determination of $\mathbf{s}$ can be facilitated by using the assumptions regarding the nature of the players. These assumptions give a certain symmetry to the payoff matrix $\mathbb{P}$ and also the the stationary vector $\mathbf{s}$. It follows (Exercise 4) that $\sigma_2 = \sigma_3$ and $\sigma_4 = 1 - 2\sigma_2 - \sigma_1$. Consequently we need only determine two of the four coördinates of $\mathbf{s}$ in order to determine the entire vector. The first two equations of $\mathbf{s}(\mathbb{P} - \mathbb{I}) = \mathbf{0}$ are

$$(x^2 - 1 - w^2)\sigma_1 + (yz + zy - 2w^2)\sigma_2 = -w^2,$$
$$(x\tilde{x} - w\tilde{w})\sigma_1 + (y\tilde{z} + z\tilde{y} - 1 - 2w\tilde{w})\sigma_2 = -w\tilde{w}.$$

Solving this, we obtain

$$\sigma_1 = \frac{w\tilde{w}(2yz - 2w^2) - (y\tilde{z} + z\tilde{y} - 1 - 2w\tilde{w})w^2}{(x^2 - w^2 - 1)(y\tilde{z} + z\tilde{y} - 1 - 2w\tilde{w}) - (x\tilde{x} - w\tilde{w})(2yz - 2w^2)}$$

and

$$\sigma_2 = \frac{w\tilde{w}(x^2 - 1 - w^2) - w^2(x\tilde{x} - w\tilde{w})}{(2yz - 2w^2)(x\tilde{x} - w\tilde{w}) - (y\tilde{z} + z\tilde{y} - 1 - 2w\tilde{w})(x^2 - w^2 - 1)}.$$

Recalling the interpretation of σ_i given in the discussion of Markov processes in Chap. 3, we note that σ_1 can be thought of as the fraction of the time the process is in state CC over a large number of trials. We denote this by $p_{CC}^{(\infty)}$. Likewise, $p_{CD}^{(\infty)}$, $p_{DC}^{(\infty)}$, and $p_{DD}^{(\infty)}$ have a similar interpretation. Thus by definition

$$p_{CC}^{(\infty)} = \frac{2w\tilde{w}(zy - w^2) - w^2(y\tilde{z} + z\tilde{y} - 1 - 2w\tilde{w})}{(x^2 - w^2 - 1)(y\tilde{z} + z\tilde{y} - 1 - 2w\tilde{w}) - 2(x\tilde{x} - w\tilde{w})(yz - w^2)}$$

and

$$p_{CD}^{(\infty)} = \frac{w\tilde{w}(x^2 - w^2 - 1) - w^2(x\tilde{x} - w\tilde{w})}{(2yz - 2w^2)(x\tilde{x} - w\tilde{w}) - (y\tilde{z} + z\tilde{y} - 1 - 2w\tilde{w})(x^2 - w^2 - 1)}.$$

These formulas are quite complicated, and they seem to contribute relatively little to the understanding of the problem. However, one can gain some insight by examining a special case.

Example. Consider a game in which the primary tendency of the players is to remain in state CC if they are in CC, but occasionally they succumb to temptation and try to take advantage of this same inclination of their opponent. In mathematical terms we take this assumption to be that $y = 0$,

$w = 1$, $z = 0$, and $0 < x < 1$, but x is close to 1. The proximity of x to 1 is a measure of the magnitude of the strength of the stabilization tendency. Thus, if x is very close to 1, then the tendency to stabilize on strategy C, and thus to remain in state CC, is very strong. On the other hand, if $1 - x$ is relatively large, then the players succumb to temptation quite often. Note that if $x = 1$, the process is no longer regular. Thus there must be some deviation from stability for our model to apply. Using the formulas for the asymptotic values of the probabilities p_{CC} and p_{CD} we have

$$p_{CC}^{(\infty)} = \frac{1}{2 + 2x - 3x^2} \quad \text{and} \quad p_{CD}^{(\infty)} = \frac{x\tilde{x}}{2 + 2x - 3x^2}.$$

These equations can be used to compute the decrease in reward which results from occasional switches in strategy from cooperation to noncooperation. For simplicity we consider only the two-parameter APDG in which $T > R > 0$ and $P = -R$, $S = -T$. The expected payoff in the totally stable case in which the players always play strategy C is obviously R. In this example the asymptotic value of the expected payoff to each player is a function of x. In view of the symmetry of the game we need consider only one function, and we denote it by

$$\begin{aligned} R_\infty(x) &= \frac{R}{2 + 2x - 3x^2} + \frac{S(x - x^2)}{2 + 2x - 3x^2} + \frac{T(x - x^2)}{2 + 2x - 3x^2} + \frac{P(1 - x^2)}{2 + 2x - 3x^2} \\ &= R \cdot \frac{x^2}{2 + 2x - 3x^2}. \end{aligned}$$

We can compare the average payoffs R and R_∞ by considering the ratio $r(x) = R_\infty(x)/R$. The function r satisfies $r(0) = 0$, $r(1) = 1$, and $r'(x) = [2x(2 + x)]/(2 + 2x - 3x^2)^2$. Thus, r is increasing on the interval $[0, 1]$ and $r'(1) = 6$.

We can interpret this result in the following way: For small deviations from total stability, i.e., $x < 1$ but x close to 1,

$$\frac{R_\infty(x)}{R} \approx 1 + 6(x - 1) \quad \text{or} \quad R_\infty(x) \approx R - 6R(1 - x).$$

Thus for small deviations the ratio of the average payoffs decreases about six times as fast as the deviation $(1 - x)$ of the parameter x from the value $x = 1$.

To obtain some indication of the range of x for which the above model is valid, it is useful to solve the equation $p_{CC}^{(\infty)} = 1/(2 + 2x - 3x^2)$ for x in terms of $p_{CC}^{(\infty)}$. We have

$$x = \frac{1}{3}\left[1 \pm \sqrt{7 - \frac{3}{p_{CC}^{(\infty)}}}\right],$$

and hence we see that the model cannot possibly be valid unless $p_{CC}^{(\infty)} > 3/7$. Therefore, for this discussion to be applicable, in the long run the game must be in state CC at least three sevenths of the time. Since there are many players and games for which this condition is not met, the model of this example is indeed a special case which is not always appropriate.

There are a number of other special cases which could be discussed. However, we feel it is more profitable to consider possible criticisms of this model and modifications which can be made. There are two natural points for criticism. First, it is unlikely that any two players are really psychologically identical. In view of this the model can only be considered to be describing the play of one average player against a second average player. The parameters of the problem can then be estimated by averaging data from many players. However, it then follows that the results obtained cannot be used to predict the play of a specific pair of contestants.

A second vulnerable point is that for most players the parameters x, y, z, and w probably change with time. They are not really constants, as this model has assumed. To meet this criticism, Rapoport and his associates have refined the model to include the variation of these parameters with repeated plays of the game. One can interpret their refinement of the model as one which has an element of learning built into it. This learning can be assumed to be simple or complicated and the complexity of the model changes accordingly. We recommend [RC] to the interested reader for the details of these more refined models.

6.2.5 Examples and the Need for Utility Theory

If the only applications of the theory of the APDG were to the operation of service stations and to the prosecution of certain pairs of criminal suspects, then a continued study of this problem would hardly seem to be justified. However, this is far from the actual situation. In this section we shall try to indicate some of the many fields in which there are problems that can be phrased as a Prisoner's Dilemma game. Also, we shall consider explicitly a question which is implicit in much of our work, namely, the treatment of games with nonnumerical payoffs. We begin with examples from a number of different areas.

Example 1: *Agriculture.* Suppose that 1001 wheat farmers each own 1000 acres which can be devoted to raising wheat. How many acres should each man sow? Consider the situation from the point of view of a single farmer. He may sow any amount from 0 to 1000 acres. The other farmers, considered as a unit, sow from 0 to 1 million acres. As the number of acres sown increases, the price of wheat decreases. Suppose that if only a small amount of wheat is planted, say 100,000 acres or less, then the farmers receive a price per bushel which provides a profit over both fixed and operating expenses. But if 1 million acres or more are sown, then the price is so low that the farmers fail to recover all their fixed expenses, although they do have an operating profit. Thus it is in the farmer's collective interest to limit production. But since each individual farmer has an operating profit, it is always more desirable for him to plant all 1000 acres. Of course, we are assuming that the farmers do not have a more lucrative alternative to planting wheat. Thus no matter how many acres the other farmers plant, if this number is considered fixed, then an individual farmer will always maximize his return by planting all 1000 of his acres. Each farmer faces a Prisoner's Dilemma. He may cooperate to limit production, or he may defect and grow as much wheat as possible.

Example 2: *Law.* Suppose that two parties sign an agreement to perform certain tasks; for example, one party will supply a certain product and the second party will buy the product. Each party has two choices. They can adhere to the agreement or violate it. The rewards and penalties for each course of action depend on the contract and the external circumstances. It may be that the supplier pays a penalty if he fails to deliver. However, even if he chooses to violate the agreement, he may be adequately compensated. For example, he may have found a better buyer for his product. Similarly, the buyer may pay a penalty if he does not accept the shipment he ordered. On the other hand, if his sales decrease, then he may not be able to sell all that he ordered, and hence he may lose in any case. It might actually be to his advantage to pay the penalty rather than pay for merchandise he cannot sell.

As a special case of this legal model we have the following example.

Example 3: *International Agreements.* Two countries sign a treaty to stop the testing of nuclear weapons. They each have two choices: adhere to the treaty or violate it. If they adhere to the treaty, then their citizens suffer less radiation danger and they maintain a certain amount of international goodwill; however, they are not able to test their new weapons and a certain subset of their population will feel insecure. If one party violates the agreement and the other does not, then the violator receives international condemnation (negative return) and new confidence in their weapons systems (positive return). If both parties violate the agreement, then everyone is subject to more radiation and neither party is relatively stronger militarily. However, they may now both be stronger relative to a third party who did not sign the original agreement and who has tested all along.

Example 4: *Student Study Tactics.* The students in a certain class know that their grade depends on their score on the final exam. They also know that the professor always curves the scores independent of how high or low they may be. Each student has a choice of studying or not studying for the exam.

If everyone studies, the effect on grades is as if no one had studied. However, each individual student does best by studying, independent of the actions of the others. The players (students) are in a Prisoner's Dilemma situation.

At this point the reader may well feel a certain uneasiness about these examples. They are certainly evidence of the widespread applicability of the Prisoner's Dilemma game. However, in each of these cases there is a certain vagueness about the exact nature of the rewards or penalities to the players. For example, what is the payoff to a country that breaks a nuclear test ban treaty? They receive adverse publicity in those countries which are not testing nuclear weapons, and they receive in return certain satisfaction from their increased knowledge of nuclear weapons. Both quantities are difficult to measure and evaluate numerically. Yet, in any APDG one must have payoffs which can be quantified. This is essential both to check the validity of the assumptions and to produce predictions which are susceptible to testing using statistical methods. This difficulty in assigning numerical payoffs is usually approached through the notion of *utility*. Briefly, in considering utility one first compares events in terms of the preferences of the players and then one says that event A has more utility than event B (to a certain player) if event A is preferred to event B (by that player). Then one proceeds to introduce a utility function which associates real numbers with events. Naturally one must make a number of assumptions in order to make these ideas rigorous. The outline of such a rigorous treatment is given in Project 6.4.1.

We conclude this section by discussing a possible numerical payoff matrix for Example 4. Since such a matrix is highly nonunique, the one given here must be considered simply as one of several possible payoff matrices.

Stan the student is contemplating his prospects for the upcoming weekend and for an exam on Monday morning in his class on mathematical models. He has two choices. He may study during the weekend, or he may go on a skiing trip to Black and Blue Lodge. He doesn't know what his classmates plan to do during the weekend, but to simplify the situation, he assumes that either they will all study or else none of them will study. He knows that the professor always curves the grades, and he also knows that in comparison with the rest of the class he is an average student. Thus in terms of his letter grade, he decides that the four possibilities correspond to the following outcomes (the outcomes are given only for Stan and not for the rest of the class):

		Others	
		Goof-off	Study
Stan	Ski	C	D
	Study	B	C

Considering these four possibilities, Stan decides that he prefers these outcomes in the following order:

1. *He studies and the others do not.* In this case he receives a grade of B. This is his first choice since it would allow him to get a grade of D in a future course. The implication is that he can take many weekends off in the future.
2. *No one studies.* Stan receives a C, and he also has a great weekend.
3. *Everyone studies.* Stan receives a C, but he does not enjoy the weekend.
4. *Stan does not study, but others do.* In this case Stan receives a D, and hence he must eventually get a B to maintain his C average. This creates long-term anxiety for Stan, which outweighs his short-term weekend bliss.

After arranging these outcomes in order according to his preferences, Stan assigns them numerical values which reflect in quantitative terms his preference ordering. He calls his units *piece of mind*, and he assigns the outcomes the values 10, 5, -5, and -10, respectively. Thus the matrix of payoffs to Stan is given by

		Others: Goof-off	Others: Study
Stan	Ski	5	-10
	Study	10	-5

In terms of the notation for an APDG we have $T > R > P > S$, and thus the necessary condition for an APDG is satisfied. Of course, it is also possible to assign payoffs in such a way that the condition is not satisfied (Exercise 3).

EXERCISES

1. Each of the following is the payoff matrix for a two-person game. Decide which of these games fits the decription for an APDG and if possible describe an *optimal strategy* for one or both players. In each case for which you give an optimal strategy, define what you mean by the term *optimal*.

(a)

(5, 5)	(5, 3)
(3, 5)	(3, 3)

(b)

(3, 5)	(5, 5)
(3, 3)	(5, 3)

(c)

(2, 3)	(4, 1)
(1, 4)	(3, 2)

(d)

(−1, 1)	(2, −2)
(−2, 2)	(1, −1)

(e)

(1, 0)	(−1, −1)
(−1, −1)	(0, 1)

2. In each of the following situations show how the problem may be phrased as a Prisoner's Dilemma game:
   (a) Labor-management negotations to avoid a strike and/or a lockout.
   (b) A husband-wife quarrel after which one or both parties may sue for a divorce.
3. Construct a payoff matrix for each of the examples of Sec. 6.2.5. Justify your choice of payoffs as much as possible and state your scale of units. Show that in some cases the payoffs can be chosen in different ways so that the game is or is not an APDG. Give examples of both cases.
4. Verify the assertion made in the computation of the fixed vector s in Sec. 6.2.4; i.e., show that $\sigma_2 = \sigma_3$ and that $\sigma_4 = 1 - 2\sigma_2 - \sigma_1$. Show that this result continues to hold in the more general model with assumption 4 replaced by 4′: The players are such that the process can move from any state to any other. *Hint:* With assumption 4′ the model is now stated in terms of an ergodic Markov process.
5. The matrix

(2, 1)	(0, 0)
(0, 0)	(1, 2)

   is the payoff matrix for a certain two-person game.
   (a) Why might this game be called The Battle of the Sexes?
   (b) If communication is possible, what is an optimal strategy?
   (c) If communication is not possible, what is an optimal strategy?
6. Develop a stochastic model for the play of the game given in Exercise 5. Compare your model with the one given for the Prisoner's Dilemma. What similarities and differences do you expect in the play of the two games, the Prisoner's Dilemma and The Battle of the Sexes?
7. Show that every two-person constant-sum game can be solved by solving a two-person zero-sum game. First define the concept of a *solution* of a constant-sum game.
8. (Equilibrium pairs) Consider a two-person game with payoff matrices $\mathbb{G}(R)$ and $\mathbb{G}(C)$ for the row and column players, respectively. Assume that $\mathbb{G}(R)$ and $\mathbb{G}(C)$ are $m \times n$. A *mixed strategy for R* is any probability vector $\boldsymbol{\sigma} \in R^m$ and a *mixed strategy for C* is any probability vector $\boldsymbol{\tau} \in R^n$. A pair of mixed

strategies $(\boldsymbol{\sigma}_0, \boldsymbol{\tau}_0)$, $\boldsymbol{\sigma}_0 \in R^m$ and $\boldsymbol{\tau}_0 \in R^n$, is called an *equilibrium pair* for the game with payoff matrices $\mathbb{G}(R)$ and $\mathbb{G}(C)$ if for every pair of mixed strategies $(\boldsymbol{\sigma}, \boldsymbol{\tau})$ we have

$$\boldsymbol{\sigma}\mathbb{G}(R)\boldsymbol{\tau}_0 \leq \boldsymbol{\sigma}_0\mathbb{G}(R)\boldsymbol{\tau}_0,$$
$$\boldsymbol{\sigma}_0\mathbb{G}(C)\boldsymbol{\tau} \leq \boldsymbol{\sigma}_0\mathbb{G}(C)\boldsymbol{\tau}_0.$$

(a) Find an equilibrium pair for the game of Exercise 5.
(b) Find an equilibrium pair for an APDG.

9. (Exercise 8 continued) Define the term optimal strategy for a general two-person game and discuss the relationship between optimal strategies and equilibrium pairs. In particular answer the following questions:
(a) Do optimal strategies yield equilibrium pairs?
(b) Do equilibrium pairs yield optimal strategies?

10. (Bargaining and threats) In any game which allows cooperation between the players, the possibility exists that certain players will negotiate a binding agreement to play in a certain manner and to exchange side payments. The negotiations which lead to the agreement traditionally involve both bargaining ("I will do this if you will do that") and threats ("If you don't do that, then I will do this"). There are mathematical theories for the study of games which include bargaining and threats (see [O], for example); however, since the techniques involved are based on mathematics which we have not assumed and which we do not wish to develop, we shall be content to study a few examples.
(a) Consider the payoff matrix of Exercise 5, and assume that the players in this game are able to bargain and use side payments. Define the concept of an optimal strategy for the players of this game and find an optimal strategy. Also discuss the possible bargaining and threats which might be used by the players to negotiate an agreement on how they will play the game.
(b) Same questions as (a) for the game with the payoff matrix

$$\begin{bmatrix} (10, 0) & (0, 10) \\ (5, 10) & (-10, -10) \end{bmatrix}.$$

6.3 COMMENTS ON MORE GENERAL GAMES AND APPLICATIONS OF GAME THEORY

Each of the games of Secs. 6.1 and 6.2 has only two players, and each player has only a finite number of pure strategies. Neither of these restrictions is necessary for the development of game theory, and there are many situations for which an appropriate mathematical model is a game in which at least one of them is violated. Thus it may be that there are n players ($n > 2$), that at least one player has infinitely many pure strategies, or both. A great deal of study has been devoted to such games, and they are currently under active investigation. Much of this work is beyond the scope of this book, and we shall not attempt to discuss it here. However, a number of special cases are considered in the Exercises and Projects. In this section we shall be content to give examples of such games and to give some general comments about the applicability of game theory.

6.3.1 Infinite Games

In the broadcast decision problem of Sec. 2.5 each political party has a strategy set consisting of 30 pure strategies. There is one strategy corresponding to the decision regarding broadcasting on each of the 30 days before the election. This problem can easily be rephrased so that each party has an infinite number of pure strategies. One method of doing this is to consider time as a continuous quantity rather than divided into units of 1 day each. For example, the strategy "broadcast 2.5 days before the polls open" would be different from the strategy "broadcast 2.25 days before the polls open." In this way each party may be considered to have an infinite number of strategies, one strategy for each time between 0 and 30 days preceding the election. One can also have an infinite number of strategies when time is not considered as continuous. For example, if we determine strategies only by the rational numbers in the interval [0, 30], then each party has a countably infinite number of strategies. On the other hand, the number of pure strategies is uncountable in the case where all real numbers between 0 and 30 are used as labels for different strategies.

We now give a formal definition of an infinite two-person zero-sum game. We shall distinguish one of the cases mentioned above (the one with uncountably many strategies), since this is an important one for applications.

Definition: An *infinite two-person zero-sum* game is a triple $\Gamma = (S, T, p)$, where S and T are sets, at least one of which is infinite, and p is a real function defined on $S \times T$. Such a game is said to be *continuous* if S and T are each intervals (possibly identical).

Remarks

1. As it is used in this definition the term continuous refers to the strategy sets (an interval is a continuum) and not to the payoff function p. It is possible (and, in fact, common) to have a continuous game with a payoff function that is not continuous. (See Exercises 2, 3, and 4.)
2. One can consider infinite two-person nonzero-sum games by simply introducing a second payoff function.

6.3.2 Multiperson Games

It is relatively easy to generalize the definition of a two-person game to a definition of a three-person game. In a two-person game one has two sets, S and T, and two functions, p_1 and p_2, defined on $S \times T$.

Similarly, in a three-person game one has three sets, say U, V, W, and three functions, p_1, p_2, p_3, defined on $U \times V \times W$. Thus, if $u \in U$, $v \in V$, and $w \in W$, then $p_1(u, v, w)$ is the payoff to the first player when he uses strategy u, player 2 uses strategy v, and player 3 uses strategy w. The functions p_2 and p_3 are defined similarly for players 2 and 3.

In the study of two-person games the assumption $p_1 + p_2 = 0$ yields the special case of zero-sum games. In that case one is able to develop a complete theory based on mixed strategies. Hence one is led to consider a similar assumption for three-person games; i.e., $p_1 + p_2 + p_3 = 0$. However, for three-person games the implications of the zero-sum assumption are not as clear as they are in the two-person case. First, it is no longer the case that this assumption places the interests of all the players in conflict with each other. Since the gain of one player is always equal to the combined losses of the other two players, it is easy to conceive of situations in which certain pairs of players would wish to combine interests. Moreover, if one keeps track of the coalitions over many plays of the game, it may be that the combinations vary from play to play. Thus in multiperson games one is forced to consider the possibility of communication between players and of permanent or temporary agreements. Even in the event that communication is not possible, one must still be aware of the possibility that a tacit coalition will form. Certain players may see that it is in their mutual interest to cooperate, and they proceed accordingly even though they have not made a formal agreement to do so. Thus even in zero-sum games it is necessary to consider a number of factors before one can begin to analyze a three-person game. The fact that there are three players is not important for these comments, and obviously the same things can be said about an n-person game for any $n > 2$. The aim of our remarks is simply to indicate some of the complexities involved in the study of n-person games. Some of these topics are pursued in Exercises 5, 6, 7, and 8. We conclude this brief discussion with an example of a three-person game.

Example. Consider a congressional committee which consists of five members: two Scoundrels, two Redeemers, and one Independent. A bill on tax reform is before the committee, and it will be reported out of committee only if it receives at least three favorable votes. If either party takes a public stand on the bill, and hence both party members vote for it or against it, then the party gets a return of $+1$ (in support) when they are in the majority and -1 when they are in the minority. If the party vote is split, then the party receives 0 no matter what the outcome of the vote. The independent member gets a return of $+2$ when he is the deciding vote; otherwise he receives 0. These payoffs can be supported in the following way: If a party votes to recommend the tax reform bill and the recommendation passes, then it receives a net gain of $+1$ in support ($+2$ comes from the taxpayers who seek tax reform and -1 comes from tax loophole fans). Similarly, if it votes against tax reform and the bill loses, then it also receives a net gain of $+1$ ($+2$ from the loophole users and -1 from the tax reform group; your friends always tend to give your more credit than your enemies). On the other hand, if one votes to approve (or disapprove) and the bill loses (or wins) despite that opposition, then that position has actually been harmful politically and one obtains a payoff of -1 (your friends give you 0 because you failed and your enemies give you -1 for resisting them). A similar type of reasoning can be applied to the other payoffs.

If the committee is expected to receive many such bills over the years, how should the members vote so as to maximize their return?

We model the above situation by means of a three-person game. The players are denoted by S, R, and I. The strategy sets U for S and V for R are the same, namely,

$$U = V = \{(2, 0), (1, 1), (0, 2)\},$$

where (a, b) represents a strategy of a votes for passage and b votes against passage. Using the same definition for (a, b), the strategy set for I is $W = \{(1, 0), (0, 1)\}$. The three payoff functions are denoted by p_S, p_R, and p_I, and they are easily computed using the definition of payoff outlined above. For example,

$$p_S((2, 0), (0, 2), (1, 0)) = +1,$$
$$p_R((2, 0), (0, 2), (1, 0)) = -1,$$
$$p_I((2, 0), (0, 2), (1, 0)) = +2.$$

Remember that the domain for each payoff function is $U \times V \times W$.

We invite the reader to consider "solving" this game (Exercise 6).

6.3.3 Concluding Remarks on Games and Applications of Game Theory

We have now introduced many of the main ideas and some of the central results of game theory. In Sec. 2.5 we solved a game by means of min-max theory, and in this chapter we showed that all two-person zero-sum games have a min-max solution in terms of mixed strategies. Thus one is tempted to conclude that one should always seek and use an optimal mixed strategy for every two-person zero-sum game. Is this really a conclusion of game theory? No, it is not. There may well be cases where one would not want to use a mixed strategy. To support this assertion, we return to the border battle game of Sec. 6.1. Suppose that C.B. knows that M.N. is going to use strategy (3, 0) on a certain night. This might be the case if his spies have infiltrated the M.N. camp. Should C.B. use strategy $\boldsymbol{\sigma}_0$? Surely not. He should use the pure strategy (4, 0) and gain a payoff of +3.

The point of these comments is that our formal mathematical study of game theory does not solve games or recommend behavior in an absolute sense. Rather, like all mathematics, game theory consists of "if . . . then . . ." assertions. If certain assumptions are true, then certain conclusions are true. If the assumptions are false, then mathematical game theory makes no statement concerning the conclusions. This point is especially important with regard to our assumption of "rational behavior." If the players are operating under a criterion other than that of maximizing their payoff, then the theory developed in Sec. 6.1 does not apply. This matter is referred to again in Sec. 10.1.

The experimental study of game-theoretic situations is another matter. Here, one is trying to determine how people actually behave in certain circumstances. Data are obtained and statistical techniques are used to analyze the data. This provides evidence for new conjectures and support for new assumptions about human behavior, but it is not technically part of what is usually meant by game theory. Rather it is part of psychology and perhaps sociology. Of course, such work may well motivate the study of new problems in game theory and its applications. Hence it is an integral part of the model-building process.

There is another matter which should be discussed in relation to two-person zero-sum games. This is the practical question, What does it really mean to use an optimal mixed strategy? According to our interpretation, it means that at each play of the game one uses a chance device to select a pure strategy. The probability that a given pure strategy will be selected is determined by the mixed strategy. But what are the benefits of such a method of play? Why not just reason out what you think your opponent will do and choose accordingly? The first reason for using an optimal mixed strategy is that this guarantees that your opponent cannot take advantage of you by correctly diagnosing your pattern of play. By adopting an optimal mixed strategy you have assured yourself of a definite long-run return, and this is the best long-run return that you can be certain of obtaining. But, what about games that are only played once? Does the use of a mixed strategy still make sense if you do not intend to play the game often enough for the good and bad results to average out? It is with this question in mind that we give a second reason for using an optimal mixed strategy. This is the advantage of having a well-defined decision process as opposed to selecting a strategy by some ad hoc method. If one seeks an alternative to the use of an optimal mixed strategy, then one is often faced with a sequence (perhaps infinite in length) of considerations of the following form: Since he thinks that I will think that he thinks that . . . , therefore it is best for me to do . . . ! There is no need for such reasoning if one uses a mixed strategy. The pure strategy is chosen by a chance device.

EXERCISES

1. Describe situations for which an appropriate mathematical model is the following type of game:
(a) Three-person zero-sum (finite).
(b) Three-person nonzero-sum (finite).
(c) Infinite, two-person zero-sum (not continuous).
(d) Infinite, two-person nonzero-sum (not continuous).
(e) Infinite, three-person zero-sum.
(f) Continuous, two-person nonzero-sum.

In Exercises 2, 3, and 4 we shall consider a game which is often used as a mathematical model for a broad class of problems known as timing problems. In these situations the participants must decide on the proper time for carrying out a certain action. In many cases it is desirable to wait as long as possible, and yet there is a severe penalty for waiting too long. The most dramatic example of such a situation is a duel with pistols between two people who are walking toward each other. As the distance between the duelists decreases, they are more likely to hit their target, and hence they have a good reason for waiting to fire. However, if they wait too long there is an obvious and severe penalty which results. Although in these terms such a game is (hopefully) only of historical interest, the same situation can occur in many different contexts. Section 2.5 provided one example. The following exercises introduce some other situations with similar features.

2. Let $S = T = [0, 1]$. Thus a pure strategy for Red is any number $s \in [0, 1]$, and a pure strategy for Cream is any number $t \in [0, 1]$. The game is zero-sum, and the payoff to Red for the strategy pair (s, t) is denoted by $p(s, t)$, where

$$p(s, t) = \begin{cases} 1 - 2s, & s > t, \\ t^2 - s, & s = t, \\ 2t^2 - 1, & s < t. \end{cases}$$

(a) Consider the situation from the point of view of Red and show that by min-max theory he should choose his pure strategy s so that

$$p(s, t) = \underset{u \in [0, 1]}{\text{supremum}} \{\underset{v \in [0, 1]}{\text{infemum}}\, p(u, v)\}.$$

Remark: See Appendix A for definitions of the terms supremum and infemum, and note that they cannot be replaced by maximum and minimum in this example.

(b) Show that

$$\underset{u \in [0, 1]}{\text{supremum}}\ \underset{v \in [0\ 1]}{\text{infemum}}\, p(u, v) = \underset{u \in [0, 1]}{\text{supremum}} \{\min\{1 - 2u, u^2 - u, 2u^2 - 1\}\}.$$

(c) Graph the following three functions:

$$y_1(u) = 1 - 2u, \qquad 0 \le u \le 1,$$
$$y_2(u) = u^2 - u, \qquad 0 \le u \le 1,$$
$$y_3(u) = 2u^2 - 1, \qquad 0 \le u \le 1,$$

and determine the point u_0 which satisfies $y_1(u_0) = y_2(u_0) = y_3(u_0)$.

(d) Use the information given in parts (a), (b), and (c) to obtain a "best" pure strategy for Red and Cream. Explain and defend your use of the term *best*.

3. Consider a duel with pistols between Mr. Honorable (H) and Mr. Scoundrel (S). Each man represents a team and the team receives $+1$ if their man wins the duel, -1 if he loses, and 0 if it is a draw. A duelist is said to *survive* if he is not hit by a bullet, and he does not survive if he is hit by a bullet fired by his opponent. If either both survive or both do not survive, then the duel is a draw. Initially the two participants are a distance D apart, and they then begin walking toward one another. Let ρ, $0 \leq \rho \leq 1$, represent the fraction of the distance between the two men which is still to be covered. Also, let $p_H(\rho)$ be the probability of a successful shot by Mr. Honorable when the distance between the men is ρD. Similarly define $p_S(\rho)$ as the probability of a successful shot by Mr. Scoundrel.
   (a) Let ρ_H and ρ_S be the values of ρ at which H and S fire their pistols. Assume that there is only one bullet in each pistol, that neither man can flee if he fires and misses, and that each knows when the other fires his pistol; i.e., the pistols are noisy. Find the expected return to Mr. Honorable's team in terms of $p_H(\rho)$ and $p_S(\rho)$. Denote this return by $E_H(\rho_H, \rho_S)$.
   (b) Let $p_H(\rho) = 1 - \rho$ and $p_S(\rho) = 1 - \rho^2$. What do these functions say about the accuracy of the combatants? Compute $E_H(\rho_H, \rho_S)$ in this case.
   (c) Use the results of Exercise 2 to obtain "best" strategies for Mr. Honorable and Mr. Scoundrel.

4. Suppose that salesman Stan (S_1) and salesman Sam (S_2) are both trying to sell a certain product to buyer Bigwig (B). The salesmen know that B will buy at or before time T; i.e., the probability that B will buy from one of the salesmen by time T is 1. The products are not identical and this influences B's decision to buy from one salesman rather than the other. Suppose that $p_1(t)$ and $p_2(t)$ denote the probabilities that B will buy from S_1 and S_2, respectively, at time t. Suppose that circumstances prevent either S_1 or S_2 from visiting B more than once and that regardless of which one visits B first, if he does not make the sale, then B will surely buy from the other salesman. What is the best time for S_1 to visit B if he is interested in maximizing the probability that he will make a sale? Also, answer the same question in the special case with $p_1(t) = t/T$, $p_2(t) = (t/T)^2$. What other criteria might S_1 use to decide when to visit B?

5. Let $U = V = W = \{1, 2, 3\}$, and Let p_1, p_2, p_3 be defined by

$$p_1(u, v, w) = \max\{v, w\},$$

$$p_2(u, v, w) = \min\{u, w\},$$

$$p_3(u, v, w) = w - u - v.$$

   (a) If each player seeks only to maximize his own return, then how should he play this game? Consider both the case where coalitions are legal and where they are illegal.
   (b) If each player seeks only to minimize the combined return of his two opponents, then how should he play this game?
   (c) If each player seeks to maximize his relative return (his gain minus the combined gain of the other players), then how should he play this game?

6. Consider "solving" the committee game of Sec. 6.3.2. First, represent the payoffs for this game by using two 3×3 matrices, one matrix for each strategy of player I. The entries in the matrices are the payoff triples representing the respective payoffs to the three players. Next, by examination of these matrices, consider the advantages and disadvantages of the various strategies for players R and S. Do they have a "best" pure strategy? Finally, consider possible coalitions which may form and evaluate the benefits to be obtained by the members of the coalition.
7. Compare three-person zero-sum games and general two-person games. How are they similar and how are they different?
8. Consider the coalitions which may form in a four-person game. Give examples of four-person games where it is natural to have
   (a) A coalition of all four members.
   (b) Two coalitions of two members each.
   (c) One coalition with three members.
   (d) Exactly one coalition of two members.
   (e) No coalitions.

6.4 PROJECTS

6.4.1 Preferences and Utility Theory

Consider a game in which one of the players has a choice between a 50–50 chance of winning or losing \$100 and a 50–50 chance of winning or losing \$100,000. In each case the expected return is 0 dollars. However, it is quite likely that the individual involved will have a definite preference between these two choices. Thus knowledge of the expected monetary return for a certain choice of play is not always sufficient to determine its place in the strategy of play of an individual. Instead, other factors such as the financial status of the player and his state of mind must be considered in determining his optimal strategy. In particular, individual preferences and individual concepts of utility must be considered. In this project we shall indicate briefly how some of these concepts can be given a sound quantitative foundation.

We begin with the basic fact that in his consideration of two events an individual may prefer one to the other or he may be indifferent toward them. If the events are denoted by E_1 and E_2, then we write $E_1 > E_2$ to denote that E_1 is preferred over E_2, and $E_1 \sim E_2$ to denote that E_1 and E_2 are equally preferable. We assume that $>$ and $\sim$ obey the following axioms:

1. For each pair of events E_1 and E_2, exactly one of the following holds:
   (a) $E_1 > E_2$.
   (b) $E_2 > E_1$.
   (c) $E_1 \sim E_2$.
2. $E \sim E$.

In axioms 3–7, the symbol $\longrightarrow$ means *implies.*

3. $E_1 \sim E_2 \longrightarrow E_2 \sim E_1$.
4. $E_1 \sim E_2, E_2 \sim E_3 \longrightarrow E_1 \sim E_3$.
5. $E_1 > E_2, E_2 > E_3 \longrightarrow E_1 > E_3$.
6. $E_1 > E_2, E_2 \sim E_3 \longrightarrow E_1 > E_3$.
7. $E_1 \sim E_2, E_2 > E_3 \longrightarrow E_1 > E_3$.

If for a given individual $E_1 > E_2$, then we say that E_1 has more utility than E_2 for this individual. The goal of utility theory is to provide a method for specifying how much more.

One type of event which is important in the study of utility theory is an event which has the form "E_1 will occur with probability p and E_2 will occur with probability $q = 1 - p$." We call such an event a lottery with outcomes E_1 and E_2 and associated probabilities p and q. We denote this lottery by $pE_1 + qE_2$. We adopt the following axioms about lotteries:

i. $pE_1 + qE_2 \sim qE_2 + pE_1$.
ii. $pE_1 + q\{rE_2 + (1 - r)E_3\} \sim pE_1 + qrE_2 + q(1 - r)E_3$.
iii. $pE + qE \sim E$.
iv. If $E_1 \sim E_3$, then for every p, $\{pE_1 + qE_2\} \sim \{pE_3 + qE_2\}$.
v. If $E_1 > E_3$, then for every p, $0 < p \leq 1$, $\{pE_1 + qE_2\} > \{pE_3 + qE_2\}$.
vi. If $E_1 > E_2 > E_3$, then there exists $p \in [0, 1]$ such that

$$\{pE_1 + qE_3\} \sim E_2.$$

Note. In each of the axioms $p + q = 1, p \geq 0$.

Problem 1. Justify each of axioms i–vi by showing that they are reasonable assumptions about an individual's preferences toward everyday events.

Problem 2. Prove that if $E_1 > E_2 > E_3$ and $\{pE_1 + qE_3\} \sim E_2$, then $0 < p < 1$ and p is unique.

With this preparation we are now ready to prove that there is a function which can be used to measure an individual's preferences. Such a function is known as a *utility function*, and its construction is the object of the following theorem.

Theorem: There exists a function u mapping the set of all events into the real numbers and satisfying the conditions

α. $u(E) > u(F)$ if and only if $E > F$.
β. $u(pE + qF) = pu(E) + qu(F)$ for every $p \in [0, 1]$ and all events E and F.

The proof of this theorem is quite long. We provide an outline of the proof and ask the reader to fill in some of the details.

We begin by considering the case where a person is indifferent to all events.

Problem 3. Show that if $E \sim F$ for all events E and F, then the function u defined by $u(E) = 0$ for all E has all the properties required by the theorem.

Next, consider the case where there are events E_1 and E_2 such that $E_1 > E_2$. Then for any other event G there are five subcases to consider. They are:

1. $G > E_1$.
2. $G \sim E_1$.
3. $E_1 > G > E_2$.
4. $G \sim E_2$.
5. $E_2 > G$.

Let $u(E_1) = 1$ and $u(E_2) = 0$. The definition of $u(G)$ will depend on which of the five subcases holds. These definitions are as follows:

1. $u(G) = 1/p$, where p is chosen so that $\{pG + qE_2\} \sim E_1$ (see axiom vi on lotteries).
2. $u(G) = 1$.
3. $u(G) = p$, where p is chosen so that $\{pE_1 + qE_2\} \sim G$.
4. $u(G) = 0$.
5. $u(G) = (p - 1)/p$, where p is chosen so that $\{pG + qE_1\} \sim E_2$.

It is now necessary to show that this definition of u satisfies conditions α and β of the theorem. There are many cases to consider, depending on the events E and F.

Problem 4. Show that with this definition u obeys conditions α and β when the events E and F are such that E satisfies subcase 2 (i.e., replace G of this subcase by E) and F satisfies subcase 4 of the definition.

Problem 5. Same as Problem 4 with E and F both in subcase 1.

The proofs for the other cases are similar to those considered in Problems 4 and 5. The carrying out of these proofs completes the proof of the theorem.

Thus a utility function can always be constructed to assign a numerical rating to events in such a way that these ratings reflect preferences. This utility function is not unique; however, if v is any other such function, then there are real numbers r and s such that

$$v(E) = ru(E) + s$$

for all events E. That is, u is unique up to a linear transformation.

6.4.2 Experimental Design

Design an experiment to determine how people act when they are faced with a Prisoner's Dilemma. State all the details of your experiment. In particular, include such information as how the subjects will be chosen, how often they will play the game, how they will be paid for participating, and how the payoff will be made. Also discuss the problems of the subjects communicating directly or using a previously agreed upon plan. Consider the possibility of a sequence of experiments in which the details of later ones depend on the initial results.

6.4.3 Sequential Play Games

In many game situations (for example, parlor games such as chess and checkers) the action proceeds in stages called moves. First, one player makes a move, then after due consideration the second player makes a move, then the first player makes a second move, and so forth until the rules of the game determine a final move. Such a game might seem to be quite different from a game in which each player has only one move and both players must indicate their move simultaneously with neither player having any knowledge of the others intentions. The purpose of this project is to show that these two types of games are not really different. We begin with an example and proceed to the general case.

An Example of Gentlemen's Russian Roulette. Consider a game of Russian roulette which is played in the following manner: Player Cream plays first and he has the two options:

1. Pass and add $10.00 to the pot.
2. Add $1.00 to the pot and gamble (in this game to *gamble* means that the player rolls a fair die and notes the outcome. If a single dot lands face up, then the player does not survive and otherwise he does survive).

Player Red receives the pot if Player Cream does not survive. If Player Cream survives, then Player Red has the options:

1. Pass and add $10.00 to the pot.
2. Add $1.00 to the pot and gamble.

If both players survive, then the pot is divided equally between them. Otherwise the survivor gets the entire pot.

Problem 1. Show that if Red takes into account the possible plays by Player Cream, then he has four different pure strategies. List these strategies. *Hint:* Let one of the strategies be "Pass in both cases."

Problem 2. Form a payoff matrix for this game using payoffs which are the expected net return to Red. Show that this game is zero-sum.

Problem 3. Solve the game with the payoff matrix given in Problem 2. Find the value of the game and the optimal strategies for Red and Cream.

Problem 4. Consider a game which consists of three moves. First a move by Cream, then a move by Red, and finally a move by Cream. Each move can be made in two ways. Show how to formulate such a game as a single move game. How big is the payoff matrix?

Problem 5. Show how to formulate any sequential-play game (two-person) as a single-move game.

6.4.4 Denumerable Games

Let S and T be denumerable strategy sets for the players Red and Cream, respectively. Thus, $S = \{s_i\}_1^\infty$ and $T = \{t_i\}_1^\infty$. Consider a generalization of the theory of finite two-person zero-sum games to denumerable games. Define a payoff function on pure strategies, define mixed strategies, and consider extending the definition of the payoff function to include mixed strategies. When can this be done in a natural way? Discuss the concept of a solution for such a game. Give an example (nontrivial) of a denumerable game for which one can obtain a solution in the same way as for a finite game. Also give an example of a denumerable game which cannot be solved in this way.

REFERENCES

The initial development of game theory was strongly influenced by the research of John von Neumann and by the publication of the pioneering work *Theory of Games and Economic Behavior* [vNM] by von Neumann and O. Morgenstern. This volume demonstrated that there is a mathematical theory of game like situations and that this theory does indeed make a contribution to the understanding of certain questions in economics. Subsequently, game theory developed in many directions, and there are now many sources available for study. Some of these are very mathematical in nature, while others stress the applications of game theory in areas such as the social sciences and the rapidly developing field of decision making.

Reference [O] is a general source for a mathmatical treatment of games. It includes both two-person and n-person games as well as both zero-sum and nonzero-sum games. The viewpoint of the social scientist toward the mathematical aspects as well as the applications of game theory is given in [LuR]. Also, in [LuR] there are numerous discussions of the limitations of game theory for the social sciences. There are a number of very interesting and frequently amusing examples of elementary game situations in [W].

The topics of two-person games, the Prisoner's Dilemma, and n-person games are treated, respectively, in [R1], [RC], and [R2]. These references provide theory as well as a discussion of such matters as how the theory might be applied, if the theory is worth trying to apply, and why one might wish to learn game theory.

[L] LAVE, L., "Factors Affecting Co-operation in the Prisoner's Dilemma," *Behavioral Science*, **10** (1965), 26–38.

[LuR] LUCE R. D., and H. RAIFFA, *Games and Decisions*. New York: Wiley, 1957.

[O] OWEN, G., *Game Theory*. Philadelphia: Saunders, 1968.

[R1] RAPAPORT, A., *Two-Person Game Theory: The Essential Ideas*. Ann Arbor: The University of Michigan Press, 1966.

[R2] RAPAPORT, A., *N-Person Game Theory: Concepts and Applications*. Ann Arbor: The University of Michigan Press, 1970.

[RC] RAPAPORT, A., and A. CHAMMAH, *Prisoner's Dilemma*. Ann Arbor: The University of Michigan Press, 1965.

[vNM] VON NEUMANN, J., and O. MORGENSTERN, *Theory of Games and Economic Behavior*, 2nd ed. Princeton, N.J.: Princeton University Press, 1947.

[W] WILLIAMS, J., *The Compleat Strategyst*. New York: McGraw-Hill, 1954.

7 Graphs as Models

7.0 INTRODUCTION

Chapters 4 and 5 are devoted to the study of a very basic model for linear optimization. It is an especially important model, with applications in many areas; however, there are important optimization problems for which this linear programming model is not sufficiently refined. The model is not capable of incorporating all the essential aspects of such problems. In this chapter we shall describe one of the many useful models which may be obtained by modifying the linear programming model. In contrast to Chap. 4, we do not attempt here to develop a complete theory for this model. Instead we use this new optimization model as an introduction to the study of graph theory. After discussing a little of the general theory of graphs, we turn to the consideration of other types of problems which can be studied by means of models which involve graphs. Throughout this chapter it is our intention to proceed only far enough with a topic to provide the reader with an introduction to the types of problems which can be considered and to some of the techniques which are used. In certain instances it is useful to develop some of the associated theory, and in others it is more convenient to proceed primarily through examples. In any case the presentation is intended to stimulate interest in a broad and rapidly developing field.

7.1 NETWORKS AND FLOWS

In this section we shall consider a class of problems which can be viewed as linear programming problems with auxiliary constraints. We begin by recalling two problems from Chap. 2: the transportation problem and the job assignment problem, Secs. 2.2 and 2.6, respectively.

In the transportation problem the goal is to find an optimal schedule for shipping mobile homes from manufacturing sites to retail outlets. This problem involves a constraint which was not explicitly taken into account in the general theory of Chap. 4 or in the development of the alogorithm of Chap. 5. This constraint is the natural requirement that the coordinates of the solution vector be integers. Such a constraint is reasonable since it would be a strange decision to consider a schedule optimal if it required a mobile home to be divided. However, with this constraint, two questions immediately arise: First, do optimal schedules exist, and second, if they exist, how can they be found? It is clear, for example, that the simplex method of Chap. 5 is not generally applicable to problems which require a solution vector having integer coordinates. Indeed, the examples and the exercises provide illustrations of problems with integer data where the coordinates of the optimal vectors are not all integers. Actually, it is the case that many integer programming problems, including the transportation problem, can be solved by the simplex method (such problems are called "natural "integer problems). However, even though the simplex method can be used for certain integer problems, it cannot be used for all of them and in those cases where it can be applied it is often not the best method. In this section, we shall introduce a new model (a network) which provides a second method for solving integer transportation problems. In addition to providing a model for the integer transportation problem, the concept of a network permits the incorporation of additional constraints, namely route capacities. Thus, the network model provides a setting for problems which are more general than those considered before. Our development of this model will provide a computational scheme for determining optimal vectors, and this scheme is often more efficient than the simplex method when both methods are applicable. Although we do not go into the computational details here, these details are thoroughly discussed in the References.

The job assignment problem of Sec. 2.6 is also a linear programming problem with auxiliary constraints, and our new model will provide a means for solving this problem and generalizations of it. To make it clear how the job assignment problem has the framework of a linear programming problem, let $\mathcal{J} = \{J_1, J_2, \ldots, J_n\}$ be the set of available jobs, and let $\mathcal{A} = \{A_1, A_2, \ldots,$

$A_m\}$ be the set of applicants. Each applicant has been interviewed, tested, and then rated as appropriate or not appropriate for each job. Recall (Sec. 2.6) that an applicant is appropriate for a job if he is both qualified for the job and willing to accept it. The rating of the set of applicants is given by the $m \times n$ matrix $\mathbb{R} = \{r_{ik}\}$, where r_{ik} is 1 if A_i is appropriate for J_k, and r_{ik} is 0 if A_i is not appropriate for J_k. The first problem is to determine if all jobs can be filled with appropriate applicants. The second is to find an actual assignment of applicants to jobs so that all jobs are filled with appropriate applicants. In Sec. 2.6 we gave a condition for determining if all jobs could be filled. However, the proof of that result was not constructive, and no algorithm was given for assigning applicants to jobs. In this section we shall provide a second proof of the condition of Sec. 2.6. and this proof will give a method for finding the desired assignment. The details of the algorithm are the topic of the project of Sec. 7.3.1.

To phrase the job assignment problem in the form of a linear optimization problem, we use $m \times n$ matrices $\mathbb{S}$ of the form $\mathbb{S} = (s_{ik})$, where

1. $s_{ik} = 0$ or 1, $i = 1, \ldots, m$, $k = 1, \ldots, n$.
2. $\sum_{k=1}^{n} s_{ik} \leq 1$, $i = 1, \ldots, m$.
3. $\sum_{i=1}^{m} s_{ik} \leq 1$, $k = 1, \ldots, n$.

These matrices correspond in a one-to-one manner with assignments of applicants to jobs: If A_i is assigned to J_k, then $s_{ik} = 1$; otherwise $s_{ik} = 0$. Condition 2 states that an applicant can be assigned to at most one job, and condition 3 states that at most one applicant is be to assigned to each job. Finally, we consider the quantity $\mathbb{R} * \mathbb{S} = \sum_{i=1}^{m} \sum_{k=1}^{n} s_{ik} r_{ik}$. Each term in this sum is either 0 or 1. If the term is a 1, then $s_{ik} = 1$ and $r_{ik} = 1$ and this means that the assignment determined by $\mathbb{S}$ fills the job J_k with an appropriate applicant A_i. Hence the quantity $\mathbb{R} * \mathbb{S}$ is a measure of the number of jobs filled by appropriate applicants under the assignment $\mathbb{S}$. The job assignment problem is then the problem of finding a matrix $\mathbb{S}$ which satisfies conditions 1–3, and which maximizes $\mathbb{R} * \mathbb{S}$ for all such matrices. In this form the problem is similar to a linear programming problem (see Exercise 1); however, it has an additional constraint, condition 1. The model of this chapter explicitly includes this new condition. It is also useful in more general assignment problems (see the project of Sec. 7.3.3).

7.1.1 Definitions and Notation

It is useful to discuss both the transportation problem and the assignment problem in a single abstract setting. The new setting is known as a network, and we obtain it by abstracting the essential ideas from the transportation problem (TP from now on) in the following manner.

First, we replace each of the factories and each of the sale areas (stores) by an abstract concept called a *vertex*. A vertex can be thought of as a point in a plane, or more generally as simply an element in a set. Next, in place of each of the routes between factories and sale areas, we introduce the notion of an *edge*. Abstractly, an edge is simply an ordered pair of vertices, and it is used to indicate a directed connection between the vertices. For example, if v and w are vertices, then the edge (v, w) represents a directed connection from vertex v to vertex w, or, in terms of the TP, a route along which goods can flow from factory v to store w. Finally, it is useful to assign a nonnegative integer to each edge to reflect the *capacity* of the edge. For example, in a TP it may be that some stores can be provided with no more than a certain maximum amount of goods in each time period. This would be the case if the factories are in Japan, the stores are in the United States and goods must be transported in ships of definite (known) size. The capacity of these ships would have to be considered in solving the TP since it would be unrealistic to call a schedule optimal if it required shipments which exceeded the available capacity. Also, in the event that there is no direct route from one of the factories (denoted by vertex v) to one of the stores (denoted by vertex w), then we assign 0 capacity to the edge (v, w). We now make these ideas precise.

Definition: A *capacitated network* (or a graph with capacity) is a pair (V, k), where $V = \{v_i\}_{i=1}^n$ is a finite set of elements called *vertices* and k is a function defined on $V \times V$ with values in the set of nonnegative integers and satisfying $k(v, v) = 0$ for $v \in V$. The elements of $V \times V$ are called the *edges* of the network, and the quantity $k(v, w)$ is called the *capacity* of the edge $(v, w) \in V \times V$.

The next step in the process of abstraction deals with the notion of a flow. Recall that in the TP one seeks a shipping schedule which is optimal among all the shipping schedules which satisfy certain constraints. In the setting of networks the appropriate generalization of a shipping schedule is a flow.

Definition: *A flow*, in the network (V, k), is a function f, defined on $V \times V$, taking *only integer values* and satisfying the conditions

1. $f(v_i, v_j) = -f(v_j, v_i)$,
2. $f(v_i, v_j) \leq k(v_i, v_j)$ for $i, j = 1, 2, \ldots, n$.

Note that a flow is constrained in three ways. The flow along an edge is *always an integer*, the flow can *never exceed the capacity* of that edge, and the flow in *one direction* along an edge *is always the negative of the flow in the other direction.*

Next, we need the idea of an originating point (the source of materials for the factories) and also that of a terminating point (the consumer who empties the stores). This leads to the following definition of two types of distinguished vertices.

Definition: A vertex $v_0 \in V = \{v_i\}_{i=1}^n$ is called an *originating* point for a flow f if $\sum_{i=1}^{n} f(v_0, v_i) > 0$. A vertex $v_t \in V$ is called a *terminating* point for a flow f if $\sum_{i=1}^{n} f(v_i, v_t) > 0$. A flow with exactly one originating point, v_0, and exactly one terminating point, v_t, is called a *flow from v_0 to v_t*. The *value* of a flow from v_0 to v_t is the number $v(f) = \sum_{i=1}^{n} f(v_0, v_i)$. A zero flow has value 0.

Remarks

1. The terms *source* and *sink* are often used in place of our terms originating point and terminating point.
2. The vertices v_0 and v_t are elements of the set $V = \{v_1, v_2, \ldots, v_n\}$. Since $k(v_i, v_i) = 0$ for every $v_i \in V$, we also have $f(v_i, v_i) = 0$ for every $v_i \in V$. In particular, $f(v_0, v_0) = f(v_t, v_t) = 0$.
3. The value of a nonzero flow is only defined for flows from a vertex v_0 to a vertex v_t. The zero flow is defined by the function which assigns zero to each edge.
4. If f is a flow from v_0 to v_t and $v_j \in V$, $v_j \neq v_0$, $v_j \neq v_t$, then $\sum_{i=1}^{n} f(v_i, v_j) = 0$.
5. If f is a flow from v_0 to v_t, then $v(f) = \sum_{i=1}^{n} f(v_i, v_t)$ (Exercise 9).

The central problem in our study of networks and flows is that of finding a flow with maximum value. One method of verifying the maximality of a flow is analogous to that used in the corresponding problem in linear programming. Recall that a vector may be shown to be optimal for a primal linear programming problem by considering the dual problem and by verifying a certain equality. In the same way a flow may be shown to be maximal by considering a second problem (a minimization problem) and by verifying a certain equality. This second problem may seem at first to be unrelated to the flow problem; however, there is a deep connection between the two problems and it will be explored below (Theorem 1). To state the appropriate minimization problem, we first need the idea of a cut in a network.

Definition: A *cut* in the network (V, k), with respect to vertices v_0 and v_t, is any partition of the set V into two sets V_0 and V_t such that $v_0 \in V_0$ and

$v_t \in V_t$. The *capacity* of the cut (V_0, V_t) is the number

$$k(V_0, V_t) = \sum_{\substack{v_i \in V_0 \\ v_j \in V_t}} k(v_i, v_j).$$

Remark. There is no flow involved in the definition of a cut. Hence the use of the symbols v_0 and v_t does not presuppose a flow from v_0 to v_t.

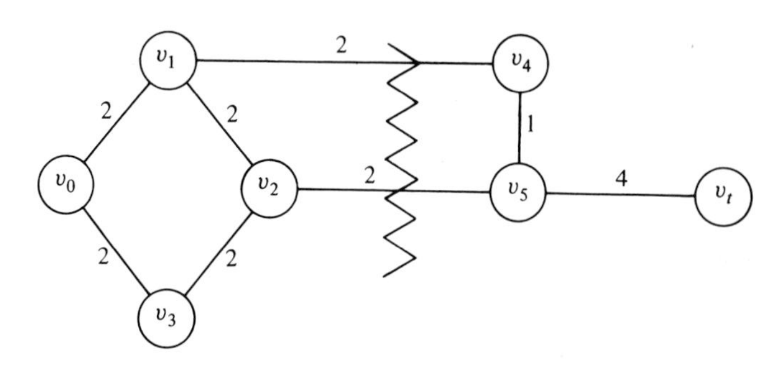

Fig. 7–1

Example. In the network shown in Fig. 7–1 the numbers on the edges indicate capacities in both directions. Edges which are not shown have zero capacity.

A flow from v_0 to v_t is given by the function f:

$$\begin{aligned} f(v_0, v_1) &= -f(v_1, v_0) = f(v_0, v_3) = -f(v_3, v_0) = f(v_1, v_2) \\ &= -f(v_2, v_1) = f(v_3, v_2) = -f(v_2, v_3) = 1, \\ f(v_2, v_5) &= -f(v_5, v_2) = f(v_5, v_t) = -f(v_t, v_5) = 2. \\ f(u, v) &= 0 \text{ for all other edges } (u, v). \end{aligned}$$

The value of f is $v(f) = f(v_0, v_1) + f(v_0, v_2) + f(v_0, v_3) + f(v_0, v_4) + f(v_0, v_5) + f(v_0, v_t) = 1 + 0 + 1 + 0 + 0 + 0 = 2$.

A cut with respect to v_0 and v_t is given by the pair of sets (V_0, V_t) where $V_0 = \{v_0, v_1, v_2, v_3\}$ and $V_t = \{v_4, v_5, v_t\}$. The cut is illustrated by the sawtooth line across the network. The capacity of this cut is given by $k(V_0, V_t) = k(v_1, v_4) + k(v_2, v_5) = 2 + 2 = 4$.

The minimization problem of interest is the problem of finding a cut with minimal capacity. The intimate connection between this problem and the maximal flow problem is crucial to our work with networks.

7.1.2 Maximal Flows and Minimal Cuts

We begin with the following important theorem (compare Theorem 2, Chap. 4).

Theorem 1 (Max Flow-Min Cut): The value of any maximal flow from v_0 to v_t is equal to the capacity of any minimal cut with respect to these vertices.

Proof: The first step in the proof is to show that $v(f) \leq k(V_0, V_t)$ for every flow f from v_0 to v_t and every cut (V_0, V_t) with respect to these vertices. Once this fact has been established, then we know that if f and (V_0, V_t) are any flow and cut such that $v(f) = k(V_0, V_t)$, then f is a maximal flow and (V_0, V_t) is a minimal cut. Also, since every maximal flow has the same value and every minimal cut has the same capacity, it follows that if we can find *one* flow f and *one* cut (V_0, V_t) for which $v(f) = k(V_0, V_t)$, then this equality holds for every maximal flow and minimal cut. We have

$$v(f) = \sum_{k=1}^{n} f(v_0, v_k) = \sum_{\substack{v_i \in V_0 \\ v_k \in V}} f(v_i, v_k).$$

This follows from the definition of a flow from v_0 to v_t. Indeed, if $v_i \in V_0$, $v_i \neq v_0$, then $\sum_{k=1}^{n} f(v_i, v_k)$ must be equal to 0. Next, we have

$$\begin{aligned} v(f) &= \sum_{\substack{v_i \in V_0 \\ v_k \in V_0}} f(v_i, v_k) + \sum_{\substack{v_i \in V_0 \\ v_k \in V_t}} f(v_i, v_k) \\ &= 0 + \sum_{\substack{v_i \in V_0 \\ v_k \in V_t}} f(v_i, v_k) \\ &\leq \sum_{\substack{v_i \in V_0 \\ v_k \in V_t}} k(v_i, v_k) = k(V_0, V_t). \end{aligned} \tag{1.1}$$

The proof is completed by exhibiting a flow f and a cut (V_0, V_t) for which equality holds. To this end, it is useful to examine inequality (1.1) to determine exactly when it is an equality. The inequality results from the condition that $f(v_i, v_k) \leq k(v_i, v_k)$ for every flow f. However, in (1.1) the pair of vertices (v_i, v_k) is such that $v_i \in V_0$ and $v_k \in V_t$. Hence, if the flow f and cut (V_0, V_t) satisfy $f(v_i, v_k) = k(v_i, v_k)$ for all $v_i \in V_0$ and $v_k \in V_t$, then we have $v(f) = k(V_0, V_t)$.

Let f be any maximal flow from v_0 to v_t. Such flows always exist because there are only a finite number of possible flows (Exercise 2). We define a cut in the following way. An edge (v_i, v_j) is said to be *saturated* if $f(v_i, v_j) = k(v_i, v_j)$. Next we introduce the notion of a *path* from vertex v to vertex v' as an ordered set of edges of the form $\{(v, w_1), (w_1, w_2), \ldots, (w_p, v')\}$. A path from one vertex to another is called *unsaturated* if no edge in the path is saturated. We now define the set V_0 to be the vertex v_0 together with all vertices that can be reached by an unsaturated path from v_0. The set V_t consists of all other vertices. We claim that (V_0, V_t) is a cut with respect to the vertices v_0 and v_t. This will be true if $v_t \in V_t$. We proceed with a proof by contradiction and assume that $v_t \notin V_t$. Then $v_t \in V_0$ and by the definition of V_0 there is an unsaturated path from v_0 to v_t. Let this path be $P = \{(v_0, w_1), (w_1, w_2), \ldots, (w_p, v_t)\}$. Also, let, $m = \min_{(w, w') \in P} [k(w, w') - f(w, w')]$. Since P is unsaturated, $m \geq 1$. We now define a flow f^* as follows:

$$f^*(w, w') = \begin{cases} f(w, w') + m & (w, w') \in P, \\ -[f(w', w) + m], & (w', w) \in P, \\ f(w, w'), & \text{otherwise.} \end{cases}$$

The flow f^* is again a flow from v_0 to v_t (Exercise 4), and since $m \geq 1$, we have $v(f^*) > v(f)$. But this is impossible because f is a maximal flow. Therefore $v_t \in V_t$ and (V_0, V_t) is a cut with respect to v_0 and v_t.

The proof is now completed by the observation that $v(f) = k(V_0, V_t)$. This equality holds because for each $v_i \in V_0$ and $v_k \in V_t$ we must have $f(v_i, v_k) = k(v_i, v_k)$ or else v_k would not be in V_t. However, as noted above, if $f(v_i, v_k) = k(v_i, v_k)$ for all $v_i \in V_0, v_k \in V_t$, then $v(f) = k(V_0, V_t)$.

Q.E.D.

The proof of Theorem 1 is constructive. It provides us with a means of finding maximal flows. The method is the following. One starts with any flow at all from v_0 to v_t. Then the sets V_0 and V_t are formed in the manner indicated in the proof. If the vertex v_t is in the set V_t, then the flow is already maximal. If $v_t \notin V_t$, then the flow is not maximal and a new flow must be constructed. The new flow is obtained in the manner used to obtain f^* from f in the proof of Theorem 1. The construction ensures that $v(f^*) > v(f)$. Either f^* is maximal and we are finished, or the process can be continued. If the process is to be continued, then we again form sets V_0 and V_t (this time for the new flow f^*) and again check the criteria: Is $v_t \in V_t$? This process can be continued until a maximal flow is obtained. Indeed, if the maximal flow has value $v(f)$, then it is clear that in at most $v(f)$ steps this process yields a maximal flow. We illustrate the process with an example.

Example. Find a maximum flow for the capacited network shown in Fig. 7–2. In the diagram the numbers on the edges indicate capacities in both directions. Edges which are not shown have zero capacity. As an initial flow we define f as follows: $f(v_0, v_1) = 1$, $f(v_0, v_3) = 1$, $f(v_1, v_2) = 1$, $f(v_1, v_4) = 0$, $f(v_3, v_2) = 1$, $f(v_2, v_5) = 2$, $f(v_4, v_5) = 0$, and $f(v_5, v_t) = 2$. The set V_0 is $\{v_1, v_2, v_3, v_4, v_5, v_0, v_t\}$ because each vertex can be reached from v_0 by an unsaturated path. Hence, $v_t \in V_0$, and this choice of f is not maximal. The path $P = \{(v_0, v_1), (v_1, v_4), (v_4, v_5), (v_5, v_t)\}$ is unsaturated, and it connects v_0 to v_t. This path can hold only one more unit of flow since $f(v_0, v_1) = 1$ and $k(v_0, v_1) = 2$. Thus we define the flow f_1 to be the same as f with the following exceptions: $f_1(v_0, v_1) = 2 = -f_1(v_1, v_0)$, $f_1(v_1, v_4) = 1 = -f_1(v_4, v_1)$,

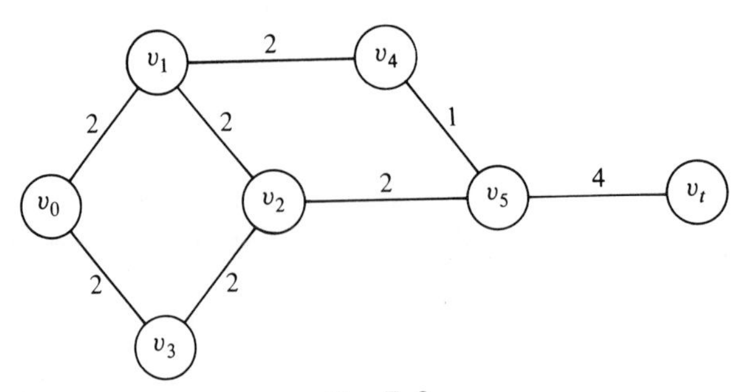

Fig. 7–2

$f_1(v_4, v_5) = 1 = -f_1(v_5, v_4)$, $f_1(v_5, v_t) = 3 = -f_1(v_t, v_5)$. Using the flow f_1 we form the sets $V_0^{(1)}$ and $V_t^{(1)}$ as in the proof of Theorem 1. We obtain

$$V_0^{(1)} = \{v_0, v_1, v_2, v_3, v_4\},$$
$$V_t^{(1)} = \{v_5, v_t\}.$$

In this case $v_t \in V_t^{(1)}$, and hence flow f_1 is a maximal flow and cut $(V_0^{(1)}, V_t^{(1)})$ is a minimal cut. The capacity of this cut is $k(V_0^{(1)}, V_t^{(1)}) = k(v_4, v_5) + k(v_2, v_5) = 2 + 1 = 3$.

The flow f_1 and the cut $(V_0^{(1)}, V_t^{(1)})$ are illustrated in Fig. 7–3.

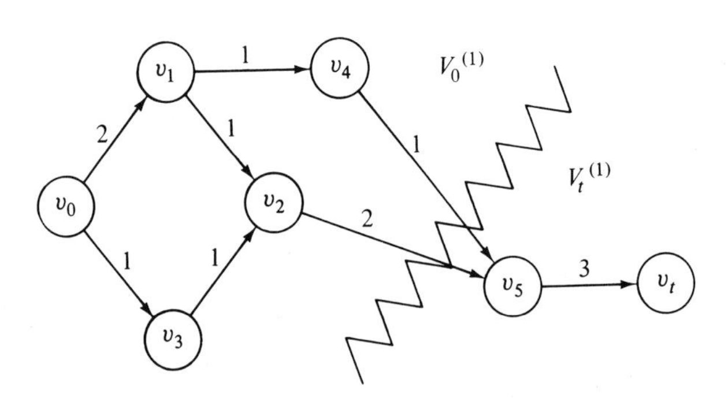

Fig. 7–3

It is not necessary to consider a very complicated example before it becomes clear that this method of finding an optimal flow is rather cumbersome. Fortunately the algorithm can be carried out more efficiently by means of a tableau. We shall not provide the details of the tableau treatment; however, the References provide numerous examples and discussions of the technique. We now turn to the use of Theorem 1 in the integer transportation problem and the job assignment problem.

7.1.3 Integer Transportations Problems

The integer transportation problem (ITP) can be phrased in the following way: Given sequences of nonnegative integers $\{s_i\}_{i=1}^m$ (supplies), $\{d_j\}_{j=1}^n$ (demands), $\{k_{ij}\}_{j=1,n}^{i=1,m}$ (route capacities), and real numbers $\{c_{ij}\}_{j=1,n}^{i=1,m}$ (costs), find a sequence of integers $F = \{f_{ij}\}_{j=1\ n}^{i=1,m}$ (the shipping schedule) such that

$$0 \leq f_{ij} \leq k_{ij}, \qquad i = 1, \ldots, m, j = 1, \ldots, n, \tag{1.2}$$

$$\sum_{j=1}^{n} f_{ij} \leq s_i, \qquad i = 1, \ldots, m, \tag{1.3}$$

$$\sum_{i=1}^{m} f_{ij} \geq d_j, \qquad j = 1, \ldots, n, \tag{1.4}$$

and $C(F) = \sum_{i=1}^{m} \sum_{j=1}^{n} c_{ij} f_{ij}$ is a minimum over all such F.

As a first step in the solution of this problem we note that it is sufficient to consider only the case in which $\sum_{i=1}^{m} s_i = \sum_{j=1}^{n} d_j$. The argument for this is as follows. First, if F satisfies (1.3) and (1.4), then $\sum_{j=1}^{n} d_j \leq \sum_{j=1}^{n} \sum_{i=1}^{m} f_{ij} \leq \sum_{i=1}^{m} s_i$, and therefore, a necessary condition for the problem to have a solution is $\sum_{j=1}^{n} d_j \leq \sum_{i=1}^{m} s_i$ (supply must equal or exceed demand). Thus if demand exceeds supply, then the problem stated above does not have a solution. However, there is a related problem involving rationing which does have a solution (see Exercise 11). Next, if $\sum_{j=1}^{n} d_j < \sum_{i=1}^{m} s_i$, then a new problem can be formed by introducing an additional store with demand $d_{n+1} = \sum_{i=1}^{m} s_i - \sum_{j=1}^{n} d_j$ and with the shipping costs to the new store being 0 from each factory. The new store is suggestively called a *dump*. This new problem satisfies the condition that supply is equal to demand, and solutions of the new problem are easily identified with solutions of the original problem (Exercise 5). Therefore it is no restriction to assume that $\sum s_i = \sum d_j$, and we do so from now on.

We have already noted that a necessary condition for an ITP to possess a solution is for supply to equal or exceed demand. This condition is not usually sufficient for the existence of a solution because of the limitations on shipping schedules which are imposed by the route capacities. A shipping schedule which satisfies (1.2), (1.3), and (1.4) will be called a *feasible schedule*. If there is a feasible schedule, then it follows that the demand at each store is less than the total capacity of the routes terminating at that store. Even this is not a sufficient condition because the supply may not be located correctly to utilize the available capacity over each route. With these preliminary comments in mind we are now ready to establish a necessary and sufficient condition for the existence of a feasible shipping schedule. As a matter of notation we let $\mathcal{W} = \{W_i\}_{i=1}^m$ and $\mathcal{S} = \{S_j\}_{j=1}^n$ be the factories and sales areas, respectively. Also, for subsets $\mathcal{W}' \subset \mathcal{W}$ and $\mathcal{S}' \subset \mathcal{S}$, we let $d(\mathcal{S}') = \sum_{S_i \in \mathcal{S}'} d_i$ and $s(\mathcal{W}') = \sum_{W_i \in \mathcal{W}'} s_i$ be the demand of the subset $\mathcal{S}'$ and the supply of the subset $\mathcal{W}'$, respectively. Also, we let $k(\mathcal{W}', \mathcal{S}') = \sum_{W_i \in \mathcal{W}', S_j \in \mathcal{S}} k_{ij}$ be the capacity between the subsets $\mathcal{W}'$ and $\mathcal{S}'$, respectively. The desired theorem has the following form.

Theorem 2: A necessary and sufficient condition for the existence of at least one feasible shipping schedule for the ITP is that for each pair of subsets $\mathcal{W}' \subset \mathcal{W}$ and $\mathcal{S}' \subset \mathcal{S}$ the relation

$$d(\mathcal{S}') - s(\mathcal{W}') \leq k(\mathcal{W} \setminus \mathcal{W}', \mathcal{S}') \tag{1.5}$$

is true.

We point out that in this case, in contrast to the more general linear optimization problems of Chap. 4, the existence of a feasible vector guarantees the existence of an optimal vector (Exercise 6). It is not necessary to consider any sort of dual problem.

Proof: We show the necessity of the condition first. Thus suppose that $F = \{f_{ij}\}$ is a feasible shipping schedule for the ITP. Then F satisfies (1.2), (1.3), and (1.4), and hence for any subsets $\mathcal{S}' \subset \mathcal{S}$ and $\mathcal{W}' \subset \mathcal{W}$ we have

$$d(\mathcal{S}') = \sum_{S_j \in \mathcal{S}'} d_j \leq \sum_{S_j \in \mathcal{S}'} \sum_{i=1}^{m} f_{ij}$$

and

$$s(\mathcal{W}') = \sum_{W_i \in \mathcal{W}'} s_i \geq \sum_{W_i \in \mathcal{W}'} \sum_{j=1}^{n} f_{ij}.$$

Therefore

$$\begin{aligned} d(\mathcal{S}') - s(\mathcal{W}') &\leq \sum_{S_j \in \mathcal{S}'} \sum_{i=1}^{m} f_{ij} - \sum_{W_i \in \mathcal{W}'} \sum_{j=1}^{n} f_{ij} \\ &= \sum_{S_j \in \mathcal{S}'} \sum_{W_i \in \mathcal{W} \setminus \mathcal{W}'} f_{ij} - \sum_{W_i \in \mathcal{W}'} \sum_{S_j \in \mathcal{S} \setminus \mathcal{S}'} f_{ij} \\ &\leq \sum_{S_j \in \mathcal{S}'} \sum_{W_i \in \mathcal{W} \setminus \mathcal{W}'} f_{ij} \\ &\leq \sum_{S_j \in \mathcal{S}'} \sum_{W_i \in \mathcal{W} \setminus \mathcal{W}'} k_{ij} \\ &= k(\mathcal{W} \setminus \mathcal{W}', \mathcal{S}'). \end{aligned}$$

Next, we establish the sufficiency of condition (1.5) by considering a specific capacitated network. The vertices of the network consist of the factories of $\mathcal{W}$, the sale areas of $\mathcal{S}$, and two special vertices, called the originating point v_0 and the terminating point v_t. The capacities of the edges are

$$\begin{aligned}
k(W_i, S_j) &= k_{ij}, \qquad && i = 1, \ldots, m, j = 1, \ldots, n, \\
k(v_0, W_i) &= s_i, && i = 1, \ldots, m, \\
k(S_j, v_t) &= d_j, && j = 1, \ldots, n, \\
k(v, w) &= 0, && \text{all other edges } (v, w).
\end{aligned}$$

We now claim that a minimal cut with respect to v_0 and v_t is given by (V_0, V_t), where $V_0 = \{v_0\} \cup \mathcal{W} \cup \mathcal{S}$ and $V_t = \{v_t\}$. To establish this claim, let (K_0, K_t) be any other cut with respect to these vertices. Then $K_0 = \{v_0\} \cup K'_0$ and $K_t = \{v_t\} \cup K'_t$, where $K'_0 \cap K'_t = \emptyset$ and $K'_0 \cup K'_t = \mathcal{S} \cup \mathcal{W}$. Hence

$$\begin{aligned}
k(K_0, K_t) &= k(v_0, K'_t) + k(K'_0, K'_t) + k(K'_0, v_t) \\
&= s(K'_t \cap \mathcal{W}) + k(K'_0 \cap \mathcal{W}, K'_t \cap \mathcal{S}) + d(K'_0 \cap \mathcal{S}).
\end{aligned}$$

Also,

$$k(V_0, V_t) = k(V_0, v_t) = d(\mathcal{S}) = d(K'_0 \cap \mathcal{S}) + d(K'_t \cap \mathcal{S}).$$

Therefore

$$k(K_0, K_t) - k(V_0, V_t) = s(K'_t \cap \mathcal{W}) + k(K'_0 \cap \mathcal{W}, (K'_t \cap \mathcal{S}) - d(K'_t \cap \mathcal{S}).$$

But, by the hypotheses of the theorem,

$$\begin{aligned}
d(K'_t \cap \mathcal{S}) - s(K'_t \cap \mathcal{W}) &\le k(\mathcal{W} \setminus (K'_t \cap \mathcal{W}), K'_t \cap \mathcal{S}) \\
&= k(K'_0 \cap \mathcal{W}, K'_t \cap \mathcal{S}),
\end{aligned}$$

and hence $k(K_0, K_t) - k(V_0, V_t) \ge 0$ and (V_0, V_t) is a minimal cut.

There is a maximal flow f from v_0 to v_t, and by Theorem 1 this flow has value $v(f) = k(V_0, V_t) = d(\mathcal{S})$. We let f determine a shipping schedule $F(f) = \{f_{ij}\}$ by setting $f_{ij} = f(W_i, S_j)$. It remains to be shown that F is feasible. First, since every flow from v_0 to v_t satisfies $f(W_i, S_j) \le k(W_i, S_j)$, we have $f_{ij} \le k_{ij}$, and thus F satisfies condition (1.2). Next, since the net flow is zero at each factory and each sales area (remember that f is a flow from v_0 to v_t), we obtain

$$\sum_{i=1}^{m} f_{ij} = \sum_{i=1}^{m} f(W_i, S_j) = f(S_j, v_t)$$

and

$$\sum_{j=1}^{n} f_{ij} = \sum_{j=1}^{n} f(W_i, S_j) = f(v_0, W_i).$$

Since $f(v_0, W_i) \le k(v_0, W_i) = s_i$, condition (1.3) is satisfied. Also, $f(S_j, v_t) \le k(S_j, v_t) = d_j$, and the fact that f is a maximal flow with value $v(f) = \sum_{j=1}^{n} d_j$ implies that $f(S_j, v_t)$ cannot be less than d_j for any j, $j = 1, 2, \ldots, n$.

Thus, $f(S_j, v_t) = d_j$ and condition (1.4) is satisfied. Therefore the shipping schedule $F = \{f_{ij}\}$ is feasible for the ITP. Q.E.D.

We now have a necessary and sufficient condition for the existence of feasible shipping schedules for the ITP. In one sense this solves the ITP because there are only a finite number of such schedules, and hence one can obtain an optimal schedule by trial and error. However, trial-and-error methods are not practical even for problems where n and m are only modestly large (say $m = n = 15$). Accordingly, considerable effort has been devoted to the development of efficient algorithms for obtaining optimal solutions. Many of these algorithms are based on the statement and proof of Theorem 1. These are often called Ford-Fulkerson methods since they first proved Theorem 1. Other methods are based on specializations of the simplex method. The References at the end of the chapter include a number of sources which treat these algorithms in detail.

7.1.4 The Assignment Problem Revisited

In Sec. 2.6 we introduced the concept of a complete set of applicants in order to state a condition under which all the available jobs could be filled by appropriate applicants. In this section we shall give a second proof of that result. More importantly, we shall give a proof which leads to a method of actually finding an assignment of applicants to jobs. We begin by introducing some new notation. We use the setting of the job assignment problem as given in the introduction to this section. For any subset $\mathcal{J}_\alpha$ of the set of jobs $\mathcal{J}$, we let $\mathcal{A}(\mathcal{J}_\alpha)$ be the set of all applicants who are appropriate for at least one of the jobs in the set $\mathcal{J}_\alpha$. Also, we let $|\mathcal{J}_\alpha|$ and $|\mathcal{A}(\mathcal{J}_\alpha)|$ be the number of elements in the sets $\mathcal{J}_\alpha$ and $\mathcal{A}(\mathcal{J}_\alpha)$, respectively. Using this notation, we say that a set of applicants $\mathcal{A}$ is *complete* for a set of jobs $\mathcal{J}$ if $|\mathcal{A}(\mathcal{J}_\alpha)| \geq |\mathcal{J}_\alpha|$ for every subset $\mathcal{J}_\alpha \subset \mathcal{J}$. We are now ready to restate and reprove a result on the existence of solutions for the job assignment problem.

Theorem 3 (Theorem 1, Sec. 2.6): The set of jobs $\mathcal{J} = \{J_1, \ldots, J_n\}$ can be filled from the set of applicants $\mathcal{A} = \{A_1, \ldots, A_m\}$ if and only if $\mathcal{A}$ is complete for $\mathcal{J}$.

Proof: If all the jobs can be filled by the applicants, then certainly $\mathcal{A}$ is complete for $\mathcal{J}$. In fact, for any subset $\mathcal{J}_\alpha$, the set of applicants who are assigned to fill these jobs is a subset of $\mathcal{A}(\mathcal{J}_\alpha)$, and hence $|\mathcal{J}_\alpha| \leq |\mathcal{A}(\mathcal{J}_\alpha)|$.

Next, suppose that $\mathcal{A}$ is complete for $\mathcal{J}$. It will be shown that all the jobs can be filled by considering a certain capacitated network. The vertices of this network consist of the elements of the set of applicants $\mathcal{A}$, the elements of the set of jobs $\mathcal{J}$, and two special vertices denoted by v_0 and v_t. Let $M = m + n$, and define the capacities of the edges of this network by the formulas

$$\begin{aligned} k(v_0, A_i) &= 1, && i = 1, \dots, m, \\ k(J_k, v_t) &= 1, && k = 1, \dots, n, \\ k(A_i, J_k) &= M, && \text{if } A_i \text{ is appropriate for } J_k, \\ k(A_i, J_k) &= 0, && \text{if } A_i \text{ is not appropriate for } J_k, \\ k(v, w) &= 0, && \text{all other edges } (v, w). \end{aligned} \tag{1.6}$$

Pictorially, we represent such a network as in Fig. 7–4.

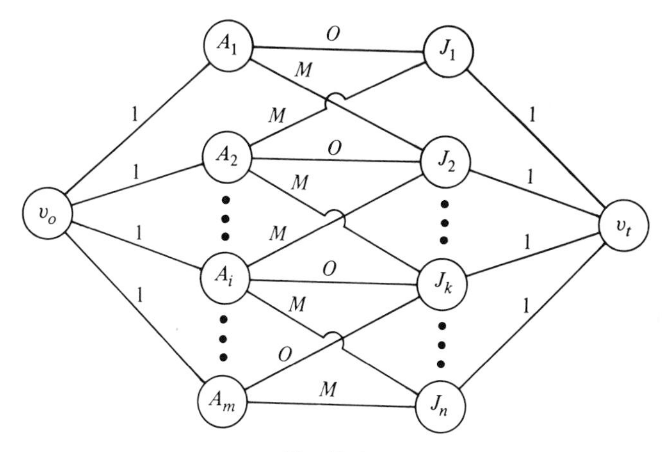

Fig. 7–4

Next, consider flows in this network, and let f be any maximal flow from v_0 to v_t. Such a maximal flow certainly exists. Indeed, the set of flows from v_0 to v_t is a finite set, and this set is nonempty since it contains the zero flow. It is useful to distinguish two cases for the value of the flow f. Case 1 is $v(f) = n$, and case 2 is $v(f) < n$. These cases cover all possibilities for $v(f)$ in view of the constraints on the capacities of the edges connected to v_t.

In case 1 ($v(f) = n$), the flow f provides an assignment of applicants to jobs which fills all available jobs. This assignment is given by the rule "assign applicant A_i to job J_k if and only if $f(A_i, J_k) = 1$." Since $v(f) = n$, it easily follows that each job is filled by exactly one appropriate applicant (Exercise 7).

The proof is concluded by showing that case 2 ($v(f) < n$) is impossible if $\mathcal{A}$ is complete for $\mathcal{J}$. To this end, note that if $v(f) < n$, then by the max flow-min cut theorem there is a minimal cut (V_0, V_t) with respect to v_0 and v_t with capacity less than n. Let $V_0 = \{v_0, A_1, \dots, A_p, J_1, \dots, J_q\}$. (There is no loss of generality in assuming this since a renumbering of the jobs and applicants will always make it possible to express V_0 in this form. It is also possible, of course, that either all the A_is or J_ks are missing from V_0). The definition of the capacity of a cut states that $k(V_0, V_t)$ is the sum of capacities of all edges of the form (v, w), where $v \in V_0$ and $w \in V_t$. Such edges have one of the following forms:

1. (v_0, A_i), $p < i \leq m$.
2. (A_i, J_k), $i \leq p$, $k > q$.
3. (A_i, J_k), $i > p$, $k \leq q$.
4. (J_k, v_t), $k \leq q$.

The edges of types 1 and 4 have capacity 1, and edges of types 2 and 3 either have capacity 0 or $M = m + n$. Since $k(V_0, V_t) < n$, no edge in the sum can have a capacity greater than n. Thus all edges of types 2 and 3 have capacity 0. Therefore we have

$$n > k(V_0, V_t) = \sum_{i=p+1}^{m} k(v_0, A_i) + \sum_{k=1}^{q} k(J_k, v_t) = m - p + q.$$

In an equivalent form, the above inequality states that $m - p < n - q$.

The proof is concluded by considering the set $\mathcal{J}_\alpha = \{J_{q+1}, \ldots, J_n\}$ and $\mathcal{A}(\mathcal{J}_\alpha)$. The above argument has shown that $k(A_i, J_k) = 0$, for $i \leq p$ and $k > q$. Therefore, $\mathcal{A}(\mathcal{J}_\alpha) \subset \{A_{p+1}, \ldots, A_m\}$, and hence $|\mathcal{A}(\mathcal{J}_\alpha)| \leq m - p$. But this in turn implies that

$$|\mathcal{A}(\mathcal{J}_\alpha)| \leq m - p < n - q = |\mathcal{J}_\alpha|.$$

Since this last inequality is a contradiction of the definition of completeness, it has been shown that the assumption $v(f) < n$ is incompatible with the assumption of completeness. This concludes the proof.

The proof of Theorem 3 shows that the job assignment problem can be solved by finding a maximal flow in a certain capacitated network. As noted earlier, the proof of Theorem 1 provides a constructive method for finding maximal flows in capacitated networks and hence a method of constructing solutions to the assignment problem. These ideas are considered in more detail in Projects 7.3.1 and 7.3.2.

EXERCISES

1. The Downjim Drug Company processes two types of ethical drugs, URP and ULP. Each drug must go through two distinct processes, one in laboratory A and the other in laboratory B. The drug URP requires 4 hours in laboratory A and 2 hours in laboratory B, while ULP must be in laboratory A for 2 hours and in laboratory B for 4 hours. Only one drug can be processed in each laboratory at a time. Also, the nature of the processing is such that once it has been begun it must be completed or else the batch in process is ruined. Thus an integer number of batches of each drug must be processed each day. Suppose that the chemists in each laboratory work for 8 hours each day. What is the maximum total number of batches of drugs that can be produced each day?

(a) Formulate this problem in linear programming terms.

(b) Try to solve this problem using the simplex method. Is your solution satisfactory?

(c) Solve the problem using another method. Describe the general features of your method.

2. Write the job assignment problem of Sec. 7.1 as a linear programming problem with auxiliary constraints.

3. Show that the function f defined on $V \times V$ by $f(v_i, v_j) = 0$ for all $v_i, v_j \in V$ is a flow in (V, k) for any k. Also, show that there are only a finite number of possible flows in a capacitated network.

4. Show that the function f^* of the proof of Theorem 1 actually defines a flow from v_0 to v_t. Note that in particular it must be shown that f^* is integral valued and f^* has a unique originating point and a unique terminating point.

5. In Sec. 7.1.3 an ITP in which supply exceeds demand is converted into an ITP in which supply equals demand. Show that there is a one-to-one correspondence between the solutions of these two problems.

6. Show that the existence of a feasible vector for the ITP implies the existence of an optimal vector.

7. Show that the assignment rule of the proof of Theorem 3 actually works. That is, show that each available job is filled by one and only one appropriate applicant.

8. In the network in Fig. 7–5 the numbers represent capacities in both directions. All edges not given numbers have zero capacity. Find a maximal flow and minimal cut with respect to v_0 and v_t. Take as your initial flow one with $f(v_0, v_i) = 1$ for all vertices v_i with $k(v_0, v_i) \neq 0$.

9. Same as Exercise 8 for the network in Fig. 7–6.

10. Show that if f is a flow from v_0 to v_t and if $V = \{v_i\}_{i=1}^n$ is the set of vertices of the network, then $v(f) = \sum_{i=1}^{n} f(v_i, v_t)$.

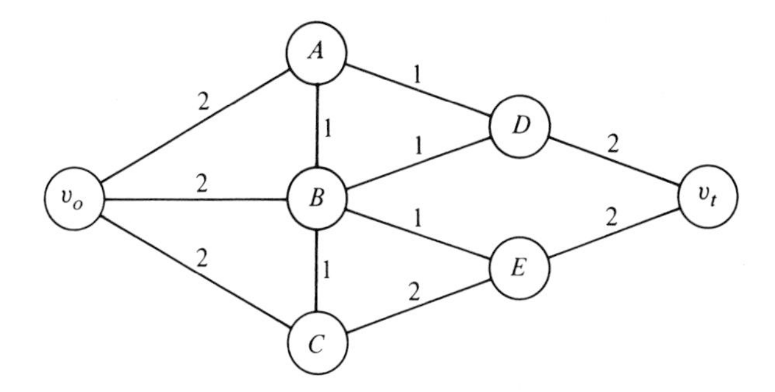

Fig. 7–5

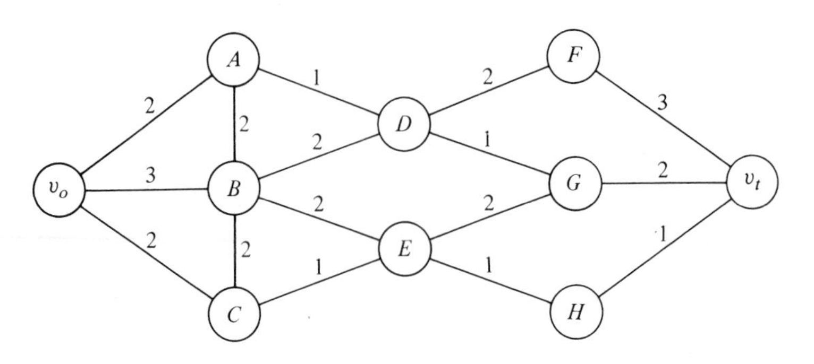

Fig. 7–6

11. Consider an integer programming problem in which supply is less than demand. Formulate an equivalent problem in which supply equals demand by introducing a dummy factory. Discuss how the solution of this new problem solves the rationing problem which results when demand exceeds supply.

7.2 DIRECTED GRAPHS

In Sec. 7.1 it was shown that the study of the job assignment problem and the integer transportation problem was facilitated by making use of the abstract setting of a capacitated network. The purpose of this section is to show that a somewhat more general concept is also useful in applications. The concept to be introduced is that of a directed graph, or, more compactly, a digraph. We shall begin the section with a short general discussion of digraphs, and then we shall consider a few of the many problems in which the notion of a digraph arises naturally. Our goal is to indicate how the study of these problems is simplified by the introduction of graph-theoretic concepts.

7.2.1 Definitions and Notation

Since the word *graph* is used in more than one way in mathematics, we begin with a discussion of the sense in which the word is used in this section. This is most easily done by contrasting the present use with the more familiar use of the term in calculus. The term graph in calculus is usually associated with a function of one or more real variables and is intended to somehow denote a "picture" of the function. In the case of a function f which associates a real number $f(x)$ with each real number x in a set X, the graph of f is the set in R^2 consisting of all pairs $(x, f(x))$ for $x \in X$. For example, if $X = \{x: -1 \leq x \leq +1\}$ and $f(x) = x^2$, then the graph of f is the set of points shown by the bold line in Fig. 7–7.

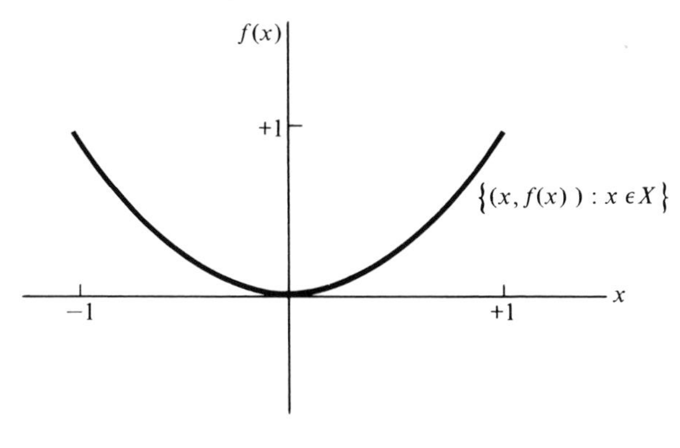

Fig. 7–7

The use of the word *graph* of interest to us in this section can be expressed intuitively as follows: A graph consists of a collection of points and a collection of lines connecting certain pairs of the points. There may be several lines connecting a given pair of points and the lines may have an assigned direction. We refer to the points as vertices and the lines as edges, but we note that these terms are by no means universal in the literature. The pictures of Fig. 7–8 illustrate some possible graphs by giving arrangements of vertices and edges.

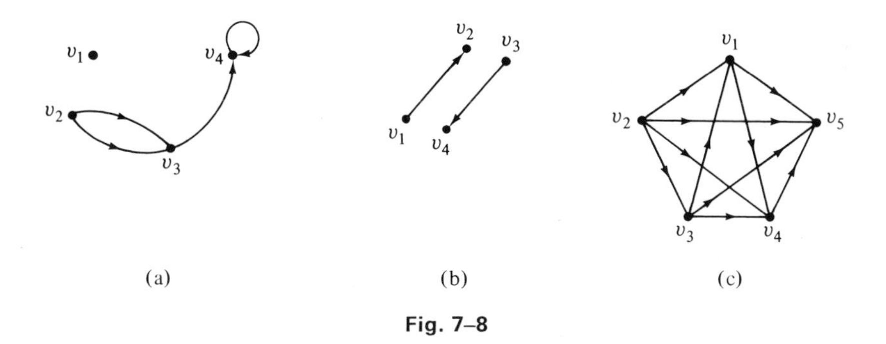

Fig. 7–8

We must, of course, make the notion of a graph more precise. We begin by concentrating on a somewhat more restrictive object, and then we note how this is related to the general notion of a graph.

Definition: Let V be a nonempty finite set and let E be a subset of $V \times V$ such that $v \in V$ implies that $(v, v) \notin E$. Then the pair (V, E) is a *directed graph* (or *digraph*). The elements of V are the *vertices* of the digraph, the elements of E are the *edges* of the digraph, and if $(v, w) \in E$, then the edge (v, w) is said to be *directed from vertex v to to vertex w.* We shall use the term *path in the digraph* (V, E) to mean a sequence of edges $P = \{(v_i, w_i)\}_{i=1}^n$ satisfying the condition $v_{i+1} = w_i$, $i = 1, 2, \ldots, n-1$. Such a path P is said to *start* at v_1, *end* at w_n, and have *length n.* If $v_1 = w_n$, the path is called a *circuit.*

Remarks

1. In Fig. 7–8 both (b) and (c) are digraphs, but (a) is not a digraph.
2. Every capacitated network is a digraph. The set of vertices V of the network is the set of vertices of the digraph, and the edges of the digraph are all elements of the set $V \times V$ except those of the form (v, v).
3. This definition is adequate for our purposes; however, for certain applications of graph theory a more general definition is preferable. For example, this definition does not allow a digraph to have more than one edge directed from one vertex to another [as between v_2 and v_3 of (a) in Fig. 7–8], nor does it allow a loop at a vertex [such as at v_4 of (a) of Fig. 7–8]. There are alternative definitions of digraphs which allow these to occur.
4. One can obtain general or undirected graphs from digraphs by adding the condition that if $(v, w) \in E$, then $(w, v) \in E$. The two directed edges (v, w) and (w, v) are then identified and considered to be the undirected edge between v and w.

There is an extensive theory of graphs and digraphs, and many major advances have been achieved in recent years. Much of the impetus for the development of the subject has come from the increasing number of applications of the theory, especially in the social sciences. It is not our purpose to develop a general theory of graphs; instead, we give an introduction to a few of the important ideas via some special problems where the setting of a graph seems especially appropriate and useful. For the reader with interests in the theoretical direction, we recommend [Be], [H], and [O].

7.2.2 Leadership and Extent of Influence

Consider the task of a sociologist who is studying the process of opinion formation by teenagers. Since this is obviously a complicated matter involving both sociological and psychological factors, the sociologist first restricts his study to an attempt to examine the effect that the opinions of one's peers have on one's own opinions on matters of mutual interest.

As a first step in the sociologist's program of study he asks each of the students in a class for the name of that fellow student whose opinions he values most. This may not be a completely reliable method for determining the effect of the second student's opinion upon the first, but it seems reasonable to assume that if student B respects the opinions of student A above all others, then it is likely that the opinions of student B are influenced to some extent by the opinions of student A. This assumption is the key to the sociologist's method of analyzing the data which he obtains. The data consist of pairs of student names, and, as such, they are conveniently represented by means of a digraph. The vertices of the digraph are the student names, and the directed edges are the pairs (v, w), where student v is the student whose opinions are most respected by student w. For example, if there are five students in the class under investigation, then the resulting digraph (called the *associated* digraph) might be as shown in Fig. 7–9.

Figure 7–9 indicates that student A respects the opinions of student B, B respects C, C and D respect A, and E respects D. Equivalently, we could say that student A influences students C and D, B influences A, C influences B, and D influences E. Thus we are led to the following definition.

Definition: In any class of students, we say that student v *directly influences* student w if student w respects the opinions of student v above all others.

The digraph of Fig. 7–9 includes all the information submitted by the students. Moreover, it also provides a source of additional information in the sense that it readily indicates possible secondary influences among students. For example, student A influences student D and student D influences student E, and hence it is likely that the opinions of student A have some effect upon the opinions of student E. Such an effect might be called a secondary influence, to distinguish it from direct influence such as that of student A upon student D. Similarly, A has a secondary influence upon B. In fact, as given in Fig. 7–9 it is clear that student A has either a direct or secondary influence upon each of the other four students. It is natural to

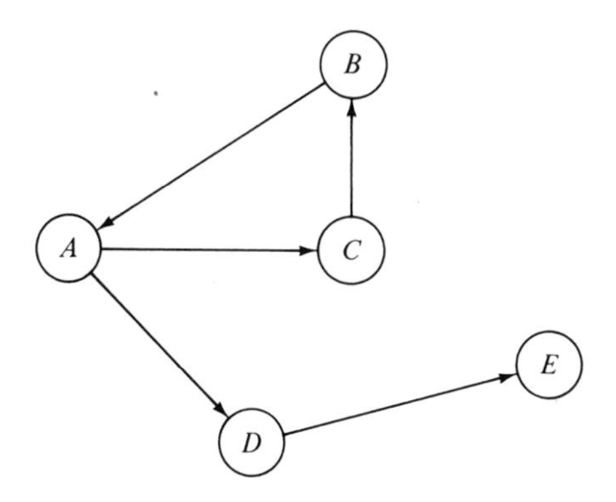

Fig. 7–9

consider such a student to be a leader of his fellow students, and we shall formalize this concept below.

Naturally, a situation with the characteristics of Fig. 7–9 need not occur in every case. It may happen that the digraph splits into disconnected parts, and then certainly there is no leader in the sense described above. Even if the digraph does not split into parts, it may not be an easy matter to simply inspect the graph and determine whether a leader exists. However, if the digraph does not split, *then a leader always exists* (Exercise 9). Knowing that a leader exists does not mean it is easy to find one. We now investigate the problem of determining who (if anyone) is a leader. We begin by describing exactly what is meant by a leader.

Definition: In a class of students, student v is said to have *influence of order* n over student w if in the associated digraph there is a path of length n with start at v and end at w and there is no shorter path with this property. We say that *v has influence over w* if v has influence of order n over w for some n. A student is said to be a *leader* if he has influence over all other students in the class. He is a *consensus leader* if he has influence of order 1 over all other students.

Although a large class will not usually have a consensus leader, it may have a leader, and there may be subclasses which do have consensus leader. How does one find the leaders and/or consensus leaders? An obvious method is that of trial and error. In turn, each vertex of the digraph is examined and a list is made of all vertices which can be reached by paths from this vertex. For large classes (i.e., a large number of vertices) this would be a tedious process, and one would have to be very careful to be sure that all paths have been considered. Hence it is desirable to develop a more efficient procedure for discovering which vertices are influenced by a given one. The procedure which we shall use is based on the incidence matrix of the digraph.

Definition: Let (V, E) be a digraph with $V = \{v_i\}_{i=1}^n$. The *incidence matrix* for this digraph is the $n \times n$ matrix $\mathbb{E} = (\epsilon_{ij})$, where $\epsilon_{ij} = 1$ if $(v_i, v_j) \in E$ and $\epsilon_{ij} = 0$ otherwise.

Example. The digraph of Fig. 7–9 has the incidence matrix

$$\begin{array}{c} \\ \\ \mathbb{E} = \end{array} \begin{array}{c} \\ A \\ B \\ C \\ D \\ E \end{array} \begin{array}{c} \begin{array}{ccccc} A & B & C & D & E \end{array} \\ \begin{bmatrix} 0 & 0 & 1 & 1 & 0 \\ 1 & 0 & 0 & 0 & 0 \\ 0 & 1 & 0 & 0 & 0 \\ 0 & 0 & 0 & 0 & 1 \\ 0 & 0 & 0 & 0 & 0 \end{bmatrix} \end{array}.$$

Henceforth the vertices A, B, C, D, and E will be referred to by 1, 2, 3, 4, and 5, respectively.

To illustrate the way in which we use the incidence matrix to study the influence of members of a class, we compute the powers of the matrix $\mathbb{E}$. For the $\mathbb{E}$ of this example we have

$$\mathbb{E}^2 = \begin{bmatrix} 0 & 1 & 0 & 0 & 1 \\ 0 & 0 & 1 & 1 & 0 \\ 1 & 0 & 0 & 0 & 0 \\ 0 & 0 & 0 & 0 & 0 \\ 0 & 0 & 0 & 0 & 0 \end{bmatrix},$$

$$\mathbb{E}^3 = \begin{bmatrix} 1 & 0 & 0 & 0 & 0 \\ 0 & 1 & 0 & 0 & 1 \\ 0 & 0 & 1 & 1 & 0 \\ 0 & 0 & 0 & 0 & 0 \\ 0 & 0 & 0 & 0 & 0 \end{bmatrix},$$

and

$$\mathbb{E}^4 = \begin{bmatrix} 0 & 0 & 1 & 1 & 0 \\ 1 & 0 & 0 & 0 & 0 \\ 0 & 1 & 0 & 0 & 1 \\ 0 & 0 & 0 & 0 & 0 \\ 0 & 0 & 0 & 0 & 0 \end{bmatrix}.$$

For an arbitrary incidence matrix $\mathbb{E}$, let $\mathbb{E}^n = (\epsilon_{ij}^{(n)})$. In this example we first consider the matrix $\mathbb{E}^2 = (\epsilon_{ij}^{(2)})$ and, in particular, the first row of this matrix. This row consists of the entries $\epsilon_{11}^{(2)} = 0$, $\epsilon_{12}^{(2)} = 1$, $\epsilon_{13}^{(2)} = 0$, $\epsilon_{14}^{(2)} = 0$, and $\epsilon_{15}^{(2)} = 1$. Thus, $\epsilon_{13}^{(2)} = 0 = \sum_{k=1}^{5} \epsilon_{1k}\epsilon_{k3}$, and since each entry ϵ_{ij} is either 0 or positive, we see that for each k, either $\epsilon_{1k} = 0$ or $\epsilon_{k3} = 0$. In terms of the digraph this means that for each k, either vertex 1 does not influence vertex k, or else vertex k does not influence vertex 3. In other words, the entry $\epsilon_{13}^{(2)} = 0$ means that vertex 1 does not have influence of order 2 on vertex 3. This is easily verified by examining the digraph of Fig. 7–9 and noting that there is no path of length 2 from vertex number 1 (A) to vertex 3 (C). Similarly, the fact that $\epsilon_{11}^{(2)} = 0$ and $\epsilon_{14}^{(2)} = 0$ is reflected in the digraph by the lack of paths of length 2 between vertices 1 and 1, and vertices 1 and 4 (D). On the other hand, $\epsilon_{12}^{(2)} = \sum_{k=1}^{5} \epsilon_{1k}\epsilon_{k2} = 1$, and hence there exists a k such that $\epsilon_{1k} = 1$ and $\epsilon_{k2} = 1$. In terms of the digraph there is a vertex which is both influenced by vertex 1 and in turn which influences vertex 2.

Such a vertex is vertex 3 (C) since the graph has a directed edge from A to C and another from C to B. In other words $\epsilon_{12}^{(2)} = 1$ implies that there is a path of length 2 with start at vertex 1 and end at vertex 2. Similarly, $\epsilon_{15}^{(2)} = 1$ indicates a path of length 2 from vertex A to vertex E. The corresponding path in the digraph is the path from A to D to E.

The above discussion is given for a special example and it is phrased in terms of the first row of $\mathbb{E}^2$, but the results apply in the general case and to all the rows. Each nonzero entry in $\mathbb{E}^2$ corresponds to a path of length 2 in the digraph. If the nonzero entry is $\epsilon_{ij}^{(2)}$, then the path starts at vertex i and ends at vertex j. In other words, we now have a simple method for finding all influences of second order between the students. We simply form the incidence matrix $\mathbb{E}$, square it, and observe the nonzero entries. With this process in mind we define the sets $I_i^{(1)}$ and $I_i^{(2)}$ as follows:

$$I_i^{(1)} = \{j: \epsilon_{ij} \neq 0\}, \qquad i = 1, 2, \ldots, n,$$
$$I_i^{(2)} = \{j: j \neq i, \epsilon_{ij} = 0, \epsilon_{ij}^{(2)} \neq 0\}, \qquad i = 1, 2, \ldots, n.$$

Thus $I_i^{(1)}$ is the set of vertices which are directly influenced by vertex i, and $I_i^{(2)}$ is the set of vertices which are influenced of order 2 by vertex i. Naturally we eliminate vertex i from the set $I_i^{(2)}$, since we are not interested in chains of influence which start and end at the same vertex.

We now consider the higher powers of the matrix $\mathbb{E}$. First, consider a nonzero entry in the matrix $\mathbb{E}^3$. For example, suppose that $\epsilon_{ij}^{(3)} = \sum_{k=1}^{n} \epsilon_{ik}^{(2)}\epsilon_{kj} \neq 0$. Then there is an index k such that $\epsilon_{ik}^{(2)} \neq 0$ and $\epsilon_{kj} \neq 0$. Equivalently, there is a path of length 2 from vertex i to vertex k, and there is a path of length 1 from vertex k to vertex j. The combination of these paths gives a path of length 3 from vertex i to vertex j. In other words, the nonzero entries of $\mathbb{E}^3$ correspond to paths of length 3 in the digraph. On the other hand, if $\epsilon_{ij}^{(3)} = \sum_{k=1}^{n} \epsilon_{ik}^{(2)}\epsilon_{kj} = 0$, then for each k, either there is not a path of length 2 from i to k (i.e., $\epsilon_{ik}^{(2)} = 0$) or else there is not a path of length 1 from vertex k to j (i.e., $\epsilon_{kj} = 0$). Hence there cannot be a path of length 3 from i to j if $\epsilon_{ij}^{(3)} = 0$. As above, we let

$$I_i^{(3)} = \{j: i \neq j, \epsilon_{ij} = 0, \epsilon_{ij}^{(2)} = 0, \epsilon_{ij}^{(3)} \neq 0\},$$

and thus $I_i^{(3)}$ is the set of vertices over which vertex i has influence of order 3.

We return to the example used above to illustrate these ideas. In $\mathbb{E}^3$ the entry $\epsilon_{25}^{(3)} = 1$, and hence there is a path of length 3 from vertex 2 (B) to vertex 5 (E). Inspection of Fig. 7–9 yields the path from B to A to D to E. For this example the sets $I_i^{(1)}$, $I_i^{(2)}$, and $I_i^{(3)}$ are

$$I_1^{(1)} = \{3, 4\}, \qquad I_1^{(2)} = \{2, 5\}, \qquad I_1^{(3)} = \emptyset,$$
$$I_2^{(1)} = \{1\}, \qquad I_2^{(2)} = \{3, 4\}, \qquad I_2^{(3)} = \{5\},$$
$$I_3^{(1)} = \{2\}, \qquad I_3^{(2)} = \{1\}, \qquad I_3^{(3)} = \{4\},$$
$$I_4^{(1)} = \{5\}, \qquad I_4^{(2)} = \emptyset, \qquad I_4^{(3)} = \emptyset,$$
$$I_5^{(1)} = \emptyset, \qquad I_5^{(2)} = \emptyset, \qquad I_5^{(3)} = \emptyset.$$

The pattern has now been set. Paths of length p are discovered by examining the pth power of the incidence matrix $\mathbb{E}$. We define

$$I_i^{(p)} = \{j: i \neq j, \epsilon_{ij} = 0, \epsilon_{ij}^{(2)} = 0, \ldots, \epsilon_{ij}^{(p-1)} = 0, \epsilon_{ij}^{(p)} \neq 0\}.$$

Then $I_i^{(p)}$ is the set of vertices which are influenced of order p by vertex i (Exercise 4). The total influence set for vertex i (assuming n vertices) is given by

$$I_i = I_i^{(1)} \cup I_i^{(2)} \cup \cdots \cup I_i^{(n-1)}.$$

Therefore we have established the following:

Theorem 1: The student represented by vertex i is a leader if and only if $I_i = V - \{v_i\}$.

In the example that we have been carrying along, we have five students, and

$$I_1^{(4)} = \emptyset, \qquad I_2^{(4)} = \emptyset, \qquad I_3^{(4)} = \{5\}, \qquad I_4^{(4)} = \emptyset, \qquad I_5^{(4)} = \emptyset.$$

Hence, $I_1 = \{2, 3, 4, 5\}$, $I_2 = \{1, 3, 4, 5\}$, $I_3 = \{1, 2, 4, 5\}$, $I_4 = \{5\}$, and $I_5 = \emptyset$. Therefore, in this special case each of students A, B, and C is a leader. In some sense A is a stronger leader because $I_1^{(1)} \cup I_1^{(2)} = I_1$; however, we shall not pursue this or any of the other possible refinements which could be made. Our purpose was simply to illustrate how a digraph model can be used for studies of this sort.

7.2.3 Traffic Patterns

Consider the problem of the newly appointed Director of City Planning. The traffic situation is a mess, and he is charged with straightening it out. He knows that at the present time there is two-way traffic on all streets, and this is a major source of the problem because many of the streets are simply too narrow to permit an efficient flow of traffic in both directions. The obvious solution is to create one-way streets. But which streets should be one-way, and which way should they be directed? The director decides to take a simple approach to the first question and make them all one-way—or at least make as many one-way as possible. At this stage he turns the problem over to his assistant with the following instructions: Design a traffic pattern for the city streets which has as many one-way streets as possible, subject to the single condition that it always be possible to travel (legally) between any two points in the city.

After some thought about his new assignment the assistant director decides that the proper setting for this problem is that of a digraph. More accurately, the initial setting is that of a general graph (streets are edges and corners are vertices), and the problem is to convert the undirected graph into a digraph in which there is always a path from an arbitrary vertex to any other vertex. Also, since tenure is rarely awarded to the assistant to the Director of City Planning, he decides to solve the problem for a general city instead of for the particular city where he happens to work now.

The first step in the solution of this problem is the realization that it is usually hopeless to try to convert all the streets to one-way traffic. Almost every city has a certain number of dead-end streets, and these can not be one-way or else entry to or exit from the street will be prohibited. Also, many cities are separated into two or more parts by physical barriers such as rivers and lowered (or elevated) railroad tracks. If there is only one bridge over the river or one overpass for the railroad tracks, then these cannot be made one-way without one part of the city losing access to another part.

With the above remarks for motivation, we now present a formal solution of this problem. We shall show that with the exception of dead-end streets and certain types of bridges, all other streets can be made one-way. Moreover, we shall show that there is a constructive procedure for deciding which way to direct the streets. We begin with a number of definitions.

Definition: A graph is said to be *connected* if for any two vertices v and w of the graph there is a sequence of vertices $\{v_i\}_{i=1}^n$ such that each of the pairs $\{v, v_1\}, \{v_1, v_2\}, \ldots, \{v_{n-1}, v_n\}, \{v_n, w\}$ is an *edge* in the graph. Such a sequence is called a *path* between v and w. (Recall that in a general graph an edge is a pair and not an ordered pair.)

We are now able to formalize the notions of dead-end streets and connecting bridges.

Definition: In a connected graph an edge is said to be a *bridge* if whenever this edge is removed from the graph, the graph is no longer connected. An edge $\{v, w\}$ is said to be a *terminal edge* if either v or w belongs to no edge other than $\{v, w\}$. A *separating edge* is an edge which is either a bridge or a terminal edge. An edge which is not a separating edge is called a *circuit edge*.

If $\{v, w\}$ is a circuit edge in a connected graph, then by definition it is not a bridge. Hence, if $\{v, w\}$ is removed from the set of edges, the graph is still connected. In other words, there is a path between v and w which does not include the edge $\{v, w\}$. This fact lies at the heart of the proof of the following key result.

Theorem 2: Let G be a connected graph. It is possible to assign a single direction to each circuit edge in such a way that when each separating edge is given a double direction there is a directed path from any vertex to any other vertex.

We shall outline the proof of this theorem below. However, we first illustrate the proof by means of examples.

Example 1. Let G be the undirected graph of Fig. 7–10.

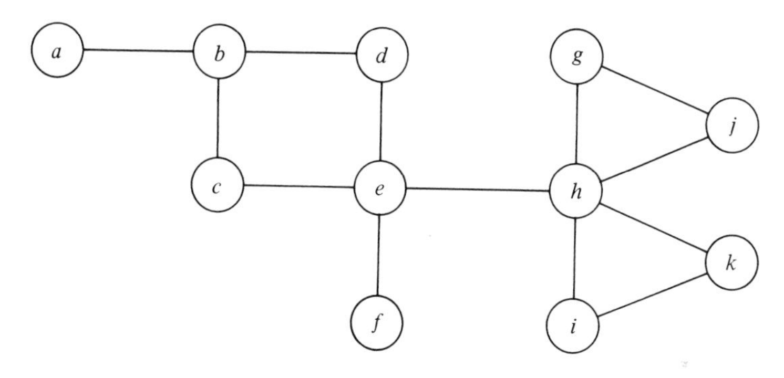

Fig. 7–10

Inspection of this graph shows that the edges $\{a, b\}$ and $\{e, f\}$ are terminal edges, while $\{e, h\}$ is a bridge. We begin by replacing one of these separating edges by a pair of directed edges. The new graph ($\{a, b\}$ replaced) is shown in Fig. 7–11.

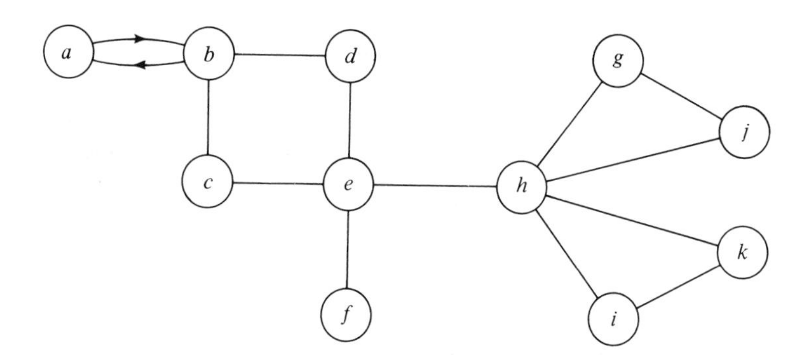

Fig. 7–11

Next we choose any undirected edge which shares a vertex with a directed edge. Either $\{b, c\}$ or $\{b, d\}$ is such an edge. We work with $\{b, c\}$ and note that it is a circuit edge. Hence there is a path between vertices b and c which does not include the edge $\{b, c\}$. Such a path is given by the edges $\{b, d\}$, $\{d, e\}$, and $\{e, c\}$. Together with edge $\{b, c\}$ this path provides a circuit from vertex b to vertex b. Pictorially, we have the circuit shown in Fig. 7–12. None of the edges in this circuit have previously been assigned a direction, and hence there are two possible ways to order the circuit to form a directed path with start at b and end at b. The clockwise direction is chosen arbitrarily, and the directed edges are (b, d), (d, e), (e, c), and (c, b). The graph at this stage is shown in Fig. 7–13.

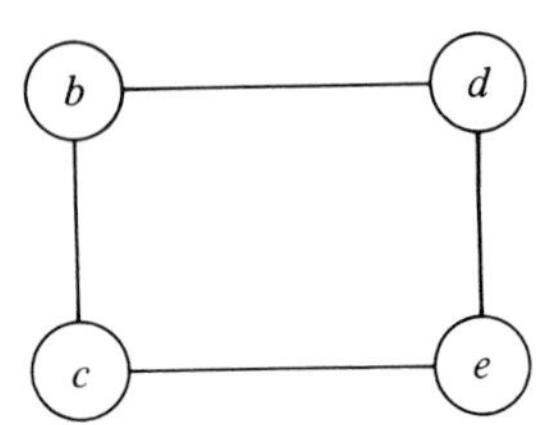

Fig. 7–12

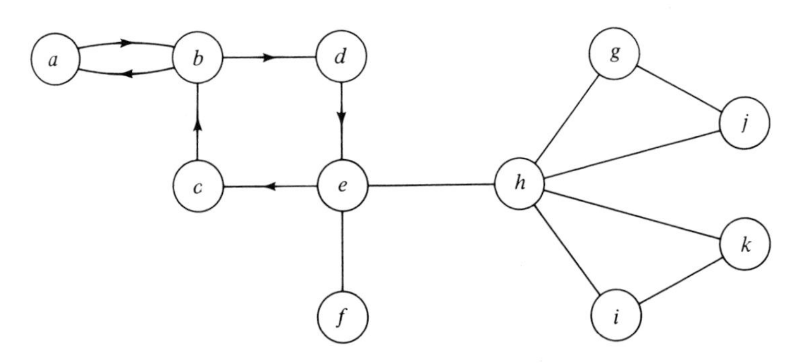

Fig. 7–13

There are still undirected edges in the graph, and thus the process must be continued. However, we note that given any vertex on a directed edge of the graph, there is a directed path to any other vertex on a directed edge of the graph. This property is to be retained as the remaining edges are assigned directions.

We continue by choosing any undirected edge which shares a vertex in common with a directed edge. Either $\{e, f\}$ or $\{e, h\}$ is such an edge, and since both are separating edges, we replace each of them by a pair of directed edges as shown in Fig. 7–14.

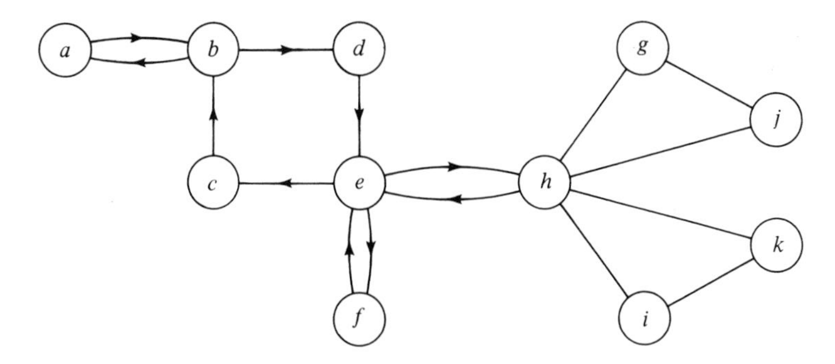

Fig. 7–14

Notice that again the construction is such that there is a directed path from any vertex on a directed edge to any other vertex on a directed edge. Since there remain undirected edges in the graph, the edge $\{g, h\}$ is chosen as an undirected edge which shares a vertex with a directed edge ($\{h, i\}$ could also be used). This is a circuit edge, and it is assigned the direction *away* from the directed portion of the graph; i.e., $\{g, h\}$ is assigned direction (h, g). Since $\{g, h\}$ is a circuit edge, there is an edge which shares vertex g with $\{g, h\}$. This is edge $\{g, j\}$, and since it is undirected, it is now assigned the direction (g, j). Continuing, $\{g, j\}$ shares vertex j with edge $\{j, h\}$, and $\{j, h\}$ is undirected, and so it is assigned the compatible direction (j, h). This yields a directed path from vertex h to vertex h using edge $\{g, h\}$, and another stage of the procedure is completed. The directions assigned to $\{g, j\}$ and $\{j, h\}$ were selected to preserve the property that at each stage there is a directed path between any two vertices on directed edges.

The final stage in making G a directed graph is to repeat this last procedure on the path $\{h, k\}, \{k, i\}, \{i, h\}$. The final directed graph is shown in Fig. 7–15.

We note that there is a directed path from any given vertex in the graph of Fig. 7–15 to any other given vertex in this graph.

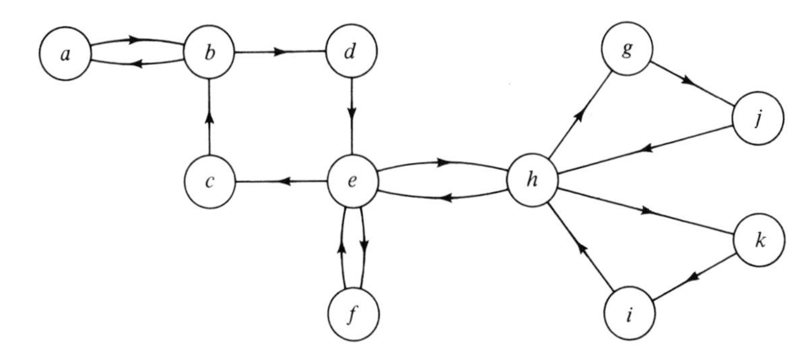

Fig. 7–15

The method used in Example 1 is applicable to any connected graph, and it is the basis for the proof of Theorem 2. We briefly discuss a second, somewhat different, example before carrying out the proof.

Example 2. Let G be the connected graph shown in Fig. 7–16.

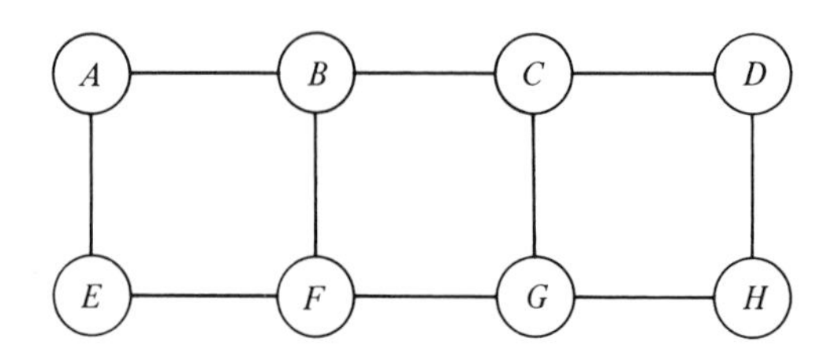

Fig. 7–16

In this graph all edges are circuit edges. We choose one at random, $\{A, B\}$, and determine a circuit from vertex A to vertex A which includes edge $\{A, B\}$. Such a circuit is $\{A, B\}, \{B, F\}, \{F, E\}$, and $\{E, A\}$. Since no edges have been assigned directions at this stage, we arbitrarily assign this circuit the clockwise direction shown in Fig. 7–17.

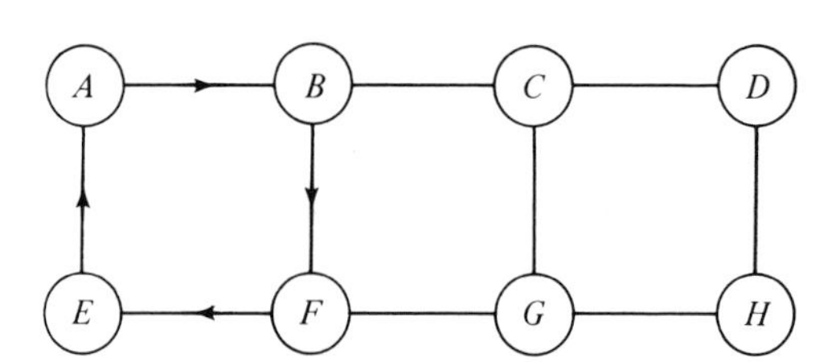

Fig. 7–17

We now mimic the technique used in Example 1. We choose an edge which is undirected but which shares a vertex with a directed edge. For example, $\{B, C\}$ has the shared vertex B. We direct $\{B, C\}$ away from the vertex which is shared by assigning it the direction (B, C), and then we choose an edge containing vertex C. The edge chosen is $\{C, G\}$, it is undirected, and it is assigned the appropriate direction (C, G). Similarly, $\{G, F\}$ is assigned the direction (G, F). The process now stops since all edges containing vertex F have been assigned a direction already. The new graph is shown in Fig. 7–18.

In this stage of the process we did not assign directions to all the edges in a circuit from vertex B to vertex B. Nevertheless, there is a directed path from any vertex on a directed edge to any other vertex on a directed edge. This is true because we did assign a direction to the edges of a path with its

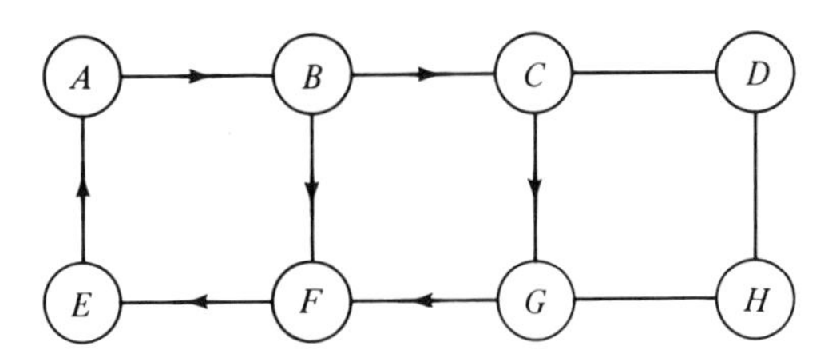

Fig. 7–18

start and end in that part of the graph which had already been assigned directions.

The process is now concluded by assigning a direction to the edges of the path $\{C, D\}$, $\{D, H\}$, $\{H, G\}$. This is done so as to give a directed path from vertex C to vertex G, and the resulting directed graph is shown in Fig. 7–19.

We note at this point that the above process (and resulting digraph) is not unique. In fact, there were two essentially different choices which were possible at each of the second and third stages of the process.

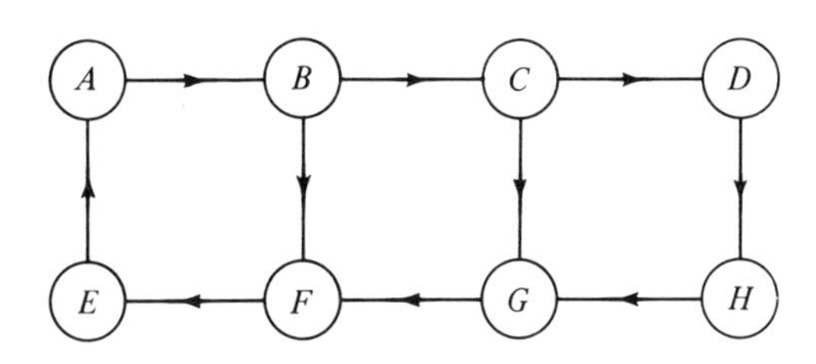

Fig. 7–19

With the above examples as motivation, we now give an outline of the proof of Theorem 2.

Outline for the Proof of Theorem 2

1. If the graph has separating edges, choose a separating edge $\{v, w\}$ at random, and replace it by the directed edges (v, w) and (w, v). Then go to step 3.
2. If the graph does not have separating edges, then choose a circuit edge $\{u, v\}$ at random, determine a circuit with start and end at u which includes the edge $\{u, v\}$, and assign a direction to the edges of this circuit to give a directed path from vertex u to vertex u. Proceed to step 3.
3. If all the edges are not directed, then choose an undirected edge $\{v, w\}$ which shares a vertex with a directed edge (u, v). If $\{v, w\}$ is a separating edge, replace it by the directed edges (v, w) and

(w, v), and begin step 3 again. If $\{v, w\}$ is a circuit edge, assign it direction (v, w) and go to step 4.

4. Since (v, w) is a circuit edge, there is another circuit edge $\{w, w_1\}$. If this edge is undirected, assign it direction (w, w_1). Again, since (w, w_1) is a circuit edge, there is a circuit edge $\{w_1, w_2\}$. If it is undirected, assign it direction (w_1, w_2). Continue in this manner until either a directed edge is reached or else an undirected edge of the form $\{w_n, v\}$ is obtained. If a directed edge is reached, return to step 3. If the edge $\{w_n, v\}$ is reached, then assign it direction (w_n, v) and return to step 3. Continue with steps 3 and 4 until all edges of the graph are directed.

Since a graph is assumed to have only a finite number of edges (at least this is assumed for the graphs of this chapter) and since each repitition of steps 3 and 4 assigns a direction to at least one new edge, the process terminates in a finite number of steps. Thus the proof is completed by showing that there is always a directed path from any given vertex to any other given vertex. This follows rather directly from the procedure used to assign the directions to the edges. We leave the details of showing this fact to the reader (Exercise 5).

As a final remark about the process of assigning directions in an undirected graph, we return to the original setting of traffic in a city. Theorem 2 states that if all connecting bridges and dead-end streets have two-way traffic, then it is possible to make all other streets one-way, while retaining the condition that one be able to legally drive between any two points in the city.

Connectedness is only one of a number of abstract concepts which are useful in the study of graphs. We introduce three more of these concepts (*Euler* and *Hamiltonian circuits* and the *valence* of a vertex) in the projects (Sec. 7.3), where this short discussion of topics in graph theory is continued.

EXERCISES

1. Draw all digraphs with two vertices and all digraphs with three vertices.

2. Two digraphs, (V_1, E_1) and (V_2, E_2), are said to be *isomorphic* if there is a one-to-one mapping of V_1 onto V_2 which preserves incidence. Thus, if the mapping is denoted by f, then $(v, w) \in E_1$ if and only if $(f(v), f(w)) \in E_2$. Give an example of two digraphs each of which has three vertices and four edges and which are not isomorphic. How many non-isomorphic digraphs are there with three vertices? *Remark:* We invite the reader to note the significant increase in the complexity of this problem in going from digraphs with three vertices to those with four vertices. There are 218 nonisomorphic digraphs with four vertices.

3. In Fig. 7–8(c), find all vertices which are "leaders" in the sense of this section.

4. Using the notation of Sec. 7.2.2, show that $\epsilon_{ij}^{(p)}$ is nonzero if and only if there is a path of length p with vertex i as the start and vertex j as the end.

5. Give an example of a class of six students in which three are leaders and three are not leaders.

6. Complete the proof of Theorem 2 by showing that the process given by steps 1–4 results in a directed graph in which there is always a directed path from any given vertex to any other given vertex. *Hint:* Show that at each stage of the process of assigning directions to the edges of the graph there is always a directed path from any vertex on a directed edge to any other vertex on a directed edge.

7. In each of the graphs in Fig. 7–20, find all the terminal edges and all the bridges.

8. Convert each of the graphs of Fig. 7–20 into a directed graph in which there is a directed path from any given vertex to any other given vertex.

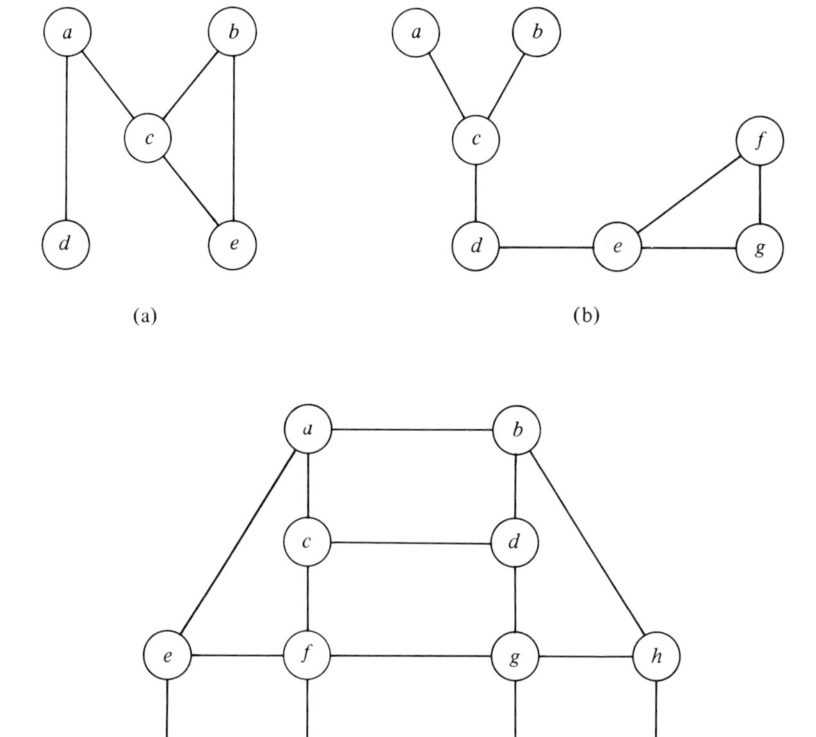

Fig. 7–20

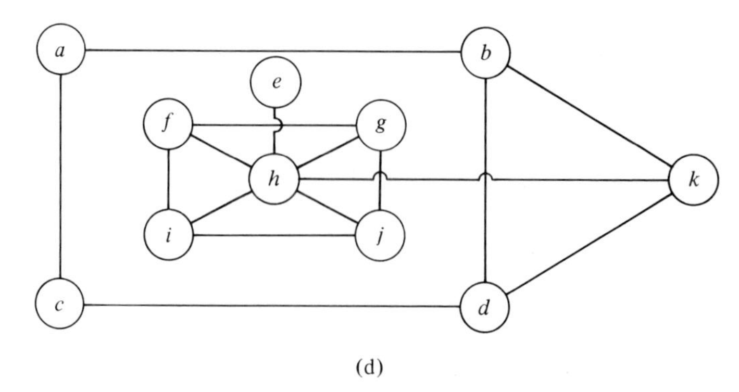

(d)

Fig. 7-20 (cont.)

9. Use the terminology of Sec. 7.2.2 and the definition of connected in Sec. 7.2.3 to show that

(a) If the influence graph for a class of students is connected, then at least one of the students in the class is a leader.

(b) If a class has a leader, then he is not unique. Thus there are always at least two leaders in a class which has leaders.

7.3 PROJECTS

7.3.1 Job Placements

Develop an algorithm for solving the job assignment problem by using the methods of networks and flows. In particular, describe how your algorithm is to start and how the process of improvement is carried out. Use the techniques given in the proof of Theorem 1 of Sec. 7.1.

Use the algorithm developed above to assign as many applicants to jobs as is possible for the following rating matrix $\mathbb{R} = (r_{ik})$. Here the convention is that A_i is appropriate for J_k if and only if $r_{ik} = 1$.

$$\mathbb{R} = \begin{array}{c} \\ A_1 \\ A_2 \\ A_3 \\ A_4 \\ A_5 \\ A_6 \\ A_7 \end{array} \begin{array}{c} \begin{array}{ccccccc} J_1 & J_2 & J_3 & J_4 & J_5 & J_6 & J_7 \end{array} \\ \begin{bmatrix} 0 & 1 & 0 & 0 & 1 & 1 & 0 \\ 1 & 1 & 1 & 0 & 0 & 0 & 0 \\ 1 & 1 & 0 & 0 & 0 & 0 & 0 \\ 0 & 0 & 1 & 0 & 1 & 0 & 1 \\ 1 & 0 & 0 & 0 & 0 & 0 & 0 \\ 0 & 0 & 1 & 1 & 0 & 0 & 0 \\ 1 & 1 & 1 & 0 & 0 & 0 & 0 \end{bmatrix} \end{array}.$$

7.3.2 The Chain of Command

Develop a graph theory model for the transfer of information in a large organization such as the military, the government, or a large company. Give definitions for positions of authority in terms of your graph. Try to reflect the subtleties which may occur in these organizations. For example, convention may dictate that certain conversations are one-way, while others are two-way.

7.3.3 The General Job Assignment Problem

We have already noted that the job assignment problem of Sec. 2.6 can be generalized. The purpose of this project is to indicate one such generalization.

1. Formulate the job assignment problem of Sec. 2.6 as a linear programming problem with auxiliary constraints. Thus, find a matrix $\mathbb{A}$ and vectors **b** and **c** so that this assignment problem is the standard maximum problem [$\mathbb{A}$, **b**, **c**] together with other constraints.
2. Develop a model for the consideration of a job assignment problem where each applicant is rated for each job on a scale from 0 to 5. Formulate this problem as a linear programming problem with auxiliary constraints, and exhibit the matrices and vectors used to write the problem as a standard maximum problem.
3. Let the matrix $\mathbb{R} = (r_{ik})$, which is given below, be the rating matrix representing the qualifications of four applicants for four jobs. Thus, $r_{ik} = 3$ means that applicant A_i is rated 3 for job J_k. Assume that higher ratings indicate more skill and experience for the job. Find an assignment of applicants to jobs which maximizes the sum of the ratings of the applicants for the jobs they are assigned to fill. Is this assignment unique?

$$\mathbb{R} = \begin{array}{c} \\ A_1 \\ A_2 \\ A_3 \\ A_4 \end{array} \begin{array}{c} \begin{array}{cccc} J_1 & J_2 & J_3 & J_4 \end{array} \\ \begin{bmatrix} 0 & 1 & 0 & 1 \\ 3 & 3 & 5 & 0 \\ 2 & 1 & 1 & 1 \\ 3 & 2 & 3 & 2 \end{bmatrix} \end{array}.$$

7.3.4 Euler and Hamiltonian Circuits

In Sec. 7.2.3 (traffic patterns) the problem is to convert a general graph into a directed graph in such a way that if v and w are arbitrary vertices of the graph, then there is a directed path from v to w. Two related and somewhat similar problems are the following:

a. Given a graph and a vertex v, find a path which starts and ends at v and which passes through each vertex of the graph exactly once.
b. Given a graph and a vertex v, find a path which starts and ends at v and which includes each edge of the graph exactly once.

Problem 1. Why might employees of the postal service be interested in problems a and b? Who else might be interested?

In both problem a and problem b there are really two separate questions to consider. First, do the desired paths exist, and second, if they exist, how does one find them? In order to (at least partially) answer these questions, we give a number of definitions, and we pose a number of problems for the reader.

Definition: A circuit in the graph (V, E) is called a *Euler circuit* (or *closed Euler path*) if it includes each edge of E exactly once. The *valence* of a vertex v is the number of edges of the form $\{v, w\}$.

Problem 2. Give two examples of graphs with five vertices. The first should contain ten edges and a Euler circuit (show the circuit), and the second should contain nine edges and no Euler circuit (explain why none exists).

The examples of Problem 2 provide the motivation for the following theorem.

Theorem A: A graph contains a Euler circuit if and only if it is connected and the valence of each vertex is an even number.

Problem 3. Show that if a graph contains a Euler circuit, then the graph is connected and the valance of each vertex is even.

Problem 4. Show that if a graph is connected and if the valence of each vertex is even, then the graph contains a Euler circuit.

Definition: A circuit in a graph (V, E) is said to be a *Hamiltonian* circuit if it includes each vertex of V exactly once.

Problem 5. Give an example of a graph with five vertices and six edges which contains a Euler circuit but no Hamiltonian circuit.

A complete theory of Hamiltonian circuits does not exist. However, there are a number of partial results concerning the existence of such circuits. One such result is the following concerning complete graphs. [A graph (V, E) is *complete* if for each $v, w \in V$, $v \neq w$ implies that $\{v, w\} \in E$.]

Theorem B: Every complete graph contains a Hamiltonian circuit.

Problem 6. Prove Theorem B for all complete graphs with three, four, or five vertices.

Problem 7. Prove Theorem B.

7.3.5 Sex Assignments and Breeding Experiments

Everyone is familiar with the pictorial representation of family trees. In these "pictures" one uses points and lines to represent family structures and interfamily relationships. Thus, if M and F are the parents of three children, two boys and one girl, then we have picture (a) of Fig. 7–21. Similarly, picture (b) of Fig. 7–21 may be thought of as representing the interfamily relationship of two families.

Problem 1. Describe the family relationship shown in Fig. 7–21(b). Also, illustrate three generations of a family by a similar graph.

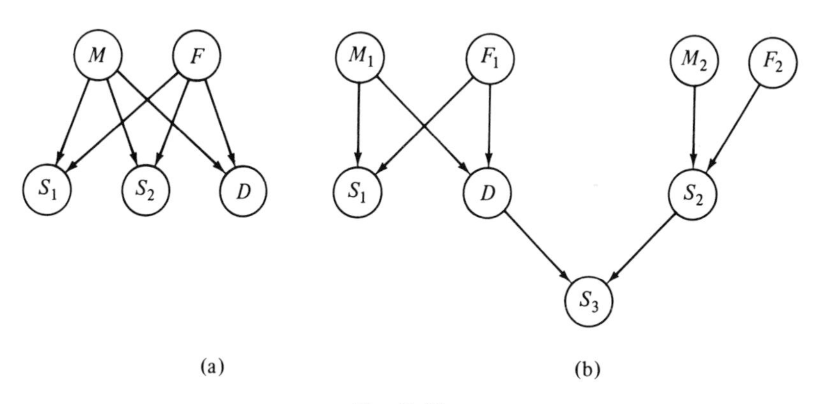

Fig. 7–21

One can conceive of a large number of such pictures. Some can represent family trees and some cannot. We consider the problem of characterizing those which can represent family trees.

We shall actually work in a somewhat simpler context than that of human relationships. Most human societies preclude family trees of the forms shown in Fig. 7–22. However, it would be quite complicated to give conditions which would eliminate pictures of this sort. Accordingly, we suppose that we are dealing with a population which is completely characterized by the condition of sexual reproduction. If the picture or graph is such that by a proper assignment of sexes to the members of the population the picture represents the results of an experiment which theoretically could take place,

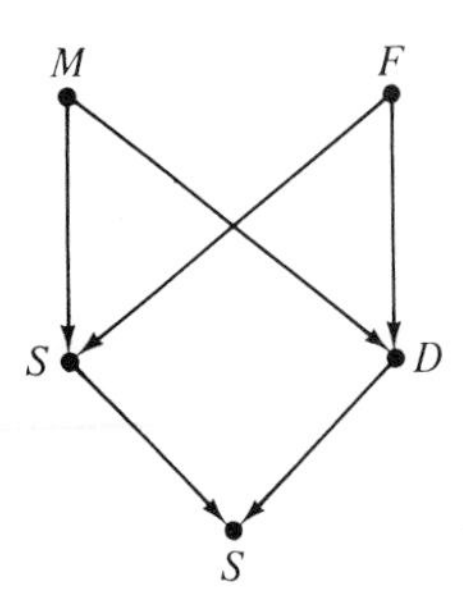

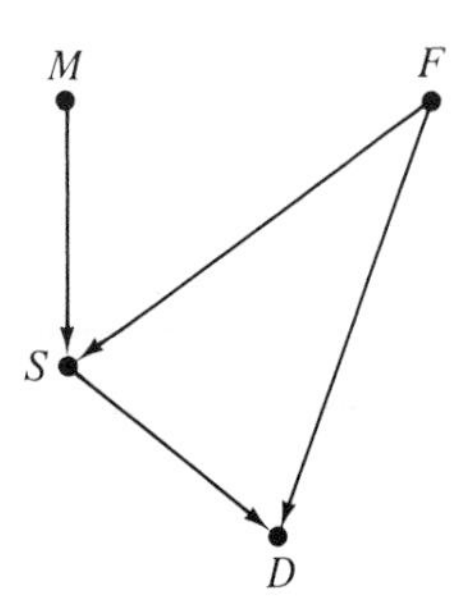

Fig. 7–22

then we say that the graph is *realizable as a breeding experiment*. We are interested in determining which graphs are realizable as breeding experiments. With this in mind we introduce the following concept.

Definition: In a directed graph (V, E), the *index* of a vertex v is the number, $\rho(v)$, of edges of the form $(u, v) \in E$.

Let us now examine the consequences of the assumption that the population under study is characterized by sexual reproduction. First, in the use of digraphs to represent breeding experiments, we adopt the convention that the presence of the directed edge (u, v) means that v is the offspring of u. Then, since each individual has exactly two parents, one male and one female, there will be exactly two edges terminating at each vertex. That is, for any $v \in V$, $\rho(v) = 2$. In pictorial terms the basic component of the graph is as shown in Fig. 7–23, where O is the offspring of M and F which are members of opposite sexes. In this connection it is necessary to point out that breeding experiments do not extend into the indefinite past. One always reaches a point where one or both of the parents of a member of the population are unknown. Thus one must replace the condition $\rho(v) = 2$ by $\rho(v) \leq 2$. We have therefore derived the following:

Property 1: If a digraph represents the results of a breeding experiment, then for all $v \in V$ one has $\rho(v) \leq 2$.

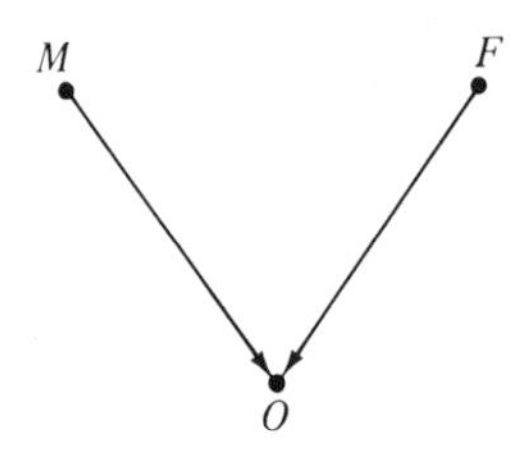

Fig. 7–23

Problem 2. Show that it is not always possible to assign sexes in a digraph with $\rho(v) \leq 2$ for all $v \in V$. More precisely, show that it is not always possible to divide the vertices into two classes M and F in such a way that for every $v \in V$ with $\rho(v) = 2$, there exists $u_1 \in M$ and $u_2 \in F$ such that $(u_1, v) \in E$ and $(u_2, v) \in E$.

Hint. Use a digraph with six vertices.

In general, suppose that v_1 is a vertex in a digraph and suppose that it is assigned sex M. It follows that if v_1 has offspring with v_2, then $v_2 \in F$. If v_2 has offspring with v_3, then $v_3 \in M$, and so forth. In this manner we obtain an alternating sequence $v_1, v_2, v_3, \ldots, v_n$ of members of M and F. In the graph they are connected by a sequence of edges alternately traversed in the direction given by the graph and the opposite direction. See Fig. 7–24. If at some stage v_n ($n > 2$) and v_1 have a common offspring—call it w_n—then we can close the path by adding (v_1, w_n). So that this assignment is legitimate in light of our assumption of sexual reproduction, we must have $v_n \in F$. That is, the number of parents occurring in the closed alternating path must be even.

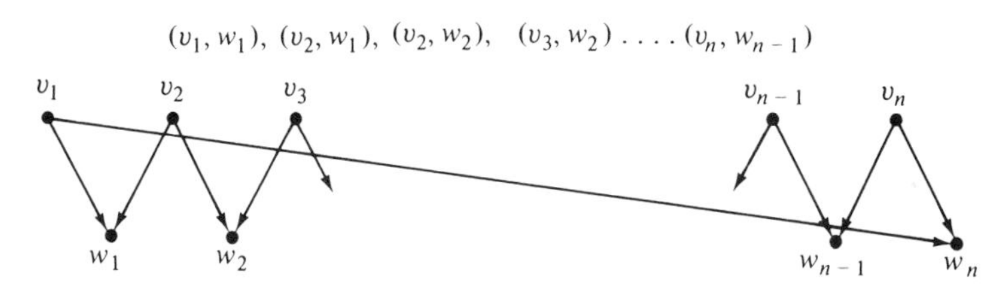

Fig. 7–24

We have therefore verified the necessity portion of the following *sex assignment theorem.*

Theorem: Let $G = (V, E)$ be a digraph with $\rho(v) \leq 2$ for each $v \in V$. It will be possible to assign sexes to all the vertices of G in a manner which is compatible with a breeding experiment, if and only if there are no closed alternating paths with an odd number of parents.

Proof: Only the sufficiency portion of the proof remains. We give a constructive proof including a part for the reader to complete. Select a vertex v_1: any vertex will do. Assign v_1 to class M. If v_1 has no offspring, or if offspring of v_1 have no other (known) parents, then the consequences of assignment of sex to v_1 are complete. If v_1 has offspring in common with other vertices, then one forms all alternating paths from v_1 and assigns sexes to the parents in a manner compatible with our assumption of reproduction. We claim that every vertex which can be reached from v_1 via an alternating path is assigned a unique sex.

Problem 3. Verify this claim by showing that it is impossible to have an alternating path from v_1 to v_n by which v_n is assigned to class M and another alternating path by which v_n is assigned to class F.

Next, if all vertices can be reached from v_1 by alternating paths, the proof is complete. If not, select a vertex which cannot be so reached, assign it sex M, and let it play the role of v_1 in the paragraph above. Since there are only a finite number of vertices, the process eventually comes to an end, and we have assigned a sex to each vertex. Q.E.D.

Note that the assignment of sexes to vertices in a graph is highly non-unique and may perhaps be accomplished in a number of different ways.

Now that we have a theorem which gives a necessary and sufficient condition for sexes to be assigned to vertices, we ask whether any digraph satisfying the hypotheses of the theorem can actually be realized in a breeding experiment. Figure 7–25 indicates that further restrictions are necessary. Indeed, $\rho(v) = 1$ for all $v \in V$, and there are no closed alternating paths. But clearly such a situation cannot arise in a real experiment.

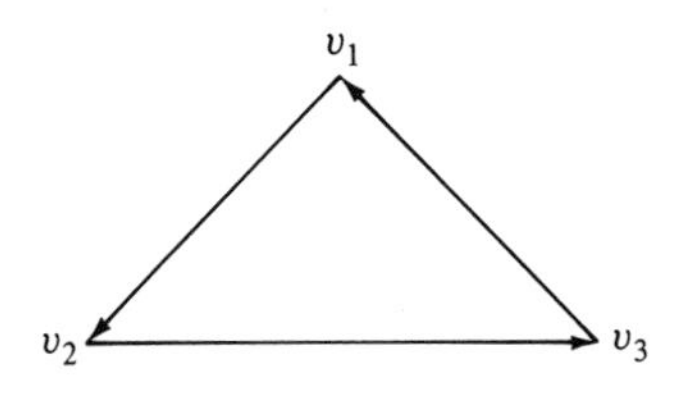

Fig. 7–25

We impose a restriction which eliminates this. Suppose that $v_1, v_2, \ldots, v_m$ constitute a series of individuals each of which is the offspring of its predecessor. In a graph this series corresponds to the directed path

$$(v_1, v_2), (v_2, v_3), (v_3, v_4), \ldots, (v_{m-1}, v_m).$$

Since the births of these individuals are sequential in time, it cannot happen that for some m we have (v_m, v_1). That is, the graph cannot contain any directed path which forms a circuit. This is a necessary condition that the digraph be realizable in an experiment.

It turns out that this condition together with the hypotheses of the sex assignment theorem are sufficient for a digraph to be realizable. Most of the proof of this fact is given here, and the reader is encouraged to put the pieces together (Problem 5).

Problem 4. In how many ways can sex assignments be made in the graph given in Fig. 7–26? First verify that it can be done at all.

Problem 5. State and prove the theorem on breeding experiments which is referred to in the concluding paragraph above.

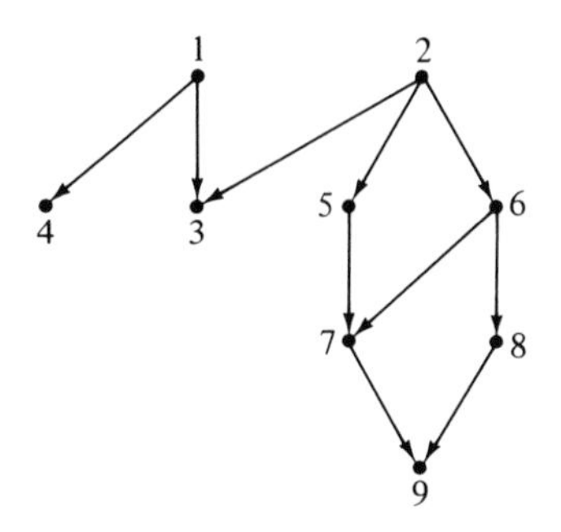

Fig. 7–26

REFERENCES

Theorem 1 of Sec. 7.1 and many of the algorithms associated with networks and flows are due to L. Ford and D. Fulkerson. These algorithms are discussed in most books on programming and networks, in particular in [BeGH], [FoFu], and [Hu]. An elementary introduction to graphs is provided by [O], and more mathematically advanced and complete treatments are given by [Be] and [H]. A number of interesting applications of graph theory are discussed in [BuS], and [HNC] treats directed graphs from the point of view of their applicability to the social sciences.

[Be] BERGE, C., *The Theory of Graphs and Its Applications.* London: Methuen, 1964.

[BeGH] BERGE, C., and A. GHOUILA-HOURI, *Programming, Games, and Transportation Networks.* London: Methuen, 1965.

[BuS] BUSACKER, R., and T. SAATY, *Finite Graphs and Networks, An Introduction with Applications.* New York: McGraw-Hill, 1965.

[FoFu] FORD, L., and D. FULKERSON, *Flows in Networks.* Princeton, N.J.: Princeton University Press, 1962.

[H] HARARY, F., *Graph Theory.* Reading, Mass.: Addison-Wesley, 1969.

[HNC] HARARY, F., R. NORMAN, and D. CARTWRIGHT, *Structual Models: An Introduction to the Theory of Directed Graphs.* New York: Wiley, 1965.

[Hu] HU, T. C., *Integer Programming and Network Flows.* Reading, Mass.: Addison-Wesley, 1969.

[O] ORE, O., *Graphs and Their Uses.* New York: Random House, 1963.

8 Models for Growth Processes

8.0 INTRODUCTION

The concept of growth, that is, the increase or decrease of some quantity, is a fundamental one which may be utilized in the study of many problems arising in the life, physical, and social sciences. In this and the following chapter we shall examine several situations in which models based on growth processes have proved useful. Since it is this process which is our particular concern, we shall emphasize those aspects of the actual biological, economic, physical, or sociological situation which are appropriately modeled as growth phenomena. It is necessary to mention, however, that the basic process is often somewhat modified or partially obscured by other phenomena which may or many not be of the same type. For example, the increase of the fox population in an area might be significantly influenced by an epidemic of rabies, which itself can be modeled as a growth process, or by the construction of a highway through the area, an extraneous event which must be accounted for by special methods. The approach taken here is to separate the interacting processes as much as possible and to concentrate on one at a time. First, we shall construct very simple models, some based on an assumption of continuous-time and some on a discrete-time approximation. These models involve no mathematical difficulties, but the results they predict are not always in good agreement with the real world. In order to reduce the inconsistencies between the predictions based on the model and the results obtained by direct observations, and also to consider more complicated systems, we shall introduce additional assumptions and consider modified models. These more involved models lead naturally to deeper mathematics. At times a complete discussion is beyond the scope of this book, and we shall provide only qualitative observations.

In Part A we shall restrict our attention to deterministic models. Recall that such models are based on the premise that if complete information on the system is known at a specific time, then its future behavior can be predicted exactly. In the situation of simple population growth, for example, the assumption of complete information means that we know the parameters of the system, the birth and death rates, and the size of the population at a specific time. In Part B we shall consider models which incorporate certain probabilistic or random behavior, i.e., stochastic models.

Although in this chapter we shall usually use the terminology of population growth or radioactive decay, the models constructed can be used in many other situations. Some of these are considered in detail in the next chapter, and others are indicated in the Exercises and Projects (Sec. 8.7).

PART A DETERMINISTIC MODELS

8.1 SIMPLE GROWTH PROCESSES

Our basic concern throughout this and the next chapter is with the change in the amount of a quantity or the size of a population with time. Specific examples to be considered include the variation in the number of units in a population, of people infected with a disease, and of radioactive atoms. In this section we consider the simplest possible models. First we introduce a discrete-time model and note its behavior as the time intervals become short. Next, we turn to continuous-time models and an economic application.

8.1.1 Discrete-Time Models

Suppose that one is interested only in determining the amount of a quantity at times belonging to a certain finite set T. The situation is simplest in the case of uniformly distributed times, and we restrict our attention to this special situation. For definiteness, suppose that time is measured from some initial time, say time 0, and that the time interval of interest is $0 \le t \le K$. For the moment, K is assumed to be an integer. Our models are based on assumptions regarding the amount present at time 0 and the increase from one time in T to the next. Specifically, we assume that

1. At time 0 there is amount N.
2. There is a constant r such that if $t_1, t_2, t_1 < t_2$, are any two consecutive times in T, then the amounts present at t_1 and t_2 are related by

$$(\text{Amount at } t_2) - (\text{Amount at } t_1) = r(t_2 - t_1)(\text{Amount at } t_1).$$

The constant r is known as the *growth rate*.

For the first model we take $T = T(1) = \{0, 1, 2, \ldots, K\}$, and we denote the amount present at time t by $A(1, t)$. The notation is intended to remind the reader that the interval between successive times in T is 1 unit. With this notation the assumptions become

$$A(1, 0) = N,$$
$$A(1, k) - A(1, k-1) = rA(1, k-1), \qquad k = 1, 2, \ldots, K.$$

It follows at once that

$$A(1, k) = A(1, k-1)(1+r), \qquad k = 1, 2, \ldots, K,$$

and consequently

$$A(1, k) = N(1+r)^k, \qquad k = 0, 1, \ldots, K. \tag{1.1}$$

As an example, suppose that the population of the world was 3 billion in 1960 and that the growth rate was 2% per year. Then the above model predicts a world population of $(1.02)^{10} \times 3 \times 10^9$ in 1970.

For construction of the next model, we take $T = T(1/2) = \{0, 1/2, 1, 3/2, \ldots, K\}$, and define $A(1/2, t)$ to be the amount present at time t for any $t \in T(1/2)$. Again the notation indicates that two consecutive times in T are $\frac{1}{2}$ unit apart. An argument similar to that used to derive (1.1) can be used to obtain

$$A(1/2, t) = N\left(1 + \frac{r}{2}\right)^{2t}, \qquad t \in T(1/2). \tag{1.2}$$

It is useful to compare $A(1, 1)$ and $A(1/2, 1)$. We have

$$A(1, 1) = N(1+r),$$
$$A(1/2, 1) = N\left(1 + \frac{r}{2}\right)^2 = N\left(1 + r + \frac{r^2}{4}\right).$$

Thus, if $N > 0$ and $r \neq 0$, then $A(1/2, 1) > A(1, 1)$. In fact, it is easy to show that if $|r| < 1$, then $A(1/2, k) > A(1, k)$ for $k = 1, 2, \ldots, K$ (Exercise 1).

We now turn to the general case where each unit interval of time is divided into n equal pieces. We set $T = T(1/n) = \{0, 1/n, 2/n, \ldots, K\}$, and $A(1/n, t)$ equal to the amount present at time t, $t \in T(1/n)$. We have

$$A(1/n, t) = N\left(1 + \frac{r}{n}\right)^{nt}, \qquad t \in T(1/n). \tag{1.3}$$

The derivation of (1.3) is similar to that of (1.1) and (1.2).

We illustrate the results predicted by these models with the example of the growth of a savings account in an institution which pays interest. Savings institutions frequently emphasize in their advertising the advantages of quarterly, monthly, or even daily compounding of interest. In Table 8–1 we

Table 8–1

n	$A(1/n, 1)$
1	1.06000
2	1.06090
3	1.06121
4	1.06136
6	1.06152
12	1.06168
24	1.06176
360	1.06183

list the amounts resulting after 1 year from an initial amount $N = 1$, an interest rate $r = 0.06$, and interest compounded at equally spaced times during the year.

An examination of Table 8–1 indicates that $A(1/n, 1)$ increases as n does, but apparently $A(1/n, 1)$ tends toward a limit as n becomes large. Since n becoming large is the same as the times in the set T becoming closer and closer together, it is useful to consider this case in more detail for comparison with models for continuous time. To this end, we rewrite (1.3) with $t = 1$ as

$$A(1/n, 1) = N\left\{\left(1 + \frac{1}{n/r}\right)^{n/r}\right\}^{r},$$

and for the remainder of this discussion we assume that $r > 0$. The rate r is fixed, it is 0.06 in the example, and therefore n/r increases as n does.

Finally, as n increases indefinitely, $A(1/n, 1)$ tends monotonically to the limit

$$N \lim_{n\to\infty} \left\{\left(1 + \frac{1}{n/r}\right)^{n/r}\right\}^r = N\left\{\lim_{n\to\infty}\left(1 + \left(\frac{1}{n/r}\right)^{n/r}\right\}^r$$
$$= N\left\{\lim_{\lambda\to\infty}\left(1 + \frac{1}{\lambda}\right)^{\lambda}\right\}^r$$
$$= Ne^r.$$

The last steps make use of the fact that $(1 + 1/\lambda)^\lambda$ is a monotonically increasing function of λ, and

$$\lim_{\lambda\to\infty}\left(1 + \frac{1}{\lambda}\right)^{\lambda} = e. \tag{1.4}$$

In the example our conclusion means that

$$\lim_{n\to\infty} A(1/n, 1) = \exp(0.06) = 1.0618365\ldots.$$

It is reasonable to expect that there is nothing special about the time $t = 1$, and consequently we are led to the conjecture that

$$\lim_{n\to\infty} A(1/n, t) = Ne^{rt} \tag{1.5}$$

for all times $t \in [0, K]$. In the natural extension of the model determined by axioms 1 and 2 to the case of continuous time, we shall encounter functions of exactly this form.

8.1.2 A Continuous-Time Model

In the real world, of course, time is not a variable taking only values in a discrete set, and for many purposes the approximation of Sec. 8.1.1 is not appropriate. Thus we introduce the natural extension of axioms 1 and 2 of Sec. 8.1.1 to continuous times and study the resulting model. Axiom 1 has nothing to do with whether time is considered to be a discrete or continuous variable, and we retain it unchanged. Let $A(t)$ denote the amount present at time t. Axiom 2 is stated in terms of the amounts present at two consecutive times of a discrete set, and since time is now a continuous variable, this must be changed. A natural replacement is

$$A(t_2) - A(t_1) \cong r(t_2 - t_1)A(t_1), \tag{1.6}$$

where $\cong$ is to be understood as "approximately equal to" and the approximation becomes better as $|t_2 - t_1|$ becomes small. For $t_2 \neq t_1$, Eq. (1.6) is equivalent to

$$\frac{A(t_2) - A(t_1)}{t_2 - t_1} \cong rA(t_1). \tag{1.7}$$

The approximate equation (1.7) involves a difference quotient for the function A and its occurrence leads to the idea of letting $t_2 \longrightarrow t_1$ and using the resulting derivative. If we suppose that the approximation in (1.7) becomes increasingly precise as $t_2 \longrightarrow t_1$, then

$$\lim_{t_2 \to t_1} \frac{A(t_2) - A(t_1)}{t_2 - t_1} = \frac{dA}{dt}(t_1)$$

exists and is equal to $rA(t_1)$. We use this heuristic discussion as a basis for an axiom system for continuous-time growth models. Let us again normalize the situation by supposing that we are concerned with times in an interval $[0, T)$. Here, T is a positive real number, or in some instances T may be ∞. The axiom system we choose for continuous-time models is

1. $A(0) = N$.

2′. A is a differentiable function and

$$\frac{dA}{dt}(t) = rA(t) \quad \text{for all } t \in [0, T). \tag{1.8}$$

For $t = 0$ the derivative in axiom 2′ is one-sided. As usual with functional notation, it is conventional to suppress the independent variable and to write (1.8) as

$$\frac{dA}{dt} = rA. \tag{1.9}$$

It is useful to look at the model defined by axioms 1 and 2′ from a heuristic point of view. The assumption embodied in axiom 2′ is that the rate of change of a quantity at time t is proportional to the amount of the quantity present at that time. The proportionality factor is assumed to be independent of time. In a study of biological population growth based on these axioms, it must be realized that there are both births and deaths occurring. Thus the growth rate r is actually the algebraic sum of two parameters: r_B, the birth rate, and r_D, the death rate, $r_B - r_D = r$. If one considers the special case of human population growth and if $r_B > r_D$, as is the case in most modern nations, then this model is that studied in detail by the English economist and demographer T. R. Malthus (1766–1834).

Returning to our mathematical model, the function A is specified by the axioms as the solution to the initial value problem

$$\frac{dA}{dt} = rA, \qquad t \geq 0, \\ A(0) = N. \tag{1.10}$$

This problem has the solution (see Appendix B)

$$A(t) = Ne^{rt}, \qquad t \geq 0. \tag{1.11}$$

It is useful to note for future application that if (1.10) is replaced by

$$\frac{dA}{dt} = rA, \qquad t \geq t_0, \\ A(t_0) = N, \tag{1.12}$$

that is, if time $t = t_0$ is taken as the reference time instead of $t = 0$, then the function A is given by

$$A(t) = Ne^{r(t-t_0)}, \qquad t \geq t_0. \tag{1.13}$$

Note that (1.13), or the special case (1.11), is an exponential function which is either increasing, constant, or decreasing with time according as $r > 0$, $r = 0$, or $r < 0$. The function defined by (1.13) is graphed in Fig. 8–1.

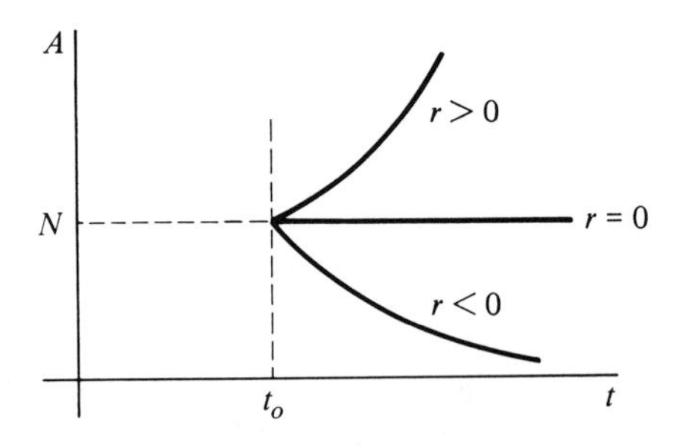

Fig. 8–1

It should be pointed out that the mathematical model constructed on the basis of the axioms of this section explicitly assumes that A is a differentiable function of time. Certain growth processes observed in the real world clearly do not satisfy this assumption. For example, if A denotes the number of units in some population, then it is obviously integer-valued, and hence it is not even continuous, let alone differentiable. In such situations this model must be viewed as an approximation whose usefulness depends on the parameters of the system. In population studies involving large populations, say several thousands of units, and for times which are not too short or too long in relation to the growth rate r, the model based on these axioms yields substantially the same results as much more complicated models not involving differentiability assumptions. If the system to be studied does not satisfy these or other criteria which justify the assumptions, then a deterministic model as a simple growth process is inappropriate and modifications must be introduced.

8.1.3 Economic Growth and Future Values

On the basis of the models discussed in Secs. 8.1.1 and 8.1.2, it is reasonable to make the following definition:

Definition: In an economy with interest rate r (per year) the future value in t years of \$1 today is e^{rt} dollars.

The question considered here is a sort of converse one: How does one assign a present value to money which becomes available at a future date? Again we assume an economy in which dollars grow at the rate r, and we ask for the number N of dollars to set aside now in order that they grow to A dollars in t years. Since we have $A = Ne^{rt}$, it is clear that $N = Ae^{-rt}$. This is the definition we adopt for the present value of dollars to be available in the future:

Definition: In an economy with interest rate r (per year) the present value of \$1 available t years in the future is e^{-rt}.

As an example of an application of these notions we derive an expression for the present value of the future profits of a business. The total future profit of an economic enterprise becomes available over a period of time, perhaps a rather long period. Let us denote the profit per unit time at the time t by $p(t)$, and we assume p to be a continuous function of time. If we take the current time as $t = 0$, then the total profit from now to a specified time T is

$$\int_0^T p(t)\,dt. \tag{1.14}$$

However, we note that the *present* value of these profits is not given by (1.14), and to provide this information additional work must be done. We begin by observing that the profit earned in a time interval (t_j, t_{j+1}) is given by

$$\int_{t_j}^{t_{j+1}} p(t)\,dt.$$

Using the mean value theorem for integrals, we see that this can be written as

$$p(t_j')(t_{j+1} - t_j),$$

where $t'_j \in (t_j, t_{j+1})$. However, this amount of money becomes available during the time interval (t_j, t_{j+1}), and, if we think of this as a small interval, we may suppose that the total amount is available at time t_{j+1}. Thus, using our notion of present value, the present value of these profits is

$$p(t'_j)(t_{j+1} - t_j)e^{-rt_{j+1}}.$$

Again assuming that the interval (t_j, t_{j+1}) is small, we conclude by the continuity of p that $p(t'_j)$ is approximately equal to $p(t_{j+1})$. Therefore we have finally that the profits earned in the time interval (t_j, t_{j+1}) have an approximate present value of

$$p(t_{j+1})e^{-rt_{j+1}}(t_{j+1} - t_j).$$

Now, consider a time interval $(0, T)$ and suppose that it is subdivided into a large number of short subintervals. We apply the above approximation to each subinterval, and we conclude that the quantity

$$\sum p(t_{j+1})e^{-rt_{j+1}}(t_{j+1} - t_j)$$

is an approximation for the present value of the profits earned before time T. Finally, we note that for very short subintervals the above sum is approximately equal to

$$\int_0^T p(t)e^{-rt}\, dt.$$

As a special case, if we let T become large, then it is reasonable to define the present value of all profits to be earned in the future by

$$\int_0^\infty e^{-rt}p(t)\, dt.$$

Note that as the interest rate increases, this quantity decreases. For example, if the profit flow is constant, say at k dollars per year, then the current value of all future profits is k/r, although the total value (when earned) of all future profits is clearly infinite.

EXERCISES

1. Prove that if $N > 0$ and $|r| < 1$, then $A(1/2, k) > A(1, k)$ for $k = 1, 2, \ldots, K$.
2. Show that f defined by $f(x) = (1 + 1/x)^x$ is an increasing function of x.
3. How much money would you need to deposit in a bank at 6% interest in order to be able to withdraw $500 per month for 20 years if the entire principal is to be consumed in that time?
   (a) Under the assumption that interest is credited and withdrawals are made continuously.

(b) Interest credited continuously, withdrawals made monthly beginning with the last day of the first month after deposit.
(c) Interest and withdrawals quarterly beginning with the last day of the first quarter after deposit.

How much do you need to deposit monthly over a period of 30 years to accumulate the fund needed for a 20-year withdrawal?
(d) Interest credited continuously at 6%.
(e) Interest credited semiannually at 6%.

Note: For the amount needed, round the figure obtained in (a) to the nearest thousand dollars.

4. Consider a model for population growth based on axiom 1 and axiom 2″: A is a differentiable function and

$$\frac{dA}{dt}(t) = r(t)A(t), \qquad t \in [0, T),$$

where r is a continuous function.

Show that if $R(t) = \int_0^t r(\tau)\,d\tau$, then $A(t) = N \exp\{R(t)\}$.

5. A culture of bacteria initially contains N_0 individuals and if left undisturbed is observed to double in size in T days. Suppose that bacteria are removed from the culture at a uniform rate of R per day. Find the number of bacteria present as a function of time. Show that this function is qualitatively different for different ratios N_0/R. Find the critical value of N_0/R and graph the various types of solutions. Answer the same question if R individuals are extracted from the population at one time at the end of each day.

6. The population of a culture doubles in 60 days. In how many days will the population triple? Support your answer.

7. Most demographers agree that a very rapid increase in population began during the seventeenth century. World population data beginning at about this time are provided in Table 8–2. Using these data, discuss the variation of r (the growth rate) with time. Estimate as closely as possible the current value of r and justify your answer.

8. Using Table 8–2, estimate the date when the world population was 3 billion. Estimate when it will be 5 billion.

Table 8–2

Date	World population (in millions)
1650	545
1750	728
1800	906
1850	1171
1900	1608
1950	2509
1960	3010
1970	3611

9. Suppose that the land area of the earth is 57.2 million square miles and that half of this area is suitable for human habitation. Using Table 8–2, estimate when there will be one human being for each square foot of habitable land.

10. Suppose that the model of Sec. 8.1.1 is used to describe human population growth and that r is not a constant but decreases as the population increases. Discuss, qualitatively, the behavior of $A(1, k)$ as k increases if $r = r_0 - cA(1, k)$ where r_0 and c are constants. Use this model and the data from Table 8–2 for 1950 and 1960 to compute the size of the population for large k.

8.2 MODIFIED MODELS OF POPULATION GROWTH: COMPETITION

Observations confirm that the model given by (1.10) is adequate to describe the growth of many populations. In general, sufficient conditions for the applicability of the model include an environment containing unlimited resources—space and food, for example—and a population consisting either of units of a single type or of units of several types each of which is described adequately by (1.10). In this section we shall examine models for growth processes not adequately described by (1.10). There are several aspects of such processes which could be incorporated in modified models, but it is sufficient for our purposes to concentrate on the impact of competition. Other possibilities are indicated in the Exercises and Projects (Sec. 8.7).

We consider two forms of competition: first, competition for limited resources within a population of individuals of a single type, and second, competition for food and survival in a predator-prey interaction. With respect to the former, any model which purports to represent a real-world process must reflect the fact that in actuality the available resources are finite. Thus when one speaks of *unlimited resources*, what is meant is that there is an ample supply and the growth of the population is not inhibited due to shortages. However, for true pure growth processes, i.e., those described by (1.10) with $r > 0$, the model predicts that the population grows beyond all bounds. Therefore it must eventually encounter constraints to its continued growth due to inadequate support for its basic requirements. At such a point the model (1.10) ceases to describe the system and modifications are necessary.

8.2.1 Models with Discontinuous Growth Rates

One of the simplest models which can be introduced to include the restriction on growth due to finite resources is based on a discontinuous growth rate. The basic assumption is that the population grows (or declines) to a certain maximum sustainable size and is constant for larger times. It is easy to make this idea precise. The total resources are examined and a decision is made as to the size of the population which can be supported on these resources. For simplicity it is assumed that the resources can support a population of constant size M indefinitely. Thus it is assumed that the consumable resources are replenished at a regular rate. Other assumptions regarding the size of the maximum sustainable population can be handled by arguments similar to those that follow (Exercise 2). With these assumptions we can take as a mathematical model for population growth the following differential equation:

$$\frac{dA}{dt}(t) = \begin{cases} rA(t), & r > 0 \text{ for those values of time for which } A(t) < M, \\ 0, & \text{if } A(t) = M, \\ sA(t), & s < 0, \text{ if } A(t) > M. \end{cases} \tag{2.1}$$

In some situations it may be that $-r = s$, but this is not necessary. Thus, if time is measured from the initial time t_0 and the initial population size is N, then the graph of the function A appears as in Fig. 8–2(a), (b), or (c) according as $N < M$, $N = M$, or $N > M$, respectively. It can be shown, for example, that if $N < M$, then

$$A(t) = \begin{cases} Ne^{r(t-t_0)}, & t_0 \leq t \leq t_0 + \frac{1}{r}[\log M - \log N], \\ M, & t > t_0 + \frac{1}{r}[\log M - \log N]. \end{cases}$$

Similar analytic expressions can be obtained in the remaining case $N > M$ (Exercise 1).

Notice that the model defined by (2.1) predicts a population with a discontinuous growth rate. When $A(t)$ reaches the value M, then the growth rate changes abruptly from rM (or sM) to 0. This is reflected in the graphs of Fig. 8–2 by the corners at the points P and Q in figures 2(a) and (c).

8.2.2 Models with Continuous (Nonconstant) Growth Rates

While the model of Sec. 8.2.1 has the merit of removing the prediction of arbitrarily large populations, it is unrealistic in that it is based on an assumption of a discontinuous growth rate. We now turn to a closer examination of the processes which tend to inhibit the growth of a population with the intention of constructing a more appropriate model.

It is useful to begin with a somewhat imprecise discussion of one possible interpretation of the growth of a population in a finite environment. Suppose that the population is such that in the absence of constraints due to

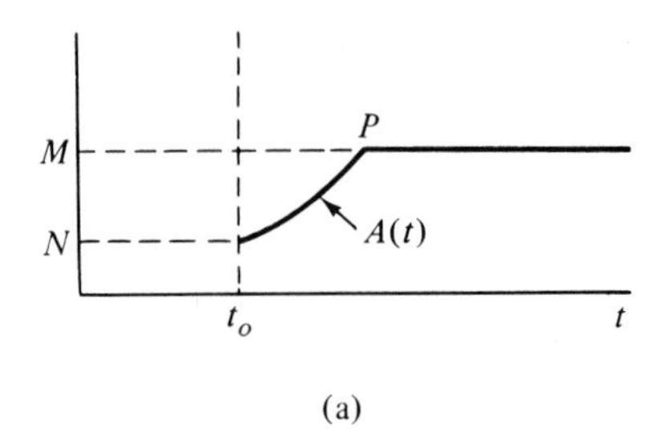

(a)

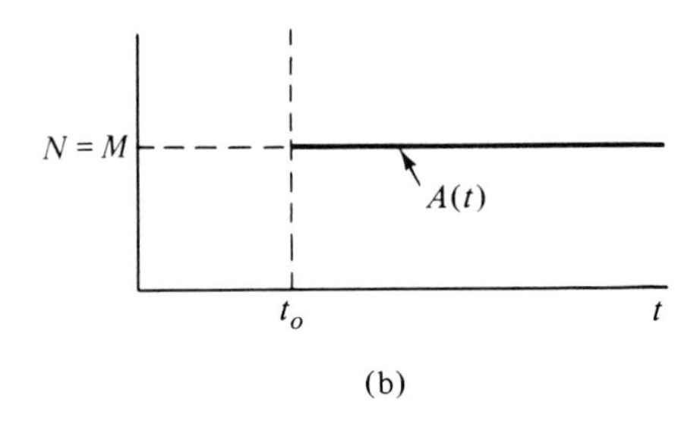

(b)

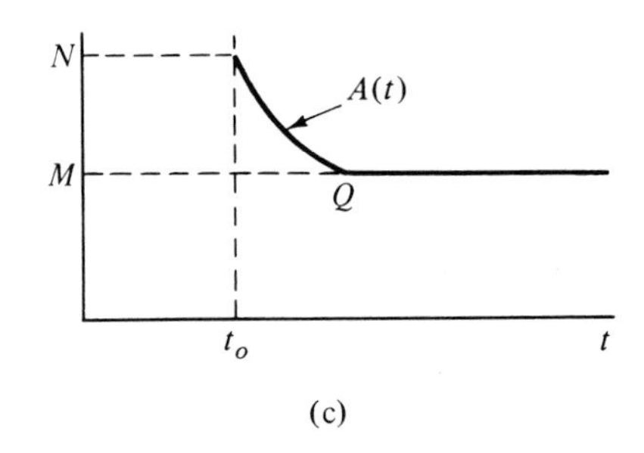

(c)

Fig. 8–2

the environment it would develop according to (1.10) with a growth rate $r > 0$. Also, suppose that this natural growth is affected by the environment in an additive manner. That is, assume that

$$\text{Total growth rate} = \text{Natural growth rate} + \text{Modification due to environment.}$$

The effect of a finite environment can be interpreted as follows. Suppose that at a specific time every individual in the population has sufficient resources to support its existence (all others having died off) and that an incoming member will compete with only one present member for its resources. If the incoming member succeeds in appropriating the resources of a current member, then the latter individual must compete with the remainder of the population, but only one at a time. If $A = A(t)$ denotes the number of individuals in the population at time t, then there are $A(A - 1)/2$ possible pairs to be engaged in competition for resources. It is reasonable to assume that this competition restricts the growth of the population, and consequently we are led to assume that the effect of competition for resources is a reduction in the growth rate by an amount proportional to $A(A - 1)$. If natural growth and competition of this form are the only two effects to be considered, then we have

$$\frac{dA}{dt} = rA - cA(A - 1),$$

$r > 0$ and $c > 0$, as the differential equation describing the growth process. It is convenient to write this in the form

$$\frac{dA}{dt} = kA - cA^2, \tag{2.2}$$

where $k = r - c$ depends both on the natural growth rate r and the competition parameter c.

We remark that this argument is based on the assumption that each member of the population has under its control a certain *package* of resources and that each incoming individual must either find an unused package or contend with exactly one present member for its package. This point of view has observational justification in the results of research on the domination of certain subterritories by members of an animal population which as a whole occupies a definite region or territory.

Equation (2.2) can also be supported by quite different arguments. For example, writing (2.2) as

$$\frac{dA}{dt} = (k - cA)A,$$

we can consider it as an example of an equation of the form

$$\frac{dA}{dt} = r(A)A; \tag{2.3}$$

that is, an equation of the form of (1.9), where the growth rate r is a function of the size of the population A. It is useful to consider (2.3) for a moment. If the population growth is measured from time t_0 and $A(t_0) = N$, then in order for the process to be a true growth one, it is necessary that $r(N) > 0$. In view of the heuristic arguments given above, it is to be expected that $r(A)$ decreases as A increases, perhaps approaching 0 as A approaches a maximum value. Obviously (2.2) is just one example, although in a sense the simplest, of equations of the form (2.3). If N is such that $k - cN > 0$, then the population increases, at least for small values of time. Since $k - cA = 0$ if $A = k/c$, it is reasonable to expect that the population cannot increase beyond this value. Thus, on the basis of these very general observations, we are led to certain predictions about the qualitative behavior of A. These predictions will be verified below.

Before proceeding to our consideration of (2.2), it should be noted that such an equation can be viewed as an analytic expression involving certain parameters, k and c in the case at hand. One can simply consider (2.2) as a given equation and attempt to determine those values of k and c which provide functions A which agree with experimental data. That is, if one has available the results of some experiments involving growth processes, then one could try to fit solutions of (2.2) to these data. Naturally, the values of the parameters c and k which provide the best fit to the data will depend on just how the notion of *fit* is defined. This is a basic question involving parameter estimation, and we shall discuss such questions in Chap. 10.

We turn now to a detailed investigation of the model defined by (2.2). Thus we consider the initial value problem

$$\begin{aligned} \frac{dA}{dt} &= kA - cA^2, \\ A(t_0) &= N, \end{aligned} \tag{2.4}$$

$k > 0$, $c > 0$. The algebraic signs of the constants k and c are taken in accordance with the heuristic discussion of competition given above. If k and c are simply viewed as parameters, then other possibilities arise. The details for the other cases are left to the reader (Exercise 3).

We define $M = k/c$ and rewrite the differential equation as

$$\frac{dA}{dt} = c(M - A)A.$$

This equation can be solved by writing it in the form

$$\frac{dA/dt}{A(M - A)} = c,$$

using a partial fraction decomposition on the left-hand side and integrating the resulting equation. If we suppose that $M - A(t) > 0$, for $t > t_0$, then proceeding as indicated we have

$$\frac{1}{M}\{\log A - \log(M - A)\} = ct + c_1$$

or

$$\frac{A}{M - A} = c_2 e^{cMt}.$$

Here c_1 and c_2 are constants to be determined by the initial conditions. After some elementary algebraic manipulations we obtain

$$A(t) = \frac{Mc_2}{c_2 + e^{-cMt}}.$$

Finally, using the initial condition $A(t_0) = N$, we find the size $A(t)$ of the population at time t to be

$$A(t) = \frac{MN}{N + (M - N)e^{-cM(t-t_0)}}, \qquad t \geq t_0. \tag{2.5}$$

We assumed that $M - A > 0$, and in particular that $M - N > 0$, in deriving (2.5). It can be verified by a calculation that *if* $N < M$, *then* the function A defined by (2.5) does indeed satisfy (2.4) and $N \leq A(t) < M$ for all $t \geq t_0$. In fact, the function (2.5) exhibits exactly the qualitative behavior which we expect. As time increases, the function A is increasing, and $A(t)$ tends to M as t increases indefinitely. Notice that M depends on the relative magnitudes of the parameters k and c but is independent of the initial size N of the population. A graph of (2.5) is given in Fig. 8–3. The S-shaped curve is one which appears frequently in experimental as well as theoretical studies and is known as the *logistic* or *saturation* curve. It was noted by the Belgian sociologist P. F. Verhurst in about 1840 in connection with the increase of human populations and was rediscovered in the 1920s by the American biologists R. Pearl and L. J. Reed. The results predicted by this model are in close agreement with observation in certain cases, for example, in studies of the growth of a population of fruit flies (drosophila) in a limited space. Many biologists feel that this model is of rather wide applicability, at least as a reasonable first approximation. It should be mentioned that as the population climbs close to its maximum value, its growth may be affected by extraneous factors, and as a result the graph of A may not approach the horizontal asymptote smoothly but instead may oscillate about it.

We note that in Fig. 8–3 the graph of A has an apparent inflection point. Denoting this point by t_I, it is easy to check that $A(t_I) = M/2$. Thus the rate of growth is increasing until the population reaches one half its

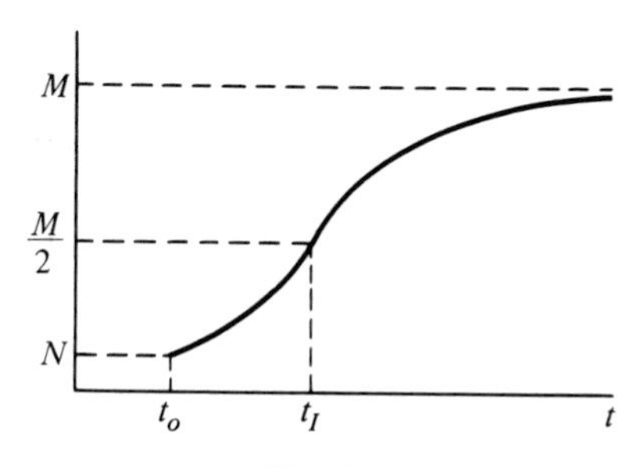

Fig. 8–3

maximum size, the rate of growth is a maximum when $A = M/2$, and the rate of growth decreases for larger values of A.

Early in the discussion of this model we assumed that $M - A(t) > 0$ for $t \geq t_0$, or, as we later proved to be equivalent, $M - N > 0$. Basically this assumption is that the initial population is smaller than the maximum sustainable one. However, the initial value problem (2.4) can be solved under the assumption that $k/c = M < N$. Real situations for which this is an appropriate model are relatively rare, and this case will not be investigated further here, however see Exercise 6.

8.2.3 Competition Involving Two Different Populations

The ideas and techniques developed in the study of the preceding models can also be applied in the formulation of mathematical models for competition involving two different populations. As an example we consider the following predator-prey or parasite-host system. The setting is an environment containing two different populations which interact with one another. There is a population K of predators (Killers) which consists of $A_K(t)$ individuals at time t. It is assumed that the existence of members of K depends in an essential way on another population. This latter population, V, consists of passive individuals (Victims) and contains $A_V(t)$ members at time t. For example, the population V might serve as the sole source of food for members of K. Let us suppose that the environment contains ample resources for the support of V and that in the absence of the population K the growth of V can be modeled by (1.10) with $r > 0$. Also, in view of our assumption regarding the dependence of K on V, we assume that in the absence of members of V, i.e., in isolation, the growth of K can be modeled by (1.10) with an appropriate $r < 0$. Thus the growth of either population in isolation is completely known. It remains to consider how the model (1.10) is affected by the predator-prey interaction. In qualitative terms the following behavior is to be expected. If at some time A_V is large relative to A_K, then a larger population K could be supported, and hence A_K tends to increase. However, as this happens, the number of individuals in V available for support per member of K declines, the growth of K slows, and ultimately A_K decreases. The decrease in A_K permits V to recover, and the process begins again. In other words, some sort of cyclic increase and decrease in both A_K and A_V is conjectured.

We proceed with a more precise discussion. As in Sec. 8.2.2 our basic assumption is that for both populations

$$\text{Total growth rate} = \text{Growth rate in isolation} \\ + \text{Modification due to interaction.}$$

In symbols this assumption becomes

$$\frac{dA_K}{dt} = -sA_K + \left(\frac{dA_K}{dt}\right)_{\text{due to interaction}},$$

where $s > 0$. The form of the first term on the right-hand side is dictated by the assumption regarding growth in isolation. There is a similar expression for dA_V/dt. It remains to produce a suitable candidate for the term $(dA_K/dt)_{\text{due to interaction}}$. If we think of the interaction as being one to one (for example, one wolf eats one rabbit), then interactions occur during contacts between pairs of members of the two populations. At time t there are $A_K(t)A_V(t)$ possible pairs of individuals, one from K and one from V. Thus it is reasonable to assume that the rate of change of A_K (or A_V) due to the predator-prey interaction is proportional to $A_K A_V$. Also, in the situation described here the effect of the interaction is to increase A_K and decrease A_V. Therefore a mathematical model for the growth of two populations in a predator-prey situation characterized by these assumptions is given by the differential equations

$$\begin{aligned} \frac{dA_V}{dt} &= rA_V - aA_V A_K, \\ \frac{dA_K}{dt} &= -sA_K + bA_V A_K, \end{aligned} \tag{2.6}$$

where r, s, a, and b are positive constants and where suitable initial values are given for A_V and A_K.

Although it is quite difficult to obtain an exact solution of this system for A_V and A_K as functions of time, it is possible to use approximation techniques to deduce some information on the qualitative relation between A_V and A_K. The idea is to look at the curves in the A_V-A_K plane parametrized with respect to t and defined by (2.6). We rewrite the system (2.6) as

$$\begin{aligned} \frac{dA_V}{dt} &= A_V(r - aA_K), \\ \frac{dA_K}{dt} &= A_K(bA_V - s). \end{aligned} \tag{2.7}$$

Notice that if A_V and A_K are constant functions, $A_V = s/b$ and $A_K = r/a$, then (2.7) is satisfied. This indicates that in the A_V-A_K plane the point $(s/b, r/a)$ is a stationary one for the system (2.7), and it is natural to ask what happens for small deviations from this stationary situation. To this end we introduce new coordinates X and Y by making the linear change of variables $X = bA_V - s$, $Y = aA_K - r$. Thus, if X is sufficiently small, then A_V is close to s/b, and if Y is sufficiently small, A_K is close to r/a. In terms of the new variables the system (2.7) is

$$\begin{aligned} \frac{dX}{dt} &= -(X+s)Y, \\ \frac{dY}{dt} &= (Y+r)X. \end{aligned} \tag{2.8}$$

If X and Y are both quite small, then this system is approximately

$$\frac{dX}{dt} = -sY,$$

$$\frac{dY}{dt} = rX,$$

where the XY terms in (2.8) have been neglected in comparison with sY and rX. The functions

$$X(t) = C \cos (rs)^{1/2}t$$

and

$$Y(t) = C\sqrt{\frac{r}{s}} \sin (rs)^{1/2}t$$

are solutions of the latter system for every value of the constant C. These functions define a family of confocal ellipses in the X-Y plane, one ellipse for each value of the constant C. Therefore one expects that the solutions of the system (2.8) have graphs which look much like ellipses near $X = 0$, $Y = 0$.

If we multiply the first equation of system (2.8) by $X(Y + r)$, the second by $Y(X + s)$, and add, then we obtain

$$X(Y+r)\frac{dX}{dt} + Y(X+s)\frac{dY}{dt} = 0.$$

If we restrict ourselves to the case that both populations are actually present, that is, $A_V > 0$ and $A_K > 0$, then $X + s > 0$ and $Y + r > 0$. Geometrically this is a restriction in the X-Y plane to values of X and Y in a quarter plane (Fig. 8–4). Dividing the above equation by $(Y + r)(X + s)$, we have

$$\frac{X}{X+s}\frac{dX}{dt} + \frac{Y}{Y+r}\frac{dY}{dt} = 0,$$

or, equivalently,

$$\frac{dX}{dt} - \frac{s}{X+s}\frac{dX}{dt} + \frac{dY}{dt} - \frac{r}{Y+r}\frac{dY}{dt} = 0.$$

The last equation can be integrated directly, and we find that

$$X - s\log(X + s) + Y - r\log(Y + r) = C_1 \tag{2.9}$$

or

$$\frac{e^{X+Y}}{(X + s)^s(Y + r)^r} = e^{C_1},$$

where C_1 is a constant to be determined by the initial conditions. The locus in the X-Y plane of the points which satisfy (2.9) near $X = 0$, $Y = 0$ is a family of closed curves, one curve of the family corresponding to each value of C_1. For values of C_1 close to $s^{-s}r^{-r}$ these curves look very much like ellipses (Fig. 8–4). The arrows indicate the directions traversed on the curves with increasing time in each case.

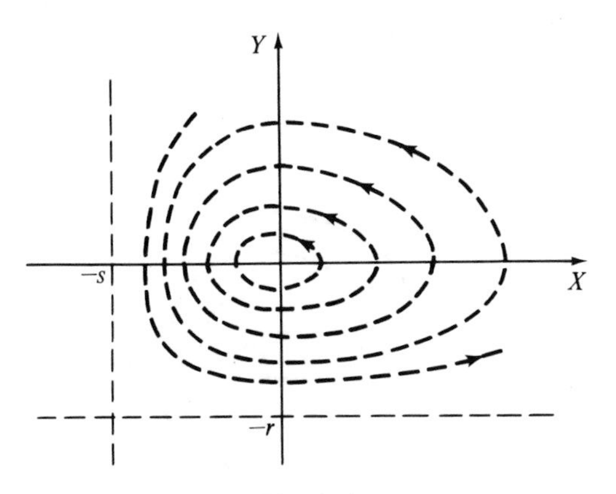

Fig. 8–4

Clearly, we have also deduced information about the relation between A_V and A_K. The change of variables from the X-Y plane to the A_V-A_K plane is a linear one, and consequently the curves in the X-Y plane transform to similar curves in the A_V-A_K plane. Our conclusion is that for A_V near s/b and A_K near r/a the curves defined by (2.7) are approximately ellipses. Moreover, if the initial values of A_V and A_K satisfy $A_V > 0$, $A_K > 0$, then the curves defined by (2.7) are closed curves in the A_V-A_K plane. Thus in any case the growth of each population can be described as that of regular increase and decrease. It follows from (2.7) that if A_V is ever 0, then it is 0 for all larger values of time. This observation together with the fact that for any set of initial conditions there is exactly one solution of (2.7) (a true statement but not proved here) shows that if A_V is ever positive, then it is always positive. Thus this model predicts that the prey population can never be completely extinguished by the predators.

It is to be emphasized that this analysis holds only for the model governed by (2.7). A more general model in which the interaction is no longer of the product form may lead to quite different results. For example, a model leading to a system of differential equations of the form

$$\frac{dA_V}{dt} = rA_V + f(A_V, A_K),$$

$$\frac{dA_K}{dt} = -sA_K + g(A_V, A_K),$$

even with f and g polynomial functions, may give different predictions. For instance, such a system could lead to a prediction that the populations each tend to a stable size. The graph of the solution of the system in the A_V-A_K plane might appear as in Fig. 8–5. In this example the size of the prey population oscillates about P_V, and the size of the predator population oscillates about P_K as time increases. However, for large times t the quantity $A_V(t)$ is close to P_V and $A_K(t)$ is close to P_K.

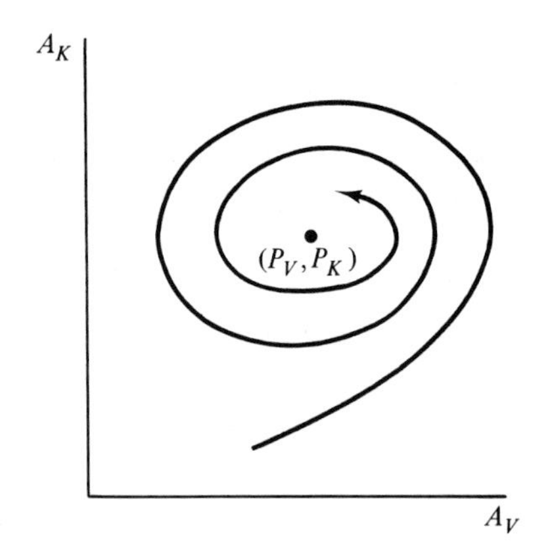

Fig. 8–5

EXERCISES

1. Derive the expression for $A(t)$ given by (2.1) in the case $N < M$. Obtain a similar expression which is valid for $N > M$.

2. Consider the following situation which is similar but slightly more general than that of Sec. 8.2.1. Suppose that at time t there are resources adequate to support a population of size Mt, $t \geq 1$. Formulate a discontinuous growth rate model for this situation. If $A(1) = N < M$, find $A(t)$ for all $t \geq 1$.

3. Investigate the solution of the initial value problem (2.4) in the case $k > 0$, $c < 0$. First give qualitative observations based only on the differential equation, and then solve the initial value problem.

4. Use the techniques of the calculus to find the point of inflection of the graph of the function A defined by (2.5). If this point has coordinates (t_I, A_I), verify that $A_I = M/2$.

5. Using the model (2.4) as a model for human population growth and the data from Table 8–2 for 1900 and 1950, estimate the maximum size of the world population. When will the population be 90% of this maximum value?

6. Solve the initial value problem (2.4) in the case $M < N$. Graph the solutions for $N = \frac{1}{4}M$ and $N = \frac{7}{4}M$ on the same coordinate system.

7. Suppose that X and Y are solutions of (2.8) which are related as in Fig. 8–4. Graph $X(t)$ and $Y(t)$ as functions of t on the same coordinate system.

8. The fact that the solutions of (2.8) which pass through points very close to $X = 0$, $Y = 0$ are approximately ellipses can also be deduced from the form of the solution, Eq. (2.9).
(a) Expand $\log(1 + \rho)$ in a Taylor series about $\rho = 0$.
(b) If $|X/s| < 1$ and $|Y/r| < 1$, expand the solution (2.9) of (2.8) in an infinite series.
(c) Show that for X and Y sufficiently small the loci (2.9) are approximately ellipses, one ellipse for each value of C_1. What are the semiaxes? How do these ellipses compare in shape with those arrived at in the text?

8.3 DECAY OF RADIOACTIVE SUBSTANCES

If the growth constant r in Eq. (1.10) is negative, then, as we noted above, A decreases with increasing time and tends to 0 as t increases beyond all bounds. One of the most important of the numerous processes with this observed behavior is the decay of radioactive nuclides. The fact that the differential equation (1.10) describes the behavior of a radioactive sample was deduced from experimental evidence by E. Rutherford and F. Soddy in the early 1900s. In this case (1.10) is to be interpreted as giving an expected value for the disintegration rate of a pure radioactive sample at an instant when A nuclides remain untransformed. It is convenient in this discussion to denote the decay constant by k instead of r. This constant is determined by experiments.

The phenomenon to be considered here is basically the spontaneous dissociation of one nuclide into another together with the emission of radiation. More precisely, it has been observed that certain nuclides are inherently unstable and that after a definite but unpredictable lapse of time after formation, but without external stimulation, they undergo a transition resulting in the emission of some sort of radiation and the formation of another nuclide. Because of the unpredictability of the dissociation, the process is inherently a probabilistic one and one would expect a stochastic model to be most appropriate. However, as with our consideration of population growth, if the sample of radioactive material contains a large number of nuclides and if the time period to be considered is not too short, then the mathematical model given by (1.10) may be an adequate approximation. Here, a *large* sample would be, for example, a sample suitable for microscopic work in a laboratory, and *long* time periods are those which are comparable to the *half-life* of the nuclide (see below).

8.3.1 Simple Decay of a Single Nuclide

A given nuclide may experience a transition a microsecond or a billion years after formation, but the presence of a measurable delay is essential for the process to be included in our discussion. The delay may be expressed in quantitative terms by giving the decay constant or the half-life of the nuclide. This important quantity is defined as the time during which, on the average, one half of a given sample of atoms will have decayed and one half will not. Note that if we know either the decay constant or the half-life, we can compute the other. Indeed,

$$A(t) = Ne^{-kt}, \qquad A(0) = N,$$

so that if T is the half-life, we have

$$\frac{N}{2} = Ne^{-kT} \qquad \text{or} \qquad kT = \log 2.$$

Remember that the differential equation is used as an approximation to describe the aggregate behavior of a large number of atoms, and therefore we should not expect to be able to deduce from this model exactly when a specific atom will decay. Indeed, it is impossible to obtain such detailed information. However, one can determine the *mean lifetime* of a specific type of nuclide.

Suppose that at the time our investigation begins, say at time $t = 0$, there are N nuclides present. Since there are only a finite number of atoms and eventually all will decay, we can list the transition times $t_1, t_2, \ldots, t_n$. Suppose that these times are listed in increasing order. Simultaneous transitions are possible and we suppose that a_j nuclides decay at time t_j, $\sum_{j=1}^{n} a_j = N$. With these conventions the mean length of time before transition, the mean lifetime (MLT), of a nuclide is

$$\frac{1}{N}\sum_{j=1}^{n} t_j a_j.$$

Let A_j denote the number of undecayed nuclides present immediately prior to time t_j, $j = 1, 2, \ldots, n$, $A_{n+1} = 0$. Then $a_j = A_j - A_{j+1}, j = 1, 2, \ldots, n$, and we have

$$\text{Mean lifetime (MLT)} = \frac{1}{N}\sum_{j=1}^{n} t_j(A_j - A_{j+1}).$$

It is useful to rewrite this last sum as

$$\frac{1}{N}\left[\sum_{j=1}^{n} t_j A_j - \sum_{j=1}^{n} t_j A_{j+1}\right] = \frac{1}{N}\left[\sum_{j=1}^{n} t_j A_j - \sum_{j=2}^{n} t_{j-1} A_j\right]$$

$$= \frac{1}{N}\left[t_1 A_1 + \sum_{j=2}^{n} A_j(t_j - t_{j-1})\right].$$

We use the last expression for the MLT to justify the assertion MLT $= 1/k$. Indeed, the sum

$$\frac{1}{N}\sum_{j=2}^{n} A_j(t_j - t_{j-1})$$

looks very much like a Riemann sum for the integral $(1/N)\int_0^{t_n} A(t)\,dt$, and we exploit this similarity. If we set $t_0 = 0$, then

$$MLT = \frac{1}{N}\sum_{j=1}^{n} A_j(t_j - t_{j-1}).$$

Using (1.10) to replace A_j by Ne^{-kt_j}, we obtain

$$MLT = \sum_{j=1}^{n} e^{-kt_j}(t_j - t_{j-1}).$$

The replacement of A_j by Ne^{-kt_j} is, of course, an approximation, and the reader is invited to consider its implications (Exercise 1).

Now, suppose that the number of nuclides increases indefinitely. It seems reasonable to assume, and the assumption is supported by observation, that the time t_n of the last transition increases indefinitely and that on any bounded time interval $[0, T]$ the time interval $t_{j+1} - t_j$ between successive transitions tends to 0. This is sufficient to show that as the number of nuclides increases

$$MLT \longrightarrow \int_0^{\infty} e^{-kt}\,dt$$

(Exercise 2). Since this integral can be evaluated explicitly, we conclude that its value, $1/k$, is a reasonable approximation to the mean lifetime for a nuclide whose decay constant is k.

We remark that among the more than 800 known radioactive nuclides no two have exactly the same decay constants. Thus a radioactive substance can be identified simply by making a sufficiently accurate measurement of its decay constant, or, equivalently, its half-life or mean lifetime. The range of variation of the decay constants of naturally occurring radioactive species is indicated by the values 1.6×10^{-18} sec^{-1} of thorium (Th^{232}) and 2.3×10^{6} sec^{-1} of polonium (Po^{212}), a range of about 10^{24} sec^{-1}.

8.3.2 Application to Radiocarbon Dating

The precise quantitative results obtainable from this and more refined models of radioactive decay have been utilized in a variety of applications. We shall give a brief discussion here of the technique for determining the age of archeological specimens by using the physical properties of an isotope of carbon, so-called radiocarbon dating. This technique was developed by the U.S. chemist W. F. Libby in the late 1940s, and he received the 1960 Nobel Prize in Chemistry for his work. The method has been applied successfully for many years and is now considered a standard tool in archeological research. Newer methods involving potassium and argon have the same theoretical basis, and techniques with a different but related foundation involving uranium have also been developed. However, it should be remembered that these methods are based on certain assumptions, and if subsequent investigations cast doubt on the assumptions, then the conclusions must also be questioned.

Carbon 14, written C^{14}, is one of three naturally occurring isotopes of the element carbon. The other two are C^{13} and C^{12}. The latter two are stable, but C^{14} is radioactive, decaying into nitrogen 14 with an emission of radiation. The half-life of C^{14} is 5570 ± 40 years. Using the model developed above and this known constant, we can compute the age of an object isolated at a specific but unknown time from a knowledge of the ratio of the number of C^{14} atoms currently present to those initially present. Age determinations with this theoretical basis have been successively carried out on materials up to 70,000 years old.

Let us consider the situation in more detail. The production of C^{14} atoms has, according to current theories, proceeded at a nearly constant rate for many thousands of years. Since the rate of decay is proportional to the number of C^{14} atoms in existence, eventually the two processes reached an equilibrium state. Carbon 14 is created in the atmosphere as a result of cosmic-ray bombardment of nitrogen. Now, since the time for carbon atoms to "mix" into the oceans and into living plants and animals is short compared to the half-life, there is essentially a uniform distribution of C^{14} atoms in carbon-bearing material on the surface of the earth. We refer to that portion of the earth having access to this carbon containing a uniform (in time and space) fraction of C^{14} atoms as the *steady-state environment*. Suppose now that an object which contains carbon has been in contact with this natural source of C^{14} atoms for a sufficient period of time for its carbon content to have the same ratio of C^{14} atoms as the steady-state environment. Upon isolation from this environment the number of C^{14} atoms present in the object begins to decrease at a rate determined by the decay constant. Thus, if the concentration of C^{14} in plant remains of an unknown age is found to be just half of the concentration in a plant living today, then we would assume that the plant was isolated from the steady-state environment, buried in the mud, for example, about 5570 years ago.

The amount of C^{14} in living plants is too small, approximately one atom in 10^{12}, to be measured by direct methods. However, since the rate of emission of radiation is proportional to the number of radioactive atoms, we can use the amount of radioactivity to deduce the concentration of C^{14} atoms.

To obtain quantitative results, let U_0 denote the number of C^{14} atoms initially present in a sample of organic material whose age is to be determined. To be more exact, we shall determine the date that it was isolated from the steady-state environment. Let U denote the number of C^{14} atoms present now in the sample of unknown age. By our assumptions, a sample of carbon of the same size extracted from a plant living at the present time contains U_0 carbon 14 atoms. Thus the current rates of decay of the contemporary sample and the sample of unknown age are, respectively,

$$r = -kU_0,$$
$$s = -kU = -kU_0 e^{-kT},$$

where T is the (unknown) age of the sample. From these equations we obtain

$$T = \frac{1}{k} \log \frac{r}{s} = \frac{5570}{\log 2} \log \frac{r}{s} \text{ years.}$$

This relation provides an expression for the age of the sample in terms of quantities observable today, namely rates of radioactive decay. Since the rate of decay is very small, on the order of 13 disintegrations per minute per gram of contemporary carbon, and since the emissions are quite weak, the laboratory work involved in making the required measurements is very exacting.

8.3.3 Radioactive Series Decay

Radioactive decay of a sequential sort is frequently observed in nature. A radioactive nuclide of type A decays into a nuclide of type B, which is also radioactive and which decays into a nuclide of type C, which we assume to be stable. Processes of this type are known to continue for as many as ten steps, but for purposes of illustration three will be sufficient.

Suppose that the decay constants are k_A for transitions from nuclides of type A to type B and k_B for transitions from nuclides of type B to type C, $k_A \neq k_B$. Also, suppose that our process begins at time $t = 0$ and that there are initially A_0 atoms of type A and none of types B and C. If we abuse the notation somewhat and let $A = A(t)$, $B = B(t)$, and $C = C(t)$ denote the number of atoms of type A, B, and C, respectively, which are present at time

t, then the equations which represent the situation mathematically are

$$\frac{dA}{dt} = -k_A A,$$
$$\frac{dB}{dt} = k_A A - k_B B, \tag{3.1}$$
$$\frac{dC}{dt} = k_B B.$$

The first equation of this system can be solved immediately, and making use of the given initial data, we have

$$A(t) = A_0 e^{-k_A t}. \tag{3.2}$$

The second equation of (3.1) therefore becomes

$$\frac{dB}{dt} = A_0 k_A e^{-k_A t} - k_B B,$$

or

$$\frac{dB}{dt} + k_B B = A_0 k_A e^{-k_A t}.$$

This is a standard first-order equation whose solution is easily obtained (see Appendix B). We have

$$B(t) = \frac{A_0 k_A}{k_B - k_A}[e^{-k_A t} - e^{-k_B t}]. \tag{3.3}$$

With (3.3), the last equation of (3.1) can be used to obtain C by a straightforward integration:

$$C(t) = \frac{A_0 k_A k_B}{k_B - k_A}\left[\frac{e^{-k_B t}}{k_B} - \frac{e^{-k_A t}}{k_A}\right] A_0.$$

One can obtain qualitative information on the behavior of the activities of the different types of nuclides even without using the explicit solutions of (3.1). Let us define the *activity* of a nuclide to be its rate of decay, i.e., the decay constant times the number of nuclides. Then activities of the nuclides of types A and B are Ak_A and Bk_B, respectively. The initial activity of nuclides of type A is $A_0 k_A$, and since $d(k_A A)/dt = -k_A^2 A$, the rate of change of the activity is negative and the activity decreases with increasing time. Since $A(t) \longrightarrow 0$ as time increases, the activity approaches 0 as time increases indefinitely. Also, the activity of nuclides of type B is initially 0, and using the second equation of (3.1) we have

$$\frac{d(k_B B)}{dt} = k_A k_B A - k_B^2 B,$$

so that

$$\frac{d(k_B B)}{dt}(0) = k_A k_B A_0,$$

which is positive. Finally, $k_B B$ approaches 0 as time increases indefinitely since eventually all atoms of types A and B will decay. Thus, $k_B B$ must have a maximum for some time $t_{\max}$, and the second equation of (3.1) gives $t_{\max}$ as a solution of

$$k_A A(t) = k_B B(t).$$

Using (3.2) and (3.3), one obtains

$$t_{\max} = \frac{\log (k_B/k_A)}{k_B - k_A}.$$

One can verify that the relation between the activities is as shown in Fig. 8–6.

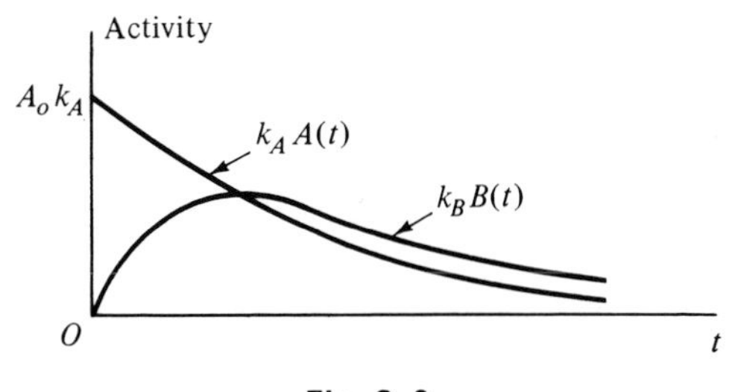

Fig. 8–6

Under certain conditions, one finds in radioactive series a phenomenon known as *long-term* or *secular equilibrium*. This arises when one element decays quite slowly and its successor very rapidly. Suppose, for example, that k_A is small and that k_B is large in comparison. This is the situation for the pair radium-radon, where the half-life of radium is 1620 years and that of radon about 4 days. As a first approximation we may therefore assume that the number A of atoms of that type is constant. With this assumption, we obtain

$$\frac{dB/dt}{k_A A - k_B B} = 1$$

from the second equation in (3.1). Hence

$$-\frac{1}{k_B}\log(k_A A - k_B B) = t + c_1$$

or

$$c_2(k_A A - k_B B) = e^{-k_B t}.$$

The constant of integration c_2 is to be obtained from the initial conditions: $B = 0$ at $t = 0$. Thus, $c_2 = 1/(Ak_A)$, and the number B of atoms of that type at time t is given by

$$B(t) = \left(\frac{k_A}{k_B}\right)A(1 - e^{-k_B t}).$$

This expression leads to the desired result. We observe that for $k_B t$ large we have $B(t) \cong (k_A/k_B)A(t)$; that is, the number of atoms of type B is approximately a fixed fraction of the number of atoms of type A. Remember that the type A atoms were assumed to be relatively inert. Looking back at (3.1), this means that for such times $dB/dt \cong 0$, and hence B is approximately constant. This allows us to compute the decay constant in cases where the half-life is very small and consequently difficult to measure. Indeed, after equilibrium has been established, k_B can be determined from the ratio B/A and the known value of k_A.

EXERCISES

The logarithms appearing in this section are to the base e.

1. Discuss the errors introduced by making the approximation $A_j \cong Ne^{-kt_j}$ in deriving the estimate MLT $= 1/k$.

2. Under the assumptions given in the text on the behavior of the sequence of transition times $\{t_j\}_{j=1}^{n}$ as the number of nuclides increases, viz., as N becomes arbitrarily large

   (a) $t_n \longrightarrow \infty$, and
   (b) For every fixed T $\max\limits_{t_j \leq T} |t_j - t_{j-1}| \longrightarrow 0,$

   show that $\sum\limits_{j=1}^{n} e^{-kt_j}(t_j - t_{j-1}) \longrightarrow \int_0^\infty e^{-kt}\,dt.$

3. (a) Given that radium has a half-life of 1620 years, find the decay constant in dimensions of sec^{-1}.
   (b) Given that thorium 232 has a decay constant of $1.6 \times 10^{-18}\ \text{sec}^{-1}$, find its half-life in convenient units.

4. An ancient refuse deposit was excavated at a site known as Fell's Cave near the Strait of Magellan in 1937 by Julius B. Bird of the American Museum of Natural History. In the 1950s, carbon 14 dating methods were applied to the samples. Emissions from C^{14} disintegrations were found to be occurring at a rate of $62 \pm 2\%$ of that of a contemporary sample of the same size. Taking the

half-life of C^{14} as 5570 ± 40 years, find the approximate age of the deposit and estimate your possible errors.

5. A baby tooth which came out on August 31, 1960, was found to contain strontium 90, whose half-life is 29 years. At the time of loss the tooth was examined and the strontium was found to be decaying at the rate of 263 atoms per minute. What was the decay rate on May 31, 1969?

6. Ten percent of a radioactive substance disintegrates in 100 years. What is its half-life?

7. Use your knowledge of radioactive decay and the following simplified situation with fictitious data to compute the age of the earth. Suppose that during some ancient cataclysm (perhaps the formation of the earth) mineral bodies of some uncontaminated radioactive element were formed at various positions throughout the earth. Suppose that one of these bodies originally consisted solely of uranium with a half-life of 4.56×10^9 years. Today this body is being mined and is found to be contaminated with lead, the ratio of uranium atoms to lead atoms being 0.9. Assume that lead is the stable end product of uranium disintegrations and find the date of the cataclysm.

8. A gram of carbon extracted from a recent cadaver has a decay rate of 18,700 atoms per day. A gram of carbon extracted from an Egyptian mummy has a decay rate of 8500 atoms per day. How old is the mummy?

9. By constructing appropriate graphs, discuss the qualitative behavior of the functions A and B which satisfy the system (3.1). In particular, compare the graphs of these functions under the following conditions: (a) $k_A > k_B$, (b) $k_A < k_B < 2k_A$, and (c) $2k_A \leq k_B$. You should construct your graphs using only the differential equations (3.1) and not their explicit solutions.

10. Same as Exercise 9 for the activities $k_A A(t)$ and $k_B B(t)$.

11. Extend the discussion of Sec. 8.3.3 to series with three unstable elements:

$$A \xrightarrow{k_A} B \xrightarrow{k_B} C \xrightarrow{k_C} D.$$

Here we assume that D is stable. Graph the activities for this case as in Fig. 8–6. Include information on as much of the qualitative behavior as you can.

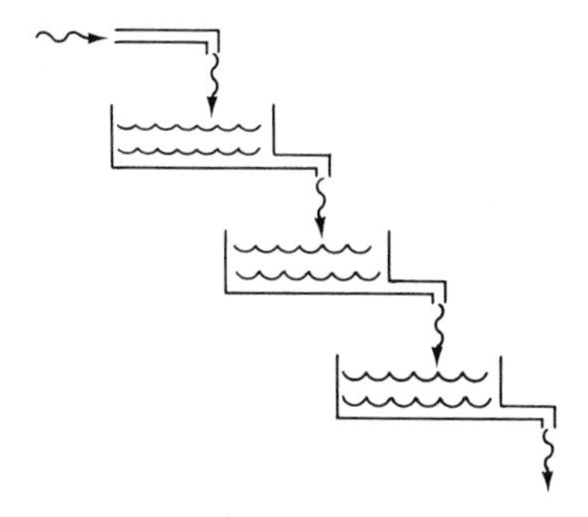

Fig. 8–7

12. Using models similar to those developed here, discuss the flow of water in a system of tanks. Assume that in a system represented schematically in Fig. 8–7 water flows into the top tank with a certain rate, from the top tank into the second with a certain (possibly different) rate, and on downward until it flows out of the last tank.

PART B STOCHASTIC MODELS

In the preceding sections of this chapter we have studied growth and decay processes from the point of view of a deterministic system. Such an analysis is usually satisfactory for the study of a reasonably large sample. The growth of a population of bacteria in a laboratory culture or the decay of a radioactive sample of laboratory size can be adequately accounted for from this point of view. Also, we could obtain more refined models of this basic type. For example, in the case of population growth, the model could be modified by differentiating between the growth constants for various age and sex groups—or, in the case of human populations, for different socioeconomic groups. Such modifications in the model usually produce more reliable conclusions. However, still referring to the study of population growth, a deterministic approach is basically incompatible with a model intending to predict changes in a small population. When one is concerned with relatively few individuals, then the growth of the population through time may be strongly influenced by chance events. If the model is to be useful in connection with the explanation and prediction of observable phenomena, then these chance events cannot be ignored, and we are led naturally to consider stochastic models. Such a model must consider the various events which can occur and incorporate them in a quantitative way into the model. The predictions provided by the two models are intrinsically different. Whereas the deterministic model provides a function giving the size of the population for any specified time, the stochastic model gives a probability distribution of population sizes for each time. Thus our goal is to construct a family of probability distributions, one for each instant in time in which we are interested.

8.4 BASIC PRINCIPLES

The process to be discussed in detail here is known as a *pure birth process*, and it is probably the simplest of the stochastic growth models. It is a special case of the more general birth-death processes which are conceptually and mathematically more difficult. Birth-death processes arise naturally in the models of epidemics and queues considered in the following chapter, and we defer their consideration until then. The purpose of the present section is to formulate carefully the basic assumptions underlying our study and to develop a useful mathematical model of the process. The study of the model will be continued in subsequent sections, and the results obtained will be compared with those based on the analogous deterministic model. Although the discussion is phrased in terms of population growth, as usual the reader should appreciate that the methods are not restricted to such problems.

To begin our discussion, it is useful to examine somewhat more closely the deterministic model for a simple growth process. Recall that the basic assumption was one involving the rate of growth. Specifically, the rate of growth of a population of A members was assumed to be proportional to the size of the population:

$$\frac{dA}{dt} = kA. \tag{4.1}$$

The constant k is positive for growth processes. Another way of expressing the same idea is to assume that in a small time interval, say Δt, the increment in the size of the population, call it ΔA, is given by

$$\Delta A = kA\,\Delta t + h(\Delta t), \tag{4.2}$$

where $h = h(\Delta t)$ is a function of Δt which satisfies $\lim_{\Delta t \to 0} [h(\Delta t)/\Delta t] = 0$. The equivalence of (4.1) and (4.2) is simply a consequence of the definition of the derivative.

We base our stochastic model on assumptions similar to (4.2), but with regard to probabilities rather than sizes of populations. We make three basic assumptions. They have to do with the probability of a single member of the population reproducing in a specific time interval of length Δt. Intervals Δt of interest to us are small. Heuristically, our assumptions are, first, that the probability of a single individual reproducing in a short time interval depends only on the length of the time interval—in fact, is proportional to this length—and not on the time at which the interval begins; second, that the probability of reproducing more than once is small; and third, that these probabilities do not depend on what happens in any other (disjoint) time intervals. Before making these assumptions precise, it is convenient to introduce a very useful notation for smallness. The idea is a simple one and the notational savings are significant. Let f be a function defined on $(0, \delta)$ for some positive δ. If $f(x) \longrightarrow 0$ as $x \longrightarrow 0$, we write $f(x) = o(1)$, $x \longrightarrow 0$. If $f(x)/x \longrightarrow 0$ as $x \longrightarrow 0$, we write $f(x) = o(x)$, $x \longrightarrow 0$; and, in general, if m is a positive real number and $f(x)/x^m \longrightarrow 0$ as $x \longrightarrow 0$, we write $f(x) = o(x^m)$, $x \longrightarrow 0$. This is a special instance of a rather general notation, but it is adequate for our purposes.

We now state the assumptions for this model in a formal way:

Assumptions

1. The probability that an individual reproduces exactly once in a time interval of length Δt is equal to $k\,\Delta t + h_1(\Delta t)$, and $h_1(\Delta t) = o(\Delta t)$, $\Delta t \to 0$.
2. The probability that an individual reproduces more than once in an interval of length Δt is equal to $h_2(\Delta t)$, and $h_2(\Delta t) = o(\Delta t)$, $\Delta t \to 0$.
3. The reproduction (or nonreproduction) of an individual in a time interval I_1, is independent of its reproduction (or nonreproduction) in any time interval I_2, $I_1 \cap I_2 = \emptyset$.

In this discussion we shall be dealing with populations consisting of an integral number of individuals, and consequently it is appropriate to use the symbol n to represent the size of the population. As the first step in the mathematical development of this model we compute the probability of the event $Q(n; \Delta t)$: The size of the population increases from n to $n + 1$ in the time interval $(t, t + \Delta t)$. Since we are dealing with a pure birth process, this increase can occur in essentially only one way. One of the n members of the population reproduces exactly once, and the other $n - 1$ members do not reproduce at all. There are n possibilities for the individual who reproduces, and consequently the desired probability is given by

$$\Pr[Q(n; \Delta t)] = n(k\,\Delta t + h_1(\Delta t))(1 - (k\,\Delta t + h_1(\Delta t)) - h_2(\Delta t))^{n-1}.$$

This formula can be simplified by using the binomial expansion on the last factor and then collecting all terms which are small in comparison to Δt. Proceeding in this way, we obtain

$$\Pr[Q(n; \Delta t)] = nk\,\Delta t + H(\Delta t),$$

where $H(\Delta t) = o(\Delta t)$, $\Delta t \to 0$. This result, which is based on the assumptions concerning the nature of the process, can now be used to derive the relevant distributions.

For a fixed time t, consider a sample space consisting of the set of all populations and define a random variable n on this sample space by assigning to each population its size. We are interested in determining the probability distribution associated with this random variable. Thus we let $P_n(t)$ denote the probability that the population consists of exactly n individuals at time t, and we look for equations which can be used to determine $P_n(t)$. An important observation is that the distribution $\{P_n(t)\}_n$ (t fixed) can be determined by investigating how the individual functions P_n change with time. We begin by considering $P_n(t + \Delta t)$, that is, the probability that the population consists of exactly n individuals at time $t + \Delta t$. The quantity $P_n(t + \Delta t)$ can be expressed in terms of $P_n(t)$, $P_{n-1}(t)$, and other terms which are small in comparison with Δt. Indeed, this probability is essentially the sum of two terms, each of which is the product of two probabilities. The first term is the product of the probability that the population consists of n members at time t and the probability that none of these n individuals reproduce in the interval $(t, t + \Delta t)$. The second term is the product of the probability that the population consists of $n - 1$ members at time t and the probability that there is exactly one individual added to the population in the interval $(t, t + \Delta t)$. The other terms are due to the possibility that a population of n members at time $t + \Delta t$ can arise by beginning with fewer than $n - 1$ individuals at time t and adding two or more individuals during the interval $(t, t + \Delta t)$. Writing this out in symbols and simplifying by collecting all terms which are small in comparison with Δt (see the Project of Sec. 2.7.1), we have

$$P_n(t + \Delta t) = P_n(t)(1 - kn\,\Delta t) + P_{n-1}(t)k(n - 1)\,\Delta t + H_1(\Delta t)$$

with $H_1(\Delta t) = o(\Delta t)$, $\Delta t \to 0$. This equation can be rewritten as

$$\frac{P_n(t + \Delta t) - P_n(t)}{\Delta t} = k(n - 1)P_{n-1}(t) - knP_n(t) + \frac{H_1(\Delta t)}{\Delta t},$$

and letting Δt approach 0, we obtain a differential equation for $P_n(t)$:

$$\frac{dP_n(t)}{dt} = k[(n - 1)P_{n-1}(t) - nP_n(t)]. \tag{4.3}$$

If we suppose that the population initially contains N individuals, then this equation holds for values of n larger than N. An equation for P_N can be derived by using an argument similar to the one given above and observing that there can be N individuals at time $t + \Delta t$ only by having N individuals at time t and no reproduction in the interval $(t, t + \Delta t)$. The relation between $P_N(t)$ and $P_N(t + \Delta t)$ is

$$P_N(t + \Delta t) = P_N(t)(1 - kN\,\Delta t) + H_2(\Delta t),$$

where $H_2(\Delta t) = o(\Delta t)$, $\Delta t \to 0$, and the desired differential equation for P_N is

$$\frac{dP_N}{dt} = -kNP_N. \tag{4.4}$$

These differential equations, (4.4) and (4.3) with $n = N + 1, N + 2, \ldots$, determine the probability distributions $\{P_n(t)\}_{n \geq N}$ and thus the evolutionary nature of the process. Some of the Projects and the models of the next chapter show that one can use the same approach in more general situations. In each case the desired probability distributions are given as the solutions to a system of differential equations. Usually such systems are difficult to study, and sometimes they are totally intractable by any but numerical methods. However, as one might expect from the simplicity of our basic assumptions, in this special case of a pure birth process with constant birth coefficient (k) an explicit representation for the probability distributions can be obtained.

EXERCISES

1. Let f and g be defined on $(0, \delta)$ for some $\delta > 0$, and suppose that there are nonnegative real numbers m and n such that $f(x) = o(x^m)$, $g(x) = o(x^n)$, $x \longrightarrow 0$.
   (a) Prove that $(f \cdot g)(x) = o(x^{m+n})$, $x \longrightarrow 0$.
   (b) Set $r = \min(m, n)$ and prove that $(f + g)(x) = o(x^r)$, $x \longrightarrow 0$.
   (c) If f is as above and g is bounded on $(0, \delta)$, prove that $(f \cdot g)(x) = o(x^m)$, $x \longrightarrow 0$.
2. Let $\{f_k\}_1^K$ be a sequence of functions each of which is defined on $(0, \delta)$ and satisfies $f_k(x) = o(x^m)$, $x \longrightarrow 0$, and let $a_1, \ldots, a_K$ be constants. Define f by $f(x) = \sum_1^K a_k f_k(x)$, $x \in (0, \delta)$. Prove that $f(x) = o(x^m)$, $x \longrightarrow 0$.
3. Give the details of the argument leading to the expression $\Pr[Q(n; \Delta t)] = nk\,\Delta t + H(\Delta t)$, $H(\Delta t) = o(\Delta t)$, $\Delta t \longrightarrow 0$.
4. Describe some settings where the pure birth model of this section might be appropriate. Also, give some of the limitations of this model.

8.5 A FORMULA FOR THE PROBABILITY DISTRIBUTIONS

In this section we shall derive an explicit expression for the probability distributions arising in the model formulated above. This expression is deduced by examining special cases; then a detailed proof is provided.

8.5.1 A Proposal

Since the initial size of the population, N, is known precisely, if we assume that the process begins at time $t = 0$, then we have $P_N(0) = 1$. Thus the determination of P_N reduces to the solution of the initial value problem

$$\begin{aligned} \frac{dP_N}{dt} &= -kNP_N, \\ P_N(0) &= 1, \end{aligned} \tag{5.1}$$

where the differential equation is that of Eq. (4.4). This initial value problem can be solved immediately, and we have

$$P_N(t) = e^{-kNt}. \tag{5.2}$$

Next, (5.2) can be used together with (4.3) with $n = N + 1$ to obtain an equation involving only P_{N+1}:

$$\frac{dP_{N+1}}{dt} = k[Ne^{-kNt} - (N + 1)P_{N+1}]. \tag{5.3}$$

Equation (5.3) can be solved using the techniques discussed in Appendix B. Indeed, the function $e^{-k(N+1)t}$ is a solution of the equation

$$\frac{dP_{N+1}}{dt} = -k(N + 1)P_{N+1},$$

and the function Ne^{-kNt} is a solution of (5.3). Thus the function

$$[ce^{-kt} + N]e^{-kNt} \tag{5.4}$$

is a solution of (5.3) for any constant c. To determine the appropriate value for c, we use the fact that the population is known to contain exactly N individuals at time $t = 0$. Thus the probability that it contains $N + 1$ individuals at that time is 0; that is, $P_{N+1}(0) = 0$. For the solution (5.4) of (5.3) to satisfy this condition, c must be taken to be $-N$. We conclude that

$$P_{N+1}(t) = Ne^{-kNt}(1 - e^{-kt}). \tag{5.5}$$

Clearly, one can continue in this manner. The next step is to use (5.5) in (4.3) with $n = N + 2$ to compute P_{N+2}. This yields the differential equation

$$\frac{dP_{N+2}}{dt} = k\{(N + 1)Ne^{-kNt}(1 - e^{-kt}) - (N + 2)P_{N+2}\}, \tag{5.6}$$

which, as above, can be solved using standard techniques (Appendix B). We find that the function

$$ce^{-k(N+2)t} + N(N + 1)[\tfrac{1}{2}e^{-kNt} - e^{-k(N+1)t}]$$

satisfies (5.6) for any value of the constant c. So that the initial condition $P_{N+2}(0) = 0$ will be satisfied, it is necessary that c be taken equal to $\frac{1}{2}N(N + 1)$. Using this value of c, we obtain

$$P_{N+2}(t) = \frac{N(N + 1)}{2}e^{-kNt}(1 - e^{-kt})^2. \tag{5.7}$$

An examination of these formulas for P_N, P_{N+1}, and P_{N+2} leads one to conjecture that

$$P_n(t) = \frac{(n-1)!}{(n-N)!(N-1)!} e^{-kNt}(1 - e^{-kt})^{n-N}, \tag{5.8}$$

for $n \geq N$ (recall the convention $0! = 1$). This is indeed true, but the details of verification are formidable.

8.5.2 A Proof

The formula for the general term in the probability distribution $\{P_n\}_{n \geq N}$ conjectured above was based on the solutions for $n = N, N+1$, and $N+2$. Here we establish the validity of this conjecture by providing a proof.

To this end, we consider the differential equation

$$\frac{dP_n}{dt} + knP_n = k(n-1)P_{n-1}, \tag{5.9}$$

which holds for $n > N$. Since the left-hand side of (5.9) can be written as

$$e^{-knt}\frac{d}{dt}(e^{knt}P_n),$$

we are led to make a change of dependent variable. We set

$$U_n = e^{knt}P_n, \qquad n \geq N,$$

and making the substitution in (5.9), we obtain

$$e^{-kt}\frac{dU_n}{dt} = k(n-1)U_{n-1}.$$

The left-hand side of this equation suggests another change of variable, the time variable in this instance. Indeed, we note that if we set $s = e^{kt}$ we obtain by the chain rule the relation

$$e^{-kt}\frac{dU_n}{dt} = k\frac{dU_n}{ds}.$$

Therefore, making this change, noting that the mapping $t \longrightarrow e^{kt}$ maps $[0, \infty)$ onto $[1, \infty)$, and writing

$$U_n(t) = U_n\left(\frac{1}{k \log s}\right) = V_n(s), \qquad s \geq 1 \quad \text{and} \quad n > N,$$

we obtain finally the equation

$$\frac{dV_n}{ds} = (n-1)V_{n-1}, \qquad n > N. \tag{5.10}$$

An entirely similar argument shows that the basic differential equation for P_N becomes

$$\frac{dV_N}{ds} = 0 \tag{5.11}$$

with the same changes of variables.

Now (5.11) obviously has the solution

$$V_N = \text{constant}.$$

Also, $U_N(0) = V_N(1) = 1$, so that the constant value of V_N must be 1; that is,

$$V_N(s) = 1, \qquad s \geq 1.$$

We set $n = N + 1$ in (5.10) and we use this to obtain

$$\frac{dV_{N+1}}{ds} = NV_N = N.$$

Therefore

$$V_{N+1}(s) = Ns + c, \qquad s \geq 1,$$

where the constant c is to be determined. The condition $U_{N+1}(0) = 0$ gives $V_{N+1}(1) = 0$ so that c must be equal to $-N$, and consequently

$$V_{N+1}(s) = N(s-1).$$

The result in the general case is contained in the following:

Theorem 1: The solution to the system

$$\frac{dV_N}{ds} = 0, \qquad V_N(1) = 1,$$

$$\frac{dV_n}{ds} = (n-1)V_{n-1}, \qquad V_n(1) = 0, \quad n > N,$$

for $s > 1$ is given by

$$V_N(s) = 1,$$

$$V_n(s) = \frac{(n-1)(n-2)\cdots N}{(n-N)!}(s-1)^{n-N}, \qquad n > N.$$

Proof: We have established the theorem for $n = N$ (and also $n = N + 1$). The proof will be completed by induction. Let us therefore assume the truth of the theorem for $n = m$, that is,

$$V_m = \frac{N(N+1)\cdots(m-1)}{(m-N)!}(s-1)^{m-N}.$$

The differential equation for $n = m + 1$ is

$$\frac{dV_{m+1}}{ds} = \frac{N(N+1)\cdots(m-1)m}{(m-N)!}(s-1)^{m-N},$$

and this has the obvious solution

$$V_{m+1}(s) = \frac{N(N+1)\cdots m}{(m-N+1)!}(s-1)^{m-N+1} + c,$$

where c is to be determined. However, $V_{m+1}(1) = 0$ requires that $c = 0$, and consequently the conclusion of the theorem also holds in the case $n = m + 1$. The proof is complete.

Equation (5.8) is easily obtained from Theorem 1. Indeed, we have $s = e^{kt}$ so that

$$U_n(t) = V_n(e^{kt}) = \frac{N(N+1)\cdots(n-1)}{(n-N)!}(e^{kt}-1)^{n-N},$$

and (5.8) now immediately follows from the change of variable $U_n = e^{knt}P_n$, $n \geq N$.

8.6 PROPERTIES OF THE PROBABILITY DISTRIBUTIONS

An investigator studying growth processes is usually less interested in an expression for the probability distributions, i.e., the equivalent of (5.8), than he is in what can be deduced about the process from the distributions. For example, if the initial size of the population is 10, then one would expect that for small times t, $P_{11}(t) > P_{12}(t) > \ldots$. Is this a prediction of the model? Also, what is the expected size of the population at time t? Examples of questions of this sort will be considered in this section.

8.6.1 Qualitative Properties of the Probability Distributions

The model introduced in Sec. 8.4 is simple enough that we were able to use elementary techniques to obtain exact solutions to the differential equations determining the probability distributions. However, it is also worthwhile to examine directly the qualitative features of the distributions. Our motives in doing this are twofold. First, the general preconceptions which one has regarding the situation are commonly qualitative ones, and it is desirable to compare them with the model at an early stage in the study. If there is substantial disagreement between what the model predicts and what one's intuition says ought to be true, then both must be examined more carefully. Second, parts of our analysis utilize only the differential equations, and thus the method, although not the details, can be applied in other models. Frequently, stochastic models for biological processes give rise to differential equations which either cannot be solved explicitly or else have such complicated solutions that little can be deduced from them. Thus it is useful to be able to obtain information about the solutions without knowing the solutions explicitly. Of course, there are also occasions when one can use expressions such as (5.8) to deduce information about the form of the distributions which is much more precise than that which can be obtained from the differential equations alone.

What are the qualitative features of the distributions which appear intuitively obvious and which we expect the model to confirm? Clearly, for each fixed n we expect that as time becomes arbitrarily large the probability that the population consists of n individuals becomes small. That is, for each n, $P_n(t) \longrightarrow 0$ as $t \longrightarrow \infty$. Next, since the model was designed to represent a growth process, if N is the initial size of the population, then we expect that $P_N(t)$ is monotonically decreasing. That is, the population is likely to increase in size as time passes. Also, with the same reasoning for $n > N$, we expect that $P_n(t)$ increases initially, has a unique maximum, at $t = t_{nM}$ say, and then decreases. Finally, we expect that the sequence $\{t_{nM}\}_{n>N}$ is a monotonically increasing sequence in n. The goal of this section is to confirm these qualitative expectations by a careful study of the differential equations. Many of the details are covered in the Exercises.

We begin our discussion by considering P_N. Since P_N satisfies the initial value problem (5.1), it is a nonnegative function (see the corollary to Theorem B.3, Appendix B). It should be noted that this would follow trivially if we knew at this point that $\{P_n(t)\}_{n \geq N}$ was, in fact, a probability distribution for each t. This fact is proved in the next subsection. It now follows from (4.4) that the slope of the graph of P_N is always nonpositive and is equal to $-kN$ at $t = 0$. Next, $P_N(t) \longrightarrow 0$ as t becomes arbitrarily large. Indeed, P_N is a nonnegative, nonincreasing function of t, and consequently either $P_N(t) \longrightarrow 0$ as $t \longrightarrow \infty$ or there is a constant c such that $P_N(t) \geq c > 0$ for all values of t. But if the latter inequality were to hold, then the fact

$$\frac{dP_N(t)}{dt} = -kNP_N(t)$$

would imply that

$$\frac{dP_N(t)}{dt} < -kNc$$

for all values of t. This last inequality is clearly inconsistent with the non-negativity of P_N. Thus we conclude that the graph of P_N appears somewhat as in Fig. 8–8.

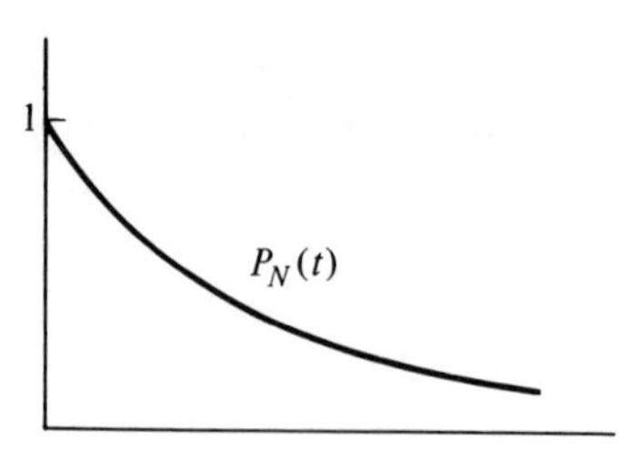

Fig. 8–8

It requires a good deal more effort to obtain as much information about P_n, $n > N$, as has been obtained about P_N. We indicate how one can proceed by considering P_{N+1} and P_{N+2}. These functions are defined by

$$\frac{dP_{N+1}}{dt} = k[NP_N - (N+1)P_{N+1}], \qquad P_{N+1}(0) = 0, \qquad (6.1)$$

and

$$\frac{dP_{N+2}}{dt} = k[(N+1)P_{N+1} - (N+2)P_{N+2}], \qquad P_{N+2}(0) = 0, \qquad (6.2)$$

respectively. Note that we have suppressed the explicit dependence on t for notational convenience. It follows from (6.1) and (6.2) that the initial portions of the graphs of P_{N+1} and P_{N+2} appear as in Fig. 8–9 (Exercise 1). To continue, we note that P_{N+1} is a nonnegative function and $P_{N+1}(t)$ is positive for $t > 0$ (Exercise 2). It has been shown above that P_N is a decreasing function, $P_N(0) = 1$, and P_{N+1} is initially increasing, $P_{N+1}(0) = 0$. Thus it seems reasonable that there is a smallest time t for which $P_{N+1}(t) = [N/(N+1)]$ $P_N(t)$. This can be made precise in the following way. Define

$$t_{1M} = \sup\{t : P_{N+1}(x) < \frac{N}{N+1} P_N(x) \text{ for } x \in [0, t)\}$$

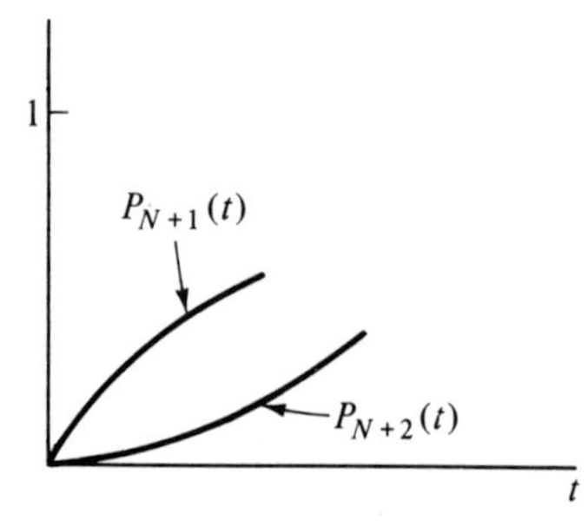

Fig. 8–9

(See Appendix A for the definition of sup.) The time t_{1M} defined in this way exists and is unique (Exercise 3). The definition of t_{1M} implies that

$$(N+1)P_{N+1}(t_{1M}) = NP_N(t_{1M}),$$

and this together with the differential equation defining P_{N+1} gives

$$\frac{dP_{N+1}}{dt}(t_{1M}) = 0.$$

With these preliminaries we can show that P_{N+1} has a global maximum at t_{1M}. In fact, we shall prove the stronger assertion that t_{1M} is the only local maximum of P_{N+1}. We begin by evaluating d^2P_{N+1}/dt^2 at an arbitrary point t^*, where $[dP_{N+1}/dt]\,(t^*) = 0$. Differentiating the differential equation for P_{N+1} and setting $t = t^*$, we have

$$\frac{d^2P_{N+1}}{dt^2}(t^*) = kN\frac{dP_N}{dt}(t^*) < 0.$$

Thus at every critical point of the function P_{N+1}, i.e., at every point t^* where $[dP_{N+1}/dt]\,(t^*) = 0$, the function has a negative second derivative. This means that each of these critical points is a local maximum. In particular, t_{1M} is a local maximum. Moreover, since P_{N+1} is continuously differentiable, it cannot have two distinct local maxima without having at least one local minimum. Since it has no local minima, t_{1M} must be the only local maximum and our assertion is proved.

Finally, since P_{N+1} has exactly one critical point, is nonnegative, and is initially increasing, it follows that it is eventually decreasing. An argument similar to that applied to P_N shows that $P_{N+1}(t) \longrightarrow 0$ as $t \longrightarrow \infty$.

Summarizing, it has been shown that P_{N+1} initially increases, $[dP_{N+1}/dt](0) > 0$; has a single local maximum; and tends to 0 as t tends to ∞. Therefore the graph of P_{N+1} appears as in Fig. 8–10.

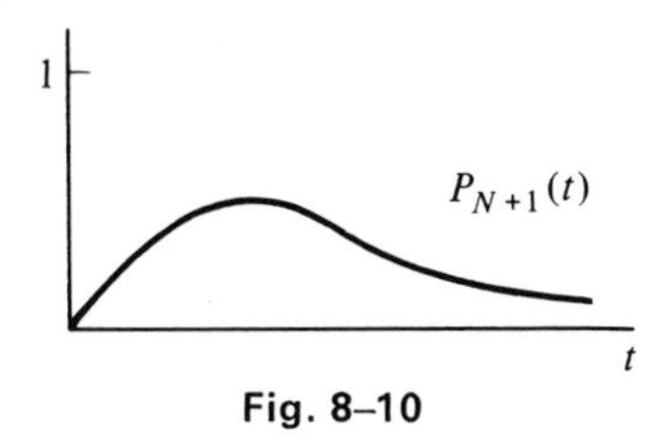

Fig. 8–10

The analysis can be continued using the same ideas. We outline one more step.

1. $P_{N+2}(0) = 0$, $[dP_{N+2}/dt](0) = 0$, $[d^2P_{N+2}/dt^2](0) > 0$. Thus, P_{N+2} is initially increasing and $P_{N+1}(t) > P_{N+2}(t)$ for sufficiently small times t.
2. Define $t_{2M} = \sup\{t : P_{N+2}(x) < [(N+1)/(N+2)]P_{N+1}(x)$ for $x \in [0, t)\}$. The quality t_{2M} exists and is unique. Moreover,

$$\frac{dP_{N+2}}{dt}(t_{2M}) = 0.$$

3. $t_{2M} > t_{1M}$.
4. The function P_{N+2} has a unique local maximum which occurs at t_{2M}.
5. $P_{N+2}(t) \to 0$ as $t \to \infty$.

Thus we conclude that the graph of P_{N+2} is initially tangent to the t axis, increasing for small times; has a unique maximum value; and decreases to 0 as t tends to ∞. The graph of P_{N+2} is qualitatively as in Fig. 8–11.

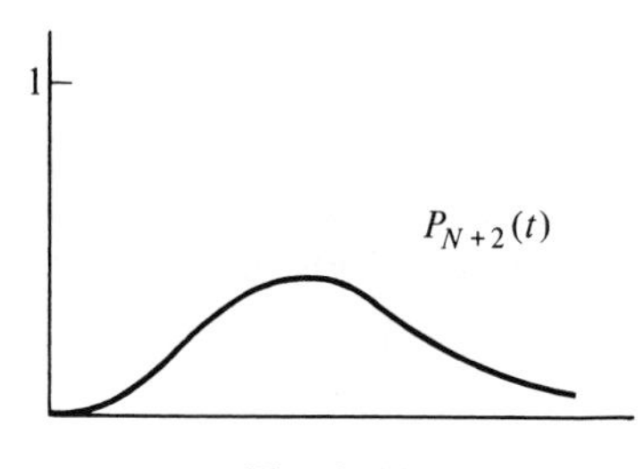

Fig. 8–11

The techniques applied in this section to a specific example are typical of those which involve only the differential equations and which have wider use. The reader will note that we have verified, at least for $n = N, N+1, N+2$, the conjectures based on intuition mentioned at the beginning of the section.

8.6.2 Quantitative Properties of the Probability Distributions

In the development of the theory of the model introduced in Sec. 8.4, it was asserted that the family of functions $\{P_n(t)\}_{n \geq N}$ constituted a family of probability distributions, one for each time t. To verify this assertion, we shall show that

$$0 \leq P_n(t) \leq 1, \tag{6.3}$$

for each n and each t, and

$$\sum_{n \geq N} P_n(t) = 1, \tag{6.4}$$

for each t. We use the fact that (5.8) gives a solution to the system of differential equations (4.3) and (4.4). Since (5.8) is clearly a nonnegative function, the first half of inequality (6.3) is obvious. We prove (6.4), which includes the second half of (6.3). Using the form of P_n given by (5.8), we see that our task is to prove

$$e^{-kNt} + \sum_{n \geq N+1} \frac{(n-1)\cdots N}{(n-N)!} e^{-knt}(e^{kt}-1)^{n-N} = 1. \tag{6.5}$$

To obtain a better idea of what we are dealing with, let us write out the first few terms on the left-hand side of (6.5). We have

$$\begin{aligned} &e^{-kNt} + Ne^{-k(N+1)t}(e^{kt}-1) + \frac{N(N+1)}{2} e^{-k(N+2)t}(e^{kt}-1)^2 + \cdots \\ &= e^{-kNt}\left[1 + Ne^{-kt}(e^{kt}-1) + \frac{N(N+1)}{2} e^{-2kt}(e^{kt}-1)^2 + \cdots\right]. \end{aligned}$$

The last expression in square brackets can be simplified by noting that it is a binomial series with a negative exponent (Appendix A). We write

$$(1-x)^{-n} = 1 + nx + \frac{n(n+1)}{2!}x^2 + \frac{n(n+1)(n+2)}{3!}x^3 + \cdots, \tag{6.6}$$

where n is positive and the series on the right converges for $|x| < 1$. Since k is positive, the inequality $1 - e^{-kt} < 1$ holds for all $t > 0$. Therefore, if we set $n = N$ and $x = e^{-kt}[e^{kt} - 1]$, the series on the right in (6.6) will converge for any $t > 0$. We now make use of this in (6.5) and we have

$$\begin{aligned} \sum_{n \geq N} P_n(t) &= e^{-kNt}[1 - x]^{-N} = e^{-kNt}[1 - (1 - e^{-kt})]^{-N} \\ &= e^{-kNt}e^{kNt} = 1. \end{aligned}$$

Thus, Eq. (6.4), and consequently the assertion, has been verified.

For each fixed time the probability distribution $\{P_n\}_{n \geq N}$ is the distribution associated with the random variable n which assigns to each population in the set of all populations its size. It is interesting to determine the expected value $E[n]$ of n and its variance $V[n]$. For notational reasons we prefer to use μ instead of $E[n]$ and σ^2 instead of $V[n]$. Thus, by definition, $\mu = \sum_{n \geq N} nP_n$. Finally, we note that since the distribution $\{P_n\}_{n \geq N}$ changes with time, so does μ and therefore $\mu = \mu(t)$. The function μ can be determined directly by using (5.8) (Exercise 8), and the result is

$$\mu(t) = Ne^{kt}. \tag{6.7}$$

It is interesting to note that in this stochastic model with growth constant k the mean size of the population at time t is exactly the size predicted by a deterministic model with growth constant k. Relationships of this sort give us confidence that our approach to the problem is soundly based.

Likewise, one can use the precise form of the solutions to derive an expression for the variance (Exercise 9). By definition

$$\sigma^2(t) = \sum_{n \geq N} (n - \mu(t))^2 P_n(t). \tag{6.8}$$

Instead of giving the details of the method described above, we choose to adopt another approach. This alternative not only avoids a considerable amount of involved, although elementary, algebra, but it also provides a method which is independent of a knowledge of the explicit solutions of the differential equations. The basic idea here is to obtain differential equations for μ and σ^2. We consider μ first.

It follows from the definition that

$$\frac{du}{dt} = \sum_{n \geq N} n \frac{dP_n}{dt},$$

and we use the basic differential equations (4.3) and (4.4) to remove the derivatives from the right-hand side. Since the form of the equations is different for dP_N/dt and dP_n/dt, $n > N$, we break up the sum on the right-hand side accordingly. We have

$$\begin{aligned} \frac{d\mu}{dt} &= N(-kNP_N) + \sum_{n > N} nk((n-1)P_{n-1} - nP_n) \\ &= -kN^2P_N + k\sum_{n > N} n(n-1)P_{n-1} - k\sum_{n > N} n^2 P_n. \end{aligned} \tag{6.9}$$

Consider the first summation on the right-hand side. We have

$$\begin{aligned} \sum_{n > N} n(n-1)P_{n-1} &= \sum_{n > N} (n-1)^2 P_{n-1} + \sum_{n > N} (n-1)P_{n-1} \\ &= \sum_{n \geq N} n^2 P_n + \sum_{n \geq N} nP_n. \end{aligned}$$

Combining this and (6.9), we obtain

$$\begin{aligned} \frac{d\mu}{dt} &= -kN^2P_N + k\sum_{n \geq N} n^2P_n + k\sum_{n \geq N} nP_n - k\sum_{n > N} n^2P_n \\ &= k\sum_{n \geq N} nP_n, \end{aligned}$$

so that

$$\frac{d\mu}{dt} = k\mu.$$

This is a first-order linear differential equation for the function μ and its solution is

$$\mu(t) = ce^{kt}.$$

Finally, since $P_N(0) = 1$ and $P_n(0) = 0$ for $n > N$, we have $\mu(0) = N$, and this gives (6.7).

The same technique can be used to derive an expression for $\sigma^2(t)$. An argument entirely analogous to that given above for μ shows that σ^2 is a solution of the initial value problem

$$\frac{d\sigma^2}{dt} = 2k\sigma^2 + k\mu, \tag{6.10}$$

$$\sigma^2(0) = 0 \tag{6.11}$$

(Exercise 10). We have an explicit expression for μ and consequently (6.10) becomes

$$\frac{d\sigma^2}{dt} - 2k\sigma^2 = kNe^{kt}. \tag{6.12}$$

This is a standard linear differential equation for σ^2, and the usual techniques show that the function

$$ce^{2kt} - Ne^{kt}$$

satisfies (6.12) for every value of c. By picking $c = N$, condition (6.11) is satisfied. Thus we conclude that the variance σ^2 is given by

$$\sigma^2(t) = Ne^{2kt}(1 - e^{-kt}). \tag{6.13}$$

From this expression for the variance we see that the standard deviation is

$$\sigma(t) = \sqrt{N}\, e^{kt}(1 - e^{-kt})^{1/2}.$$

One deduction that can be made from this formula is that for large values of time the standard deviation is approximately a constant times the mean. Precisely,

$$\sigma(t) \sim (N)^{-1/2}\mu(t)$$

in the sense that

$$\lim_{t\to\infty} \frac{\sigma(t)}{\mu(t)} = (N)^{-1/2}.$$

8.6.3 Concluding Remarks

The model proposed in Sec. 8.4 is probably the simplest stochastic model for the growth of a population. As a result it is subject to the same criticisms as the very primitive deterministic models discussed in

Part A of this chapter. There are, of course, situations in which the predictions of this model fit well with observations. This is particularly true when it is viewed as predicting a spatial rather than temporal distribution of events. That is, the time dimension t is replaced by a space dimension x and the model predicts the distribution of events (e.g., numbers of plants) occurring along one spatial dimension [Ba].

Just as in the deterministic situation, various modifications of this basic model have been proposed. For example, the predator-prey interaction described in Sec. 8.2 from a deterministic point of view can also be modeled as a stochastic process. Some of these modifications are the topics of Projects (Sec. 8.7), and many others may be found in the References.

Perhaps the most important point to be gleaned from this discussion is that the probabilistic features of biological processes can be incorporated into mathematical models. Assumptions regarding the biological situation have been made precise, and specific conclusions have been drawn from these assumptions. A check of the validity of the conclusions frequently involves statistical work.

EXERCISES

1. Use (6.1) and (6.2) to show that the initial portions of the graphs of P_{N+1} and P_{N+2} appear as in Fig. 8–9. In particular, show that
   (a) $[dP_{N+1}/dt](0) > 0$, $[d^2P_{N+1}/dt^2](0) < 0$.
   (b) $[dP_{N+2}/dt](0) = 0$, $[d^2P_{N+2}/dt^2](0) > 0$.
   What do (a) and (b) say about the values of these derivatives for small values of t?

2. Show that $P_{N+1}(t) > 0$ for $t > 0$. *Hint:* Suppose that there are some values of $t \neq 0$ for which $P_{N+1}(t) = 0$, and set $\tau = \sup\{t: P_{N+1}(x) > 0 \text{ on } (0, t)\}$. Show that $P_{N+1}(\tau) = 0$, that $[dP_{n+1}/dt](\tau) > 0$, and that this leads to a contradiction of the definition of t. (See Appendix A for the sup notation.)

3. Show that the quantity t_{1M} defined in this section exists and is unique. *Hint:* If t_{1M} does not exist, then $P_{N+1}(t) < [N/(N+1)]P_N(t)$ for all values of t. Deduce that there are two times t_1 and t_2 such that $P_{N+1}(t_1) = P_{N+1}(t_2)$, and show that this is incompatible with the nonexistence of t_{1M}.

4. Verify assertions 1–5 concerning P_{N+2} (Sec. 8.6.1).

5. Using the precise solutions for the model of Sec. 8.4, find approximate values for t_{1M} and t_{2M} in terms of the parameters of the system: k and N. What restrictions (if any) are imposed on the model by your approximation?

6. Find exact values for $\max_{0 \leq t < \infty} P_{N+1}(t)$ and $\max_{0 \leq t < \infty} P_{N+2}(t)$ for the model discussed in Secs. 8.4 and 8.5. How do these values behave as the size of the initial population increases?

7. Using the discussion of the model of Sec. 8.4 given in this section, graph P_N,

P_{N+1}, and P_{N+2} on the same coordinate system. How do the values of the various probabilities compare at different times?

8. Derive Eq. (6.7) for the mean μ directly from (5.8). *Hint:* Make use of Eq. (6.6).

9. Derive Eq. (6.13) for the variance σ^2 directly from (5.8).

10. Show that the variance σ^2 of the probability distributions $\{P_n\}_{n \geq N}$ satisfies the initial value problem:

$$\frac{d\sigma^2}{dt} = 2k\sigma^2 + k\mu,$$

$$\sigma^2(0) = 0.$$

11. Assuming an initial population of N members, construct a stochastic model for a pure decay process. State your assumptions as completely as you can.

12. Find the family of probability distributions which result from the model constructed in Exercise 11. Find the mean and variance of the family as functions of time.

13. Discuss the relevance of the model constructed in Exercise 11 as a model for radioactive decay. Compare the results of deterministic and stochastic theory for this problem.

8.7 PROJECTS

8.7.1 Preyed Upon Predators

Consider a closed ecological system consisting of three interacting populations: fish, seals, and Eskimos. Suppose that the Eskimos are entirely dependent on seals for food, that the seals eat only fish, and that the fish have a large food supply, inexhaustible for the purposes of this discussion. Formulate a model for this system and develop as much of the theory as you can. State your assumptions carefully and provide some justification. Comment on your conclusions.

Consider the same situation under the assumption that Eskimos can and do catch both fish and seals. Proceed as in the first part.

8.7.2 Other Stochastic Models

The stochastic model constructed in this chapter was based on several assumptions. The most important can be stated easily by considering the possible reproductive behaviors of an individual during the time interval $(t, t + \Delta t)$.

Definition: Let $E_1(t, \Delta t)$ be the event that an individual reproduces exactly once in the interval $(t, t + \Delta t)$, and let $E_2(t, \Delta t)$ be the event that an individual reproduces more than once in the interval $(t, t + \Delta t)$.

With this terminology, the assumptions on which the stochastic model of this chapter was based can be stated as

i. $\Pr[E_1(t, \Delta t)] = k\,\Delta t + h_1(\Delta t)$, where $h_1(\Delta t) = o(\Delta t)$, $\Delta t \to 0$, and k is a positive constant which is independent of the size of the population, and
ii. $\Pr[E_2(t, \Delta t)] = h_2(\Delta t)$, where $h_2(\Delta t) = o(\Delta t)$, $\Delta t \to 0$.

The reader can appreciate that in many cases the assumption on k in axiom i may not be a reasonable one. Indeed, as the population grows to a size where the resources are no longer adequate, then one expects the constant k to decrease as n increases. Also, taking a cue from human reproduction, one expects that k may not be a constant at all, but rather a function of time. In this project the reader is asked to explore these possibilities.

Part I. Construct a stochastic model for population growth under the following more general assumptions:

i′. $\Pr[E_1(t, \Delta t)] = k_n\,\Delta t + h_1(\Delta t)$, where $h_1(\Delta t) = o(\Delta t)$, $\Delta t \to 0$, and k_n is a positive constant which depends on n, but not on time, and
ii′. Same as ii above.

Problems

1. Find the differential equations characterizing the probability distributions.
2. If the population has size N at time $t = 0$, determine $P_N(t)$ and $P_{N+1}(t)$.
3. Without solving the infinite system of differential equations found in Problem 1, determine as much of the qualitative behavior of the probability distributions as you can.

Part II. Construct a stochastic model for population growth under the following assumptions:

i″. $\Pr[E_1(t, \Delta t)] = k(t)\,\Delta t + h_1(\Delta t)$, where $h_1(\Delta t) = o(\Delta t)$, $\Delta t \to 0$, and $k(t)$ is a positive function of time, which is independent of n, and
ii″. Same as ii above.

To make the analysis simpler, assume that the function $k(t)$ of axiom i″ does not depend on the specific individual. That is, the same function $k(t)$ applies to every member of the population.

Problems

1. Find the differential equations characterizing the probability distributions.
2. Suppose that we are dealing with a fairly homogeneous population

over a short enough time period so that it is reasonable to suppose that each individual reproduces exactly once. Let $G(t)$ be the probability that an individual reproduces before time t. What is the relation between $G(t)$ and $k(t)$? How would the graphs of $G(t)$ and $k(t)$ look over the lifespan of human individuals having exactly one offspring.

3. Returning to assumption i'' of Part II, suppose that $k(t)$ depends on the individual in the following way. If an individual is born at time τ (we suppose that members of the original population are born at time 0), then the associated function $k(t)$ is given by

$$k(t) = \begin{cases} 0, & 0 \leq t \leq \tau, \\ K(t - \tau), & \tau < t, \end{cases}$$

where $K(s)$ is a fixed positive function defined for $s > 0$. Construct a model for this situation and develop as much of the theory as you can.

8.7.3 Refined Deterministic Models

Part I. Formulate a deterministic model for population growth which takes into account age and sex differences. State your assumptions carefully and make them as plausible as possible.

Part II. Consider a population in which each individual belongs to exactly one of two subpopulations. We call these subpopulations Male and Female and refer to them as the two sexes. Let $M(t)$ denote the number of males in the population at time t, $F(t)$ the number of females in the population at time t, and $P(t) = M(t) + F(t)$. We suppose that births result from random encounters between individuals of the opposite sex and that the death rate is constant. To simplify the situation, we suppose that the ratio of males to females is independent of time. Precisely, we make three assumptions:

i. The ratio $M(t)/F(t)$ is independent of time. That is, there is a constant λ, $0 \leq \lambda \leq 1$, such that $M(t) = \lambda P(t)$, $F(t) = (1 - \lambda)P(t)$, for all times t.

ii. The birth rate is proportional to the number of random encounters between males and females. That is, the birth rate is given by

$$k_b M(t) F(t) = k_b \lambda(1 - \lambda)P(t)^2, \qquad k_b > 0.$$

iii. The death rate is given by $k_d P(t)$.

Problems

1. Formulate and solve the differential equation which governs the growth of the population under these assumptions.

2. Discuss the asymptotic behavior of $P(t)$ under various assumptions on the relative magnitudes of the parameters.
3. Since the birth rate increases as the size of the population increases, the population becomes arbitrarily large provided that k_d is not too big. Thus for low death rates and environments with limited resources, our model must be modified. Show that if we observe a stable population size for large times, that is, if

$$\lim_{t\to\infty} P(t) = P_\infty, \qquad 0 < P_\infty < \infty,$$

then the relative birth rate,

$$\frac{\text{Birth rate}}{\text{Population size}},$$

cannot be a linear function of $P(t)$. Investigate the situation with the assumption that the relative birth rate is a quadratic function of the population size.

8.7.4 Competing Populations

Suppose that two populations cohabit the same environment and compete for limited resources. An example of this situation would be two kinds of fish with a common food supply in the same lake. Let the two types be A and B and suppose that $P_A(t)$ denotes the number of individuals of type A present at time t, and likewise for $P_B(t)$.

Problems

1. Give a detailed argument justifying the use of the following differential equations as a mathematical model for the system described above:

$$\frac{dP_A}{dt} = aP_A - bP_A^2 - cP_AP_B,$$

$$\frac{dP_B}{dt} = dP_B - eP_B^2 - fP_AP_B,$$

where a, b, c, d, e, f are positive constants.
2. Discuss the asymptotic behavior of P_A and P_B under various assumptions concerning the magnitudes of the parameters. In particular, show that if $a > d$, then P_A increases to a limit and P_B decreases to 0.

REFERENCES

Elementary growth processes are frequently considered in the later chapters of calculus textbooks, for example, [LS], and in the early chapters of books on elementary differential equations. We recommend the Preface and parts of Chap. 1 of [BN] as especially appropriate for readers of this book. The content of [BN] is more related to the physical sciences, and the treatment is more concerned with the development of the mathematics, but the reader will find the discussion useful. A popular account of radiocarbon dating which surveys the origins of the technique can be found in [De], and a more substantial discussion, including many examples, is found in [Al]. Potassium-argon methods are covered in [DL].

Deterministic models for population growth were first extensively investigated by A. J. Lotka and V. Volterra. Their work is discussed in [L], [D'A], and [Lo]. The deterministic models considered here are treated in more detail in [D'A]. The situation is discussed from an experimental point of view in [AB], which also considers the question of estimating the parameters in biological problems.

Stochastic treatments of growth processes can usually be found in textbooks including applications of probability theory under the designations *birth and death processes* or *Furry-Yule processes*. The original application of probability theory of this type to birth processes was given by Yule in [Y]. Our model is based on the relevant sections of Chap. XVII of [F]. This book is a classic in probability theory and its applications, and every student should at least browse through it. The model studied here as well as several others are presented in [Ba].

[Al] Allibone, T. E., ed., *The Impact of The Natural Sciences on Archaeology.* London: Oxford University Press, 1970.

[AB] Andrewartha, H. G., and L. C. Birch, *The Distribution and Abundance of Animals.* Chicago: University of Chicago Press, 1954.

[Ba] Bartlett, M. S., *Stochastic Population Models.* London: Methuen, 1960.

[BN] Brauer, F., and J. A. Nohel, *Ordinary Differential Equations.* Reading, Mass.: Benjamin, 1967.

[D'A] D'Ancona, U., *The Struggle for Existence.* Leiden, Netherlands: E. J. Brill, N. V., 1954.

[De] Deevey, E. S., Jr., "Radiocarbon Dating," *Scientific American,* **186** (Feb. 1952), 24–28.

[DL] Dalrymple, G. B., and M. A. Lanphere, *Potassium-Argon Dating.* San Francisco: Freeman, Cooper, 1969.

[F] Feller, W., *An Introduction to Probability Theory and Its Applications,* 3rd ed. New York: Wiley, 1968.

[L] Leigh, E. R. "The Ecological Role of Volterra's Equations," in *Lectures on Mathematics in the Life Sciences: Some Mathematical Problems in Biology* (M. Gerstenhaber, ed.). Providence, R. I.: American Mathematical Society, 1968.

[Lo] LOTKA, A. J., *Elements of Mathematical Biology*. New York: Dover, 1956.

[LS] LOWENGRUB, M., and J. G. STAMPFLI, *Topics in Calculus*. Waltham, Mass.: Ginn/Blaisdell, 1970.

[Y] YULE, G. U., "A Mathematical Theory of Evolution, Based on the Conclusions of Dr. J. C. Willis, F.R.S.," *Philosophical Transactions of the Royal Society of London*, Ser. B, **213** (1925), 21–87.

9 Growth Models for Epidemics, Rumors, and Queues

9.0 INTRODUCTION

This chapter can be viewed as a continuation of Chap. 8, and although little explicit use is made of the results obtained there, the reader will benefit from a familiarity with the problems and methods of that chapter. In particular, Secs. 8.1, 8.2, and 8.4 have a bearing on matters discussed here.

The primary objective of this chapter is to show that growth and decay models can be used in many ways other than the population studies considered above. In Sec. 9.1, we shall use such models to study the spread of an infection through a population. We shall construct both deterministic and stochastic models with the same sort of basic assumptions, and we shall identify some common features of these models. There are similarities, as in Chap. 8, and some surprising differences.

In Sec. 9.2 we shall investigate the spread of rumors, a topic that has significant overlap with Sec. 9.1. However, in this study we shall emphasize not a comparison of deterministic and probabilistic models but rather models based on continuous and discrete time. Models based on an assumption of discrete time, that is, that time occurs in chunks instead of being a continuously divisible quantity, are especially important because of their relation to computation. To utilize the advantages of electronic computing machines, it is usually necessary to formulate the model in terms of discrete time, and these formulations may have other advantages as well.

It may seem reasonable to the reader that the spread of epidemics and rumors is an example of a growth (or decay) process. It is perhaps less obvious that the familiar phenomenon of a waiting line is another example of such a process. The theory of queues, the term we prefer for waiting lines, is exceptionally rich and interesting. Section 9.3 is devoted to the study of queues. However, both for reasons of space and difficulty we shall give only the briefest introduction to this important topic. Models for queues involve the most general birth-death processes studied in this book.

9.1 THE SPREAD OF SIMPLE EPIDEMICS

In this section we shall investigate the diffusion of a disease through a population. Throughout the discussion we shall adopt terminology which on occasion indicates that we are concerned with human populations. The models which we shall construct are not so restricted, and if the appropriate conditions are present, then they will apply to mosquitos as well as to men. However, even if we were to restrict ourselves to human populations, there is ample justification for our study. It is difficult for most people who have the advantages of modern health and medical and sanitary aids to obtain the proper perspective of the role of disease in human history. We shall mention here only one of the numerous references to plague and pestilence found in recorded history. In the fourteenth century a series of devastating epidemics swept Europe and England. The first major outbreak occurred near the middle of the century and successive waves of lesser magnitude followed at irregular intervals for a period of approximately 50 years. Mortality rates varied from region to region and figures from 15% in some areas to 60% in others can be supported. It has been estimated that Europe as a whole lost 25 million people, that is, one quarter of the total population, in the 50-year period beginning about 1340. Attrition at these rates in populations of comparable sizes has not occurred recently. However, this does not necessarily mean that widespread epidemics have been eliminated by increased knowledge and technical skills. Indeed, in this century and in so-called advanced nations there have been several significant outbreaks of influenza. In addition, there is serious concern that even more massive epidemics may develop in the future. One group of eminent biologists feels that there is increasing danger of epidemics of extreme severity in some underdeveloped nations. Here one has, as a result of the unrestricted population growth and the tendency of people to concentrate in urban areas, a phenomenon almost ready-made for the rapid spread of infection: a huge, centralized group of people often with inadequate nourishment and sanitary conditions.

Although the concern for a precise description of the progress of an epidemic is an old one, the idea of developing a mathematical theory which can predict the behavior of an epidemic, as compared to the collection and analysis of data, is a very recent one. With the growth of pathology in the last century it began to appear that the complex questions encountered in a study of the spread of infection were amenable to mathematical study. As a result, what might be called mathematical epidemology was born. The task of constructing a mathematical model for the spread of a disease through a population leads to many interesting questions, some of which are biological in nature and others mathematical. Since this presentation uses only elementary mathematics and almost no biology, we are restricted to models which include only the most basic aspects of the situation. We shall find, however, that these simple models will give insight into the basic processes at work. In particular, they can be used to deduce qualitative results which are compatible with observations. With regard to the mathematical techniques to be used, we shall find that approximation techniques are somewhat more important than they have been in previous work.

Let us now turn to a more precise statement of our problem. We shall consider the spread of an infectious disease through a population, and we shall be primarily concerned with simple models which display the right qualitative features. The term *infectious* is used here to indicate that the disease can be transmitted at some stage from an infected individual to an uninfected but susceptible member of the population. This transmission can either be direct or through an intermediary who may or may not himself contract the disease. We shall not elaborate on the nature of the biological processes involved, and the reader should appreciate that one can construct more realistic (and more complicated) models which reflect the biology of the situation more completely. For example, the question of the existence and nature of a transmission vector is not considered here even though in certain cases it may be an obviously important consideration. In general, we shall discuss only those aspects of the various processes which will be incorporated in the models. It should be understood by the reader that the omission of a point does not necessarily mean we feel it to be unimportant. It is, in fact, more likely that its neglect means that the models resulting from its inclusion are more complicated than our study permits.

In their broad features the models constructed for the spread of a disease display some similarities to those constructed in Chap. 8 to study the growth of populations. As in that situation we shall develop both deterministic and stochastic models. Many of the comments made in Chap. 8 concerning the validity of the models hold here as well. In particular, if one is concerned with large populations and a model reflecting only the overall spread of the disease, then a deterministic model is frequently adequate. In any case, such a model is usually a reasonable first approximation. On the other hand, if one is concerned with a small population or with the spread of a disease through a small group contained in a larger population, then the results are likely to be more satisfactory if a stochastic model is used. In the latter situations the chance elements which are always present in the problem can no longer be safely ignored.

Let us now consider in detail the assumptions concerning the population and the disease of the epidemic which will serve as a basis for the models to be constructed. The term *epidemic* is used here to indicate a change in the composition of a population in which at least initially the number of individuals having the disease increases. More precise definitions are given later. With regard to the population, we first assume that the size of the population is constant throughout the epidemic. This is a reasonable assumption for populations with moderate birth and death rates and for diseases of relatively short duration. Next, we assume that we are dealing with a homogeneous and uniformly mixing population. Such an assumption would clearly be justified if we were concerned with, say, the spread of measles through an elementary school. However, it would appear that the assumption is on much weaker ground if it is applied to the same disease in even a modest-sized city. It is therefore somewhat unexpected to find that data collected from observations indicate that the assumption is not excessively restrictive. What apparently happens in many cases is that an epidemic can be viewed as consisting of several smaller subepidemics occurring in separate geographical or social subdivisions of the whole population. In general, these smaller epidemics are not at the same stage of development at a given time, but they are not completely independent phenomena either. Thus the global development of the epidemic can be considered as the development of several subepidemics occurring in slightly different but usually overlapping time intervals in several subpopulations. The interaction of these two simultaneous developments, the spread of the disease in each subpopulation and the spread through the collection of subpopulations, produces results consistent with an assumption of homogeneity and uniform mixing.

Additional assumptions will be made in connection with the construction of specific models, and it is to these models that we now turn. As a final introductory comment, however, it is appropriate to reemphasize the guidelines of our study. We are interested in learning something in general terms about the development of epidemics. The results will be mostly of a qualitative nature and there will be obvious common features in the several simplified models to be studied.

9.1.1 Deterministic Models

To begin our work, we develop two deterministic models for the spread of a disease. The first is extremely simple and serves to introduce the basic ideas and to illustrate in a very rough way some qualitative features of the development of the epidemic. The second is more refined and leads to a well-known empirical law.

Let us consider the composition of the population at an arbitrary time t. We suppose that at each fixed time the population is composed of individuals of four types which can be specified as follows:

S: That subpopulation composed of individuals who are uninfected and susceptible to the infection.
E: That subpopulation composed of individuals who are infected but who are not yet capable of spreading the infection.
I: That subpopulation composed of individuals who are infected and actively spreading the disease.
R: That subpopulation composed of individuals who are not susceptible or who have been infected and either subsequently cured or removed from possible contact with members of S.

Our use of the term *cured* in the definition of the subpopulation R is in the sense that no reinfection is possible. Thus the R subpopulation is an absorbing state; once an individual enters the R group, he remains there for all future times. We refer to individuals in subpopulation S as *susceptibles* and denote their number at time t by $s(t)$; individuals in group E are known as *exposed* and their number at time t will be $e(t)$; individuals in group I are known as *infectives* and their number at time t will be $i(t)$; and individuals in group R will be called *removals* and their number at time t will be $r(t)$. The functions s, e, i, and r are actually integer-valued, i.e., step functions. However, as is usual with deterministic models, we treat them as differentiable functions. For large populations this is not an unreasonable approximation.

The populations involved in most real epidemics contain individuals in all four of these classes. That is, there is a definite period of incubation, which may vary with the disease and from individual to individual, during which the infected individual is not capable of passing the disease to a susceptible. Also, there is usually a definite duration, which again varies from disease to disease and to a much lesser extent from one individual to another, after which the infected individual is no longer capable of communicating the disease to a susceptible. After an individual who has been infected ceases to be so, then, depending on the nature of the disease, he may rejoin group S or he may be added to group R. Consequently in the most general situation an individual would proceed either from S to E to I to R or from S to E to I to S. In the latter case he could conceivably contract the disease a number of times. The epidemics considered here are considerably less than the most general. First we neglect passage through group E. That is, we assume that every individual who contracts the disease is immediately capable of communicating it to others. With this assumption we introduce a population composition vector

$$\mathbf{p}(t) = (s(t), i(t), r(t))$$

which contains the relevant information on the composition of the population at time t. The previous assumption that the population is of fixed size gives the equation

$$s(t) + i(t) + r(t) = \text{constant},$$

and consequently a knowledge of two of the coordinates of the population vector automatically gives the third. The subpopulations S, I, and R are assumed to be pairwise disjoint and their union is the entire population. We assume there is no transition from subpopulation I to subpopulation S. Further assumptions will be introduced as we proceed.

The qualitative behavior of $\mathbf{p}(t)$ as time increases is obvious. Suppose for the moment that we begin at time $t = 0$ with a population of $n + 1$ individuals. For an epidemic to develop, there must be at least one infected individual, i.e., $i(0) \geq 1$, and for definiteness we assume that $i(0) = 1$. The other members of the population are assumed to be in S. That is, $\mathbf{p}(0) = (n, 1, 0)$. The function s is nonincreasing since as time progresses the members of S become infected, or at least some of them do. If we assume that the disease has a finite term, then the function i is initially increasing as more individuals become infected and ultimately decreasing as individuals pass from I to R and the size of the S subpopulation declines. This behavior of the function i can be used to give a precise mathematical meaning to the *end* of such an epidemic: An epidemic will be said to end at time t_e if

$$i(t_e) = 0 \qquad \text{and} \qquad i(t) > 0 \text{ for all } t < t_e.$$

Indeed, if $i(t_e) = 0$, then there are no infected individuals at time t_e, and therefore no one is available to continue spreading the infection. Note, in particular, that the end of an epidemic is not defined in terms of the number of susceptibles.

We continue our model building by looking more closely at the simplest possible nontrivial situation. We consider an epidemic in which the infection is spread by contact between members of a population and in which no individuals are ever in group R. Intuitively, this can be thought of as a highly contagious disease which is not serious enough for infected individuals to be quarantined and whose duration is long compared to the time required for it to spread through the population. If in the general case one thinks of an individual as passing from S to I to R, then in this special case no one makes the last step. With only this information we can deduce something about the behavior of **p**. Indeed, since an individual remains infected and able to communicate the disease to others for the entire period after his initial infection, it follows that for large times the population vector will have its first and last coordinates 0. Implicit here is the assumption that over a long time period every possible contact between members of the population will take place.

The detailed behavior of the vector **p** depends on the assumptions we make concerning the way in which the disease spreads through the population. For the present model let us assume that the rate of growth of the infected subpopulation is proportional to the number of contacts per unit time between members of groups S and I. In turn, using our previous assumption regarding homogeneity and uniform mixing, the number of contacts per unit time at time t between members of S and members of I is proportional to $s(t) \cdot i(t)$. This assertion can be justified in the same manner as a similar statement in the predator-prey model discussed in Sec. 8.2. In precise terms this assumption becomes

$$\frac{ds}{dt} = -k \cdot s(t) \cdot i(t) = -k \cdot s(t) \cdot (n + 1 - s(t)), \tag{1.1}$$

where the second equality results from the basic assumption of this model, namely, that the entire population (of $n + 1$ members) is contained in the two groups S and I. Here k is a nonnegative constant determined by the physical parameters. Of course, an obvious generalization is to admit the possibility that k actually depends on time; $k = k(t)$. This refinement will be pursued in the Exercises. If we retain our previous assumption that at time $t = 0$ there is exactly one infected individual, then the condition to be satisfied by the solution $s = s(t)$ of (1.1) at time $t = 0$ is

$$s(0) = n. \tag{1.2}$$

Equations (1.1) and (1.2) form a mathematical model for the simple situation characterized by the assumptions discussed above. The reader will recognize the similarity between this model and the growth model of Sec. 8.2.

We now investigate this model with mathematical tools and see what information can be obtained. Rewriting (1.1) as

$$\frac{ds/dt}{s(n + 1 - s)} = -k,$$

we can expand the left-hand side using partial fractions to obtain

$$\frac{ds/dt}{(n+1)s} + \frac{ds/dt}{(n+1)(n+1-s)} = -k. \tag{1.3}$$

From Eq. (1.3) we have

$$\frac{1}{n+1}[\log s - \log(n+1-s)] = -kt + c,$$

or

$$\frac{s}{n+1-s} = c_1 e^{-k(n+1)t}.$$

It is convenient for future work to reduce this still further. Straightforward algebraic manipulations give

$$s(t) = \frac{n+1}{1 + c_2 e^{k(n+1)t}} \tag{1.4}$$

as a useful form of a solution of Eq. (1.1). Using condition (1.2) the constant c_2 in (1.4) can be evaluated. We find $c_2 = n^{-1}$. Thus the unique solution of (1.1) for $t > 0$ which satisfies (1.2) for $t = 0$ is

$$s(t) = \frac{n(n+1)}{n + e^{k(n+1)t}}, \qquad t > 0. \tag{1.5}$$

The asserted uniqueness is not obvious, nor does it follow from the discussion in Appendix B. Fortunately, in this special case it can be proved directly without too much effort (Exercise 3). Of course, (1.5) also implies that

$$i(t) = \frac{n+1}{1 + ne^{-k(n+1)t}}, \qquad t > 0.$$

The last two equations give precise and total information on the behavior of the functions s and i. However, it is also useful to have a geometrical picture of what is happening. In fact, as an aid for giving a clear impression of the shift of individuals from group S to group I, a graph of these functions is often more useful than the analytic expression. This task is left to the reader (Exercise 4).

In studies of this sort it is conventional to pay more attention to the *epidemic curve*, that is, the graph of $-(ds/dt)$ as a function of time, than to the function s itself. This is perhaps due to our natural interest in knowing how fast a disease is spreading as opposed to knowing the exact number of people who have the disease. Heuristically, we expect the rate of change of size of subpopulation S to be small initially (after all at the very beginning there is only one infected individual), to increase as the size of I increases, and to decrease as almost all individuals become infected. This behavior can be deduced from our results. Equation (1.1) and the expressions for s and i obtained above lead us to

$$-\frac{ds}{dt} = ks(t)i(t) = \frac{kn(n+1)^2 e^{k(n+1)t}}{[n + e^{k(n+1)t}]^2}, \tag{1.6}$$

and the epidemic curve is given in Fig. 9–1. Notice that

$$-\frac{ds}{dt}(0) = k \cdot s(0) \cdot i(0) = kn,$$

$$-\frac{ds^2}{dt^2}(0) = k\left(s(0)\frac{di}{dt}(0) + i(0)\frac{ds}{dt}(0)\right) = k^2 n(n-1),$$

and that ds/dt approaches 0 as t increases to infinity. It follows that the epidemic curve has at least one relative maximum. To obtain the coordinates of one of these maxima, we differentiate (1.6) and look for the zeros of the resulting expression. Following this procedure and noting that the denominator of (1.6) is always positive, we conclude that there is exactly one relative maximum and the time associated with the maximum value of (ds/dt) is determined by the equation

$$n - e^{k(n+1)t} = 0.$$

Thus the time t_M at which the size of the S subpopulation is changing most rapidly is

$$t_M = \frac{1}{k(n+1)} \log n.$$

Observe that this time of the maximum rate of infection varies inversely with the parameter k and is a decreasing function of the size of the total population. It is interesting to determine the size of the group of susceptibles at time

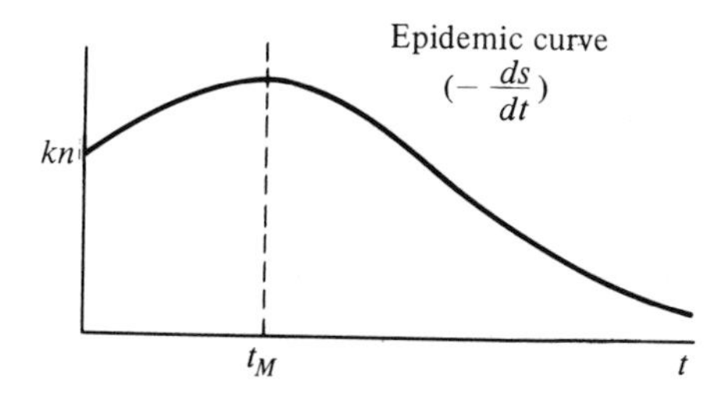

Fig. 9–1

t_M and also the maximum rate of infection, $-(ds/dt)|_{t=t_M}$. We obtain

$$s(t_M) = \frac{n+1}{2}$$

and

$$-\frac{ds}{dt}(t_M) = \frac{k(n+1)^2}{4}.$$

There is still useful information to be obtained from the epidemic curve, and we continue our discussion a little further. Let us replace t by $t_M + \tau$ in the right-hand side of (1.6) and denote the resulting function by f. We have

$$f(\tau) = k(n+1)^2 \frac{e^{k(n+1)\tau}}{[1 + e^{k(n+1)\tau}]^2}$$

for $-t_M < \tau < \infty$. Next, note that if we multiply the numerator and denominator of the right-hand side of this equation by $e^{-2k(n+1)\tau}$, then we obtain

$$\begin{aligned} f(\tau) &= k(n+1)^2 \frac{e^{-k(n+1)\tau}}{[e^{-k(n+1)\tau} + 1]^2} \\ &= k(n+1)^2 \frac{e^{k(n+1)(-\tau)}}{[1 + e^{k(n+1)(-\tau)}]^2} \\ &= f(-\tau) \end{aligned}$$

for $-t_M < \tau < t_M$. The meaning of this is, of course, that the graph of $-(ds/dt)$ is symmetric about the time value t_M. In Sec. 9.2 we shall compare this epidemic curve with a related curve obtained from a stochastic model constructed with similar simple assumptions.

It is clear that a great deal has been ignored in constructing the simple model discussed above, and one expects that it is of very restricted applicability. To obtain more useful models, one must reduce the assumptions imposed. Many of the more general models involve mathematical techniques of considerable depth and are not appropriate for discussion here. However, we can generalize the situation in one important direction while confining ourselves to elementary mathematics. In particular, we shall admit the possibility that some individuals are ultimately members of subpopulation R. In this case one expects the development of the epidemic to be highly dependent on the relation between the rate of infection and the rate of removal. For example, we expect that if members of S become members of I much more rapidly than members of I are removed to R, then at least initially the epidemic develops much like the simpler situation discussed above. Alternatively, if the infection rate, i.e., the rate at which members of S become members of I, is much smaller than the removal rate, the rate at which members of I move to R, then it seems likely that no epidemic will develop. Our goal in the discussion to follow is to make this argument more precise. In particular, we shall obtain an important result known in the literature as the threshold theorem. This result, which was first stated by W. O. Kermack and A. G. McKendrick, gives quantitative information on the occurrence and size of certain epidemics. The form of the result derived here is one of the simplest—indeed, it is close to the original form—and it admits numerous extensions. For example, versions are known for models in which the infection is transmitted by an intermediary and for stochastic models.

We begin, just as in the first model considered in this section, with the basic assumption that the disease is passed from members in I to members in S through contacts between such individuals. Thus once again we have the equation

$$\frac{ds}{dt} = -ks(t)i(t),$$

which relates the rate of infection to the sizes of the two subpopulations. Next, it is necessary to make a specific assumption about the removal rate, that is, the rate of growth of the R subpopulation. Our assumption is that the removal rate at time t is proportional to the number of infected individuals at that time. We have

$$\frac{dr}{dt} = l \cdot i(t)$$

where l, like k, is an appropriate nonnegative constant determined by the physical parameters of the situation. Since the total population always consists of $n + 1$ individuals, these assumptions involving ds/dt and dr/dt also determine di/dt. Thus, the rate of change of the population vector $\mathbf{p}$ is given by the following system of (first-order nonlinear) differential equations:

$$\begin{aligned} \frac{ds}{dt} &= -k \cdot s(t) \cdot i(t), & k > 0, \\ \frac{di}{dt} &= k \cdot s(t) \cdot i(t) - l \cdot i(t), & l > 0, \\ \frac{dr}{dt} &= l \cdot i(t). \end{aligned} \tag{1.7}$$

These equations can be solved exactly for s, i, and r as functions of time, but such a solution is rather involved and for our purposes an approximate solution will be sufficient. The approximation will involve restrictions on the values of the parameters and it will be necessary to pay some attention to this matter.

To proceed with our study of (1.7), we introduce another constant $\rho = l/k$ which will be referred to as the relative rate of removal of infected individuals, or for short, the *relative removal rate*. With this notation the second equation in (1.7) can be written

$$\frac{di}{dt} = ki(t)[s(t) - \rho]. \tag{1.8}$$

We can use (1.8) to make some of our earlier comments precise. Let us assume that we begin measuring time from $t = 0$ and that initially the subpopulation R is empty, $r(0) = 0$. Therefore the initial population vector is $\mathbf{p}(0) = (s(0), i(0), 0)$. It is now necessary to be more precise about our use of the term epidemic. We say that an *epidemic develops* if di/dt is positive for some values of t. This usage is a technical one and is quite different from the common use of the terms. In day-to-day conversation one usually restricts these terms to the widespread occurrence of a reasonably dangerous disease.

Theorem 1: A necessary condition for an epidemic to develop is that the parameters k and l and the initial population vector $\mathbf{p}(0)$ be such that $s(0) - \rho > 0$, $\rho = l/k$.

Proof: The conclusion follows at once from the definitions and Eq. (1.8). If di/dt is to be positive for any values of t it must be positive at $t = 0$ since s is a decreasing function. Q.E.D.

Theorem 1 can be viewed in two different ways. If we assume that the initial population vector $\mathbf{p}(0)$ is given, then the theorem asserts that there is a certain critical value of the relative removal rate $\rho_0 = s(0)$ such that for values of ρ larger than ρ_0 an epidemic cannot begin. Thus artificial means of raising ρ, such as quarantine, which increases l, serve to impede the development of an epidemic. On the other hand, if we assume the parameter ρ to be given, then an epidemic can develop only if the initial population vector $\mathbf{p}(0)$ is such that $s(0)$ is larger than ρ. In this respect the model leads to predictions which agree with observation. A disease may be common in a locality for many years, never developing into an epidemic but suddenly increasing dramatically with the influx of new members of the population. The size of the population has exceeded the threshold value. Theorem 1 is one part of the *threshold theorem*; a second part is given below in Theorem 2. We turn next to an approximate solution of the system (1.7).

We begin by observing that the third equation of the system gives

$$ki(t) = \frac{1}{\rho}\frac{dr}{dt},$$

and substituting this into the first equation, we obtain

$$\frac{ds}{dt} = -\frac{1}{\rho}s(t)\frac{dr}{dt}$$

or

$$\frac{1}{s}\frac{ds}{dt} = -\frac{1}{\rho}\frac{dr}{dt}.$$

The latter differential equation can be solved by a straightforward integration. Taking into account the initial condition $r(0) = 0$, we obtain an equation connecting the functions s and r:

$$s(t) = s(0)e^{-r(t)/\rho}.$$

Since $s(t) + i(t) + r(t) = n + 1$ for all values of t, we have

$$i(t) = n + 1 - r(t) - s(0)e^{-(1/\rho)r(t)}.$$

Substituting this into the third equation of (1.7), we obtain a differential equation involving only the function r:

$$\frac{dr}{dt} = l[n + 1 - r(t) - s(0)e^{-(1/\rho)r(t)}]. \tag{1.9}$$

Although it is possible to solve this equation exactly for r as an explicit function of time, such a solution is complicated and we choose instead to solve an approximate equation.

From now on we shall suppress the explicit dependence of r on time and write simply r instead of $r(t)$. The difficulty encountered in solving (1.9) is due to the occurrence of the exponential term $e^{-r/\rho}$. One method of coping with this difficulty is to expand the exponential in a power series in r and retain only the first few terms. We expect the solution of the new differential equation to be useful in predicting the behavior of a solution to the original one. Expanding the exponential, we have

$$e^{-r/\rho} = 1 - \frac{r}{\rho} + \frac{1}{2}\left(\frac{r}{\rho}\right)^2 - \frac{1}{3!}\left(\frac{r}{\rho}\right)^3 + \cdots,$$

and substituting this expression for $e^{-r/\rho}$ in (1.9) results in

$$\frac{dr}{dt} = l\left[n + 1 - s(0) + \left(\frac{s(0)}{\rho} - 1\right)r - \frac{s(0)}{2}\left(\frac{r}{\rho}\right)^2 + \cdots\right].$$

It is now a matter of deciding how many terms of the infinite series are to be retained, that is, how good our approximation is to be. The retention of many terms clearly improves the approximation, but it also increases the technical difficulties of obtaining a solution. The final decision must result from a comparison of the advantages and inconveniences of retaining additional terms. Certainly we wish to retain at least the linear term in r. Otherwise the resulting equation could be solved immediately giving r as a linear function of t. Thus, r would increase indefinitely as time increases, and this is clearly incompatible with what we know of r from the real situation. If in the differential equation we discard all terms containing r to a power higher than 1, then the result is a nonhomogeneous linear differential equation. If an epidemic develops, i.e., if $s(0) > \rho$, then the magnitude of r increases indefinitely as time increases. This follows from the form of the solution (Appendix B). However, this behavior is incompatible with the definition of r. Thus the simplest approximation consistent with the real situation is one which includes at least the terms involving r^0 (constants), r, and r^2, and we proceed to use this approximation. That is, we solve the differential equation

$$\frac{d\tilde{r}}{dt} = l\left[n + 1 - s(0) + \left(\frac{s(0)}{\rho} - 1\right)\tilde{r} - \frac{s(0)}{2\rho^2}\tilde{r}^2\right], \tag{1.10}$$

where $\tilde{r}$ is a function which will be considered to be an approximation to the function r. We seek a solution valid for $t > 0$ and satisfying $\tilde{r}(0) = 0$. Actually there is exactly one such solution.

Although the algebra is somewhat involved, the solution of (1.10) presents no unusual problems. We introduce new constants

$$a = l(n + 1 - s(0)),$$

$$b = l\left(\frac{s(0)}{\rho} - 1\right),$$

$$c = \frac{ls(0)}{2\rho^2},$$

and we note that $c > 0$, since otherwise we have a linear equation. Also $a = (d\tilde{r}/dt)(0)$ must be positive if an epidemic develops, and we have already argued that $b > 0$ in the case of interest. We have

$$\frac{d\tilde{r}}{dt} = a + b\tilde{r} - c\tilde{r}^2$$

or

$$\frac{d\tilde{r}/dt}{a + b\tilde{r} - c\tilde{r}^2} = 1.$$

Since $\tilde{r}$ is a monotonically increasing function, we are interested only in $\tilde{r}$'s such that $a + b\tilde{r} - c\tilde{r}^2 > 0$, and with this restriction the equation can be integrated at once to obtain

$$-\frac{2}{q}\tanh^{-1}\left[\frac{-2c\tilde{r}+b}{q}\right]=t+c_1.$$

Here we use q for the quantity $(b^2+4ac)^{1/2}$, and the constant c_1 is to be chosen so that $\tilde{r}(0)=0$. Using the definition of the inverse hyperbolic tangent, we solve this for $\tilde{r}$ as a function t. Thus

$$\tilde{r}(t)=\frac{1}{2c}\left[b-q\tanh\left(-\frac{q}{2}t+c_2\right)\right],$$

or, in terms of exponential functions,

$$\tilde{r}(t)=\frac{1}{2c}\left[b+q\frac{1-e^{-qt+c_3}}{1+e^{-qt+c_3}}\right]. \tag{1.11}$$

In these equations the constants c_2 and c_3 are to be chosen so that $\tilde{r}(0)=0$. Since $q>b$ and $\tanh x$ increases monotonically from -1 to $+1$ as x increases from $-\infty$ to $+\infty$, it follows that both c_2 and $c_3=2c_2$ exist and are unique (as real constants). The precise values of the constants are of no particular significance in the present study.

We are now ready to turn to the main point of this discussion: the study of the asymptotic behavior of $\tilde{r}$, that is, the behavior of $\tilde{r}(t)$ for large values of t. Since every individual who is ever infected eventually winds up in R, this will give an estimate for the total size of the epidemic. We proceed by letting t become very large in (1.11). Then $\tilde{r}(t)$ tends to $\tilde{r}_\infty$, where

$$\tilde{r}_\infty=\frac{b+q}{2c}.$$

Writing this directly in terms of the parameters of the model, we have

$$\tilde{r}_\infty=\frac{\rho(s(0)-\rho)+\rho[(s(0)-\rho)^2+2s(0)i(0)]^{1/2}}{s(0)}.$$

If we now make some additional assumptions regarding the relative sizes of these parameters, we can obtain the second part of the threshold theorem mentioned earlier. In particular, it is usual to suppose that the epidemic is generated through the addition of a small number of infected individuals to a population consisting of members of S, i.e., $s(0)>\rho$ and $i(0)>0$. We use the quantity

$$\lim_{i(0)\to 0}\tilde{r}_\infty=\frac{2\rho(s(0)-\rho)}{s(0)}$$

to denote the asymptotic size of an epidemic resulting from the introduction of a *trace* of disease into a group S. We make one final approximation. Suppose that the initial value $s(0)$ is close to the *threshold* value ρ (ρ is assumed given). If $s(0) < \rho$, then no epidemic develops (Theorem 1). If $s(0) > \rho$, we can write $s(0) = \rho + \sigma$, σ small and positive. Then

$$\lim_{i(0)\to 0} \tilde{r}_\infty = \frac{2\sigma}{1 + (\sigma/\rho)} \cong 2\sigma.$$

That is, the asymptotic size of the epidemic is approximately 2σ. We state this formally as the remaining part of the threshold theorem.

Theorem 2: The total size of an epidemic resulting from the introduction of a trace infection into a population of susceptibles whose size $s(0)$ is close to the threshold value ρ is approximately $2(s(0) - \rho)$.

Another way of stating this result is that in these circumstances the number of uninfected susceptibles remaining after the epidemic is just as far *below* the threshold value ρ as the initial population size was *above* ρ.

We return to the system (1.7) for another comment. The second equation can be written

$$\frac{di}{dt} = ki(t)[s(t) - \rho],$$

and since in nontrivial situations s is a nonincreasing function and i is positive, we conclude that the number of infected individuals reaches a maximum when $s(t) = \rho$, the threshold value. In qualitative terms, the number of infected individuals initially increases until the size of the S subpopulation decreases to the threshold value. Then, as $s(t)$ becomes smaller than ρ, the number of infected individuals begins to decrease. Indeed, for values of t for which $s(t) < \rho$, any infected individual is more likely to become a member of R than to infect a member of S. One of the conclusions of this section, and a very interesting one, is that an epidemic satisfying these assumptions may run its course before the subpopulation S is exhausted. The precise size of the subpopulations at the end of the epidemic is, of course, determined by the parameters k and l and the sizes of the initial subpopulations.

As the final point in this discussion we reiterate the fact that the function $\tilde{r}$ is only an approximation to r. Thus the conclusions of Theorem 2 are not conclusions about r; they are only conclusions about an approximation to r. If the approximation is a bad one, then one cannot expect these results to yield useful information about r.

9.1.2 A Stochastic Model

The models considered in the preceding section were deterministic ones, and consequently they have the inherent limitations of such models. In particular, they proceed on the assumption that randomness, which is apparently an essential part of the process of infection transmission, can be safely ignored. We recall a point mentioned in the qualitative discussion preceding the construction of the models of Sec. 9.1.1, namely, that the assumption of uniform mixing is likely to be strictly valid only for relatively small groups, that is, for exactly those populations for which we have the most doubts about the validity of the deterministic approach. Thus it is time for us to pay more attention to the probabilistic nature of the process, and we turn to the construction of stochastic models. Since the theory developed for deterministic models seems to agree in a general way with our common-sense expectations of the development of an epidemic, we anticipate that the results predicted by deterministic and stochastic models will be qualitatively similar. Moreover, this agreement should become more and more precise as the size of the population increases. In the related question of population growth considered in Chap. 8, the stochastic theory led to a family of probability distributions for a random variable (the size of the population) whose mean value was identical to the solution of the associated deterministic problem. Thus we might be led to conjecture that the same situation prevails here. However, stochastic means and deterministic values do not usually coincide for epidemic models. Also, in the context of epidemics we shall find that stochastic models lead to more difficult mathematical problems, and consequently our discussion will not be as complete as that of Chap. 8.

The stochastic model constructed here will be based on the very simple assumptions of the first model discussed in the preceding section. That is, if S, I, and R have the same meaning as before, then we assume that the set R is empty for all times. With a stochastic model we do not attempt to predict the number of individuals in S (or I) at a given time; we view this number as a random variable, and we derive the family of probability distributions. That is, for each time t we derive a set of probabilities each of which can be interpreted as the probability that at time t the subpopulation S (or I) consists of a certain number of individuals. Even with the restrictive assumptions on the nature of the epidemic on which our model is based, a complete solution of the resulting mathematical problem is beyond the scope of this book. A study of more realistic models involves commensurately more difficulties. The interested reader can find details in the articles cited in the References.

We now turn to the construction of our model. As in the deterministic situation, we assume that initially, say at time $t = 0$, our population consists of n individuals in subpopulation S and one individual in I. We define probability functions P_m for $m = 0, 1, \ldots, n$ and $t \geq 0$ by

$$P_m(t) = Pr[s(t) = m].$$

Here, and in what follows, the functions s and i have the same meanings as in Sec. 9.1.1. The model is based on the following assumptions regarding the nature of the infection process (compare Sec. 8.4):

1. A member of subpopulation S is infected upon contact with a member of subpopulation I.
2. The probability that there is exactly one contact between a member of S and a member of I in the time interval $(t, t + \Delta t)$ is proportional to $s(t)i(t)\,\Delta t + h_1(\Delta t)$, where $h_1(\Delta t) = o(\Delta t)$, $\Delta t \longrightarrow 0$. The constant of proportionality is independent of time.
3. The probability that there is more than one contact between individuals from different subpopulations in the time interval $(t, t + \Delta t)$ is given by $h_2(\Delta t)$, where $h_2(\Delta t) = o(\Delta t)$, $\Delta t \longrightarrow 0$.
4. Once an individual is infected, he remains in subpopulation I for all future times.
5. If I_1 and I_2 are two disjoint time intervals, then the number of contacts between individuals from different subpopulations in interval I_1 has no effect on the number of such contacts in the interval I_2.

Let us now consider how the subpopulation S can arrive at time $t + \Delta t$ with a size m. There are the following possibilities: either the group S contained m individuals at time t and there were no contacts between individuals of different groups, or this group contained $m + 1$ individuals at time t and there was one contact between individuals from groups S and I, or this group contained $m + 2$ individuals at time t and there were two contacts, etc. Utilizing assumptions 1–5, we have

$$\begin{aligned}&\Pr[\text{No contacts between subpopulations } S \text{ and } I\\ &\qquad \text{in time interval } (t, t + \Delta t) \mid s(t) = m]\\ &= 1 - km(n + 1 - m)\Delta t - kh_1(\Delta t) - h_2(\Delta t),\end{aligned}$$

where k is the constant of proportionality of condition 2. Applying standard arguments from elementary probability theory (compare Sec. 8.4), we obtain

$$\begin{aligned}P_m(t + \Delta t) = {}& [1 - km(n + 1 - m)\Delta t - kh_1(\Delta t) - h_2(\Delta t)]P_m(t)\\ &+ k[(m + 1)(n - m)\Delta t + h_1(\Delta t)]P_{m+1}(t) + e(\Delta t),\end{aligned}$$

where $e(\Delta t)$ is a term which takes into account the possibility of more than one contact between members of different subpopulations. It follows from assumption 3 and the fact that there are only finitely many different possible contacts between pairs of individuals that $e(\Delta t) = o(t)$, $\Delta t \longrightarrow 0$. Collecting all terms in $P_m(t + \Delta t)$ which are small in comparison with Δt, we have

$$\begin{aligned}P_m(t + \Delta t) = {}& [1 - km(n + 1 - m)\Delta t]P_m(t)\\ &+ k(m + 1)(n - m)\Delta t P_{m+1}(t) + H(\Delta t),\end{aligned}$$

where $H(\Delta t) = o(\Delta t)$, $\Delta t \longrightarrow 0$. It follows at once that

$$\frac{P_m(t + \Delta t) - P_m(t)}{\Delta t} = -km(n + 1 - m)P_m(t) + k(m + 1)(n - m)P_{m+1}(t) + \frac{H(\Delta t)}{\Delta t},$$

and, finally, letting $\Delta t \longrightarrow 0$, we obtain the differential equation

$$\frac{dP_m}{dt} = -km(n + 1 - m)P_m + k(m + 1)(n - m)P_{m+1}. \tag{1.12}$$

This equation holds for $m = 0, 1, \ldots, n - 1$ but not for $m = n$. Indeed, in deriving Eq. (1.12), we considered the possibility that S contained $m + 1$, $m + 2, \ldots$ members, and this is clearly not possible if $m = n$. The corresponding equation for $m = n$ can be derived in a similar manner and is

$$\frac{dP_n}{dt} = -knP_n. \tag{1.13}$$

Note that (1.13) follows formally from (1.12) with $m = n$ if we define $P_{n+1} = 0$. Recall that at time $t = 0$ the size of the subpopulation S is known to be exactly equal to n, and this means that

$$P_n(0) = 1, \qquad P_m(0) = 0, \qquad m = 0, 1, 2, \ldots, n - 1. \tag{1.14}$$

The differential equations (1.12) and (1.13) and the initial conditions (1.14) provide a mathematical model based on assumptions 1–5. We base our further study on this mathematical formulation of the situation.

An explicit closed-form solution to this system can be obtained in several ways. We could, for example, solve (1.13) subject to (1.14) and substitute the resulting P_n into (1.12) with $m = n - 1$. In this way we obtain a differential equation for P_{n-1}. In turn we could solve this equation subject to (1.14) to obtain P_{n-1}. Continuing, substitution of the function P_{n-1} into (1.12) with $m = n - 2$ gives a differential equation for P_{n-2}, and so forth. In a moment we shall proceed in this way for a few steps so that the reader may appreciate the computational difficulties which arise when n is even modestly big, say $n > 10$. Another approach (see [B3]) allows a direct computation of P_m but involves more sophisticated mathematical techniques. In any case the resulting general expressions are quite complicated and not particularly illuminating.

We illustrate the first of the approaches mentioned above. The usual technique applied to Eq. (1.13) and the initial condition $P_n(0) = 1$ gives immediately

$$P_n(t) = e^{-knt}. \tag{1.15}$$

We obtain a differential equation for P_{n-1} by substituting this in (1.12) with $m = n - 1$:

$$\begin{aligned}\frac{dP_{n-1}}{dt} &= knP_n - 2k(n-1)P_{n-1} \\ &= kne^{-knt} - 2k(n-1)P_{n-1}.\end{aligned}$$

Techniques developed in Appendix B show how this equation can be solved, and taking into account $P_{n-1}(0) = 0$, we obtain

$$P_{n-1}(t) = \frac{n}{n-2}[1 - e^{-k(n-2)t}]e^{-knt}. \tag{1.16}$$

Again substitution of (1.16) into (1.12) gives rise to the differential equation

$$\begin{aligned}\frac{dP_{n-2}}{dt} &= 2k(n-1)P_{n-1} - 3k(n-2)P_{n-2} \\ &= \frac{2k(n-1)n}{n-2}[1 - e^{-k(n-2)t}]e^{-knt} - 3k(n-2)P_{n-2},\end{aligned}$$

whose solution satisfying the condition $P_{n-2}(0) = 0$ is

$$\begin{aligned}P_{n-2}(t) = {} & \frac{n(n-1)}{(n-2)(n-3)(n-4)}[(n-4) - 2(n-3)e^{-k(n-2)t} \\ & + (n-2)e^{-2k(n-3)t}]e^{-knt}.\end{aligned} \tag{1.17}$$

The computational difficulties inherent in this approach are becoming obvious, and it is not worthwhile to proceed further.

For the final topic in this discussion we return to an earlier comment regarding deterministic values and stochastic means. In the introductory remarks preceding the construction of our stochastic model, we asserted that for models of epidemic processes the stochastic means are in general different from the solutions of the corresponding deterministic equations. It is time to offer some support for this statement. For each time t we view the number k of individuals in subpopulation S as a random variable with corresponding probability distribution $\{P_m\}_{m=0}^{n}$. Therefore the expected value $E[k]$ of k, or the mean μ, is given as

$$E[k](t) = \mu(t) = \sum_{m=0}^{n} mP_m(t). \tag{1.18}$$

For notational convenience we prefer to use μ instead of $E[k]$. Although we were unable to obtain a workable expression for P_m, there is still hope that techniques similar to those used in Sec. 8.6 might enable us to obtain a useful expression for μ. We encourage the reader to make the effort (Exercise 13) and to note the source of the difficulty. More sophisticated techniques can be used to obtain an explicit formula for μ and also for the variance σ^2.

However, for our limited objective much less detailed information will be sufficient. We proceed by comparing only the functional values and the values of the first two derivatives of μ and the determinstic solution s at time $t = 0$. Recalling the deterministic equation for s, which we reproduce here for convenience,

$$\frac{ds}{dt} = -ks(n + 1 - s), \tag{1.1}$$

$$s(0) = n, \tag{1.2}$$

we have

$$\frac{ds}{dt}(0) = -kn, \tag{1.19}$$

$$\frac{d^2s}{dt^2}(0) = -k^2n(n - 1). \tag{1.20}$$

Equation (1.19) follows from (1.1) by using (1.2), and (1.20) arises by differentiating (1.1) and using (1.2) and (1.19). We next compute the corresponding quantities for the stochastic mean μ. It follows from (1.18) and (1.14) that

$$\mu(0) = n,$$

and consequently $s(0) = \mu(0)$. Differentiating (1.18) and using the differential equations for P_m, we obtain

$$\begin{aligned}\frac{d\mu}{dt} &= \sum_{m=0}^{n} m\frac{dP_m}{dt} \\ &= \sum_{m=0}^{n-1} m\{-km(n + 1 - m)P_m + k(m + 1)(n - m)P_{m+1}\} - kn^2P_n.\end{aligned}$$

Again using the conditions at $t = 0$, we find that

$$\frac{d\mu}{dt}(0) = kn(n - 1) - kn^2 = -kn,$$

and therefore $(ds/dt)(0) = (d\mu/dt)(0)$. We continue by comparing the second derivatives. It is useful to note from (1.12) that

$$\frac{dP_m}{dt}(0) = 0, \qquad m = 0, 1, 2, \ldots, n - 2,$$

and consequently, if the second derivative

$$\frac{d^2\mu}{dt^2} = \sum_{m=1}^{n-1} m\left[-km(n + 1 - m)\frac{dP_m}{dt} + k(m + 1)(n - m)\frac{dP_{m+1}}{dt}\right] - kn^2\frac{dP_n}{dt}$$

is evaluated at $t = 0$, then the sum on the right-hand side contains only four nonzero terms. In fact,

$$\begin{aligned}\frac{d^2\mu}{dt^2} &= (n-2)k(n-1)2\frac{dP_{n-1}}{dt}(0) - (n-1)k(n-1)2\frac{dP_{n-1}}{dt}(0)\\ &\quad + (n-1)kn\frac{dP_n}{dt}(0) - kn^2\frac{dP_n}{dt}(0)\\ &= 2k(n-1)(n-2)(kn) - 2k(n-1)^2(kn)\\ &\quad + kn(n-1)(-kn) - kn^2(-kn)\\ &= k^2n\{2(n-1)(n-2) - 2(n-1)^2 - n(n-1) + n^2\}\\ &= -k^2n(n-2),\end{aligned}$$

and obviously $(d^2s/dt^2)(0) \neq (d^2\mu/dt^2)(0)$. We conclude that the functions s and μ are not identical. Notice, however, that for large populations the ratio of the quantities $(d^2s/dt^2)(0)$ and $(d^2\mu/dt^2)(0)$ is very close to 1. Thus for large n and for small times the graphs of s and μ are quite similar (see also [H]).

We commented earlier on the importance of the deterministic epidemic curve, i.e., the graph of $-(ds/dt)$. There is some justification (see [B3]) for regarding the graph of $-(d\mu/dt)$ as the appropriate stochastic analog of this deterministic concept. For relatively small values of n, $n = 10$ or 20 say, these two curves appear much as shown in Fig. 9–2. It was shown earlier that the deterministic epidemic curve was symmetric about the time at which its maximum occurred. This not true for the stochastic curve, and, moreover, the latter decreases more slowly than it originally increased. The times at which the maxima are assumed in the two cases are close.

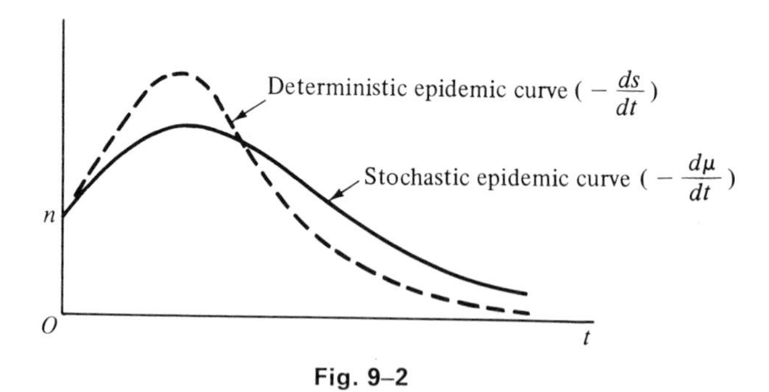

Fig. 9–2

The models developed in this section are not sufficiently refined to expect that predictions based on them will agree in detail with observations. However, they do provide a qualitative description of the course of an outbreak of disease under various assumptions. Refinements such as the recognition of an incubation period and the use of other distributions for the infectious period will provide models which agree more closely with most diseases. Several such extensions are considered, although not in great detail, in [B4].

EXERCISES

1. Consider the model associated with Eq. (1.1). Give a heuristic argument justifying the symmetry of the epidemic curve about the time t_M.
2. Solve (1.1) and (1.2) in the case where the proportionality constant k is not a constant but instead is a function of time: $k = k(t)$, $t \geq 0$. Assume that k is continuous.
3. Prove that the only solution of (1.1) which satisfies (1.2) is the function s defined by (1.5). *Hint:* If x is any other solution, write $x = sy$ and prove that the function y must be identically 1.
4. Graph the functions s and i given by (1.5) and (1.6) as functions of time.
5. Discuss in detail the qualitative nature of a disease which has a mathematical model characterized by Eq. (1.7)
6. Justify in detail the assertion that one must retain at least the first three terms in the Taylor series expansion of $e^{-r/\rho}$ in (1.9) for the resulting equation to be consistent with the real situation.
7. Show that the epidemic curve, i.e., the graph of $-(ds/dt)$, for the model given by (1.7) is not symmetric about any time.
8. For the model given by (1.7), show that the graph of dr/dt is a bell-shaped curve which is symmetric about t_M, where t_M is the unique value of t which satisfies
$$\frac{dr}{dt}(t_M) = \max_{0<t<\infty} \frac{dr}{dt}(t).$$
9. Prove that for the model governed by Eq. (1.10) one always has $\tilde{r}_\infty \geq 2\rho[s(0) - \rho]/s(0)$.
10. With regard to the discussion given in the derivation of the expressions for $\lim \tilde{r}_\infty$ as $i(0) \longrightarrow 0$, suppose that we wish to take into account first-order effects due to the term
$$\epsilon = \frac{2i(0)s(0)}{(s(0) - \rho)^2}.$$
Use the definition of $\tilde{r}_\infty$ and obtain expressions analogous to those of the text which contain the appropriate term linear in ϵ. Comment on this result in terms of its significance for the real problem.
11. Graph the functions P_n, P_{n-1}, P_{n-2} given by (1.15), (1.16), and (1.17) and discuss the behavior of these functions in detail.
12. Derive Eq. (1.13) using the same sort of arguments as were used in the derivation of (1.12).
13. In an analogous situation in Chap. 8 we were able to derive a differential equa-

tion for μ which did not contain any of the functions P_k. Try the same technique for the μ of this section and note the source of the difficulty.

14. Suppose that the population of infectives initially consists of q individuals and that the total population consists of $n + q$ individuals. Using a model similar to that of this section, derive a differential equation analogous to (1.12) for this situation.

15. Formulate a stochastic model for an epidemic in which the duration of infection is not long in comparison with the length of the epidemic. In particular, in addition to subpopulations S and I considered in this section, you should introduce a group R of removals. Derive differential equations for the probabilities P_{lm}, where $P_{lm}(t) = \Pr[i(t) = l, s(t) = m$ at time $t]$, $l = 0, 1, 2, \ldots, n + 1$, $m = 0, 1, 2, \ldots, n$.

9.2 DISCRETE-TIME MODELS FOR THE SPREAD OF RUMORS

Let us begin by reiterating a comment made earlier in this chapter concerning the applicability of growth models. A moment's thought should convince the reader that the models developed in the preceding section are not restricted to the spread of disease through a population. They are applicable to many other diffusion processes and, in particular, to the spread of a rumor. The validity of any of these models as representations of the rumor-spreading process depends, as always, on the degree to which our basic assumptions are true for this situation. Although there are many obvious similarities in the diffusion processes, there are also some differences. For example, a rumor is obviously spread through contacts between individuals, provided that we enlarge our definition of contact to include written and other "at a distance" communications. Thus, making the obvious changes in notation, an equation of the form (1.1) is a reasonable description of the rate at which a rumor spreads under assumptions similar to those of Sec. 9.1.1. On the other hand, once they are contracted, many diseases have a quite precisely known duration, whereas there will probably be a wide variation between individuals in their persistence at spreading a rumor. Any discussion which takes into account such factors must involve in an essential way the psychology and sociology of the process, and consequently it is beyond the scope of this study. Therefore in this section we shall proceed in another direction, and we shall use certain questions of rumor diffusion to consider another type of model. Our subject involves what are known generically as discrete-time models (see also Sec. 8.1.1).

9.2.1 Rumors with Limited Communication

There are circumstances in which it is either necessary or more convenient to consider time as a discrete rather than a continuous variable. We might, for example, consider an experimental study involving public opinion in which the population is sampled periodically, say daily or weekly. In such a case we are interested in constructing a model which predicts the composition $\mathbf{p}(t)$ of the population only for certain times t. That is, if we assume the composition known for some initial time t_0, then we wish to predict $\mathbf{p}(t)$ for t belonging to some discrete set T. We take $T = \{t_1, t_2, \ldots\}$. One can assume that the times t_j in the set T are selected either according to some definite scheme or by some random device. Frequently an appropriate choice of T will be indicated by the nature of the process to be modeled. This is true of the example which forms the major part of this section.

We consider a set of $N + 1$ villages in the rather thinly populated arid region of East Nowher. These villages are isolated from each other and the only communication is by means of a primitive telephone system. This system is such that any two villages can call one another, but only two villages can use the system at a time. We suppose that each village contains exactly one telephone. We use the term *village* here in the sense of a unit in our population. The population is, of course, the collection of $N + 1$ villages. We speak of one village calling another or having heard the rumor when we actually mean an individual in one village calls an individual in another or an individual hears the rumor. However, what happens internally in each village is of no concern for the present problem, so it is sufficient to consider each village as an indivisible unit. This convention is not only appropriate but very convenient for exposition.

Suppose that at time t_0 a rumor is introduced into one village. At some later time an individual in this village calls another village to pass the rumor along. Next, either of these villages can call another, and so forth; the rumor can be spread only through telephone communication. Even though it is not as suggestive in this example, we retain the notation of the preceding section. Thus we introduce

- S: The set of villages which have not yet heard the rumor but which would be interested in spreading it.
- I: The set of villages which have heard the rumor and are interested in spreading it to other villages.
- R: The set of villages not in the set $S \cup I$.

Note these that sets vary with time, and denote the number of villages in sets S, I, R at time t by $s(t)$, $i(t)$, $r(t)$, respectively. The model to be studied here is based on the following assumptions.

1. Whenever a village makes a call, it is equally likely to call each of the other N villages.

We have stated this assumption in terms of the callers only for convenience. It could just as well have been stated as a characteristic of the communication system. That is, the following is an equivalent assumption:

1′. The telephone system is such that all calls are connected at random. That is, if one village makes a call, then it is equally likely to reach each of the remaining N villages.

The remaining assumptions have to do with the effect of the telephone calls.

2. If a village of type I calls a village of type S, then the latter becomes a village of type I and the former remains type I.
3. If a village of type I calls a village of type I or R, then the village originating the call becomes a village of type R while the village receiving the call does not change its type.
4. Telephone calls originating in villages of type S or R do not alter the types of any of the villages.

As a result of assumptions 2–4 it follows that only telephone calls originating in villages of type I are of any significance insofar as our problem is concerned. Such calls will be referred to as *meaningful calls.*

Assumptions 2 and 3 are especially crucial ones in that they specify the reactions of the villages with respect to the rumor. In more descriptive language, the second assumption asserts that as soon as a village which is inclined to spread rumors hears one, it begins to spread it. Likewise, the third assumption says that as soon as a village which is actively spreading the rumor calls a village that already knows it or is not interested, then the village doing the calling loses interest. Interest in rumor spreading once lost cannot be rekindled through further telephone calls.

Our next task is to select the set T for which we shall predict the composition of the set of villages. Since it is only meaningful calls which are important, we shall take

$$t_k = \text{Time of conclusion of } k\text{th meaningful call}, \qquad k \geq 1.$$

An alternative would be to assume that meaningful calls occur at regular intervals and then select the t_k's in a regular manner and such that exactly one meaningful call occurs in each interval $t_j < t \leq t_{j+1}, j = 0, 1, 2, \ldots$. Recall that the composition of the set of villages at time t_0 is known. If $\mathbf{p}(t) = (s(t), i(t), r(t))$, then $\mathbf{p}(t_0) = (N, 1, 0)$.

The randomness with which rumor dissemination takes place cannot usually be completely ignored, and purely deterministic models are only useful as first approximations. There is a blend of determinism and probability in the model considered here. More precisely, Assumption 1′ means that the village originating the call does not know in advance the type of village it will reach, while Assumptions 2 and 3 specify exactly how the villages react once the call is completed. A more elaborate model would admit the possibility of several responses to a meaningful call with appropriate probability distributions for the alternatives. Returning to the case at hand, the partially stochastic nature of our model makes it impossible to compute $\mathbf{p}(t) = (s(t), i(t), r(t))$, $t \in T$, precisely. The most we can hope to determine is a sort of expected value for $\mathbf{p}(t)$ instead of $\mathbf{p}(t)$ itself. We pursue this line of thought. The vector $\mathbf{p}(t_1)$ can be determined exactly since the first meaningful call must go to a village of type S. However, at the very next step Assumption 1 introduces uncertainty into the picture. The second meaningful call goes to a village of type S with probability $(N - 1)/N$ and to a village of type I with probability $1/N$. The composition vector $\mathbf{p}(t_2)$ is therefore not well defined. In a situation of this sort it is natural to turn to expected values, and the expected value $E[\mathbf{p}](t_2)$ can easily be obtained. We take this as our definition of the *composition* of the set of villages at time t_2. If we now take $E[\mathbf{p}](t_2)$ as the actual composition at time t_2, then we can compute the probabilities needed to obtain the expected value at t_3, and so forth. This procedure can obviously be continued step by step for all times t_k in the set T. We should not, however, continue to denote the result by $\mathbf{p}$ or even by $E[\mathbf{p}]$. We make the following definition.

Definition: The *composition* $\mathbf{p}_k$ of the set of villages at time t_k is defined recursively as follows:

$$\mathbf{p}_0 = (N, 1, 0),$$

$$\mathbf{p}_1 = (N - 1, 2, 0),$$

$$\mathbf{p}_2 = (s_2, i_2, r_2) \text{ is defined to be } E[\mathbf{p}](t_2),$$

and $\mathbf{p}_k = (s_k, i_k, r_k)$, $k = 3, 4, \ldots$, is defined as the expected value of $\mathbf{p}$ at t_k given that $\mathbf{p}(t_{k-1}) = \mathbf{p}_{k-1}$.

This definition is adequate for our investigations. It provides a unique set of vectors which takes into account in a reasonable way the stochastic elements of our model and which contains information on the composition of the set of villages at each time in the set T.

Let us now see what conclusions can be drawn from a model based on these definitions and assumptions. If $\mathbf{p}_{k-1} = (s_{k-1}, i_{k-1}, r_{k-1})$, then the vector $\mathbf{p}(t_k)$ must be either

$$(s_{k-1} - 1, i_{k-1} + 1, r_{k-1}) \tag{2.1}$$

or

$$(s_{k-1}, i_{k-1} - 1, r_{k-1} + 1). \tag{2.2}$$

The vector (2.1) results whenever the kth meaningful call reaches a village of type S, and this happens with probability s_{k-1}/N. The vector (2.2) results from meaningful calls to villages of type I or R, and consequently occurs with probability $(N - s_{k-1})/N$. Thus, since s_k is the first coordinate of the expected value of $\mathbf{p}(t_k)$ given $\mathbf{p}(t_{k-1}) = \mathbf{p}_{k-1}$, we have

$$\begin{aligned} s_k &= \frac{s_{k-1}}{N}(s_{k-1} - 1) + \frac{N - s_{k-1}}{N}(s_{k-1}) \\ &= \left(\frac{N-1}{N}\right)s_{k-1}. \end{aligned} \tag{2.3}$$

Similarly,

$$\begin{aligned} i_k &= \frac{s_{k-1}}{N}(i_{k-1} + 1) + \frac{N - s_{k-1}}{N}(i_{k-1} - 1) \\ &= i_{k-1} + \frac{2}{N}s_{k-1} - 1. \end{aligned} \tag{2.4}$$

Again, as a consequence of the equality $s_k + i_k + r_k = N + 1$, it is unnecessary to compute r_k (Exercise 1). To obtain some feeling for the variation of s_k and i_k with increasing values of k, i.e., with increasing time, we evaluate the first few vectors $\mathbf{p}_k$:

$$\begin{aligned} \mathbf{p}_0 &= (N, 1, 0), \\ \mathbf{p}_1 &= (N - 1, 2, 0), \\ \mathbf{p}_2 &= \left(N - 2 + \frac{1}{N}, 3 - \frac{2}{N}, \frac{1}{N}\right). \end{aligned}$$

The explicit solution of (2.3) for $k = 1, 2, \ldots$ is straightforward (Exercise 2), and we obtain

$$s_k = N\left(\frac{N-1}{N}\right)^k, \qquad k = 0, 1, 2, \ldots. \tag{2.5}$$

If we use this result in (2.4), then we have

$$i_k = i_{k-1} + 2\left(\frac{N-1}{N}\right)^{k-1} - 1,$$

or, in a form which is more convenient for our intentions,

$$i_k - i_{k-1} = 2\left(\frac{N-1}{N}\right)^{k-1} - 1.$$

This equation is a *pure difference equation* for i_k. That is, the value of the difference $i_k - i_{k-1}$ is known for every value of k, $k = 1, 2, 3, \ldots$. Such a system of difference equations can be solved simply by writing them for

$k = 1, 2, \ldots, m$ and adding. Carrying this out, we have

$$i_1 - i_0 = 2 - 1$$
$$i_2 - i_1 = 2\left(\frac{N-1}{N}\right) - 1$$
$$\cdots\cdots\cdots\cdots\cdots\cdots$$
$$i_m - i_{m-1} = 2\left(\frac{N-1}{N}\right)^{m-1} - 1,$$

and adding both sides of these m equalities, we obtain

$$i_m - i_0 = 2\sum_{k=0}^{m-1}\left(\frac{N-1}{N}\right)^k - m.$$

The simple form of the left-hand side results from the cancellation of all i_j's except for i_0 and i_m. Finally, $i_0 = 1$ and $\sum_{k=0}^{m-1} x^k = (1 - x^m)/(1 - x)$, and so

$$i_m = 2N\left[1 - \left(1 - \frac{1}{N}\right)^m\right] + 1 - m. \tag{2.6}$$

Equation (2.6) holds with $m = 0, 1, 2, 3, \ldots$. All further discussion of this model will be based on these expressions for s_k and i_k [$k = m$ in (2.6)].

First, let us consider briefly the qualitative behavior of s_k and i_k. Clearly, s_k is a monotone decreasing function as k increases and $\lim_{k\to\infty} s_k = 0$. The behavior of i_k is not quite so evident. As k increases, the expression $[1 - (1 - 1/N)^k]$ increases, but $-k$ obviously decreases, and it therefore becomes a question of relative magnitudes. The usual techniques of the calculus, motivated by appropriate pictures, lead us to conclude that for small values of k the quantity $2N[1 - (1 - 1/N)^k]$ increases more rapidly than k, but for large values of k the situation is reversed (Exercise 3). Thus we expect the graph of i_k to appear somewhat as in Fig. 9–3. The data in Fig. 9–3 actually represent the sequence $\{i_k\}$ for $N + 1 = 30$. We have also included the sequence $\{s_k\}$ for comparison. In fact, making some straightforward approximations, it can be shown that i_k (as a function of k) has a relative maximum which is also a global maximum. This maximum value may in certain cases be achieved for two distinct but consecutive values of k (Exercise 4).

9.2.2 The Life Span of a Rumor

One of the most important predictions provided by our model concerns the length of time the rumor is in circulation. Once we know

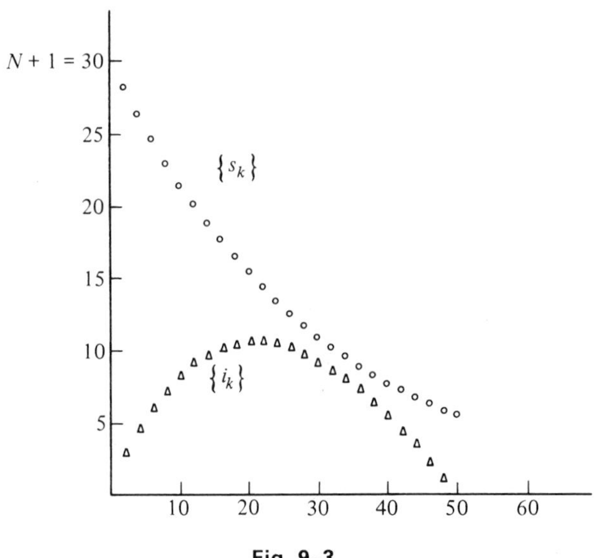

Fig. 9–3

this, we should be able to estimate the total number of villages which the rumor reaches. The first step here is to define what is meant by a rumor being in circulation. With the villages classified into types as in Sec. 9.2.1, it is natural to say that the rumor is in circulation as long as the set of villages of type I is nonempty and that the rumor is dead as soon as this set is empty. More precisely, we say that the rumor is dead as soon as no more meaningful calls are possible. Thus it dies after the kth meaningful call if k is the last integer for which $i_k \geq 0$. The mathematical problem is, therefore, given the number $N + 1$ of villages, find the integer k_0 for which $i_{k_0} \geq 0$ and $i_j < 0$ for $j > k_0$. To begin, we consider the equation $i_{k_0} = 0$, although, of course, this equation may not have an integer solution. We return to Eq. (2.6), and we observe that solving $i_{k_0} = 0$ involves solving a transcendental equation. We expect this to require other than purely algebraic methods, and we again find approximation techniques useful. We shall proceed with the help of reasonable approximations to obtain an estimate for k_0. Finally, we shall compute k_0 exactly for some specific values of N and compare the two results.

To justify our approximations, we reason as follows. If N is quite large, then we expect the integer k_0 to be large as well, probably at least comparable to N in magnitude. Let us write $k_0 = \lambda N$ and see if we can obtain a useful estimate on λ. Actually, k_0 can always be written as $\lambda(N)N$, but we hope to do this with a constant λ rather than a λ which depends on N. From the condition $i_{k_0} = 0$ and Eq. (2.6) we obtain

$$0 = 2N\left[1 - \left(1 - \frac{1}{N}\right)^{\lambda N}\right] - \lambda N + 1$$

or

$$\lambda = 2 - 2\left[\left(1 - \frac{1}{N}\right)^N\right]^\lambda + \frac{1}{N}. \tag{2.7}$$

Now, if N is very large, then $1/N$ can be neglected and $(1 - 1/N)^N$ will be very close to e^{-1}. Therefore we expect that the λ which satisfies (2.7) will be approximately equal to λ^*, where λ^* is the solution of

$$\lambda^* = 2 - 2e^{-\lambda^*}. \tag{2.8}$$

This equation has exactly one solution so it makes sense to refer to *the* λ^* which satisfies it. Since (2.8) is again a transcendental equation, for λ^* instead of k_0, the reader might wonder if we have made any real progress. Indeed, we have simplified the situation and in the process have obtained two new pieces of information. First, since (2.8) does not contain N explicitly, we conclude that for large values of N, k_0 is approximately equal to $\lambda^* N$, where λ^* is a constant. Second, Eq. (2.8) is of a technical form which is relatively easy to solve, or at least solve approximately. Let us denote the right-hand side of (2.8) by $g(\lambda^*)$; that is, we define a function g by

$$g(\lambda^*) = 2 - 2e^{-\lambda^*}.$$

Then Eq. (2.8) can be written

$$\lambda^* = g(\lambda^*), \tag{2.9}$$

and the problem is seen to be one of finding a fixed point of the function g. Equation (2.9) can be solved graphically or by an iterative procedure.

We consider a graphical method first. The graph of the function g is given by the solid curve in Fig. 9–4, and the graph of the identity function is given by the dashed line. Thus the point of intersection, denoted by P in the figure, provides the desired value of λ^* (Exercise 5).

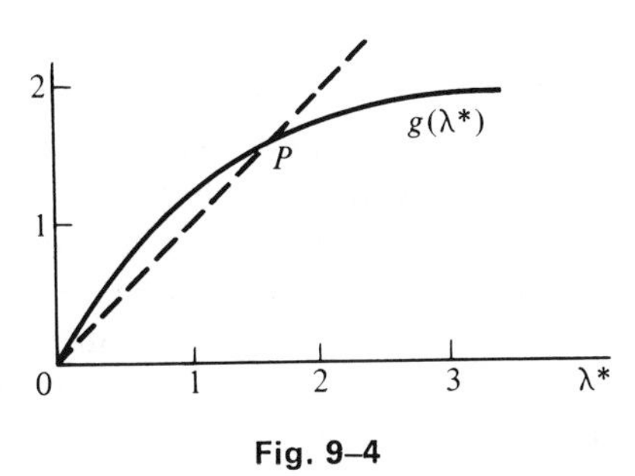

Fig. 9–4

An algorithm for the solution of equations of the form (2.9) is given in Appendix A. Adapting the notation introduced there to the present situation, we define f by $f(x) = x - g(x)$. We have $f''(x) = 2e^{-x}$ and consequently $f''(x)$ is positive for all values of x. Also, $f(2) = 2e^{-2}$ is positive. Thus the iterative method described in Sec. A.2 can be applied with $x_1 = 2$. In this case, the successive approximations x_n to a solution of the equation $f(x) = 0$ are

$$x_1 = 2.000$$
$$x_2 = 1.629$$
$$x_3 = 1.594$$
$$x_n = 1.594 \qquad n > 3,$$

to three decimal place accuracy. The result stated in Appendix A asserts that the sequence $\{x_n\}$ converges to a solution λ^* of the equation $f(x) = 0$. We conclude that Eq. (2.9) is satisfied with $\lambda^* = 1.594$. Here we have used Newton's method; another iterative method of computing λ^* is the topic of Exercise 7.

If N is sufficiently large, then k_0 is approximately equal to $1.594N$, and we conclude that the rumor dies after there have been $1.594N$ meaningful calls. Of course, the number of meaningful calls must be an integer, so this conclusion must be modified in a minor way. We take k_0 to be the largest integer which is no larger than $1.594N$ (why?). Note that k_0 may not be the closest integer to $1.594N$.

How many villages never hear the rumor? To answer this question, we must determine the number of villages of type S remaining when the rumor dies. This number, s_{k_0}, is approximately equal to $s_{\lambda N}$, and according to Eq. (2.5),

$$s_{\lambda N} = N\left(\frac{N-1}{N}\right)^{\lambda N}.$$

Again assuming that N is large enough to justify replacing $(1 - 1/N)^N$ by e^{-1}, we claim that $s_{\lambda N}$ is about the same as $s_{\lambda^* N}$,

$$s_{\lambda^* N} = Ne^{-\lambda^*} = 0.238N.$$

Thus we conclude that when the rumor ceases to circulate there will be almost 24% of the villages which have not heard it.

Whenever approximations of this sort are introduced, one should either provide an analysis of the errors or at least do enough computational work to provide numerical support for their validity. Table 9–1 summarizes some numerical data for comparison with the results of this section. The integer k_0 is as above. We define an integer M (or a pair of consecutive integers) by

$$i_M = \max_{1 \le j \le k_0} i_j.$$

In this model the quantity i_M gives the maximum number of villages spreading the rumor at any one time, and M is such that after M meaningful calls the maximum number of villages is spreading the rumor. We selected five values of N and computed the corresponding values of M, i_M, k_0, and k_0/N. The latter should be close to λ or λ^*, at least for a large number of villages. In Table 9–1 notice that the number i_M is not an integer, although the number

Table 9–1

Number of villages	M	i_M	k_0	k_0/N
10	7	4.4	17	1.700
50	35	16.7	81	1.620
100	69	32.0	161	1.610
500	347	154.8	799	1.598
1000	693	308.2	1595	1.595

of villages of type I at any time is an integer. This discrepancy is, of course, a result of the method of analysis adopted here.

EXERCISES

1. With $\mathbf{p}_k = (s_k, i_k, r_k)$ as defined in this section, derive a formula for r_k and prove that $r_k + i_k + s_k = N + 1$.
2. Give a formal proof using mathematical induction that if $s_0 = N$ and $s_k = ((N-1)/N)s_{k-1}$, $k \geq 1$, then $s_k = N((N-1)/N)^k$, $k \geq 1$.
3. Prove that for fixed N the quantity

$$2N\left[1 - \left(1 - \frac{1}{N}\right)^k\right] - k + 1$$

   is increasing for small values of k and decreasing for large values.
4. Prove that (2.6) has exactly one maximum value which is achieved either for one value of m or for two consecutive values. Remember that m is an integer.
5. Refine the graphical technique introduced in Sec. 9.2.2 and estimate the location of the fixed point of g as closely as you can.
6. How large must N be for the solutions of (2.7) and (2.8) to differ by less than 0.05?
7. Another iterative method of solving an equation of the form $x = g(x)$ which can be applied to Eq. (2.9) is given by the following theorem: Suppose g is a continuous function defined on an interval $[a, b]$ with range contained in $[a, b]$ and for which there is a constant $k < 1$ such that $|g'(x)| \leq k$ for all $x \in [a, b]$. Take $x_0 \in [a, b]$ and define a sequence $\{x_n\}$ by $x_n = g(x_{n-1})$, $n = 1, 2, \ldots$. Then the sequence $\{x_n\}$ converges to a solution x^* of the equation $x = g(x)$.

(a) Verify that the theorem can be applied to Eq. (2.9) and the interval [1, 2].
(b) Take $x_0 = 1$ and evaluate the first five terms in the sequence $\{x_n\}$.
Note that in this example the convergence of the sequence $\{x_n\}$ generated using this method is much slower than the convergence of the sequence generated by using Newton's method.

8. The model developed in this chapter has many of the features of a Markov chain. Pursue this idea. In particular, develop as thoroughly as you can a Markov chain model for the spread of a rumor under the assumptions of this section.

9. Formulate a continuous-time analog for the model of this section. How must the assumptions be modified?

9.3 MODELS FOR QUEUES

We conclude the present chapter with an abbreviated discussion of the questions and methods involved in the construction of models for situations involving waiting lines, or, to use the term we prefer, *queues*. One's first impression may be that queues do not obviously display the same growth characteristics as the epidemics and rumors of the preceding sections. However, once an effort is made to describe the development of a queue in precise terms, then the connection becomes more apparent. The relation of this to the earlier sections will become even clearer when we begin the detailed model building. We caution the reader that what follows is not to be interpreted as a survey of queuing theory but rather as a sampling of those aspects of the subject which are likely to be of most interest to the individual seeking to acquire a knowledge of the basic notions for applications in model building.

Real-world situations involving queues are very common, and consequently there has been substantial interest in constructing mathematical models with these characteristics. Certain obvious instances come immediately to mind: lines at supermarket checkout counters, theater box offices, and college registrations; airplanes waiting to land at Kennedy Airport on Friday afternoon; and automobiles at a tollway ticket booth. Less obvious examples might be the accumulated mail waiting to be processed at the local postoffice or the stack of examinations waiting to be typed in the department office. A little thought is sufficient to identify the features common to all these situations. A unit (person, airplane, automobile, letter, examination) arrives at a service facility (checkout counter, airport, tollbooth, postoffice, typist) to be processed, but due to the presence of other units, a delay is encountered. The sequence arrival, delay, service is the object of our study.

We shall begin by considering further the basic concepts common to all queues, and we shall introduce a simple but hopefully illuminating example. In the next subsection we shall develop a model for a special but important type of queue, and we shall continue with comments on generalizations in Sec. 9.3.3. Since many problems arising in applications lead to queuing models for which the mathematical analysis is either formidable or perhaps even impossible, we continue in the next chapter with Monte Carlo techniques and simulation.

9.3.1 Basic Concepts

It will be useful in the discussion which follows to have an example in mind which gives concrete meaning to the concepts we introduce. Although this example is quite simple, it displays the essential characteristics of much more general processes. Our setting is the shop of the neighborhood hair stylist Mr. Ur Soot. We note that many traditionalists still think of Mr. Soot as a barber and his shop as a barbershop, and since this setting is an appropriate one for our purposes we use these terms. We suppose that Mr. Soot is the only barber working in the shop and that he cuts hair whenever he has a customer: There are no lunch or coffee breaks in his barber business. The first point to be considered is the arrival of customers: How many are expected and when will they arrive? It has been observed that the rate of arrivals varies with the day of the week, the time of the day, the weather, and many other factors. One can take these variations into account, but for our purposes, and as a first approximation to the more complicated (and more realistic) situation, we shall assume that the arrivals are randomly distributed throughout the day. If we make this assumption, then Mr. Soot gives us an estimate of an average of three customers per hour as a reasonable numerical value. As the customers enter the shop, they take a number and hair cuts are given in order. Finally, the length of time required for Mr. Soot to cut a customer's hair is again a variable quantity. Among other factors, it depends on the length of the hair, the style of the cut, and how well Mr. Soot knows the customer (the care he uses and how much they talk). Again, Mr. Soot provides a numerical estimate. He states that it is reasonable to assume that he spends an average of 15 minutes on each haircut.

With this information we can develop a mathematical model for Mr. Soot's barbershop, and this is what we shall do in detail in a moment (Sec. 9.3.2). An important part of our task is to make the ideas introduced in a naive way above much more precise. After we have agreed on exactly what the terms mean, we shall be able to answer in an appropriate sense such questions as, how long should a customer expect to wait for a haircut? On the average, how many customers are waiting for haircuts at a given time? How does the situation change if Mr. Soot does less talking and averages 12 minutes per haircut instead of 15? The answers to questions such as these may be more than curiosities to Mr. Soot. As an illustration, we cite the possibility of expansion. The question of opening another chair in the shop is directly related to average waiting times. Customers will rarely patronize a barbershop where the expected waiting time is more than one half or three quarters of an hour. Consequently, if the rate of arrivals increases and Mr. Soot is unable to further reduce service time, then a knowledge of the average waiting times will assist him in deciding whether it is better to hire an assistant or to lose a few customers.

Before we begin the development of our detailed model for the barbershop, however, it is useful to consider general queues and to identify some of their important features. What aspects of a queuing process give it its distinctive characteristics? There are three essential subprocesses to be considered: the arrival of units at a service facility, the selection of units to be serviced from among the arrivals, and finally the serivce activity itself. Here we adopt terminology of units, arrivals, service, etc., as general words which encompass all sorts of special cases—in particular, all those mentioned above. Thus insofar as we are concerned, the queuing process will be completely determined once we specify how the arrivals appear—in short, the *arrival scheme;* how the selection from among the arrivals is made—the *selection scheme* or *queue discipline;* and the schedule involved in administering the service—the *service scheme.* In many cases, indeed probably in most cases arising in applications, the total group of arrivals is composed of several subgroups, and the service facility consists of several subfacilities each capable of providing the complete service. For example, on a tollway there are usually several lanes and several tollbooths, each of which can handle any automobile. The arriving automobiles can come in on any lane and go through any booth. Also, in the context of Mr. Soot's barbershop, it might be that the barbershop has more than one entrance, and if he hires an assistant, there will be two service facilities. For the present discussion we consider all arrivals as a single homogeneous group, and we suppose that units are handled by the service facility one at a time. Some relaxations of these assumptions will be considered later in the text and in the Exercises. For the present we shall consider only this setting.

The easiest of the three schemes for us to discuss is the second. We formulate models only under the first come-first served selection scheme. That is, we suppose that as each unit arrives it is given a number and that the units are then serviced in the order specified by the numbers. Other selection methods are, of course, possible. For example, a man may leave the barbershop because he is tired of waiting, or priorities may be assigned to arrivals, as with regular customers who have standing appointments. Another illustration of the latter is encountered in the airline reservation system. Individuals traveling on youth standby or military standby status may lose their position in the list for seat assignment if a full-fare-paying passenger arrives. As one expects, other selection schemes lead to significantly more complicated models, and the answers to some of the questions posed above, such as those involving average waiting times, may be quite involved.

Next we consider arrival schemes, and we identify two quite different types. It is convenient to introduce the notion of *interarrival interval.* If units arrive at times $t_1, t_2, t_3, \ldots$, then the time intervals $t_2 - t_1, t_3 - t_2, \ldots$ are known as interarrival intervals. We classify arrival schemes into two fundamentally different groups.

1. *Deterministic Arrivals.* In this case the times of successive arrivals—or, what is the same thing, the time of the first arrival and the interarrival intervals—are precisely known. The lengths of the interarrival intervals may be equal or unequal but they must be distributed according to a known rule which gives definite values. A mechanized production line is often set up so that units arrive at certain points on an exact schedule.

2. *Stochastic Arrivals.* Here the interarrival intervals are not precisely known and the arrivals obey no fixed schedule. If the arrivals are completely haphazard, that is, if nothing at all is known about the distribution of the lengths of the interarrival intervals, then nothing more can be said about the behavior of the queuing process. To continue we must make some assumptions concerning the interarrival intervals. There are obviously many possible assumptions and whichever one is made must be justified in terms of the real-world situation on which the model is based.

Similarly, one can develop models with either deterministic or stochastic service schemes. The model as a whole can therefore be described as completely deterministic, completely stochastic, or mixed.

A model for a completely determinstic queuing process can easily be constructed. The details are simpler for the case of regular arrivals and constant service time, but the same techniques are applicable even if this is not true. We proceed under the simpler assumption. In particular, we suppose that the arrivals are regular at a rate of λ per hour and that the first unit arrives at the moment the service facility is available for use. Finally, we suppose that the time to service one unit is $1/\mu$ hours. Thus, if the service facility is in continuous use, then the rate of service is μ units per hour. There are two quite different situations according as $\lambda \leq \mu$ or $\lambda > \mu$. If $\lambda \leq \mu$, then there is never any queue at all. The service on one unit is completed before (in the case $\lambda < \mu$) or at the same time as (in case $\lambda = \mu$) another unit arrives. If $\lambda > \mu$, then a queue does form, and in fact it grows indefinitely long as time passes. If N hours have passed since the first arrival, then there are approximately $(\lambda - \mu)N$ units waiting to be served. Additional information about this and related deterministic models can be obtained by devising appropriate representations and alogorithms (Exercises 1 and 2).

9.3.2 A Typical Model

We proceed now with the construction of a model for the barbershop example introduced in Sec. 9.3.1. This model is special in the sense that specific assumptions are made concerning interarrival intervals and service times, but it is typical in that the mathematical techniques used can be applied under a reasonably wide class of assumptions. Also, the results are in general indicative of the kind of information one obtains even in more elaborate studies. To be more precise, we investigate here a single-station queuing process where the interarrival intervals and service times are random with known mean or expected values. We assume that the queue discipline is first come-first served. Our first task is to make explicit just what is meant by random arrivals.

Let $n = n(t)$ be the number of units in the queue (individuals waiting for a haircut) at time t. We specifically include in this figure the unit (if any) which is actually receiving service at that time. Thus, if at time t there are no units waiting and none being serviced, then $n(t) = 0$, while if there are no units waiting but there is one being serviced, then $n(t) = 1$. Let $P_n = P_n(t)$ be the probability that at time t the queue length is n units. The basic assumptions made here concerning the notion of randomness are as follows.

1. There is a constant λ (> 0) such the probability of exactly one new unit arriving in the time interval $(t, t + \Delta t)$ is given by

$$\lambda \Delta t + h_1(\Delta t),$$

where

$$h_1(\Delta t) = o(\Delta t), \qquad \Delta t \longrightarrow 0.$$

Note that h_1 is assumed to be independent of the initial time t of the interval $(t, t + \Delta t)$; that is, it depends only on the length of the interval in question. The constant λ is known as the *mean arrival rate.*

2. The probability of more than one unit arriving in the time interval $(t, t + \Delta t)$ is $h_2(\Delta t)$, where

$$h_2(\Delta t) = o(\Delta t), \qquad \Delta t \longrightarrow 0.$$

3. If I_1 and I_2 are two nonoverlapping time intervals, then the number of units arriving in interval I_1 has no bearing on the number of arrivals in interval I_2.

A process for which the occurrence of an event (arrival of a unit) satisfies conditions 1–3 is known in probability theory as a *Poisson process.* Thus it is common to refer to our assumptions as the assumption of *Poisson arrivals.* The reader should note the similarity of these assumptions and those of Secs. 8.4 and 9.1.2. All these systems are examples of probabilistic processes known as birth-death processes. We make similar assumptions concerning service times. Specifically, we assume that

1′. There is a constant μ (> 0) such that if there is a unit being serviced at time t, then the probability that the service is completed in the interval $(t, t + \Delta t)$ is

$$\mu\Delta t + k_1(\Delta t),$$

where

$$k_1(\Delta t) = o(\Delta t), \qquad \Delta t \to 0.$$

Again we assume that the function k_1 depends only on the length Δt of the time interval. The constant μ is known as the *mean service rate.*

2′. The probability of service being completed on more than one unit in the time interval $(t, t + \Delta t)$ is $k_2(\Delta t)$, where

$$k_2(\Delta t) = o(\Delta t), \qquad \Delta t \to 0.$$

3′. If I_1 and I_2 are two nonoverlapping time intervals, then the number of units whose service is completed in interval I_1 has no effect on the number of units whose service is completed in I_2.

These are our fundamental assumptions. It will be necessary to make additional ones as we proceed with our study. The reader should note that by definition the processes of arrival and service are independent, subject, of course, to the condition that service can take place only if there are units to be serviced.

There is a small but important point that crops up frequently in the arguments that follow: What is the probability that the queue at time t is empty and that there is exactly one arrival in the interval $(t, t + \Delta t)$ whose service is completed in this interval? By our assumptions, the probability of exactly one arrival is $\lambda\,\Delta t + h_1(\Delta t)$, and the probability of this arrival having its service completed in the interval is no larger than $\mu\,\Delta t + k_1(\Delta t)$. Indeed, if service is begun after time t, then the probability of completion must be no larger than if it began at time t. Thus the probability of arrival and departure in the interval $(t, t + \Delta t)$ is no larger than

$$l(\Delta t) = [\lambda\,\Delta t + h_1(\Delta t)][\mu\,\Delta t + k_1(\Delta t)] = o(\Delta t), \qquad \Delta t \to 0.$$

Let us now consider how we can arrive at time $t + \Delta t$ with a queue consisting of n units. For the moment suppose that $n > 0$ and recall that the queue length includes the unit being serviced. First, we could have begun at time t with a queue of n units, and there were no arrivals and no departures during the interval $(t, t + \Delta t)$. Second, we could have begun at time t with a queue of $n - 1$ units, and there was one arrival and no departures in the interval. Third, at time t there was a queue of $n + 1$ units, and there was one departure and no arrivals in the interval. In addition, there are all the other possibilities corresponding to queue lengths at time t of $1, 2, 3, \ldots, n - 2,$

$n + 2, n + 3, \ldots$. Using our assumptions, we have

$$\begin{aligned} P_n(t + \Delta t) = P_n(t)&(1 - \lambda\,\Delta t - h_1(\Delta t) - h_2(\Delta t))(1 - \mu\,\Delta t - k_1(\Delta t) - k_2(\Delta t)) \\ &+ P_{n-1}(t)(\lambda\,\Delta t + h_1(\Delta t))(1 - \mu\,\Delta t - k_1(\Delta t) - k_2(\Delta t)) \\ &+ P_{n+1}(t)(\mu\,\Delta t + k_1(\Delta t))(1 - \lambda\,\Delta t - h_1(\Delta t) - h_2(\Delta t)) \\ &+ H_1(\Delta t), \end{aligned} \tag{3.1}$$

where $H_1(\Delta t) = o(\Delta t)$, $\Delta t \longrightarrow 0$. Equation (3.1) follows from our assumptions and standard arguments involving probabilities. The term $H_1(\Delta t)$ includes the probabilities corresponding to all those possibilities involving at least two events (arrivals or departures). The estimate on the behavior of $H_1(\Delta t)$ as $\Delta t \longrightarrow 0$ follows from a rather complicated but not particularly illuminating mathematical argument involving infinite series with binomial coefficients. It is certainly reasonable in light of assumptions 2 and 2′ to expect this behavior. We do not give the proof, but we invite the reader to construct one of his own. After rearranging and collecting terms in (3.1), we obtain

$$\begin{aligned} P_n(t + \Delta t) - P_n(t) = -(\lambda + \mu)P_n(t)\,\Delta t + \lambda P_{n-1}(t)\,\Delta t + \mu P_{n+1}(t)\,\Delta t \\ + H_2(\Delta t), \end{aligned}$$

where H_2 is the sum of H_1 and other terms which have similar limiting behavior as $\Delta t \longrightarrow 0$. It follows therefore that

$$H_2(\Delta t) = o(\Delta t) \qquad \text{as} \qquad \Delta t \longrightarrow 0.$$

Dividing both sides of the above equation by Δt and taking the limit as $\Delta t \longrightarrow 0$, we arrive at the following differential equation for P_n:

$$\frac{dP_n}{dt} = \lambda P_{n-1} + \mu P_{n+1} - (\lambda + \mu)P_n. \tag{3.2}$$

We observe that by using the definition of λ and μ as mean arrival and mean service rates, respectively, one can easily give a purely formal argument leading to this result (Exercise 3).

The computation is slightly different in the case where we are concerned with arriving at time $t + \Delta t$ with a queue of length 0. This can happen in the following ways: The queue at time t was of length 0 and there were no arrivals (and obviously no departures); the queue at time t was of length 1 and there was one departure and no arrival in the interval. In view of our assumptions, all other cases lead to probabilities which tend to 0 as $\Delta t \longrightarrow 0$. It follows that

$$\begin{aligned} P_0(t + \Delta t) = P_0(t)&(1 - \lambda\,\Delta t - h_1(\Delta t) - h_2(\Delta t)) \\ &+ P_1(t)(\mu\,\Delta t + k_1(\Delta t))(1 - \lambda\,\Delta t - h_1(\Delta t) - h_2(\Delta t)) \\ &+ H_3(\Delta t), \end{aligned}$$

where $H_3(\Delta t) = o(\Delta t)$. The argument justifying this estimate on H_3 is again technical and is omitted. The above equation involving $P_0(t)$ and $P_0(t + \Delta t)$ can be used in an obvious way to obtain a first-order differential equation for P_0 which contains P_1. We have

$$\frac{dP_0}{dt} = -\lambda P_0 + \mu P_1. \tag{3.3}$$

Let us suppose that we measure time from some reference, which we may take to be $t = 0$, and at this time the queue consists of exactly N units. Then the model based on our assumptions is completely described by Eqs. (3.2) and (3.3) together with the initial conditions

$$P_n(0) = \begin{cases} 0, & n \neq N, \\ 1, & n = N. \end{cases} \tag{3.4}$$

Solutions of the initial value problem specified by Eqs. (3.2), (3.3), and (3.4) are not easy to obtain. The difficulty is inherent in the problem; the functions P_n are complicated and the techniques available to us are not adequate for their determination.

Let us think back to the barbershop example which generated the idea for this model. We know that arrivals and departures are random events in a sense made precise in our assumptions. Given this randomness, it appears that after the elapse of a certain period of time the system will have "forgotten" the original queue length. That is, whether the original queue was of length 0, N, or $2N$ will not matter in the long run. For long time periods it will be the parameters λ and μ which determine the length of the queue and not its initial length. With this observation for support, it is reasonable to suppose that solutions of (3.2) and (3.3) behave differently for times close to 0 and for large times. The solution for small times will depend strongly on the initial condition, whereas the behavior of the solution for large times will depend much more on the parameters λ and μ and very little on the initial conditions. In a situation such as this, the solution of the initial value problem which is valid for small times but whose relative importance declines as t increases is known as a *transient solution.* A solution of the differential equations (3.2) and (3.3) which is valid for large times and which is independent of the initial conditions is known as a *steady-state* solution. As a final bit of notation, a solution of (3.2) and (3.3) which is independent of time, that is, for which $dP_n/dt = 0$, $n = 0, 1, \ldots,$ is known as an *equilibrium solution.* In most applications it is the steady-state solutions which are of most interest, and we proceed with a study of them. A detailed investigation of the transient solution, even in this simple example, requires a substantial mathematical discussion, and it will not be undertaken here. It is an important observed fact that in many problems of this sort the steady-state solutions vary quite slowly in time. That is, they are approximately equilibrium solutions. We formalize this in our model as an assumption.

4. The steady-state solutions to (3.2) and (3.3) are actually equilibrium solutions; i.e., there is a time t^* such that if $t > t^*$, then P_n is independent of time, $n = 0, 1, 2, \ldots$.

In terms of the differential equations, assumption 4 gives us for all sufficiently large times

$$-\lambda P_0 + \mu P_1 = 0, \tag{3.5}$$

$$\lambda P_{n-1} - (\lambda + \mu)P_n + \mu P_{n+1} = 0, \qquad n > 0. \tag{3.6}$$

Note that (3.5) is the steady-state (equilibrium) version of (3.3) and that (3.6) is the steady-state version of (3.2). In this case the initial conditions are no longer of any interest.

In a mathematical situation such as that described by Eqs. (3.5) and (3.6), the function P_n can be thought of as defined on the nonnegative integers; i.e., $P_n = P(n)$, $n = 0, 1, 2, \ldots$. Equations (3.5) and (3.6) then relate the values of the function P for successive values of the argument n and are known as difference equations (see also Sec. 9.2). These equations are a discrete analog of the differential equations we have encountered previously. They can be solved successively for $P_1, P_2, \ldots$ in terms of P_0, and then P_0 itself can be determined by the condition $\sum_{k \geq 0} P_k = 1$. The latter condition follows from the fact that the P_k's represent probabilities. First, Eq. (3.5) gives

$$P_1 = \frac{\lambda}{\mu} P_0.$$

Then, (3.6) gives successively

$$P_n = \left(\frac{\lambda}{\mu}\right)^n P_0, \qquad n = 2, 3, \ldots. \tag{3.7}$$

The straightforward induction proof of (3.7) is left as an exercise (Exercise 4). We note from (3.7) that the asymptotic behavior of the probabilities P_n depends strongly on the ratio λ/μ. This ratio is known as the *traffic intensity*, the *intensity ratio*, or the *utilization factor*, and it is a measure of the number of units arriving per unit departing. It plays a fundamental role in the remaining discussion of this and related models. We note that if $\lambda > \mu$, then there are on the average more arrivals than departures. In this case the quantities P_n defined by (3.7) are either all 0 or else grow beyond all bounds as n increases and thus cannot represent probabilities. Therefore, so that the model given by assumptions 1–4 will be nontrivial, it is necessary to make a final assumption:

5. The mean arrival rate λ is smaller than the mean service rate μ.

With assumption 5, we can proceed from (3.7) to a complete determination of the P_k's. In fact, since $\sum_{k\geq 0} P_k = 1$, we have

$$\sum_{k\geq 0} P_k = P_0 \sum_{k\geq 0} \left(\frac{\lambda}{\mu}\right)^k = 1,$$

and consequently $P_0 = 1 - \rho$, where $\rho = \lambda/\mu$ is the traffic intensity. Here we have used the formula for the sum of a geometric series: If $|\rho| < 1$, then $\sum_{k\geq 0} \rho^k = 1/(1 - \rho)$. From this and (3.7) we have finally

$$P_n = \rho^n(1 - \rho), \qquad n = 0, 1, 2, \ldots. \tag{3.8}$$

Summarizing this argument, we have shown that under the stated assumptions the steady-state probability P_n (that the queue is of length n) is given by equation (3.8), $n = 0, 1, 2, \ldots$. The reader should compare this with related formulas for the probabilities associated with a geometric random variable with parameter ρ. It can be shown that if $P_n = P_n(t)$, $n = 0, 1, 2, \ldots$, denote the solutions to the original initial value problem [(3.2), (3.3), (3.4)], then

$$\lim_{t\to\infty} P_n(t) = \rho^n(1 - \rho), \qquad n = 0, 1, 2, \ldots.$$

Thus the transient solutions tend to those given by (3.8), and consequently assumption 5 has some a posteriori justification.

As is frequently the case in problems of this sort, the probabilities P_n themselves are not as useful as certain other information which can be deduced from them. In most applications quantities such as the mean or expected length of the queue or the mean time between arrival and departure are more interesting and more easily compared with data obtained by observation of the real world. We now turn to questions of this type. Let $E[n]$ denote the expected value of the queue length. Then by definition

$$E[n] = \sum_{k\geq 0} kP_k,$$

and consequently by (3.8) we have

$$\begin{aligned} E[n] &= \sum_{k\geq 0} k\rho^k(1 - \rho) \\ &= (1 - \rho)\frac{\rho}{(1 - \rho)^2} \\ &= \frac{\rho}{1 - \rho}. \end{aligned} \tag{3.9}$$

In the above we have used the fact that for $|\rho| < 1$ we have $\sum_{k \geq 0} k\rho^k = \rho/(1-\rho)^2$ (Appendix A).

Equation (3.9) is a useful piece of information. For example, it provides a particularly easily checked estimate which may be used as a test of the validity of the model. As a predictor it indicates that as the traffic intensity ρ increases toward 1 (remember that ρ is always less than 1), the expected queue length increases rapidly. To be more specific, we note that if $\rho = 1 - 1/M$, then $E[n] = (1 - 1/M)/(1/M) = M - 1$.

We continue our investigation of the consequences of (3.8). If n is the number of units in a queue and m is the number waiting for service, then $n = m + 1$, and we can easily compute the expected number of units waiting for service. We have

$$
\begin{aligned}
E[m] &= \sum_{k \geq 2} (k-1)P_k \\
&= \sum_{k \geq 2} kP_k - \sum_{k \geq 2} P_k \\
&= E[n] - P_1 - (1 - P_0 - P_1) \\
&= E[n] - 1 + P_0.
\end{aligned}
$$

Using (3.8), with $n = 0$, and (3.9), we conclude that

$$
\begin{aligned}
E[m] &= \frac{\rho}{1-\rho} - 1 + (1-\rho) \\
&= \frac{\rho^2}{1-\rho}.
\end{aligned}
$$

An easy consequence of this is that $E[m] = \rho E[n]$ or, writing this equation slightly differently, that

$$\lambda E[n] = \mu E[m].$$

We shall find this relation useful in computing the expected time that a unit spends in the system, that is, the mean time between arrival and departure. The reader will note that much of the following argument is independent of any specific assumptions concerning arrival and service times and consequently holds in more general situations. In any process having an equilibrium solution to the equations for the probabilities there cannot be more units arriving than leaving the system, at least for large times. Since the mean arrival rate is λ, this means that the mean departure rate must also be λ. The reader should note that the mean departure rate is not the mean service rate μ, which is larger than λ. Let T denote the expected length of time between the arrival of a unit and the beginning of service on it, and let T^* denote the expected length of time between the arrival of a unit and its departure from the system. It follows from these definitions that

$$T = \frac{1}{\lambda} E[m] \qquad \text{and} \qquad T^* = \frac{1}{\lambda} E[n].$$

Now, using $\lambda E[n] = \mu E[m]$, we can also write

$$T = \frac{1}{\mu} E[n].$$

As a check on this reasoning we can use these formulas for T and T^* to compute the mean service time:

$$\text{Mean service time} = T^* - T = \frac{1}{\mu}.$$

This agrees with our definition of μ as the mean service rate.

As our final topic of this discussion, let us consider the distribution of interarrival intervals. Since λ is the mean arrival rate, we have $1/\lambda$ as the mean interarrival period. Let i (≥ 0) denote the length of an interval, and set

$$p(i) = \Pr[\text{No arrivals in an interval of time of length } i]. \tag{3.10}$$

It is a consequence of our basic assumptions that $p(i)$ is well defined. In particular, the right-hand side of (3.10) does not depend on the end points of the interval but only on its length. Set $I = [t, t + i]$, $\Delta I = [t + i, t + i + \Delta i]$, and Q equal to the probability that there is no arrival in the interval I and at least one arrival in the interval ΔI. Then

$$\begin{aligned} Q = p(i)(1 - p(\Delta i)) &= p(i) - p(\Delta i)p(i) \\ &= p(i) - p(i + \Delta i). \end{aligned}$$

Here we have used the independence of the events: no arrivals in I and no arrivals in ΔI to write $p(i)p(\Delta i) = p(i + \Delta i)$. But by assumptions 1 and 2 the probability Q can also be written as

$$p(i)[\lambda \Delta i + h_1(\Delta i) + h_2(\Delta i)].$$

Consequently, we have

$$\frac{p(i) - p(i + \Delta i)}{\Delta i} = \lambda p(i) + p(i)\frac{h_1(\Delta i) + h_2(\Delta i)}{\Delta i},$$

and taking the limit as $\Delta i \longrightarrow 0$, we have a differential equation for $p(i)$:

$$-\frac{dp}{di} = \lambda p.$$

Solving this differential equation, we have

$$p(i) = ce^{-\lambda i}.$$

The constant c can be evaluated by noting the behavior of p as $i \longrightarrow 0$. Assumptions 1 and 2 guarantee that the probability of an arrival in a very short interval approaches 0 with the length of the interval. Thus, $p(i) \longrightarrow 1$ as $i \longrightarrow 0$. This implies that $c = 1$, and

$$p(i) = e^{-\lambda i}. \tag{3.11}$$

It is sometimes useful to look at this result somewhat differently. Equation (3.10) can be stated as

$$p(i) = \text{Pr[Interarrival interval is larger than } i],$$

and it is in this sense that (3.11) gives the distribution of interarrival intervals. Using the terminology of probability theory, we say that the length of the interval between successive arrivals is an exponential random variable with parameter λ. The probability density function for this random variable is $\lambda e^{-\lambda i}$.

9.3.3 More General Models

The preceding subsection contains a fairly detailed study of a particular example, and a natural question at this point is, what can be said in other situations? There is, in fact, a substantial and rapidly increasing literature on more general models. It is our intention to use two examples to illustrate the sort of results which can be obtained in situations not fitting the model of Sec. 9.3.2. The first example leads to an interesting estimate which is valid under a rather wide range of assumptions, and the second is another special case analogous in many respects to the model of Sec. 9.3.2 but whose importance justifies its inclusion here.

Let us consider first a queuing process with Poisson arrivals and an arbitrary distribution of service time intervals. More precisely, we assume the arrivals satisfy assumptions 1–3 of the preceding subsection; i.e., the probability distribution for the interarrival times is a Poisson distribution, and the service intervals are random with an unknown probability distribution. We shall continue to assume that $\lambda/\mu < 1$, where λ is the mean arrival rate and μ is the mean service rate. As pointed out earlier, such an assumption is necessary if the process is to reach and maintain an equilibrium. To begin this discussion, it is convenient to consider the situation as a unit departs from the service facility. At the instant a unit departs suppose that the length of the remaining queue (i.e., not counting the departing unit but counting the one which immediately moves into the service facility) is q units. Suppose that the next unit is serviced in a time interval of length s and during this period r new units join the queue. We view r as a random variable, and we note that its conditional distribution is a Poisson distribution with mean value λs; and we view s as a random variable whose distribution is the unknown service distribution. At the instant the next unit leaves the service facility, let q' be the size of the remaining queue. Although the following assumption can be justified by a more detailed study of the notion of equilibrium, it is suitable for our purposes to take it as an axiom.

Axiom: After the process reaches equilibrium, the random variables q and q' have the same (marginal) distribution, and consequently their expected values and standard deviations are the same. We assume that $E[q] < \infty$ and $V[q] < \infty$.

Subsequent development of this model will make essential use of this assumption. Here, as usual, $E[q]$ denotes the expected value of the random variable q and $V[q]$ its variance.

According to the definitions, we have

$$q' = \begin{cases} q - 1 + r, & q \neq 0, \\ r, & q = 0. \end{cases} \tag{3.12}$$

Indeed, if $q = 0$, then the next service interval does not begin until one unit arrives and r additional units arrive during the service interval. Thus after the unit being serviced departs, there still remain r units in the queue. It is convenient to introduce a function δ

$$\delta(q) = \begin{cases} 0, & \text{if } q \neq 0, \\ 1, & \text{if } q = 0, \end{cases}$$

and with this terminology Eq. (3.12) can be written compactly as

$$q' = q + r - 1 + \delta. \tag{3.13}$$

It follows from the definition of $\delta(q)$ that

$$\delta^2 = \delta, \qquad q\delta = 0,$$

and these relations play an important role in what follows. Our axiom contains the assumption $E[q] = E[q']$, and from (3.13) we obtain

$$E[q'] = E[q] + E[r] - 1 + E[\delta],$$

and consequently

$$E[\delta] = 1 - E[r] = 1 - \frac{\lambda}{\mu}.$$

This equation is also based on the facts that the expected length of a service interval is $1/\mu$ and the length of a service interval is independent of arrivals. We remark that

$$\begin{aligned} E[\delta] &= 0 \cdot \Pr[\delta = 0] + 1 \cdot \Pr[\delta = 1] \\ &= \Pr[\delta = 1] \\ &= \Pr[q = 0]; \end{aligned}$$

that is, $1 - \lambda/\mu$ is the probability that in equilibrium a departing unit leaves no queue at all. As expected, this probability is small if μ is close to λ in size.

We consider next the variance of q and q'—or, what is equivalent, given a knowledge of $E[q]$, the expected value of the square of q. Proceeding in this direction, we square Eq. (3.13) and obtain

$$(q')^2 = q^2 - 2q(1 - r) - 2\delta(1 - r) + 2\delta q + (1 - r)^2 + \delta^2.$$

Also, making use of the properties of δ noted above, we have

$$\delta^2 + 2\delta q - 2\delta(1 - r) = \delta(2r - 1).$$

Thus from these two equations we have

$$(q')^2 = q^2 - 2q(1 - r) + (r - 1)^2 + \delta(2r - 1).$$

It is a consequence of our assumption of Poisson arrivals that r and q are independent random variables. Thus on forming the expected value of $(q')^2$ and using the axiom again to equate $E[(q')^2]$ and $E[q^2]$, we obtain after a little algebraic manipulation

$$E[q] = E[r] + \frac{E[r(r - 1)]}{2\{1 - E[r]\}}. \tag{3.14}$$

Indeed, we write the above equation for $(q')^2$ as

$$(q')^2 - q^2 + 2q(1 - r) = r(r - 1) - (r - 1) + \delta(2r - 1),$$

and taking expected values gives

$$2E[q]E[1 - r] = E[r(r - 1)] - E[r - 1] + E[\delta]E[2r - 1]$$

or

$$2E[q](1 - E[r]) = E[r(r - 1)] + (1 - E[r]) + E[\delta](2E[r] - 1),$$

from which (3.14) follows directly. For our purposes it is preferable to write it in a more suggestive form. The following is frequently associated with the names of D. G. Kendall, A. Y. Khintchine, and F. Pollaczek, and it is the primary goal of this discussion.

Kendall's Formula: Let q, s, λ, and μ be as defined above and let $V[s]$ denote the variance of the random variable s. Then

$$E[q] = \frac{\lambda}{\mu}\left[1 + \frac{\lambda}{2(\mu - \lambda)} + \frac{\lambda\mu^2}{2(\mu - \lambda)}V[s]\right].$$

The proof of this equality is almost immediate. From (3.14) and the relation $E[r] = \lambda/\mu$ we have at once

$$E[q] = \frac{\lambda}{\mu} + \frac{E[r^2] - \lambda/u}{2(1 - \lambda/\mu)}. \tag{3.15}$$

The only matter remaining is the evaluation of $E[r^2]$. We proceed using notation appropriate for the case where s assumes values in a discrete set. We leave it to the reader to make the obvious modifications which are necessary if this is not a valid assumption. It was pointed out above that for each fixed s the function r is a random variable with mean or expected value λs. Denote this mean value by $E_s[r]$. We have

$$E_s[r] = \sum_R r \Pr[R = r \mid S = s],$$

where

$$\Pr[R = r \mid S = s] = \text{Probability that } r \text{ units arrive during an interval of length } s.$$

Since r has a Poisson distribution, it follows that

$$E_s[r^2] = \lambda s + (\lambda s)^2$$

(Exercise 6). Therefore

$$\begin{aligned} E[r^2] &= \sum_S \{\sum_R r^2 \Pr[R = r \mid S = s]\} \Pr[S = s] \\ &= \sum_S E_s[r^2] \Pr[S = s] \\ &= \lambda \sum_S s \Pr[S = s] + \lambda^2 \sum_S s^2 \Pr[S = s] \\ &= \lambda E[s] + \lambda^2 E[s^2]. \end{aligned}$$

Also, the expected value of s is $1/\mu$, and for any random variable $E[s^2] - (E[s])^2 = V[s]$. Consequently

$$E[r^2] = \frac{\lambda}{\mu} + \lambda^2\left[V[s] + \left(\frac{1}{\mu}\right)^2\right],$$

and we can use this in (3.15) to obtain the desired result. Q.E.D.

An interesting and important corollary of Kendall's formula is that the minimum of $E[q]$ over all service time distributions is achieved when $V[s] = 0$. That is, the shortest expected queue length at the times at which a unit departs, i.e., the times of the expected maximum queue length, is achieved for deterministic service facilities. In this case the expected maximum queue length is

$$E[q] = \frac{\lambda(2\mu - \lambda)}{2\mu(\mu - \lambda)},$$

and this is $(1 - \lambda/2\mu)$ times the expected maximum queue length with the random service intervals of the preceding subsection. The discussion of this model is continued in the Exercises.

As our second illustration of possible modifications of the model of Sec. 9.3.2 we consider a process with Poisson arrivals and service times depending on the length of the queue. In the barbershop illustration considered above, it might be argued that Mr. Soot works more quickly when he has a shop full of customers than when he has his only customer in the chair. On the other hand, service in a restaurant is likely to be slower when the restaurant is crowded than when it is relatively empty. It is clear from this sort of example that the dependence of the service rate on the length of the queue may be quite complicated. The particular choice made here will be appropriate for some situations and not for others. The techniques are of wider applicability than the specific example. We suppose that the service rate is proportional to the length of the queue, as might be the case in our barbershop example. Precisely, we assume that if there is a queue of length q, then the service rate is μq, where μ is a constant. If we state this in a form comparable to the assumptions of Sec. 9.3.2, we have

1″. There is a constant μ (> 0) such that if there is a unit being serviced at time t and the queue is of length q at that time, then the probability that the service is completed in the interval $(t, t + \Delta t)$ is

$$\mu q \, \Delta t + k_1(\Delta t),$$

where

$$k_1(\Delta t) = o(\Delta t), \qquad \Delta t \longrightarrow 0.$$

Using this assumption and the others of Sec. 9.3.2 in the same manner as in the discussion of that subsection, we are led to the following differential equations for the probabilities:

$$\frac{dP_0}{dt} = -\lambda P_0 + \mu P_1, \tag{3.16}$$

$$\frac{dP_n}{dt} = -(\lambda + n\mu)P_n + \lambda P_{n-1} + (n + 1)\mu P_{n+1}, \qquad n = 1, 2, \ldots. \tag{3.17}$$

We shall see as we proceed with our discussion that the condition $\lambda \leq \mu$ need not be imposed in this case. Let us suppose that there exists an equilibrium solution to these equations, i.e., a solution $\{P_n\}_{n=0}^{\infty}$ for which $dP_n/dt = 0$, $n = 0, 1, 2, \ldots$. Then from the first of the above equations we have

$$P_1 = \left(\frac{\lambda}{\mu}\right)P_0, \tag{3.18}$$

and from the second for $n = 1, 2, \ldots$, we have

$$(n+1)\mu P_{n+1} = (\lambda + n\mu)P_n - \lambda P_{n-1}. \tag{3.19}$$

We can now determine P_n for $n \geq 2$ by solving (3.19) successively for $n = 1, 2, \ldots$. The first few of these equations are

$$2\mu P_2 = (\lambda + \mu)P_1 - \lambda P_0, \tag{3.20}$$
$$3\mu P_3 = (\lambda + 2\mu)P_2 - \lambda P_1, \tag{3.21}$$
$$4\mu P_4 = (\lambda + 3\mu)P_3 - \lambda P_2.$$

Using (3.18) in (3.20), we have

$$2\mu P_2 = (\lambda + \mu)\left(\frac{\lambda}{\mu}\right)P_0 - \lambda P_0,$$

and hence

$$P_2 = \frac{(\lambda/\mu)^2}{2}P_0. \tag{3.22}$$

Next, using (3.18) and (3.22) in (3.21), we have

$$\begin{aligned} 3\mu P_3 &= (\lambda + 2\mu)\left(\frac{\lambda}{\mu}\right)^2\frac{1}{2}P_0 - \lambda\left(\frac{\lambda}{\mu}\right)P_0 \\ &= \frac{\lambda^3}{2\mu^2}P_0, \end{aligned}$$

and therefore

$$P_3 = \frac{(\lambda/\mu)^3}{2 \cdot 3}P_0.$$

This justifies the conjecture that the form of P_n for arbitrary n is

$$P_n = \frac{(\lambda/\mu)^n}{n!}P_0, \qquad n = 1, 2, \ldots. \tag{3.23}$$

A straightforward induction proof verifies that the P_n's so defined do indeed satisfy (3.19). However, this does not yet determine the P_n's completely.

Indeed, since each P_n represents a probability, we must have $0 \leq P_n \leq 1$, $n = 0, 1, 2, \ldots$, and since the queue is certain to be of some length, $\sum_{n=0}^{\infty} P_n = 1$. If we use (3.23) and the latter normalization condition, then we obtain

$$P_0 + \sum_{n=1}^{\infty} \frac{(\lambda/\mu)^n}{n!} P_0 = 1$$

or, equivalently,

$$P_0 + (e^{\lambda/\mu} - 1)P_0 = 1.$$

From this we conclude that $P_0 = e^{-\lambda/\mu}$, and therefore

$$P_n = \frac{1}{n!}\left(\frac{\lambda}{\mu}\right)^n e^{-\lambda/\mu}, \qquad n = 0, 1, 2, \ldots.$$

This formula gives the probability that (in equilibrium) the queue contains n units for any integer n. Notice that the formula is meaningful for any value λ/μ of the traffic intensity. This example will be pursued further in the Exercises, where additional properties of the probabilities $\{P_n\}$ will be obtained.

EXERCISES

1. Construct a model for a completely regular deterministic queuing process with arrival rate λ and service rate μ. Assume that there is a single service facility and that there is no limit to the possible length of the queue.
   (a) Determine the exact number of units waiting for service at time t.
   (b) Find the waiting time before service for a unit which arrives at time t.
2. Construct a model for a deterministic queuing process in which the times of successive arrivals are $\{t_1, t_2, \ldots\}$ and the service is regular with rate μ. Find a formula which gives the number of units waiting for service at time t.
3. Let λ and μ be mean arrival and service rates, respectively. Give a formal argument leading to (3.2).
4. Using (3.5), (3.6), and an induction argument, prove (3.7).
5. Show that Kendall's formula reduces to the corresponding result of Sec. 9.3.2 in the case of service distributions which satisfy assumptions 1′–3′.
6. Let r be a Poisson random variable with mean value σ. Prove that $E[r^2] = \sigma + \sigma^2$.
7. Consider a queuing process with a single service facility, Poisson arrivals with mean arrival rate λ, and an unknown service time distribution with mean service rate μ. Suppose that as a unit departs there are q units remaining in the queue, and let w and s denote, respectively, the waiting time before service and service time for the departing unit.
   (a) Show that $E[q] = \lambda\{E[w] + E[s]\}$.

(b) Using Kendall's formula and part (a), prove that the expected waiting time before service for a newly arrived unit is

$$E[w] = \frac{\lambda}{\mu}\left[\frac{1 + V[\mu s]}{2(\mu - \lambda)}\right].$$

(c) Use part (b) to obtain an expression for $E[w]/E[s]$. This ratio is a useful efficiency parameter. It is the ratio of the mean time spent waiting before service to the mean time of service waited for. Justify the following assertion: A necessary and sufficient condition for maximum efficiency is that there be no variation in the service times.

8. The purpose of this exercise is to consider the transient behavior of the probabilities P_n for the second model of Sec. 9.3.3. In particular, consider the initial value problem defined by (3.16), (3.17), and the condition $P_0(0) = 1$, $P_n(0) = 0, n > 0$. Set

$$P_n(t) = \frac{1}{n!}\left[\frac{\lambda}{\mu}(1 - e^{-\mu t})\right]^n \exp\left[-\frac{\lambda}{\mu}(1 - e^{-\mu t})\right], \qquad n = 0, 1, 2, \ldots.$$

(a) Prove that P_0 and P_1 are solutions of the initial value problem. (Note that you need to use P_0, P_1, and P_2 to do this.)

b) Investigate $\lim_{t\to\infty} P_n(t)$, and compare this limit with the equilibrium probabilities P_n computed in Sec. 9.3.3.

9. (Continuation of 8) Find the average number of units waiting in line at time t, $E[n](t)$. *Hint:* Show that $a(t) = E[n](t)$ satisfies the differential equation $(da/dt) + \mu a = \lambda$.

(a) Use this formula for $E[n](t)$ to show that for small times t the average length of the line is approximately λt.

(b) Find an estimate similar to that of (a) involving t and t^2.

(c) Show that $\lim_{t\to\infty} E[n](t) = A$, where A is the average length of the line in equilibrium. *Note:* First find A in terms of λ and μ using the results of Sec. 9.3.3.

10. Consider the queuing process of Sec. 9.3.2 and suppose that $\lambda = \mu$. Deduce as much as you can about $\{P_n(t)\}$ and $E[n](t)$.

9.4 PROJECTS

9.4.1 The Spread of an Epidemic through Two Communities

The purpose of this project is to develop a model for the spread of a disease through a population consisting of a collection of groups of individuals, each group forming an identifiable subpopulation. For definiteness we assume that there are two such groups, and we refer to them as communities. If the communities are completely isolated from each other, then the disease spreads through each independently. On the other hand, if the two groups are completely mixing, then the epidemic develops as if the union of the two groups were actually a single group. The primary concern of this model is the intermediate case, that is, the situation in which there is some mixing between the subpopulations, but this mixing is less than the mixing within either population individually. To be more precise, suppose the following:

i. The population of each community is large enough to justify ignoring the random features of the situation.
ii. The infection is spread only through contacts between individuals.
iii. The population of each community is homogeneous and uniformly mixing. The contacts between individuals from separate communities satisfy like conditions but occur less frequently.
iv. No infected individuals are removed from either community.

Let $N_1 + 1$ and N_2 denote the number of individuals in communities 1 and 2, respectively, and let $s_i(t)$, $i = 1, 2$, denote the number of susceptibles in community i at time t.

Problems

1. Find the differential equations which specify the rate of change of s_i as a function of time. Indicate how each of assumptions i–iv is used in deriving these equations.
2. Discuss the dependence of s_i, $i = 1, 2$, on the internal and cross (i.e., intercommunity) infection rates. Give an argument which provides some information on s_1 and s_2 "near" the extreme cases. By extreme cases we mean those in which (a) the internal and cross infection rates are nearly equal, and (b) the cross infection rate is small in comparison with the internal rate.
3. If community 1 initially contains exactly one infected individual and community 2 contains none, discuss as completely as you can the behavior of $s_1(t)$ and $s_2(t)$.

9.4.2 Recurrent Epidemics

Formulate a deterministic model for a recurrent epidemic in a closed population. That is, in the terminology of Sec. 9.1, assume that transitions between subpopulations S and I and between I and R are possible. In addition, assume that there is a positive number L such that if an individual makes the transition from subpopulation I to R at time T, then it makes the transition from R to S at time $T + L$.

Show that as L becomes very large this model predicts the same population distributions $\mathbf{p}(t)$ as the model of Sec. 9.1. What is the situation for very small values of L?

9.4.3 Another Rumor Diffusion Model

Consider an alternative model for the rumor diffusion situation discussed in Sec. 9.2 in which assumption 3 is replaced by 3′.

3′. If a village of type I calls a village of type I or R, then both villages become (or remain) of type R.

Adopt the other assumptions of Sec. 9.2 without change.

Problems

1. Formulate and analyze a discrete-time model for this rumor diffusion problem.
2. Provide an estimate for the percentage of villages that never hear the rumor.
3. Formulate and analyze a continuous-time model based on these assumptions. Compare your results with those obtained from the discrete-time model.

9.4.4 Refinements of the Barbershop Model

In the barbershop model developed in Sec. 9.3.2 we introduced a number of assumptions which, while making the resulting model relatively simple, are likely to require some modification in practice. Here we consider somewhat more realistic assumptions and the associated models.

Problems

1. Suppose that the barbershop is served only by Mr. Soot and that the other assumptions of Sec. 9.3.2 hold. In addition, suppose that there are M seats for customers and no customers will wait for a haircut without a seat. Derive the steady-state (equilibrium) equations analogous to (3.5) and (3.6) for this model. Solve these equations for the probability distributions P_n, $n = 0, 1, 2, \ldots$. Determine the expected number of customers in the barbershop at time t. Discuss the difficulties encountered in defining the expected time a customer waits for a haircut. Decide on an appropriate definition for this expected waiting time and compute it.
2. Consider next a model for a multiple-chair shop. Specifically, suppose that there are Q barbers with chairs and the arrival and

service times (for each barber) are subject to the conditions of Sec. 9.3.2. Thus each barber provides service subject to these assumptions and we assume that each barber operates independently. Also, suppose that the shop is capable of seating any customer who comes in, and that the customers have no preferences for certain barbers. This means that whenever a barber is available, the first customer in line is serviced by this barber. Formulate and solve the equations for P_n analogous to (3.5) and (3.6) for this model. Compute the expected number of customers waiting for a haircut and the expected waiting time for an individual customer.

3. Combine the models of problems 1 and 2. That is, assume that there are Q barbers and M seats for customers, with no customer willing to join the queue without a seat. Analyze this model as the preceding ones.

9.4.5 Discrete-Time Queuing Models

A model for a queuing process in which time is considered to be a discrete rather than continuous quantity is developed in this project. The basic axiom which distinguishes this model from that developed in Sec. 9.3.2 is the following:

i. All events (arrivals, departures, and the beginning of service) take place at definite and regularly spaced times.

Assume that the first event takes place at time $t = 0$. Then there is a time h such that all subsequent events take place at times in the set $T = \{h, 2h, 3h, \ldots\} = \{kh\}_{k=1}^{\infty}$. However, every time in this set need not be the time that an event takes place. Additional axioms for this model are the following:

ii. No more than one customer arrives at any time $t \in T$.
iii. The arrival of a unit at time $t \in T$ is an event which is independent of the arrival of customers at times $t < T$, $t \in T$. There is a number p, $0 \leq p \leq 1$, such that $\Pr\{\text{arrival at } t \in T\} = p$. It is explicitly assumed that p is independent of t.
iv. The length of the service interval for one customer does not affect the length of the service interval for any other customer. That is, the lengths of the service intervals are independent random variables. Assume that these random variables all have the same probability distribution.
v. Queue discipline is first come-first served.

In your development of this model you should solve the following problems:

1. Let I_k denote a general time interval containing k consecutive elements of the set T. Find a formula for $a_{km} = \Pr[m \text{ arrivals in } I_k]$.
2. If the number of arrivals m in I_k is interpreted as a random variable, find its expected value and variance.
3. Find the mean number of arrivals per unit time.
4. A service time s can only assume values which are integral multiples of h. Set $c_k = \Pr[s = kh]$, $k = 1, 2, 3, \ldots$. If one defines the traffic intensity ρ as the product of the mean rate of arrivals and the mean service time (see Sec. 9.3), then prove that

$$\rho = p \sum_{k=0}^{\infty} kc_k.$$

5. Observe that the formula for ρ derived in problem 4 does not depend explicitly on h, and thus it presumably remains valid as h becomes small. What does the formula become as $h \longrightarrow 0$?
6. Consider a model for the process determined by axioms i–v above which is based on a denumerable Markov process.

REFERENCES

Discussions of mathematical models for the spread of rumors and epidemics are not common in undergraduate textbooks, and consequently the list of references for this material contains more journal articles and specialized books than usual. The pioneering work of W. O. Kermack and A. G. McKendrick referred to in the text is reported in [M] and [KeM]. The topic of this paper is continued, again with attention to the biological aspects of the process, in [K2]. Here the notion of a stochastic threshold theorem is introduced, and this idea is explored further in [K3]. The latter article contains data from several artificial stochastic epidemics and the associated epidemic curves. The work of Norman T. J. Bailey reported in several papers beginning in 1950 and his books contains significant contributions to the development of a theory of epidemics ([B1]–[B4]). In general, the work of Bailey utilizes mathematical techniques more advanced than those of this book, but frequently the results are understandable even if their proofs are not. Another instance of this is the article [Wi], where the graphs summarizing part of the results can be appreciated independently of the analysis leading to them. A historical survey (to 1952) of work in this area may be found in [Se].

It is common in the literature to view models for the spread of rumors and epidemics as being interchangeable. Thus there is not a large literature directed specifically at models for rumor diffusion. The idea of building discrete-time models for diffusion processes based on difference equations was suggested to us by H. O.

Pollak and G. S. Young. A diffusion model developed along these lines but with a different set of basic assumptions is contained in [PY].

Queuing processes are now recognized as a rich source of model-building ideas and techniques, and accounts are available at many levels of difficulty. Most recent textbooks in management science or operations research contain chapters on queuing theory written from a modeling point of view, for example, [Ka] and [W]. There are also books devoted entirely to queues with varying emphasis on applications ([Kh] and [S]). The reference [S] is particularly appropriate for those interested in using queuing theory without learning all the mathematical details. It consists of a summary of many results together with a huge number of practical applications. Many standard texts on probability theory present the mathematical foundations of the subject. Our discussion of processes involving unknown service time distributions is based on [K1].

[B1] Bailey, Norman T. J., "A Simple Stochastic Epidemic," *Biometrika*, **37** (1950), 193–202.

[B2] Bailey, Norman T. J., "The Total Size of a General Stochastic Epidemic," *Biometrika*, **40** (1953), 177–185.

[B3] Bailey, Norman T. J., *The Mathematical Theory of Epidemics*. New York: Hafner, 1957.

[B4] Bailey, Norman T. J., *The Mathematical Approach to Biology and Medicine*. New York: Wiley, 1967.

[H] Haskey, H. W., "A General Expression for the Mean in a Simple Stochastic Epidemic," *Biometrika*, **41** (1954), 272–275.

[Ka] Kaufmann, A., *Methods and Models of Operations Research*. Englewood Cliffs, N.J.: Prentice-Hall, 1963.

[KeM] Kermack, W. O., and A. G. McKendrick, "A Contribution to the Mathematical Theory of Epidemics," *Proceedings of the Royal Society of London*, Ser. A, **115** (1927), 700–721.

[K1] Kendall, David G., "Some Problems in the Theory of Queues," *Journal of the Royal Statistical Society*, Ser. B, **13** (1951), 151–173.

[K2] Kendall, David G., "Deterministic and Stochastic Epidemics in Closed Populations," *Proceedings of the Third Berkeley Symposium on Mathematical Statistics and Probability*, Vol. IV. Berkeley: University of California Press, 1956.

[K3] Kendall, David G., "Mathematical Models of the Spread of Infection," *Mathematics and Computer Science in Biology and Medicine*. London: Medical Research Council, H. M. Stationery Office, 1965.

[Kh] Khintchine, A. Y., *Mathematical Methods in the Theory of Queueing*, trs. D. M. Andrews and M. H. Quenouille. New York: Hafner, 1960.

[M] McKendrick, A. G., "The Application of Mathematics to Medical Problems," *Proceedings of the Edinburgh Mathematical Society*, **44** (1926), 98–130.

[PY] POLLAK, H. O., and G. S. YOUNG, eds., *Applications of Mathematics for Secondary School Teachers*, unpublished manuscript.

[S] SAATY, Thomas L., *Elements of Queueing Theory, with Applications.* New York: McGraw-Hill, 1961.

[Se] SERFLING, R. E., "Historical Review of Epidemic Theory," *Human Biology*, **24** (1952), 145–166.

[W] WAGNER, HARVEY M., *Principles of Management Science.* Englewood Cliffs, N.J.: Prentice-Hall, 1970.

[Wi] WILLIAMS, TREVOR, "The Simple Stochastic Epidemic Curve for Large Populations of Susceptibles," *Biometrika*, **52** (1965), 571–579.

10 Practical Aspects of Model Building

10.0 INTRODUCTION

A mathematical model may be very useful to a mathematician and quite useless to a scientist. The value of the model to a mathematician lies in the structure and mathematical beauty of the model and in its relationships to other mathematical theories. The value of the model to a scientist lies in the utility of the model in explaining and predicting the events and phenomena being studied. We have formulated and developed a number of different models; however, in each case we stopped short of a full discussion of the important problem of evaluating the model. In this chapter we shall try to fill this gap by considering the art and/or science of testing, evaluating, and using the mathematical models developed in the text. The treatment given here is not intended as a thorough study or even a general introduction to the subject but rather as a natural part of the model-building activity which has been deferred until now. Thus there are frequent references to models constructed in earlier chapters. Our point of view is that of a scientist who is considering a particular mathematical model which he is interested in using. What should he do in order to decide if his model is a good one, a model which is worthy of further study and use? This question is difficult to answer completely; however, we do hope to introduce some of the tools and techniques which have been useful in the past and which are likely to be used in the future.

The chapter is organized in three sections. We shall begin with a discussion of some of the heuristics of model evaluation. Next, we shall consider the use of the statistical analysis of data in building and assessing models, and finally we shall consider computer simulation.

10.1 INTUITIVE EVALUATIONS

A scientist who builds a model to aid him in his studies usually has a good idea concerning what he wants his model to do. He may only be interested in a convenient method of consolidating his data, in which case it is likely that a number of different models are acceptable to him. However, in most cases the scientist wants his model to do more; it should also explain, and predict, and perhaps contribute to the understanding of previously unanswered questions. In such circumstances different models may give dramatically different results. The scientist must then decide which models he wishes to use and which ones he chooses to reject. There are many ways for him to make this decision. Some of the methods are mathematical and others are not. One of the most important nonmathematical techniques is based on the intuition of the model builder himself. This is the so-called *eyeball* technique, and it consists of the intuitive feelings of the scientist about the assumptions and consequences of the model. He must decide for himself if they seem to be correct, or at least have a reasonable chance of being correct.

Recall the model which was given for the spread of an epidemic in Chap. 9. The model was developed under certain assumptions about the spread of the disease, and then it was used to make predictions about the expected increases and decreases in the sizes of certain subsets of the population. These predictions are consistent with one's intuitive feeling of the nature of the spread of a disease. An epidemic may well end with a sizable portion of the population still being susceptible to the disease, and this is in accordance with the predictions of the model. Thus, at least in one sense, the model seems "right," and it passes an initial eyeball test.

As a second example of intuitive evaluations, consider the transportation problem of Sec. 2.2. One businessman may examine the model for the transportation problem and conclude that the model is useless. His reason could be that the model ignores the usual fluctuations in supply and demand. For example, he may know from experience that it rarely happens that supply and demand are constant and supply is exactly equal to demand, and hence he will have no faith in a model based on these assumptions.

On the other hand, a second businessman may find the model for the transportation problem very attractive. Perhaps he is interested in transporting oil through pipelines within his own plant. For him, all the assumptions of the model are very reasonable, and the model seems to present an adequate picture of the situation.

As a final example recall the game theory model of Sec. 6.1. The model is designed for two-person zero-sum games, and it used min-max reasoning to show how to obtain an optimal strategy for all such games. The mathematics here is sound and beautiful; however, a social scientist might well question the usefulness of the model. He may feel that most people do not use min-max reasoning and hence that this model does not really tell him how the subjects play. He may have observed that many people play so that they will receive an occasional high payoff even though this strategy results in a long-term loss which is higher than the loss incurred by using another strategy. For subjects with this method of play a strategy may be optimal even though it does not maximize the long-term financial return to them. Thus the social scientist needs a new definition of optimal and a new model which fits his observations more closely.

As the above examples indicate, an eyeball evaluation is a very personal matter. Each scientist carries out his own intuitive evaluation of models which he uses, and the results of these evaluations may well differ from individual to individual. Also, we note that the eyeball method can lead to two types of errors: first, the model may seem "right" and yet data show that the model is useless, and, second, the model may seem "wrong" because of unreasonable assumptions and yet it may produce useful insights and results.

10.2 STATISTICS FOR THE MODEL-BUILDING PROCESS

The use of statistical techniques plays an important role in several aspects of model building. However, a thorough study of techniques for the treatment of data would be a major undertaking, and it would take us too far afield. Thus we shall discuss very briefly only a few of the applications of statistics in our work.

It is sufficient for us to think of mathematical statistics as a discipline concerned with extrapolating from data which describe a subset of a given population to obtain information about the entire population. Normally one concentrates on a single characteristic of the population. For example, the population might be 20-year-old college students attending large state universities, and the subset might be 300 such students selected in some way from six such universities. The characteristic might be the probability that a subject uses the strategy *cooperate* in a Prisoner's Dilemma game if both he and his opponent used *cooperate* on the immediately preceding play of the game (see Sec. 6.2.2). In a typical situation an experiment would be conducted and a value p of the probability would be obtained for each student. These values could be used to obtain a mean value $\hat{p}$ for the subset of 300 students. A reasonable statistical question is, What is the relationship between the statistic $\hat{p}$ for the subset and the corresponding mean value $\bar{p}$ for the entire population? Problems of this sort are those of *parameter estimation.* They arise naturally because in order to make numerical predictions based on a model of a situation, one must have numerical values for the parameters occurring in the model.

Another important role of statistical methods in mathematical modeling is in testing the model for correctness and accuracy: Does the model really do what it purports to do? For example, one can ask whether and in what sense the model for learning (Sec. 2.4) provides a description of how a subject learns a simple concept. A usual approach to answering such questions is to accumulate data either from the results of controlled experiments or through selected observations of the real world and then evaluate the model by comparing the results predicted by the model with those resulting from a statistical analysis of the data. There are several standard statistical tests with may variants which are useful in this connection. Some of them are discussed in this section.

Obviously the two problems mentioned just above are not independent. One would hardly expect a model to predict accurate results if the parameters have been carelessly chosen. Thus, when evaluating the predictions based on a particular model, it is essential to keep in mind the processes used to determine the parameters and the accuracy that can be expected. Also, it is common that one set of data is used both to determine the parameters and to test the model, a part of the data being used for each purpose. In these situations systematic bias in the data will be reflected in both the parameters and the testing, and even though the model may predict results consistent with the data, the defective data may obscure deficiencies in the model.

In discussing the testing of models, it is important to note that different models are by nature tested in different ways. For some models it is natural that the axioms of the model be directly tested. For other models it is very difficult to test the axioms; however, the theorems or conclusions of the model can be easily tested. In such cases, of course, one is indirectly testing the axioms of the model. The basic assumption here is that the theorems are true whenever the axioms are true (i.e., the proofs of the theorems are valid), and hence if the theorems (predictions) are not verified in the real world, then the axioms cannot be correct statements about the real world.

The linear programming model of Chaps. 4 and 5 is a good example of a model which is usually checked by considering the hypothesis of the model as opposed to the conclusion. The conclusion of the linear programming model is that under certain conditions an algorithm (the simplex method) will yield a vector which makes a given inner product a minimum (we are thinking of the restricted minimum problem). In general one would not test this model by comparing the values of the inner product for a number of different vectors. Instead one would check the validity of the linear programming model for a real-world situation by checking that the assumptions of the model (in particular, the linearity assumption) are true (or in some sense nearly true) for that situation. On the other hand, models of population growth are often tested by comparing the conclusions of the model (the predicted growth) with the actual growth which is observed.

It is also important to keep in mind that most testing of models is not an *either-or* situation which results in the acceptance or rejection of a model on the basis of a single test. Instead much of the testing is for purposes of comparison to determine how well the model explains and predicts the real world. Frequently there are a number of competing models, and experiments are conducted in an effort to determine which one is the "best" at explaining and predicting the phenomena under study. For example, in learning theory there are several different models which may be used in an attempt to explain the paired-associate learning discussed in Sec. 2.4. Each model makes certain predictions about the responses which will be observed, and these predictions can be (and have been) statistically compared with the actual responses obtained in experiments. In [ABC], [Bu], and [ReG] there are discussions of these models and the methods of comparing them.

A statistical evaluation of the accuracy of a model is usually carried out with the use of a standard measure of the discrepancy between the predicted and the observed data. In this way one obtains a numerical measure of the *goodness of fit* for each model. Naturally, if one model gives a consistently better fit than any other model, then this model will be accepted and the others rejected. However, it often happens that one model will be the best to explain and fit certain sets of data, while another model will be the best at explaining and predicting other sets. Neither model can be rejected, since each is better under certain circumstances. Likewise, neither model should be completely accepted since in certain cases each model is not the best available. They can be conditionally accepted, studied, and used in those circumstances where they are the appropriate choice. Naturally a scientist would like to have a single model which is the best at explaining all the known experimental results. However, such a model is not always available, and the scientist must work with the models at hand until better ones are developed.

10.2.1 Estimation of Parameters

Many of the models constructed in this book cannot be tested until certain parameters are determined. We now consider the problem of estimating parameters, and we give two useful methods: the method of maximum likelihood and the method of minimum discrepancy. It should be emphasized that there are many other techniques which have been successfully used, and some of these can be found in the References for this chapter, in particular, [ReG].

The philosophy behind the method of maximum likelihood is the following. Appropriate data are collected, and it is hypothesized that the data depend on certain parameters in a specific way. Next, the probability (or density function in the nondiscrete case) that these particular data would occur in a study of this sort is determined, and then this probability is expressed as a function of the parameters. Finally, the possible parameter values are examined to determine if there are values of the parameters which maximize this probability. That is, one searches for specific parameter values such that the probability of obtaining these data is greater than or equal to the probability of obtaining the same data from a system in which the parameters take on other values. Parameters selected in this way are said to be determined by the *maximum likelihood principle*.

The following examples illustrate the method.

Suppose that a model of learning uses a parameter p which is the probability that a student in psychology P007 at Big State University can identify a concept in a certain limited time. We shall assume that all the students are equal in the ability to identify this concept. Also, suppose that 100 students are randomly selected and that 20 of them are able to identify the concept in the alloted time. Theoretically the parameter p may take any value in the interval $[0, 1]$. For each choice of p the probability that exactly 20 of 100 students will *independently* make the identification is equal to

$$\binom{100}{20} p^{20}(1-p)^{80}.$$

This formula defines a function f which depends differentiably on p, $0 \le p \le 1$. The maximum value of f on this interval can be found by examining the values of $f(\tilde{p})$ for $\tilde{p} = 0, 1$ and $\tilde{p} \in \{p : df/dp(p) = 0\}$. The equation $df/dp = 0$ is

$$\binom{100}{20}(20p^{19}(1-p)^{80} - 80p^{20}(1-p)^{79}) = 0$$

or

$$p^{19}(1-p)^{79}[(1-p) - 4p] = 0.$$

Thus $\{p : (df/dp)(p) = 0\} = \{1/5, 0, 1\}$, and we need to compare $f(0)$, $f(1)$, and $f(1/5)$. Since $f(0) = f(1) = 0$ and $f(1/5) > 0$, it is clear that the maximum value of f on $[0, 1]$ is assumed for $p = 1/5$. Thus, $1/5$ is the estimate for p

determined by the method of maximum likelihood. Note that this is also the sample proportion since $20/100 = 1/5$.

The next example is somewhat more complicated. Consider the learning model of Sec. 2.4 with the learning parameter c. Here c is the probability of learning a concept on a single trial of the experiment, and it is supposed to be independent of both the concept and the position of the trial in the sequence. Of course, we suppose that the concepts have been selected so that such an assumption is reasonable. The particular situation which we consider is that of the sequential presentations of stimuli involving six concepts. Suppose that the subject learns one concept on the second and fifth presentation and two concepts on each of the third and sixth presentations. It follows from the discussion in Sec. 2.4 that if the probability of learning on a single presentation is c, then the probability of learning on the jth presentation is $c(1-c)^{j-1}$. Therefore, using the multinomial distribution, we find that the probability of a subject responding in the manner indicated is

$$\binom{6}{1, 2, 2, 1}[c(1-c)][c(1-c)^2]^2[c(1-c)^4][c(1-c)^5]^2$$
$$= \binom{6}{1, 2, 2, 1}c^6(1-c)^{19}.$$

As above, the maximum of this function f occurs either for $c = 0$, $c = 1$, or c equal to a solution of $df/dc = 0$. We obtain 6/25 as the estimate of the parameter c given by the method of maximum likelihood. It is worth noting that this is not the same as the estimate obtained by averaging the estimates obtained by applying this method to each of the six individual concept learning situations (Exercise 1).

These two examples are, perhaps, deceptively simple. This is due in part to the fact that in both cases only a single parameter is to be determined and in part to the simple formulas which could be obtained for the relevant probabilities. In most cases of interest in model building there will be a number of parameters to be determined and the formulas for the probabilities are either very complicated or impossible to obtain. In such cases it is not reasonable to expect to solve the problem by analytic means. Indeed, it is often necessary to determine the optimal values of the parameters or approximations to them by numerical techniques using a computer. This will be illustrated by an example in the next section.

In the general case the method of maximum likelihood is used in the following way. Consider a model which has n parameters $p_1, \ldots, p_n$ to be determined. Let $L(\mathbf{x}; p_1, \ldots, p_n)$ be the function which associates with each point $(p_1, p_2, \ldots, p_n) \in \mathcal{A} \subset R^n$ the probability of obtaining given data, $\mathbf{x}$. Here $\mathcal{A}$ is the set of admissible (i.e., possible) values for the parameters. Note, also, that we are using the context of a discrete problem (i.e., a countable set of possible data vectors $\mathbf{x}$). If the context is that of a continuous problem, then L is a probability density function. The object is to find a maximum of L for $(p_1, \ldots, p_n) \in \mathcal{A}$. If L satisfies certain differentiability conditions, then the method of maximum likelihood leads to a consideration of the values of L for those points $\tilde{\mathbf{p}} = (\tilde{p}_1, \ldots, \tilde{p}_n)$ which satisfy the system of equations

$$\frac{\partial L}{\partial p_i} = 0, \qquad i = 1, 2, \ldots, n, \tag{2.1}$$

or which lie on the boundary of the set $\mathcal{A}$.

We turn now to a second method of estimating parameters, the *method of minimum discrepancy*. It is similar in spirit to the maximum likelihood technique. As above, data are collected by conducting experiments or by other suitable means, and these data are compared with the results predicted by the model. However, in the present method the parameters are to be chosen so that an appropriate measure of the difference between the data and the predictions of the model is as small as possible. The criteria used to measure this difference will vary from model to model, but the *least-square* and *chi-square* measures are common. We shall illustrate these measures and the method by examples.

Consider a model in which the following relation between age and learning ability arises. Subjects of age 4, 6, 8, and 10 years are asked to perform as many tasks of a certain type as possible in a unit time. It is hypothesized that at these ages the number of tasks performed is a linear function of age, that is,

$$\text{Number of tasks} = m \times (\text{age}) + b,$$

where m and b are parameters which cannot be determined within the model. Let n be the number of tasks performed by a subject in a unit time, and let t be the age of the subject. Suppose that the observed data are

Age	Observed n
4	15
6	17
8	29
10	34

The problem is to determine values of m and b so that the function $n = n(t)$, defined by $n(t) = mt + b$, "best" fits the data obtained from observation. For the method of least squares the best function means the one for which the sum of the squares of the differences between the observed and predicted values is minimized. Thus we seek m and b such that $\sum$ (observed value $-$ predicted value)^2 is as small as possible.

The predicted values of n are

Age	Predicted n
4	$4m + b$
6	$6m + b$
8	$8m + b$
10	$10m + b$

Thus we consider the sum of the squares of the differences,

$$S(m, b) = (15 - 4m - b)^2 + (17 - 6m - b)^2 + (29 - 8m - b)^2 + (34 - 10m - b)^2,$$

and select m and b so that $S(m, b)$ is a minimum. Since m and b are unrestricted and since S is large for $|m|$ and $|b|$ large, the desired minimum is obtained by solving the system of equations $\partial S/\partial m = 0$, $\partial S/\partial b = 0$. This system has the form

$$28m + 4b = 95,$$

$$108m + 14b = 367,$$

and the solution is $m = 3.45$ and $b = -0.40$. Thus the linear function which agrees best with the observed data in the sense of least squares is

$$n(t) = 3.45t - 0.40. \tag{2.2}$$

We leave it as an exercise for the reader to graph the data points and the function $n = n(t)$ defined by (2.2) (Exercise 2).

This method can be used to fit functions other than linear ones to observed data. For example, one could use this technique to determine growth constants in models for population growth. Simple assumptions regarding growth lead to models predicting that the size of the population at time t is given by ae^{rt}, where a and r are parameters. Experimental data taken at several times can be used to determine the values of a and r which are best in the least-squares sense (Exercise 4).

The sum of the squares of the differences between the observed values and the predicted values is one measure of the discrepancy between actual and predicted results. Another useful measure is the chi-square measure, written χ^2. Briefly, this is a weighted measure of the squares of the differences between observed and predicted frequencies. If the predicted frequencies depend on parameters, then the values assigned to the parameters influence the size of the χ^2 measure. Parameters selected so as to make the χ^2 measure as small as possible are said to have been selected according to the method of minimum discrepancy with the χ^2 measure.

As an example of this technique, consider a situation in botany in which one is concerned with the frequence of occurrence of flowers of a certain color. We suppose that it is possible to classify all flowers of the sort being investigated as to color, and we take the classifications to be the desired color and all others. Let r be the (unknown) proportion of flowers of the desired color. To estimate r, we examine three populations of flowers of the given type, each population consisting of 50 flowers. Suppose that the observed ratio of desired colors to other colors is 33:17 for the first population, 30:20 for the second, and 27: 23 for the third. Since $50r$ is the predicted frequency, the model predicts a ratio of $50r:50(1 - r)$ in each case. The χ^2 measure is defined by

$$\sum_{\substack{\text{all} \\ \text{observations}}} \frac{(\text{Observed frequency} - \text{Predicted frequency})^2}{\text{Predicted frequency}}.$$

Thus in the present case we have

$$\chi^2(r) = \frac{(33 - 50r)^2}{50r} + \frac{(30 - 50r)^2}{50r} + \frac{(27 - 50r)^2}{50r}.$$

This expression can be minimized by using the standard techniques of the calculus. Thus, since a minimum clearly does not occur at $r = 0$ or $r = 1$, we compute $d\chi^2/dr$ and find those values of r for which $(d\chi^2/dr)(r) = 0$ (Exercise 5). We obtain

$$r^2 = \frac{(33)^2 + (30)^2 + (27)^2}{3(50)^2}$$

or

$$r = 0.602,$$

where the last figure is correct to three decimal places. Note that this is close to, but not identical to, the average of the three sample values of r.

10.2.2 Testing Hypotheses and Comparing Models

The usual method of evaluating a model is to compare the results of experiments or observations with predictions based on the model. Frequently this evaluation is concerned with the parameters of an assumed probability distribution for a random variable. Occasionally it is concerned with the form of the distribution. In these cases statistical tests are appropriate. We are interested both in the test to be used to obtain a comparison and in the criteria used to measure the results of the test. The following discussion is quite selective, but the selection was made so as to give a sample which is fairly representative of the techniques in common use.

To test some feature of a model, we test a specific statement regarding this feature. By a *test* of a statement we mean a determination of whether the statement should be accepted or rejected. The method of determination makes use of the results of experiments or observations, and it relates these results to the predictions of the model. The test also gives probability statements concerning the likelihood of errors in accepting or rejecting the statement. The specific statement being considered is known as a *hypothesis based on the model* or a *hypothesis of the model.* If the hypothesis specifies completely and exactly both the form of the probability distribution and all its parameters, then it is said to be a *simple hypothesis;* otherwise it is said to be *composite.*

There are often several aspects of a model which can be tested. For example, as noted earlier, one can test the underlying assumptions (the axioms) directly, or one can test the predictions (theorems) directly and thereby test the underlying assumptions indirectly. Notice the nature of this second method. The theorems are true statements in the mathematical model; there is no question of their validity, and therefore there is no sense in testing them in the model. However, the theorems or predictions can be tested in the real world. There is no guarantee that the predictions of the model match exactly with reality. This is what we hope, but we have no right to assume it. To illustrate these notions, let us return to the model for learning of Sec. 2.4. We took as one of our basic assumptions that the learning parameter c is independent of trial and of stimulus but that it may vary from individual to individual. This assumption could be tested by designing an experiment which determines whether the parameter is, in fact, independent of trial or whether it varies in some systematic way with the trial number. Alternatively, we could accept this assumtpion and test the prediction regarding the probability of a correct response on the nth trial (Theorem 2, Sec. 2.4.2 of Chap. 2). Our hypothesis would be that the subject gives a correct response to the stimulus on the nth trial with probability $1 - \frac{1}{2}(1 - c)^{n-1}$. Such a test presupposes that the learning parameter has been determined. As part of our testing procedure we would have to establish criteria for accepting or rejecting the hypothesis. There are a number of standard tests and criteria in common use.

Let us consider the question of testing a hypothesis in more detail. We begin by noting that there are different types of errors which can occur in evaluating the results of a test of a hypothesis. First, it is possible to conclude that the hypothesis is false even though it is actually true, and, second, it is possible to conclude that the hypothesis is true even though it is actually false. Errors of the former type are often called Type I errors, and those of the latter type Type II errors. The characteristics of the experiment, the statistical test to be used, and the criteria for accepting or rejecting the hypothesis all contribute to the occurrence of the two types of errors. Ideally it is desirable to use a test which simultaneously minimizes the chances for both types of errors. In general it is not possible to find a test and reasonable criteria for acceptance which accomplish this. It is customary to proceed by agreeing in advance to an acceptable probability of a Type I error—0.05 and 0.01 are commonly accepted values—and then selecting a test which gives a tolerable Type II error.

It is useful to consider an example which illustrates some of these concepts. First, however, we introduce one of the tests which is often used to check the accuracy of a statistical hypothesis having to do with a mathematical model, the chi-square test of predicted frequencies. This test is used to determine whether predictions of the model concerning the frequencies of certain events are supported by experimental evidence. The object of the present discussion is simply to indicate to the reader how such a test can be used. A discussion of its probabilistic and statistical basis is properly in the domain of statistics, and we suggest that the interested reader consult the References at the end of this chapter. Suppose that the model predicts that a set of mutually exclusive and exhaustive events will occur with frequencies e_i, $i = 1, 2, \ldots, n$ (these are called the *expected frequencies*) and that in an appropriate experiment the observed frequencies of these same events are o_i, $i = 1, 2, \ldots, n$. The chi-square (χ^2) measure of the difference between the observed and expected frequencies was defined above, and we reproduce the definition here for convenience:

$$\chi^2 = \sum_{i=1}^{n} \frac{(o_i - e_i)^2}{e_i}.$$

Consider the genetics model of Chap. 2. Suppose that we are concerned with two distinct attributes, say color and texture, each of which occurs in two forms. If the two colors are green and yellow, yellow dominant, and the two textures are smooth and wrinkled, smooth dominant, then the phenotype distribution of the first filial generation resulting from a dihybrid cross is in the ratio 9:3:3:1, with 9/16 of the members of phenotype yellow-smooth, 3/16 of phenotype green-smooth, 3/16 of phenotype yellow-wrinkled, and 1/16 of phenotype green-wrinkled. A dihybrid cross is a mating of two individuals each of which is a hybrid with respect to each of the two genes. Thus, if the two alternative forms of each of the two genes are denoted by A, a and B, b, respectively, then the mating is a cross between individuals of genotype $AaBb$. The ratios 9:3:3:1 are predicted on the basis of a model similar to that constructed in Sec. 2.1 (see Exercise 5 of Sec. 2.1). Suppose that an experiment is conducted in which two individuals of genotype *AaBb* are crossed and the resulting descendants are classified according to color and texture. Assume that the following table summarizes the experiment by giving the observed and predicted phenotypes of 480 individuals:

Phenotype	Observed numbers	Predicted numbers
Yellow-smooth	257	270
Green-smooth	86	90
Yellow-wrinkled	103	90
Green-wrinkled	34	30

According to the definition, we have

$$\begin{aligned}\chi^2 &= \frac{(257 - 270)^2}{270} + \frac{(86 - 90)^2}{90} + \frac{(103 - 90)^2}{90} + \frac{(34 - 30)^2}{30} \\ &= 0.626 + 0.178 + 1.87 + 0.533 \\ &= 3.22.\end{aligned}$$

What interpretation is to be given to the result $\chi^2 = 3.22$? Recall that the original question was to determine whether or not the model provided an accurate description of the real-world situation or, alternatively, whether the predictions of the model were consistent with observations. One expects that a small value of χ^2 corresponds to a good fit between experiment and prediction, but this is not precise enough for our purposes. One prefers probability statements concerning the likelihood of obtaining discrepancies at least as large as the value of χ^2 actually obtained. Such probability statements can be obtained from the derived value of χ^2 and standard statistical tables. In this example the probability associated with the value 3.22 of χ^2 is 0.36. That is, the statistical test implies that if the predicted values of the frequencies were in fact the actual ones and if the experiment described above were repeated many times, then 36% of the time, purely by chance, one would obtain deviations at least as large as those just observed. On this basis one must decide whether it is reasonable to conclude that the observed data are in agreement with the predicted 9:3:3:1 ratio. In this connection it is useful to have in mind a clear statement of the hypothesis which is being tested and a criterion for evaluating the results of the test. In this example, it is natural for the hypothesis to be the statement that the four phenotypes occur with the ratios 9:3:3:1. A possible criterion is that the hypothesis will be rejected if $\chi^2 \geq \alpha$, where α is chosen so $\chi^2 \geq \alpha$ will occur by chance with probability ≤ 0.05 when the hypothesis is correct. If $\chi^2 < \alpha$, then there are two options to consider. We may decide to accept the hypothesis, or we may conclude that although the experimental results do not warrant rejection of the hypothesis, judgment on accepting the hypothesis is better reserved until further testing is complete. The critical value of α is 7.8, and hence with this criterion the above experiment would indicate that we should not reject the hypothesis. Instead, the hypothesis would be conditionally accepted pending the results of additional tests.

The situation just discussed illustrates an evaluation of the predictions of the model. One can sometimes also use the chi-square method to test the validity of assumptions. In Exercise 6 we shall consider testing the assumption on the demand made in the linear programming problem of Sec. 2.2.

The examples above are illustrations of the process of testing hypotheses where the intent is to accept or reject the hypothesis on the basis of the test. We have already noted in the introduction to this section that most model testing is not of this form. Instead the test is usually designed to obtain a quantitative measure of the ability of the model to explain the phenomena under study. In this way the model can be compared with other models which are designed to study the same phenomena. As an illustration of the process of comparing models we now turn to an example of the use of the least-squares measure to compare three different models for the growth of a population of fruit flies.

Example. Let time be measured in days and let $P(t)$ be the size of a population of fruit flies on day t. The three models which we shall consider are the linear model, the exponential model, and the logistic model. Each of these models depends on certain parameters, and in terms of these parameters the models can be expressed as follows:

Linear: $$P_1(t) = mt + b, \qquad m, b \text{ parameters,}$$

Exponential: $$P_2(t) = Ae^{kt}, \qquad A, k \text{ parameters,}$$

Logistic: $$P_3(t) = \frac{MP_0}{P_0 + (M - P_0)e^{-cM(t-t_0)}},$$

$$t_0, P_0, c, M \text{ parameters.}$$

The exponential and logistic models are discussed in Chap. 8, and the linear model is based on the assumption that the size of the population is a linear function of time.

From the discussion in Sec. 8.2 we know that the logistic curve is such that M is the limiting size of the population, t_0 is the starting time for the study, P_0 is the original population size, and c is the growth parameter. We shall specify t_0 and P_0, and hence each of our three models will have two undetermined parameters. These parameters will be determined by the data of the study. We use the following representative data for fruit fly growth:

t (in days)	Number of observed flies
8	10
16	50
24	160
32	300
40	330

Using these data we have $t_0 = 8$ and $P_0 = 10$ in the logistic model. We estimate the other two parameters for the logistic model and the two parameters in the linear and exponential models by the method of minimum discrepancy (see Sec. 10.2.1). Since we intend to use the least-square measure to compare the models, we also use this measure to estimate the parameters for each model. In this way each parameter of each model will be assigned that value which makes the least-square measure of the discrepancy of the model as small as possible.

We denote by L_i, $i = 1, 2, 3$, the least-square measure of the discrepancy between the observed data and the data predicted by the function P_i. Then L_1 depends on the parameters m and b, L_2 depends on the parameters A and k, and L_3 depends on the parameters M and c. We first estimate the parameters of the function L_1. Our method requires the computation of the partial derivatives of L_1 with respect to m and b, respectively. We have

$$\begin{aligned} L_1(m, b) = {} & (8m + b - 10)^2 + (16m + b - 50)^2 + (24m + b - 160)^2 \\ & + (32m + b - 300)^2 + (40m + b - 330)^2, \end{aligned}$$

and hence

$$\frac{\partial L_1}{\partial m} = 7040m + 240b - 55040$$

and

$$\frac{\partial L_1}{\partial b} = 240m + 10b - 1700.$$

The solution of the system of equations $\partial L_1/\partial m = \partial L_1/\partial b = 0$ is given by $m = 11.08$ and $b = -95.68$ (accurate to two decimal places). Thus these values of m and b are candidates for the values which make L_1 as small as possible. It is easily shown that indeed these values do give a minimum for L_1 (Exercise 7). With this choice of the parameters m and b the function L_1 has the value 3390.

We now turn to the exponential model and estimate the parameters A and k in the function L_2. Once again we compute the partial derivatives of the function L_2 and set these partial derivatives equal to 0. We have

$$L_2(k, A) = (Ae^{8k} - 10)^2 + (Ae^{16k} - 50)^2 + (Ae^{24k} - 160)^2 + (Ae^{32k} - 300)^2 + (Ae^{40k} - 330)^2.$$

Therefore,

$$\frac{\partial L_2}{\partial A} = 2e^{8k}\{A[e^{8k} + e^{24k} + e^{40k} + e^{56k} + e^{72k}] - [10 + 50e^{8k} + 160e^{16k} + 300e^{24k} + 330e^{32k}]\},$$

and

$$\frac{\partial L_2}{\partial k} = 16Ae^{8k}\{A[e^{8k} + 2e^{24k} + 3e^{40k} + 4e^{56k} + 5e^{72k}] - [10 + 100e^{8k} + 480e^{16k} + 1200e^{24k} + 1650e^{32k}]\}.$$

Since inspection shows that L_2 does not have a minimum when A and k are such that $Ae^{8k} = 0$, the system of equations $\partial L_2/\partial A = \partial L_2/\partial k = 0$ can be solved by finding those values of A and k such that

$$A = \frac{10 + 50e^{8k} + 160e^{16k} + 300e^{24k} + 330e^{32k}}{e^{8k} + e^{24k} + e^{40k} + e^{56k} + e^{72k}}$$

and

$$A = \frac{10 + 100e^{8k} + 480e^{16k} + 1200e^{24k} + 1650e^{32k}}{e^{8k} + 2e^{24k} + 3e^{40k} + 4e^{56k} + 5e^{72k}}.$$

If we let $x = e^{8k}$ and equate these expressions for A, we obtain the following algebraic equation for the variable x:

$$30x^{10} - x^9 + 15x^8 - 46x^7 - 20x^6 - 96x^5 - 55x^4 - 146x^3 - 90x^2 - 31x - 5 = 0.$$

We solve this equation numerically (by digital computer) and obtain 1.595 as an approximate value for the single positive real solution of the equation (Exercise 9). This gives A the value 35.29, and for these values of x and A the function L_2 has the value $L_2(k, A) = 10{,}310$. Notice that we also have a value for k: $k = \frac{1}{8}\log(1.595) = 0.058$. At this stage we have not shown that these values of k and A are the ones which make L_2 as small as possible. We have only shown that they are values for which the partial derivatives of L_2 are 0. It is still necessary to show that a minimum exists at this point. The proof of this fact is somewhat involved, and we leave the details to the Exercises (specifically, Exercises 8 and 9).

We note that the linear model gives a much smaller least-squares value than the exponential model. However, neither model is able to predict the observed data with any great degree of accuracy. We next turn to a somewhat more sophisticated model, the logistic model, and based on the discussion of Sec. 8.2 we expect to obtain better results. The unspecified parameters for the logistic model are M and c, and the least-square measure of the discrepancy between the actual and predicted data is given by the function

$$\begin{aligned} L_3(M, c) = (10 - 10)^2 &+ \left(\frac{10M}{10 + (M - 10)e^{-8cM}} - 50\right)^2 \\ &+ \left(\frac{10M}{10 + (M - 10)e^{-16cM}} - 160\right)^2 \\ &+ \left(\frac{10M}{10 + (M - 10)e^{-24cM}} - 300\right)^2 \\ &+ \left(\frac{10M}{10 + (M - 10)e^{-32cM}} - 330\right)^2. \end{aligned}$$

In this case, and in most instances arising in practice, it is difficult to solve for the values of M and c which make L_3 as small as possible. The system of equations $\partial L_3/\partial M = \partial L_3/\partial c = 0$ is a complicated system of nonlinear equations and consequently very difficult to solve. Moreover, it is difficult to show that L_3 has a minimum when evaluated at the solution of this system. In these cases the easiest method is often a search procedure which systematically tries out different values of the parameters to find the best (or nearly best) ones. This procedure is almost always carried out on a computer, and there are many prepared programs for such searches. We have used a very direct search procedure in this example. First a value of M is selected, and then the best value of c for that M is determined by comparing the values of L_3 for different choices of c and selecting that c which gives the smallest value of L_3. This process is carried out for a selection of values of M, and it yields a sequence of pairs (M, c), where the value of c is the best for that M. Finally, the values of L_3 are compared for these pairs to obtain the pair which gives the smallest value for L_3. The method is discussed in more detail in Exercise 10.

Using the search method discussed above, we obtain the pair $(M, c) = (347, 0.00062)$ as the pair which gives the smallest value for L_3. The L_3 value is

$$L_3(347, 0.00062) = 161.$$

Thus we see that for these data the logistic model is much better at "fitting" the observed growth of the population of fruit flies. The graphs of the func-

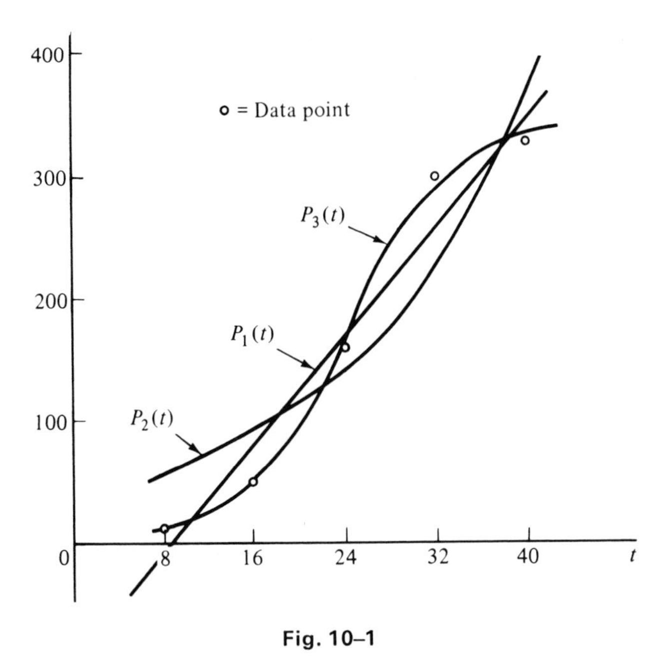

Fig. 10–1

tion $P_1(t)$, $P_2(t)$, and $P_3(t)$ are shown in Fig. 10–1, together with original data points.

EXERCISES

1. Estimate the parameter c in the learning theory model of Sec. 2.4 for the following situations:
   (a) An individual is tested and he learns a simple task on the nth trial. Use the method of maximum likelihood to estimate c.
   (b) An individual learns six simple tasks in the following way: One task is learned on the second trial, two tasks are learned on the third trial, one task is learned on the fifth trial, and two tasks are learned on the sixth trial. Use the results of part (a) to estimate c for each task, and then use the average of these six values of c to estimate c in general. Compare this estimate with the maximum likelihood estimate of c obtained in Sec. 10.2.1.
2. Show that if any line is chosen which goes through two of the four points (4, 15), (6, 17), (8, 29), (10, 34), then at least one of these four points is farther from this line than it is from the line determined by these points using the least-squares method. Graph the seven lines.
3. Use the method of least squares to find the parabola which best fits the points (1, 2.2), (2, 4.1), (3, 5.8), (4, 4.4), (5, 2.4).

4. Use the method of least squares to estimate the parameters a and r in the function $P(t) = ae^{rt}$ so as to give the best fit to the following data:

t	Observed value
5	5
10	12
15	30

5. Show that for $0 < r \leq 1$, the minimum of the function

$$\chi^2(r) = \frac{(33 - 50r)^2}{50r} + \frac{(30 - 50r)^2}{50r} + \frac{(27 - 50r)^2}{50r}$$

is attained for

$$r = \left[\frac{(33)^2 + (30)^2 + (27)^2}{3(50)^2}\right]^{1/2}.$$

6. Suppose that over a 10-week period the demand for mobile homes is as follows:

Week	1	2	3	4	5	6	7	8	9	10
Demand	9	10	8	20	21	12	11	20	20	19

Use the chi-square measure to test the assumption that demand is constant on a week-to-week basis. Note that although the assumption of constant demand can be tested by assuming that demand is any natural number, there are relatively few natural numbers which are reasonable for these data. In particular, compute the chi-square measure of discrepancy for assumed demands for 14, 15, 16, 17, and 18 mobile homes per week. For ten degrees of freedom and the same 0.05 criterion used in the examples of this section, we have $\alpha = 18.3$. Which, if any, of the choices for a constant demand should be rejected and which accepted? Which choice minimizes the χ^2 measure of discrepancy?

7. Show that the minimum of the function

$$L_1(m, b) = (8m + b - 10)^2 + (16m + b - 50)^2 + (24m + b - 160)^2 + (32m + b - 300)^2 + (40m + b - 330)^2$$

is attained for values of m and b given by the solution of the system of equations $\partial L_1/\partial m = 0 = \partial L_1/\partial b$. Show that this system has a unique solution. *Hint:* Consider (in three dimensions) the graph of the function L_1 and examine the behavior of L_1 as $|m|$ and/or $|b|$ become very large.

8. Show that the minimum of the function

$$L_2(A, k) = (Ae^{8k} - 10)^2 + (Ae^{16k} - 50)^2 + (Ae^{24k} - 160)^2 + (Ae^{32k} - 300)^2 + (Ae^{40k} - 330)^2$$

is attained for values of A and k given by a solution of the system of equations

$$\frac{\partial L_2}{\partial A} = 0 = \frac{\partial L_2}{\partial k}.$$

Hint: Consider the graph of the function L_2. In particular, note the behavior of L_2 for negative values of A and for large values of A and $|k|$. Finally, argue that given any large rectangle in the A-k plane (with one side along the $A = 0$ axis, including the origin, and lying in the $A \geq 0$ half-plane) the minimum of L_2 will be taken at a point (A, k) in the interior of the rectangle.

9. For the function L_2 of Exercise 8 show that the system of equations $\partial L_2/\partial A = 0 = \partial L_2/\partial k$ has a unique solution. *Hint:* Let $x = e^{8k}$ and find the equation for x which must be satisfied if A and k are a solution of the system of equations. Then show that this equation in x has a single positive real root.

10. To find the minimum of the function $L_3(M, c)$ of Sec. 10.2.2, a direct computer search method was used. The method found a best value of c for each M by testing a range of values of c. Then the pairs (M, c) were compared to find a best pair. Construct a flow chart for this search procedure. Also discuss the problem of finding the proper step size to use in the search for the best c to go with a given M.

10.3 SIMULATION

In this text we have studied models in which the mathematical problems were ones which could be solved by analytic or numerical means. Thus in the Markov chain models for small-group decision making we obtained explicit formulas for the transition probabilities in terms of the parameters of the system. Also, although we were unable to provide formulas for the solution of general linear optimization problems, we developed an algorithm which could be used to find solutions to these problems numerically. It frequently happens, however, that neither of these approaches (analytic or numerical) is a reasonable one. To use these methods, it may be necessary to impose restrictions and to make approximations to such an extent that the resulting model bears little relation to the real situation. An example of this might be a nonlinear stochastic optimization problem, say a transportation problem with uncertain supply and demand governed by known but not simple probability distributions and a system of bulk discounts for shipping large quantities over specific routes. A study of this situation is beyond the scope of the methods developed in this text.

Another consideration is the expenditure of effort involved. The numerical solution of the system of differential equations resulting from a rather general epidemic model may require a great deal of programming effort and computer time and still not provide answers to many interesting questions. For example, a numerical solution to the differential equations will not in general tell how the characteristics of the epidemic change with variations in the parameters. Thus there are instances in which an appropriate approach is to return to the basic processes which determine the behavior of the system and to attempt to simulate them.

The term *simulation* is usually taken to mean the creation of an artificial system which displays either the same behavior as the one being studied or behavior related in some simple way to it. Thus it might mean the creation of an electrical or mechanical model which in a well-defined way mimics the operation of a system of social or biological interest. For example, the study of real electrical networks may provide valuable information about certain types of situations which can be modeled as graphs. Such physical simulations are frequently quite useful. However, we intend here to consider only simulations carried out on a computer or, in exceptional circumstances, by hand with the aid of a desk calculator.

In general, the use of simulation permits quantitative work with much more complex systems than can be studied by analytic or numerical methods. Anticipating the example to be considered in Sec. 10.3.2, it is worthwhile to draw specific attention to the roles of time and the parameters of the system in simulation processes.

The appropriate choice of a time unit for a real-world situation may be a nanosecond or an eon. Either very long or very short time units make direct observation difficult or impossible. Computer simulation gives the investigator considerable flexibility in choosing his time scale. Events which take place in weeks, years, or centuries in the real world can be made to take place in seconds in the computer. Also, regular time intervals in the computer may be determined by the occurrence of events which do not occur periodically in the real world. That is, a completely artificial time scale can be introduced.

It is usual for a program written for a simulation to permit relatively easy variation of the parameters. Thus the model can be tested with regard to the sensitivity of the results to changes in the parameter values. Systematic variation of parameters in a simulation may provide useful information even in cases in which the system can be studied partially or completely by analytic means.

Simulation is a broad subject and there are many techniques which are primarily useful in certain special situations. Also, there are special programming languages developed particularly for simulation, e.g., GPSS (General Purpose Simulation System) and SIMSCRIPT, and the reader with more than a casual interest in simulation should become familiar with them. They are powerful tools which aid in formulating the model as well as reducing programming time. Rather than to attempt a short survey of the topic as a whole, we propose instead to treat a single example in some depth. We select a queuing problem to use as an illustration both because many of the questions that arise here are common to other simulation problems and because it is the sort of problem for which one frequently must resort to simulation. The results described here were obtained with the aid of a FORTRAN program written by Albert Hart and a CDC 6600 computer.

Since simulation is simply an alternative to other types of models, the results obtained by modeling a situation through simulation must be evaluated in the same way as those arising through analytic means. Therefore the computer output resulting from simulation should, whenever possible, be compared with data from the real situation using the same sort of statistical tests as in Sec. 10.2. Also, simulation may occasionally be applied in situations where real data cannot be obtained, either because of cost or excessive risk. In such circumstances other methods of evaluation must be applied.

Frequently one combines analytic and simulation models and uses the latter as an aid in testing the basic principles of the model. In such a procedure one formulates the assumptions in the usual way and translates them into mathematical terms. However, instead of proceeding to analyze the situation using analytic tools, the resulting mathematical situation is simulated. A comparison of the results of the simulation with the real world then provides a test of the validity of the assumptions.

Simulation is an important modeling tool. It has established its usefulness as a tool in the engineering and management sciences, and it is finding increasing applications in the life and social sciences. In the References the reader will find sources which proceed into the subject on various levels and with various audiences in mind.

10.3.1 Random Numbers for Monte Carlo Simulation

All mathematical models contain quantities which are particularly relevant for the real situation being considered. In stochastic models these quantities are usually random variables, and typically the model will be based on an assumed probability distribution for each random variable. The nature of the assumed distribution is frequently basic to the predictions based on the model, and different distributions may lead to quite different predictions. A specific distribution may be selected on the basis of experimental evidence, or it may simply be a reasonable conjecture subject to future verification. Consequently, an important part of the simulation process is the production of sequences of numbers to be viewed as the values of a random variable distributed according to an assumed probability distribution or, as is often the case, a number of such sequences with different associated probability distributions. The simplest case is that in which a random variable can assume only finitely many different values and each occurs with equal probability. The probability distribution associated with such a random variable can be generated by several different physical devices. One of these is a fair spinner on a disc divided into an appropriate number of equal-sized wedges. In view of the similarity of this device to a roulette wheel, the name *Monte Carlo* has come to be associated with the generation and application of random numbers in simulations and similar situations.

A spinner similar to the one described above can be used to handle slightly more complex random variables. Indeed, if a random variable can take only finitely many values and if the probabilities of taking the respective values are known, then a spinner on a disc with unequal wedges can be used to generate a sequence of random numbers with the appropriate distribution. Suppose that v_i, $i = 1, 2, \ldots, n$ are the values taken by a random variable and suppose that v_i is taken with probability p_i. Assign v_i to a wedge centered at the spinner with a central angle of $2\pi p_i$ radians.

It is clear that if simple techniques such as these are to be useful in simulation, then a means must be found to implement them on a computer. The point of view which we pursue here is one which allows the treatment of these and also much more complex situations. Therefore the discussion will be continued in a somewhat more general setting. In particular, in the present discussion we shall consider simultaneously those random variables which take values in a discrete (finite or infinite) set and those which take values in an interval.

A random variable f which takes values in an interval $[a, b]$ is said to have a *uniform distribution* if

$$\Pr[\alpha \leq f \leq \beta] = \frac{(\beta - \alpha)}{(b - a)}$$

for all intervals $[\alpha, \beta] \subseteq [a, b]$. A random variable with values in an interval $[a, b]$ can easily be transformed into one whose values lie in $[0, 1]$ and whose associated probability distribution has the same features. Thus, if f has a uniform distribution and takes values in $[a, b]$, then the random variable $(f - a)/(b - a)$ has a uniform distribution and takes values in $[0, 1]$. A sequence $\{x_j\}$ of numbers in the unit interval $[0, 1]$ is said to be *random with the uniform distribution*, or simply *random*, if it is a set of values taken by a random variable with a uniform distribution on $[0, 1]$. There are several methods of generating sequences of random numbers and extensive tables of such numbers have been published [R]. It is possible to read tables into a computer in order to have random numbers available for use in simulation. However, even modestly sized simulations may require quite a large set of random numbers, and consequently the storage of the necessary tables would require large amounts of memory. As an alternative, computer programs have been developed for generating sequences of random numbers as they are needed. These programs are short and require little programming effort and have acceptably short execution time. The requirements to be imposed on a particular set of random numbers will usually vary with the intended application in simulation. Statistical tests can be used to verify that the set of numbers produced is satisfactory for the intended use (see the introduction to [R]).

Most stochastic models involve random variables whose associated distribution is nonuniform and a sequence of whose values is to be generated. As an example consider again the case of a random variable assuming finitely many values, say $1, 2, \ldots, n$ for simplicity, and suppose that the value j is assumed with probability p_j, $1 \leq j \leq n$. A sequence of values with this distribution can be generated in the following manner. Let y be a random variable taking values in $[0, 1)$ and whose associated distribution is uniform. Define a new random variable x by setting $x = j$ if y is contained in the interval $[\sum_{i<j} p_i, \sum_{i \leq j} p_i)$. Thus, if $y \in [0, p_1)$, then $x = 1$; if $y \in [p_1, p_1 + p_2)$, then $x = 2$; and so forth. The procedure is illustrated geometrically in Fig. 10–2. The sequence $\{y_i\}$ which is randomly distributed on $[0, 1)$ with the uniform distribution has been used to obtain a sequence $\{x_i\}$ randomly distributed on the set $\{1, 2, \ldots, n\}$ with a distribution given by the probabilities $\{p_i\}_{i=1}^n$.

The related procedure for random variables whose associated distribution function is continuous is more complex, but the basic idea introduced above continues to be a useful one. Let ρ be the density function for a random variable and let $F(x) = \int_{-\infty}^{x} \rho(t)\, dt$ be the (cumulative) distribution function. Suppose that the density function is piecewise continuous, zero on $(-\infty, a]$, nonzero on (a, b), and zero on $[b, \infty)$. We admit the possibility that $a = -\infty$, $b = +\infty$. Then F is strictly increasing on (a, b), $\lim_{x \to a} F(x) = 0$, $\lim_{x \to b} F(x) = 1$. When restricted to the interval (a, b), the function F has an inverse which we denote by F^{-1}. The domain of F^{-1} is the interval $(0, 1)$. If $\{y_i\}$ is a sequence of random numbers uniformly distributed on the domain of F^{-1}, then the sequence $\{x_i\}$, $x_i = F^{-1}(y_i)$, is a sequence of random numbers distributed on the interval $(-\infty, \infty)$ whose associated cumulative distribution

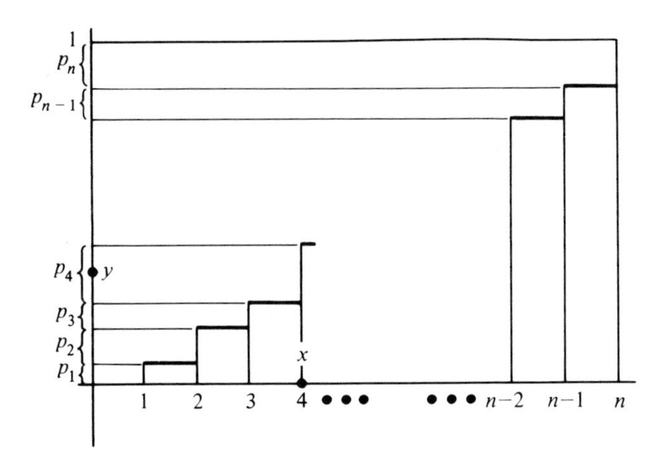

Fig. 10–2

function is F. If the density function ρ is 0 outside some interval I, then $x_i \in I$ for all i.

If the distribution function (or the density) can be described analytically, then it may be possible to find an analytic representation for F^{-1} and thus a means of determining $\{F^{-1}(y_i)\}$.

As an example, suppose that the density function ρ is given by

$$\rho(x) = \begin{cases} 0, & x \leq a, \\ \dfrac{2(x-a)}{(b-a)^2}, & a < x < b, \\ 0, & x \geq b, \end{cases}$$

so that the cumulative distribution function is

$$F(x) = \begin{cases} 0, & x \leq a, \\ \left(\dfrac{x-a}{b-a}\right)^2, & a < x < b, \\ 1, & x \geq b. \end{cases}$$

In this case the inverse function F^{-1} can be computed directly. Indeed, since $y = F(x) = (x^2 - 2ax + a^2)/(b-a)^2$, x can be expressed as a function of y using the quadratic formula. The details are simpler if the change of variable $x \longrightarrow x + a$ is made. This has the effect of replacing a by 0 and b by $b - a$ in the above definition of F. Thus, after making the indicated change,

$$F(x) = \begin{cases} 0, & x \leq 0, \\ \dfrac{x^2}{(b-a)^2}, & 0 < x < b - a, \\ 1, & x \geq b - a, \end{cases}$$

and

$$F^{-1}(y) = (b-a)y^{1/2}, \qquad 0 < y < 1.$$

We conclude that if $\{y_i\}$ is randomly distributed on (0, 1) with a uniform distribution, then $\{(b-a)y^{1/2}\}$ is distributed on $(0, b-a)$ with distribution function F.

As a second example of a nonuniform distribution which can be expressed analytically, we cite the exponential distribution. Such a distribution occurs as the cumulative distribution function of a random variable defined by the interarrival times for Poisson arrivals to a queue. The associated density function is

$$\rho(x) = \begin{cases} 0, & x \leq 0, \\ \lambda e^{-\lambda x}, & x > 0, \end{cases}$$

and consequently the cumulative distribution function is

$$F(x) = \begin{cases} 0, & x \leq 0, \\ 1 - e^{-\lambda x}, & x > 0. \end{cases}$$

Thus, if $\{y_i\}$ is a sequence of random numbers distributed on (0, 1) with the uniform distribution, then the sequence $\{x_i\}$ defined by

$$x_i = -\frac{1}{\lambda} \log(1 - y_i)$$

is randomly distributed on $(0, \infty)$ with the exponential distribution. Notice that if $\{y_i\}$ is randomly distributed on (0, 1) with the uniform distribution, then so is $\{1 - y_i\}$. It follows that the sequence $\{x_i\}$ defined above could just as well be defined by

$$x_i = -\frac{1}{\lambda} \log y_i$$

in the sense that the two sequences have the same distribution properties.

10.3.2 A Sample Simulation

We now use an example to illustrate some of the ideas and methods discussed earlier in this section. The situation to be studied is a version of the barbershop problem which was introduced in Chap. 9. Although the situation considered here is in certain respects more complicated than those considered in Chap. 9, the basic ideas remain the same. It is because of these additional complications that the methods of analysis introduced in Chap. 9 are inadequate for this situation and that resorting to simulation is necessary. We consider a shop with several chairs, and we orient our study toward the following problem: Given a knowledge of the arrival rates and service times, determine the optimal number of chairs for the shop.

In the language of Chap. 9 it is necessary to assume an arrival scheme, a queue discipline, and a service scheme. A first come-first served queue discipline corresponds to actual practice in most barbershops. With regard to the service times, we assume that each haircut requires the same time. Although there are likely to be small variations from one customer to the next and more importantly, from one barber to the next, this is an assumption which has substantial empirical support. The nature of the appropriate assumptions concerning arrivals is not nearly so clear. Shops in certain locations may have customer arrivals spaced nearly uniformly throughout the day, while other shops may have definite periods of heavy and light traffic. We assume an arrival pattern somewhat more complex than uniform arrivals all day long but one which is still simple enough that the model does not become excessively cluttered. In particular, we assume that there is a single period of heavy traffic with varying density and during the remainder of the day the traffic is uniform. Finally, we assume that there is a relatively simple sort of reneging. If a customer arrives and finds the shop "too full," in a sense to be made precise, then he leaves without joining the queue. Any customers who join the waiting line remain until serviced.

We now proceed to make these ideas more precise, and we turn first to the pattern of arriving customers. Suppose that the shop is to be open for 8 hours, 9 A.M. until 5 P.M., for example, and that traffic is distributed throughout the day as follows. Arrivals are uniformly randomly distributed with a mean arrival rate of 9 customers per hour except during the 2 hour period from 11:30 A.M. to 1:30 P.M. During the latter period customers again arrive randomly, but the underlying distribution is no longer uniform. We assume that 36 customers are expected in this period with a peak arrival rate of 27 customers per hour occurring at 12:30 P.M. Suppose that the increase in mean arrival rate from 9 customers per hour at 11:30 to 27 customers per hour at 12:30 is uniform and that the decrease back to 9 customers per hour at 1:30 is also uniform. Our simulation is a minute-by-minute evaluation of the system, and consequently it is convenient to translate the arrival rates to customers per minute. Clearly, the uniform rate of 9 customers per hour is equivalent to 0.15 customers per minute. If time is measured in minutes from 9:00 A.M., then the period of heavy traffic described above is one in which customers arrive at the rate of

$$r(t) = \begin{cases} 0.15 + 0.005(t - 150), & 150 \leq t \leq 210, \\ 0.45 - 0.005(t - 210), & 210 \leq t \leq 270, \end{cases}$$

customers per minute. For the simulation these arrival rates are to be interpreted as follows. The probability that a customer arrives during any specific minute in the period 9–11:30 A.M. or 1:30–5:00 P.M. is 0.15, and the probability that a customer arrives during minute $(t, t + 1)$, $150 \leq t < 270$, is

$$\int_t^{t+1} r(\tau)\, d\tau.$$

Thus our basic assumption concerning arrivals is that they are random throughout the day with density function given by

$$p(t) = \begin{cases} 0.15, & 0 \leq t \leq 150,\ 270 \leq t \leq 480, \\ r(t), & 150 \leq t \leq 270. \end{cases}$$

The simulation proceeds minute by minute, and for each time interval a suitable random device is used to determine whether a customer enters the shop. It is assumed that during each minute at most one customer will enter the shop. There is one exception to be considered, that of reneging. If a customer enters the shop when there are 25 customers there, either receiving haircuts or waiting, then he leaves. That is, the total number of customers in the shop never exceeds 25. In practice it may be that this number should depend also on the number of barber chairs in operation. This slight modification will not be considered further here. We arrange our simulation so that if a customer enters the shop during some minute, then he enters at the beginning of the minute, whereas if a customer leaves the shop, i.e., if a haircut is completed, then he leaves at the end of the minute. Thus it cannot happen, for example, that there are 25 customers in the shop at the beginning of a minute, service is completed on one, he leaves and another enters the shop within a single minute.

With regard to the service time, we suppose that each haircut requires 20 minutes. It is necessary to specify exactly how this is to be computed. If a customer "enters" a barber chair during a specific minute, then he leaves the chair, his haircut completed, at the end of the nineteenth following minute.

Using the sequence of events described above, the simulation proceeds as follows. For each minute a random device, in our case a random number generator, determines whether a customer enters the shop. If a customer enters the shop and there are fewer than 25 customers already there, he remains; otherwise he leaves. If the customer remains, then he joins the waiting line. If there is an empty chair and no existing queue, his stay in the waiting line may be artificial, but it is convenient to introduce it nevertheless. Next, we examine the barber chairs to see if there is an empty one. If so, the customer at the head of the queue moves into the empty chair. Finally, it must be determined whether any of the customers then being serviced are in their twentieth minute of service, i.e., whether any haircuts will be completed during the minute under consideration. If a haircuit is completed, then that individual leaves the shop and the chair he occupied is empty at the beginning of the next minute.

The schematic diagram in Fig. 10–3 should aid in understanding the proposed simulation. Such a diagram is a very useful first step in the simulation process. Subsequent steps are the refinement of the diagram into a flow chart and then into a computer program. We emphasize that the diagram gives the sequence of events to be examined *during each minute*.

In expanding the diagram into a flow chart for the entire process, one of the important matters to be settled is that of keeping track of information. The decision as to exactly what data are to be recorded must, of course, depend on the purpose of the simulation. In this example we are interested in finding the right number of chairs for the shop. The term *right* remains to be interpreted. Our interpretation of "the right number of chairs" is that this number makes the average waiting time acceptable, a necessary condition if customers are to continue coming to the shop, and also yields an acceptable level of chair usage, a necessary condition if the barbers are to continue to

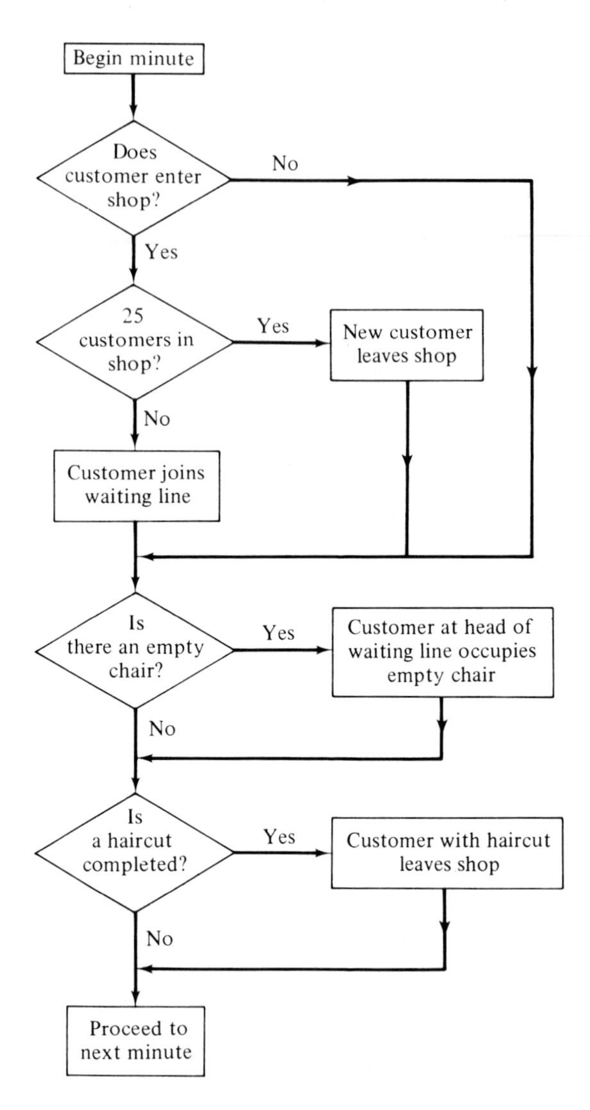

Fig. 10–3

work in the shop. Other considerations will be assumed secondary to these. Notice that in a sense the selected criteria exert pressures in opposite directions. To make the average waiting times small in periods of heavy traffic, one needs more chairs than can be used in periods of light traffic, and to have high chair utilization in periods of light traffic, it is necessary to have fewer chairs than are needed in periods of heavy traffic.

The simulation described here was carried out for 100 days, i.e., 100 separate runs, for shops with three through nine barber chairs. The selection of three chairs as an appropriate lower bound on the number of chairs which will be best was based on the following observation. Even with a uniform rate of nine arrivals per hour randomly distributed all day long, the ratio of arrival rate to service rate, the traffic intensity of Chap. 9, will be equal to 1, and there will be a logarithmic buildup of waiting customers. Certainly the heavier traffic during the 11:30 A.M.–1:30 P.M. period will aggravate the situation. The selection of nine chairs as an upper bound was an arbitrary one subsequently justified by the outcome of the simulation. The number of individuals in the shop was recorded 5 times each hour for a total of 40 times for each 8-hour day. Also, a record was kept of the number of haircuts and the average waiting time for each day. Finally, the average waiting time and the average number of haircuts was computed over all 100 days for a shop of each of the sizes considered. The necessary data were collected by inserting counters, as shown in the flow chart, Fig. 10–4.

The flow chart is somewhat incomplete in that several of the decision boxes are not completely specified and require more detail for programming. For example, consider the box having to do with whether a customer enters a shop during the minute T. Let P be the probability that a customer enters the shop in minute T, and let R be a random number selected from a set uniformly distributed on $[0, 1)$. Of course, P and the random number R may vary with T. In programming the simulation the decision box in Fig. 10–5 is to be interpreted as shown in Fig. 10–6.

We turn next to a summary of the results of the simulation. First, the average over 100 days of the daily average waiting times and the number of haircuts is given in Table 10–1. Since the time required for a haircut is 20 minutes, the average total service time is 20 minutes plus the waiting time in each case. It is immediately apparent that three and four chairs yield unreasonably long average waiting times and should be eliminated as possibilities on that basis.

There are several ways in which chair utilization can be evaluated. A very simple one is to check the number of empty chairs at each of the 40 check times on each of the 100 days. The results of such a count are presented

Table 10–1

Number of chairs	Average over 100 days of daily average waiting times (min)	Average number of haircuts
3	79.7	84.1
4	36.5	89.2
5	13.0	89.9
6	5.7	88.4
7	2.6	91.1
8	1.3	89.8
9	0.4	89.7

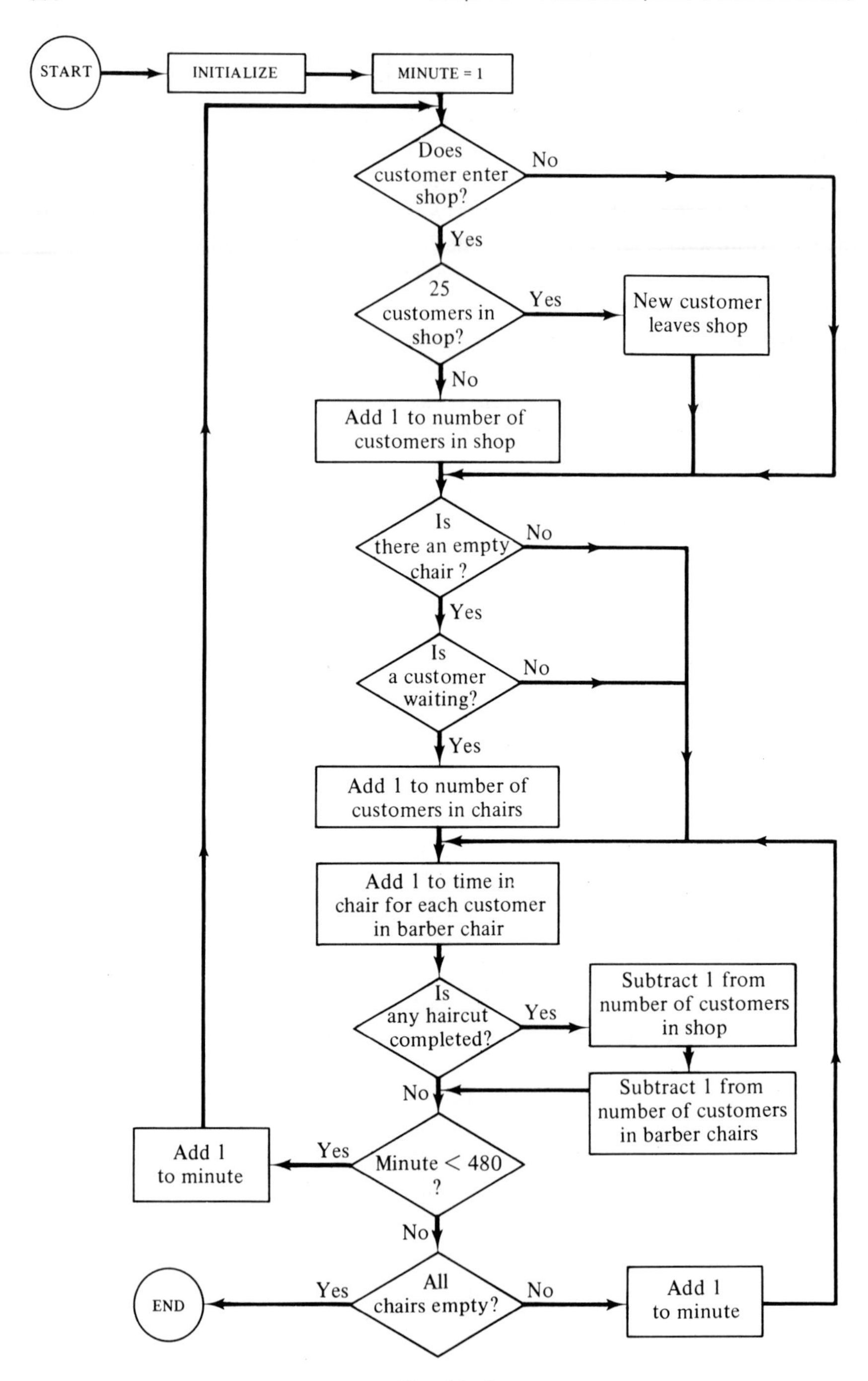

START
INITIALIZE
MINUTE = 1
Does customer enter shop?
No
Yes
25 customers in shop?
Yes
New customer leaves shop
No
Add 1 to number of customers in shop
Is there an empty chair ?
No
Yes
Is a customer waiting?
No
Yes
Add 1 to number of customers in chairs
Add 1 to time in chair for each customer in barber chair
Is any haircut completed?
Yes
Subtract 1 from number of customers in shop
Subtract 1 from number of customers in barber chairs
No
Add 1 to minute
Yes
Minute < 480 ?
No
END
Yes
All chairs empty?
No
Add 1 to minute

Fig. 10–4

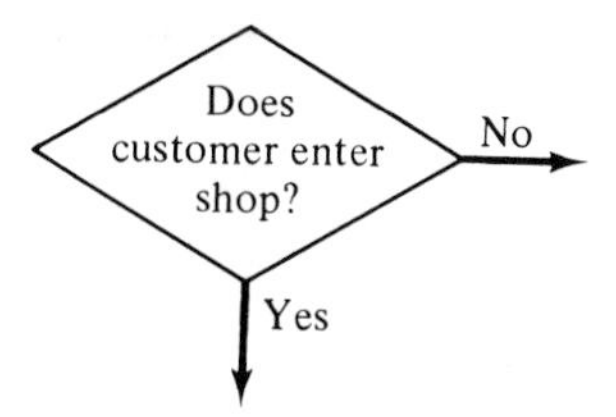

Fig. 10–5

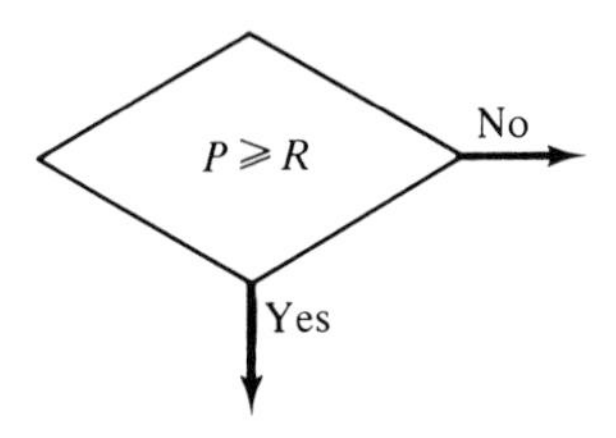

Fig. 10–6

in Table 10–2. The entries in this table give the number of check times (out of 40) at which there were *at least* the indicated number of empty chairs in the shop on each of the 100 days. When there are only a few chairs, the single check time at which there is an empty chair is the one at 12 minutes past

Table 10–2

Number of empty chairs	Number of chairs: 3	4	5	6	7	8	9
1	1	1	1	1	1	2	20
2	0	1	1	1	1	1	5
3	0	0	1	1	1	1	2
4		0	0	1	1	1	1
5			0	0	1	1	1
6				0	0	1	1
7					0	0	1
8						0	0
9							0

opening, i.e., the first check time. Every entry of 1 in the table is associated with this time. Table 10–2 was constructed to give an idea of how many empty chairs could be expected, but it might be argued that the test of 100% occupancy is too strong. Table 10–3 gives the number of check times (out of

Table 10–3

Number of chairs

Number of empty chairs	3	4	5	6	7	8	9
1	1	1	2	23	30	32	34
2	1	1	1	2	28	31	34
3	0	1	1	1	2	29	32
4		0	1	1	1	2	29
5			0	1	1	1	3
6				0	1	1	1
7					0	1	1
8						0	1
9							0

40) at which there are *at least* the indicated number of empty chairs on at least 75 of the 100 days. On the basis of the information in this table, we conclude that in a shop with eight or nine chairs on 75 days out of 100 there will be at least three empty chairs approximately three quarters of the time (actually 29 times out of 40). This is interpreted as underutilization, and the possibility of having eight or nine chairs will not be considered further. Also, on the basis of Table 10–1 we eliminate three- and four-chair shops from further consideration.

To select which number of chairs—five, six, or seven—is most desirable, it is necessary to establish a criterion for desirability. Such a criterion is, of course, somewhat arbitrary in that it requires a subjective judgment on the part of the proprietor of the shop. It should simultaneously reflect the customer's desire for quick service times and the shop owner's desire for efficient use of the facilities. Since there are 40 check times in each day and the simulation is run for 100 days, there are 4000 check times altogether. We select as the test criterion for optimality the following: The optimal shop size (five, six, or seven chairs) is that which maximizes the fraction of the check times (the ratio of the number of check times to 4000) at which

1. Every barber chair is occupied, and
2. There is no waiting line.

This information is contained in Table 10–4. Although we are interested only in shops with five, six, or seven chairs, we have included the data for four and eight chairs for comparison. It is evident from the table that according to this criterion the optimal number of chairs is five. It is also clear that for fewer than five chairs there is an imbalance due to too many customers relative to the capacity of the shop, while for more than five chairs there is an excess of capacity relative to the number of customers.

Table 10–4

		Percent of check times for which		
		1 holds	2 holds	1 and 2 hold
Number of chairs	4	84.4	23.9	8.2
	5	62.2	49.7	11.8
	6	38.0	71.2	9.2
	7	24.4	82.1	6.4
	8	15.1	89.1	4.2

EXERCISES

1. Consider a single-server queue in which 200 observations of service times have been taken and tabulated. Time is divided into units of one minute each. Suppose that the data are as follows:

Length of service (minutes)	Number	Relative frequency	Probability density	Cumulative distribution
9 or less	0	0	0	0
10	8	0.04	0.040	0.040
11	14	0.07	0.070	0.110
12	26	0.130	0.130	0.240
13	33	0.165	0.165	0.405
14	41	0.205	0.205	0.610
15	31	0.155	0.155	0.765
16	25	0.125	0.125	0.890
17	12	0.060	0.060	0.950
18	7	0.035	0.035	0.985
19	3	0.015	0.015	1.000
20 or more	0	0	0	1.000

(a) Consider the length of service time to be a random variable and find the points on the graph of its cumulative distribution function as given by the table.
(b) Approximate the graph of the cumulative distribution by a piecewise linear curve and thereby find an approximate analytic expression for the cumulative distribution function.
(c) Use the analytic expression obtained in part (b) to find a sequence $\{x_j\}$ which is randomly distributed on $(-\infty, \infty)$ with this distribution function. Be specific and find an expression which relates the sequence $\{x_j\}$ to a sequence of random numbers uniformly distributed on (0, 1).

2. Suppose that the reneging scheme of the example of Sec. 10.3.2 is modified to depend on the number of barbers in the shop as follows: If a customer enters the shop and if the number of customers waiting for service is 4 times as large as the number of barbers, then he leaves without being serviced.
(a) How is the flow chart of Fig. 10–4 changed?
(b) How are the results of the simulation changed?

3. Construct a simulation model for the passage of genetic characteristics from parents to offspring described in Sec. 3.5.3. Consider populations consisting of 100 individuals at each generation, and continue the simulation for ten generations. Compare your results with those predicted by the project cited above.

4. Construct a simulation model for the spread of a rumor as discussed in Sec. 9.2.

REFERENCES

Problems of parameter estimation and the evaluation of models play an important role in the mathematical theory of learning. A number of models are discussed and compared in [ABC], [BuM], and [ReG].

The books [Ba], [Ch], and [G] provide an introduction to simulation. The presentation in [Ba], in particular, is elementary and qualitative. Simulation and evaluation of queuing models are discussed in detail in [P], which also contains a number of case studies and applications to real problems.

The statistical questions raised in this chapter are studied in detail in [H], and are discussed at a more advanced level in [Cr].

[ABC] ATKINSON, R. C., G. H. BOWER, and E. J. CROTHERS, *An Introduction to Mathematical Learning Theory*. New York: Wiley, 1965.

[Ba] BARTON, R., *A Primer on Simulation and Gaming*. Englewood Cliffs, N.J.: Prentice-Hall, 1970.

[BuM] BUSH, R. R., and F. MOSTELLER, "A Comparison of Eight Models," in *Readings in Mathematical Social Sciences* (P. Lazerfeld and N. Henry, eds.). Cambridge, Mass.: M.I.T. Press, 1968.

[Ch] CHU, Y., *Digital Simulation of Continuous Systems*. New York: McGraw-Hill, 1969.

[Cr] CRAMÉR, H., *Mathematical Methods of Statistics*. Princeton, N.J.: Princeton University Press, 1946.

[G] GORDON, G., *System Simulation*. Englewood Cliffs, N.J.: Prentice-Hall, 1969.

[H] HOEL, P., *Introduction to Mathematical Statistics*, 4th ed. New York: Wiley, 1971.

[P] PANICO, J. A., *Queueing Theory*. Englewood Cliffs, N.J.: Prentice-Hall, 1969.

[R] RAND CORPORATION, *A Million Random Digits with 100,000 Normal Deviates*. New York: Free Press, 1955.

[ReG] RESTLE, F., and J. GREENO, *Introduction to Mathematical Psychology*. Reading, Mass.: Addison-Wesley, 1970.

APPENDICES

A Topics in Calculus

Throughout the text we assume that the reader has already studied calculus, at least to the extent of having a working knowledge of real functions, sequences, limits, continuity, differentiation, and integration. The aim of this appendix is to collect together a few special results used in the text. These are facts which are not always covered in a first course in calculus, and when they are covered, they may not be in the form we need. In particular, we shall consider topics in infinite series, Newton's method of approximation, and extrema of real functions.

A.1 INFINITE SERIES

Let $\{x_n\}_0^\infty$ be a sequence. The new sequence $\{s_n\}_0^\infty$ defined by setting $s_n = x_0 + x_1 + \cdots + x_n$, $n = 0, 1, 2, \ldots,$ is called the *infinite series with summands* $\{x_n\}_0^\infty$. This infinite series is said to *converge to sum s* if

$$\lim_{n\to\infty} s_n = s.$$

An infinite series which does not converge is said to diverge. We write $\sum_{n=0}^{\infty} x_n$ for the infinite series $\{s_n\}_0^\infty$.

Examples

1. *Geometric series.* Let r be a fixed real number and define $\{x_n\}_0^\infty$ by $x_n = r^n$, $n = 0, 1, 2, \ldots$. Then
   a. If $0 \leq r < 1$, $\sum r^n$ converges, and
   $$\sum_0^\infty r^n = \frac{1}{1-r}.$$
   b. If $r \geq 1$, $\sum r^n$ diverges.
2. Let $x_n = nr^{n-1}$, $n = 1, 2, \ldots, 0 \leq r < 1$. Then $\sum nr^{n-1}$ converges and
   $$\sum_{n=1}^\infty nr^{n-1} = \frac{1}{(1-r)^2}.$$
3. *Binomial series.* Let r and m be fixed real numbers. Define $\{x_n\}_0^\infty$ by $x_0 = 1$, $x_n = [m(m-1)\cdots(m-n+1)/n!]r^n$, $n = 1, 2, \ldots$. If $0 \leq |r| < 1$, then $\sum x_n$ converges and
   $$(1+r)^m = \sum_0^\infty x_n = 1 + mr + \frac{m(m-1)r^2}{2!} + \cdots + \frac{m(m-1)\cdots(m-n+1)r^n}{n!} + \cdots.$$

Note. In Example 3 above, if m is a positive integer, then $x_n = 0$ for $n > m$. Hence

$$(1+r)^m = 1 + mr + \frac{m(m-1)r^2}{2!} + \cdots + r^m.$$

A.2 NEWTON'S METHOD

In this text we encounter a number of problems which ultimately involve the determination of a real number r such that $f(r) = 0$, where f is some given function defined on a subset of the real numbers. Often it is quite difficult to determine the exact value of the desired r. In such circumstances it may be sufficient to approximate r to a certain degree of accuracy. In this section we present a commonly used method for carrying out such an approximation process. The method is a widely applicable one, known as Newton's method. First we shall discuss the computational aspects of the technique, and then we shall use a specific example to illustrate some geometric aspects.

Let f be defined on $[a, b]$, and assume that f' and f'' are continuous and never 0 on (a, b). Thus, f is either increasing or decreasing on $[a, b]$. Finally, we assume that $f(a)\cdot f(b) < 0$. We seek $r \in (a, b)$ such that $f(r) = 0$. The steps in Newton's method for approximating the value of r are

1. Choose a point $x_1 \in (a, b)$ such that if $f'' > 0$, then $f(x_1) > 0$, and if $f'' < 0$, then $f(x_1) < 0$.
2. Define a point x_2 by the equation

$$x_2 = x_1 - \frac{f(x_1)}{f'(x_1)}.$$

3. For $n = 2, 3, \ldots$, define x_{n+1} by the equation

$$x_{n+1} = x_n - \frac{f(x_n)}{f'(x_n)}.$$

The conclusion is that the sequence $\{x_n\}_1^\infty$ converges to the desired value of r:

$$\lim_{n\to\infty} x_n = r.$$

Example. Let $f(x) = x^3 - 2x^2 - x + 2$ and $[a, b] = [\frac{3}{2}, 3]$. Then $f'(x) = 3x^2 - 4x - 1$ and $f''(x) = 6x - 4$, and hence $f''(x) > 0$ for $x \in [a, b]$. Also $f(\frac{3}{2})f(3) = (-\frac{5}{8})(+8) < 0$. We seek $r \in [\frac{3}{2}, 3]$ such that $f(r) = 0$. As step 1 of Newton's method we select the point $x_1 = 5/2$. This selection satisfies the conditions of step 1 since $f'' > 0$ and $f(x_1) = 21/8 > 0$. To illustrate certain geometric aspects of Newton's method, we graph the function f and indicate the point $(x_1, f(x_1))$ on this graph (Fig. A–1). Also, we draw the tangent line to the graph of f at the point $(x_1, f(x_1))$. This line has the slope $f'(x_1)$, and hence the equation of this line is

$$y = f(x_1) + f'(x_1)(x - x_1).$$

We now determine the point $\tilde{x}_2$ where this tangent line intersects the x axis. Using the above equation of the tangent line, with $y = 0$, $x = \tilde{x}_2$, we obtain

$$\tilde{x}_2 = x_1 - \frac{f(x_1)}{f'(x_1)}.$$

The point $\tilde{x}_2$ is also shown in Fig. A–1. At this point we continue with Newton's method and apply step 2. This defines the number x_2 by

$$x_2 = x_1 - \frac{f(x_1)}{f'(x_1)},$$

and we immediately see that $x_2 = \tilde{x}_2$. Hence, the second approximation obtained by Newton's method is the x intercept of the tangent line to the curve at the point whose x coordinate is the first approximation. This geo-

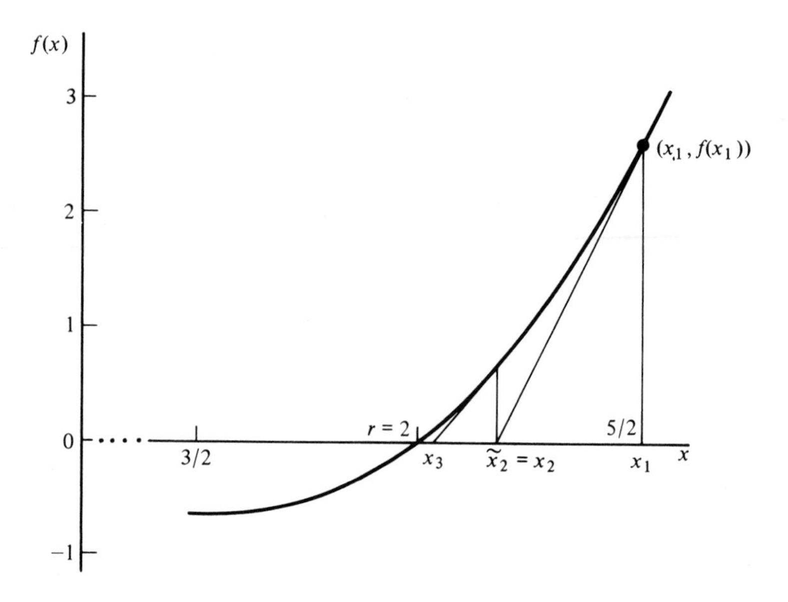

Fig. A–1

metric aspect of Newton's method continues with the higher approximations. The nth approximation, x_n, is the x intercept of the line tangent to the graph of f at the point $(x_{n-1}, f(x_{n-1}))$.

If we compute the first few approximates for the above example, we obtain

$$x_2 = \frac{5}{2} - \frac{21/8}{31/4} = \frac{67}{31} = 2.161,$$

$$x_3 = \frac{67}{31} - \frac{17640}{130138} = 2.025,$$

and

$$x_4 = 2.025 - 0.024 = 2.001.$$

Since $f(x)$ can be factored to $f(x) = (x - 1)(x - 2)(x + 1)$, we see that the exact answer is $r = 2.000$. Thus already the fourth approximation to r given by Newton's method is quite accurate. In general this is characteristic of Newton's method, as often only a few steps are needed to obtain the desired accuracy.

A.3 EXTREMA OF REAL FUNCTIONS

Let S be a subset of R^n, and suppose that f is a real function defined on S. We say that $M \in S$ is a *maximum* point for f if $f(s) \leq f(M)$ for all $s \in S$.

Similarly, a point $m \in S$ is a *minimum* point for f if $f(m) \leq f(s)$ for every $s \in S$. A point which is either a minimum point or a maximum point is called an *extreme* point. It is easy to see that not every real function has extrema on every set $S \subset R^n$ [take $n = 1$, $S = (0, +\infty)$, and $f(x) = 1/x$]. We shall quote one theorem which guarantees the existence of extrema for certain sets S and functions f. We first need two additional definitions.

We say that a subset S of R^n is *bounded* if there exists a number B such that for every $\mathbf{x} = (\xi_1, \xi_2, \ldots, \xi_n) \in S$ we have

$$|\mathbf{x}| = \sqrt{\xi_1^2 + \cdots + \xi_n^2} \leq B.$$

Also, we say that a set S in R^n is *closed* if for every sequence $\{\mathbf{x}_n\}$ in S ($\mathbf{x}_n \in S$, $n = 1, 2, \ldots$) we have $\lim_{n \to \infty} \mathbf{x}_n = \mathbf{x}$ implies that $\mathbf{x} \in S$.

Theorem A.1: Let f be a continuous real function defined on the closed bounded set $S \subset R^n$. Then f has both maximum and minimum points on S.

As a special case of the above theorem we consider $n = 1$ and $S = [a, b]$. Also we assume that f' exists on (a, b). In this case we can give a stronger statement than that of Theorem A.1. In addition to the existence of extrema, we can also make a statement about their location. We have

Theorem A.2: Let f be continuous on $[a, b]$ and assume that f' exists on (a, b). Then f has both maximum and minimum points on $[a, b]$, and, moreover, a point e is an extreme point only if at least one of the following conditions holds:

1. $e = a$ or $e = b$.
2. $f'(e) = 0$.

A.4 SUPREMA AND INFIMA

The domain of a function may be a set whose elements are subsets of the set of real numbers. Two important functions with domains of this type are the supremum and infimum functions. In order to define these functions carefully, we need a number of definitions. Let R be the set of real numbers. A subset S of R is said to be *bounded above* if there is a number u such that $s \leq u$ for all $s \in S$. The number u is an *upper bound* for S. Similarly, S is *bounded below* if there is a number ℓ (called a *lower bound* for S) such that $\ell \leq s$ for all $s \in S$. S is said to be *bounded* if it is bounded above and below. Let S be bounded above. A number $\bar{u}$ is a *supremum* for S if $\bar{u}$ is an upper bound for S and $\bar{u} \leq u$ for every upper bound u of S. Similarly, $\bar{\ell}$ is an *infimum* for a set S which is bounded below if $\bar{\ell}$ is a lower bound for S and

$\bar{\ell} \geq \ell$ for every lower bound ℓ of S. The following is taken as a basic property of the set R of real numbers:

Assumption: Every subset S of R which is bounded above has a unique supremum.

The assumption above is easily shown to be equivalent to the following:

Assumption: Every subset S of R which is bounded below has a unique infimum.

The *supremum function* (sup) is defined on the set of all subsets of R which are bounded above. For such a subset S, $\sup(S)$ is the supremum of S. Similarly, the *infimum function* (inf) is defined on the set of all subsets of R which are bounded below, and for such a set S, $\inf(S)$ is the infimum of S.

B Topics in Differential Equations

The task of solving an algebraic equation is a familiar one: Given such an equation involving an unknown, usually denoted by x, the problem is to find numerical values $n_1, n_2, \ldots, n_k$ such that when x is replaced by any one of these values consistently throughout the expression, the result is a numerical identity. For example, when we say that the equation

$$3x^2 + 5x - 2 = 0$$

has solutions $x = \frac{1}{3}$ and $x = -2$, we mean that if we replace x by $\frac{1}{3}$ consistently in the above expression, then the left-hand side becomes a real number which is equal to 0, and likewise if we begin with $x = -2$. The problem of solving a differential equation is analogous, but before we can hope to convince the reader of this, we must agree on some terminology.

B.1 DEFINITIONS AND NOTATION

Let f be a continuous function defined on a rectangle R in the t-x plane. R may be an ordinary rectangle, a strip, a half plane, or the entire t-x plane.

We proceed on the assumption that R is an ordinary rectangle and leave the obvious extensions to the remaining geometric situations to the reader. Let $R = \{(t, x): \beta_1 < t < \beta_2, \alpha_1 < x < \alpha_2\}$. We consider the *differential equation*

$$\dot{x} = f(t, x), \tag{B.1}$$

where we use the overdot to indicate differentiation with respect to t. This equation is to be interpreted as the defining relation for a function x. That is, we are looking for a function $x = x(t)$ which when substituted into (B.1) produces an identity in t. This notion can be made precise, and we turn to a formal definition. By a solution of the differential equation (B.1) on the interval $\beta'_1 < t < \beta'_2$, we mean a function $x = x(t)$ with the following properties:

1. x is defined and continuously differentiable on the interval $\beta'_1 < t < \beta'_2$.
2. For each t, $\beta'_1 < t < \beta'_2$, the point $(t, x(t))$ is contained in the rectangle R.
3. The function x satisfies the functional identity (B.1), that is,

$$\frac{dx}{dt}(t) = f(t, x(t)), \tag{B.2}$$

   for each t, $\beta'_1 < t < \beta'_2$. Of course, for a specific t, each side of (B.2) is simply a number.

It should be pointed out that in general Eq. (B.1) has many solutions. We shall refer to any specific one of these as a *particular solution*. Frequently one is interested in identifying which solution (if any) has a graph which contains a certain point P of R. If $P = (t_0, x_0) \in R$, then the condition

$$x(t_0) = x_0 \tag{B.3}$$

is known as an *initial condition*. The problem of finding that solution of (B.1) which satisfies (B.3) is known as an *initial value problem*, and it is typical of many problems in this book. Each of the initial value problems arising in the models considered in the text has the very nice property of possessing exactly one solution.

To illustrate these concepts, we consider an example. Let R be the rectangle $-1 < t < 1, 0 < x < 1$, and define f by $f(t, x) = x^2 t$ for $(t, x) \in R$. Clearly, f is a continuous function on R. The differential equation $\dot{x} = f(t, x)$ can be solved by elementary techniques (write it as $\dot{x}/x^2 = t$ and

If we wish to find a particular solution of (B.5) which satisfies an initial condition (B.3), then the constant c occurring in (B.10) or (B.11) can be evaluated by using this additional information. Indeed, by (B.10)

$$x(t_0)e^{-at_0} = x_0 e^{-at_0} = c,$$

and using this together with (B.11), we obtain

$$x(t) = x_0 e^{a(t-t_0)}. \tag{B.12}$$

It is trivial to verify that the function defined by (B.11) satisfies (B.5) and that the function defined by (B.12) satisfies both (B.5) and (B.3). The fact that (B.12) is the *only* such solution follows from the uniqueness theorem, Theorem B.3 in Sec. B.2.4.

B.2.2 The Equation $\dot{x} = ax + be^{kt}$

One of the reasons for using the particular constructive technique of the preceding section is that it generalizes easily to nonhomogeneous equations. Indeed, proceeding as above, the equation to be studied can be written as

$$\frac{d}{dt}(xe^{-at}) = be^{(k-a)t}. \tag{B.13}$$

Thus, if x is any solution of

$$\dot{x} = ax + be^{kt}, \tag{B.14}$$

then the function

$$X(t) = x(t)e^{-at}$$

has a derivative which is equal to $be^{(k-a)t}$. However, if we recall the fundamental theorem of calculus, then we can write down another function with the same derivative. In fact,

$$\int_{t_0}^{t} be^{(k-a)\tau}\, d\tau$$

is such a function. Therefore, the derivative of the difference

$$X(t) - \int_{t_0}^{t} be^{(k-a)\tau}\, d\tau$$

is 0, and the difference is a constant. The integral can be evaluated explicitly,

integrate both sides), and we find that for any constant c, $-1 < c < 0$, the function x

$$x(t) = \frac{-2c}{ct^2 + 2}$$

gives a solution to the differential equation on $-1 < t < 1$. Indeed, x is defined and continuously differentiable on $-1 < t < 1$; for each $t \in (-1, 1)$ the point $(t, x(t)) \in R$; and for each $t \in (-1, 1)$ the function x satisfies the differential equation. Thus properties 1–3 above are fulfilled. The function $x(t) = 2/(4 - t^2)$ is a solution (in fact, the only solution) of the initial value problem:

$$\dot{x} = x^2 t, \qquad -1 < t < 1,$$
$$x(0) = \tfrac{1}{2}.$$

B.2 A LINEAR DIFFERENTIAL EQUATION

Most of the models studied in the text lead to differential equations of a very special type. The particular form of these equations is a consequence of the basic assumptions on which the model is based. Adopting the terminology of the preceding section, in this section we shall consider equations of the form

$$\dot{x} = ax + b_1 e^{k_1 t} + \cdots + b_m e^{k_m t}, \tag{B.4}$$

where $a, b_1, k_1, \ldots, b_m, k_m$ are given constants. We include in our discussion the possibility that one of the k's may be 0, and consequently the expression on the right-hand side of (B.4) may contain a constant term. Note that the right-hand side of (B.4) defines a continuous function on the entire t-x plane. Usually we are interested in a solution of (B.4) which is valid for t larger than some prescribed value, t_0 say, and which satisfies an initial condition at t_0

$$x(t_0) = x_0,$$

where x_0 is a given constant. As above, this simply means that we wish to identify that solution of (B.4) whose graph contains the point (t_0, x_0) of the t-x plane.

Our approach to this problem is constructive. That is, we provide a technique for actually writing down a solution, and then we show that the solution so obtained is the only one. It is convenient to consider several simpler equations and then derive a general result which shows how the solutions to these simpler equations can be combined to yield a solution of (B.4).

B.2.1 The Equation $\dot{x} = ax$

If $a = 0$ in (B.4), then a solution can be obtained by a straightforward integration. Therefore we restrict ourselves to an investigation of the more interesting case $a \neq 0$. The simplest such equation is

$$\dot{x} = ax, \qquad a \neq 0. \tag{B.5}$$

However, there is cause for interest in this equation quite apart from its simplicity. In fact, we shall see that it plays a unique role in what follows. This role can be illustrated by the following fundamental results.

Theorem B.1: Let x_h be any solution of (B.5) and x_p be a particular solution of (B.4), both defined on an interval $(\beta'_1, \beta'_2) = I$. Then for any constant c the function

$$x = x_p + cx_h \tag{B.6}$$

is also a solution of (B.4) on the interval I.

Proof: The fact which underlies the proof of this theorem is that the operation $x \longrightarrow \dot{x} - ax$ which maps a (differentiable) function x to the function $\dot{x} - ax$ is a linear operation. This notion is exploited further in Theorem B.2. To proceed, let x_h and x_p be as in the hypothesis of the theorem, that is,

$$\dot{x}_h = ax_h, \tag{B.7a}$$

$$\dot{x}_p = ax_p + b_1 e^{k_1 t} + \cdots + b_m e^{k_m t}. \tag{B.7b}$$

If we multiply (B.7a) by any constant c and add it to (B.7b), then we have

$$\dot{x}_p + c\dot{x}_h = a(x_p + cx_h) + b_1 e^{k_1 t} + \cdots + b_m e^{k_m t}, \tag{B.8}$$

and using this it is easy to show that the function x defined by (B.6) is a solution of (B.4). Indeed, since the right-hand side of (B.4) is a continuous function on the whole t-x plane, properties 1–2 of the definition of a solution follow for x immediately from the fact that x_h and x_p are solutions to Eqs. (B.5) and (B.4), respectively. Property 3 is a direct consequence of Eq. (B.8).

Q.E.D.

Theorem B.2: If x_1 and x_2 are solutions of (B.5) on an interval I, then for any constants c_1 and c_2 the function

$$x = c_1 x_1 + c_2 x_2$$

is also a solution of (B.5) on I.

Proof: Since x_1 and x_2 are solutions of (B.5),

$$\dot{x}_1 = ax_1,$$

$$\dot{x}_2 = ax_2.$$

Multiplying the first equation by c_1, the second by c_2, and adding, we obtain

$$\frac{d}{dt}(c_1 x_1 + c_2 x_2) = a(c_1 x_1 + c_2 x_2).$$

The remainder of the proof follows as in the proof of Theorem B.1.

We encourage the reader to think of these two theorems from the point of view of linear transformations on a certain vector space. Such a point of view is often helpful in interpreting the results. For example, Theorem B.2 asserts that solutions of (B.5) form a linear space, or, using different terminology, these solutions satisfy a superposition principle.

Let us return now to the construction of a solution of (B.5). We expect that with such a simple function f, $f(t, x) = ax$, there ought to be a simple solution, and this is, in fact, the case. It is convenient to write the equation as

$$\dot{x} - ax = 0, \tag{B.9}$$

and to refer to it as a *linear homogeneous equation*. The adjective *linear* is justified by the comments above, and the term *homogeneous* is used to indicate that all the b_j's in (B.4) are 0. If a function x is a solution of (B.9), then it is also a solution of

$$(\dot{x} - ax)e^{-at} = 0.$$

Next, observe that the left-hand side of this is the derivative with respect to t of the function xe^{-at}, and consequently the equation can be written as

$$\frac{d}{dt}(xe^{-at}) = 0.$$

From this last equation we conclude that the function

$$X(t) = x(t)e^{-at}$$

must be a constant. Thus

$$x(t)e^{-at} = c \tag{B.10}$$

or

$$x(t) = ce^{at}, \tag{B.11}$$

where c is a constant.

integrate both sides), and we find that for any constant c, $-1 < c < 0$, the function x

$$x(t) = \frac{-2c}{ct^2 + 2}$$

gives a solution to the differential equation on $-1 < t < 1$. Indeed, x is defined and continuously differentiable on $-1 < t < 1$; for each $t \in (-1, 1)$ the point $(t, x(t)) \in R$; and for each $t \in (-1, 1)$ the function x satisfies the differential equation. Thus properties 1–3 above are fulfilled. The function $x(t) = 2/(4 - t^2)$ is a solution (in fact, the only solution) of the initial value problem:

$$\dot{x} = x^2 t, \qquad -1 < t < 1,$$
$$x(0) = \tfrac{1}{2}.$$

B.2 A LINEAR DIFFERENTIAL EQUATION

Most of the models studied in the text lead to differential equations of a very special type. The particular form of these equations is a consequence of the basic assumptions on which the model is based. Adopting the terminology of the preceding section, in this section we shall consider equations of the form

$$\dot{x} = ax + b_1 e^{k_1 t} + \cdots + b_m e^{k_m t}, \tag{B.4}$$

where $a, b_1, k_1, \ldots, b_m, k_m$ are given constants. We include in our discussion the possibility that one of the k's may be 0, and consequently the expression on the right-hand side of (B.4) may contain a constant term. Note that the right-hand side of (B.4) defines a continuous function on the entire t-x plane. Usually we are interested in a solution of (B.4) which is valid for t larger than some prescribed value, t_0 say, and which satisfies an initial condition at t_0

$$x(t_0) = x_0,$$

where x_0 is a given constant. As above, this simply means that we wish to identify that solution of (B.4) whose graph contains the point (t_0, x_0) of the t-x plane.

Our approach to this problem is constructive. That is, we provide a technique for actually writing down a solution, and then we show that the solution so obtained is the only one. It is convenient to consider several simpler equations and then derive a general result which shows how the solutions to these simpler equations can be combined to yield a solution of (B.4).

B.2.1 The Equation $\dot{x} = ax$

If $a = 0$ in (B.4), then a solution can be obtained by a straightforward integration. Therefore we restrict ourselves to an investigation of the more interesting case $a \neq 0$. The simplest such equation is

$$\dot{x} = ax, \qquad a \neq 0. \tag{B.5}$$

However, there is cause for interest in this equation quite apart from its simplicity. In fact, we shall see that it plays a unique role in what follows. This role can be illustrated by the following fundamental results.

Theorem B.1: Let x_h be any solution of (B.5) and x_p be a particular solution of (B.4), both defined on an interval $(\beta'_1, \beta'_2) = I$. Then for any constant c the function

$$x = x_p + cx_h \tag{B.6}$$

is also a solution of (B.4) on the interval I.

Proof: The fact which underlies the proof of this theorem is that the operation $x \longrightarrow \dot{x} - ax$ which maps a (differentiable) function x to the function $\dot{x} - ax$ is a linear operation. This notion is exploited further in Theorem B.2. To proceed, let x_h and x_p be as in the hypothesis of the theorem, that is,

$$\dot{x}_h = ax_h, \tag{B.7a}$$

$$\dot{x}_p = ax_p + b_1 e^{k_1 t} + \cdots + b_m e^{k_m t}. \tag{B.7b}$$

If we multiply (B.7a) by any constant c and add it to (B.7b), then we have

$$\dot{x}_p + c\dot{x}_h = a(x_p + cx_h) + b_1 e^{k_1 t} + \cdots + b_m e^{k_m t}, \tag{B.8}$$

and using this it is easy to show that the function x defined by (B.6) is a solution of (B.4). Indeed, since the right-hand side of (B.4) is a continuous function on the whole t-x plane, properties 1–2 of the definition of a solution follow for x immediately from the fact that x_h and x_p are solutions to Eqs. (B.5) and (B.4), respectively. Property 3 is a direct consequence of Eq. (B.8).

Q.E.D.

Theorem B.2: If x_1 and x_2 are solutions of (B.5) on an interval I, then for any constants c_1 and c_2 the function

$$x = c_1 x_1 + c_2 x_2$$

is also a solution of (B.5) on I.

Proof: Since x_1 and x_2 are solutions of (B.5),

$$\dot{x}_1 = ax_1,$$
$$\dot{x}_2 = ax_2.$$

Multiplying the first equation by c_1, the second by c_2, and adding, we obtain

$$\frac{d}{dt}(c_1x_1 + c_2x_2) = a(c_1x_1 + c_2x_2).$$

The remainder of the proof follows as in the proof of Theorem B.1.

We encourage the reader to think of these two theorems from the point of view of linear transformations on a certain vector space. Such a point of view is often helpful in interpreting the results. For example, Theorem B.2 asserts that solutions of (B.5) form a linear space, or, using different terminology, these solutions satisfy a superposition principle.

Let us return now to the construction of a solution of (B.5). We expect that with such a simple function f, $f(t, x) = ax$, there ought to be a simple solution, and this is, in fact, the case. It is convenient to write the equation as

$$\dot{x} - ax = 0, \tag{B.9}$$

and to refer to it as a *linear homogeneous equation*. The adjective *linear* is justified by the comments above, and the term *homogeneous* is used to indicate that all the b_j's in (B.4) are 0. If a function x is a solution of (B.9), then it is also a solution of

$$(\dot{x} - ax)e^{-at} = 0.$$

Next, observe that the left-hand side of this is the derivative with respect to t of the function xe^{-at}, and consequently the equation can be written as

$$\frac{d}{dt}(xe^{-at}) = 0.$$

From this last equation we conclude that the function

$$X(t) = x(t)e^{-at}$$

must be a constant. Thus

$$x(t)e^{-at} = c \tag{B.10}$$

or

$$x(t) = ce^{at}, \tag{B.11}$$

where c is a constant.

If we wish to find a particular solution of (B.5) which satisfies an initial condition (B.3), then the constant c occurring in (B.10) or (B.11) can be evaluated by using this additional information. Indeed, by (B.10)

$$x(t_0)e^{-at_0} = x_0 e^{-at_0} = c,$$

and using this together with (B.11), we obtain

$$x(t) = x_0 e^{a(t-t_0)}. \tag{B.12}$$

It is trivial to verify that the function defined by (B.11) satisfies (B.5) and that the function defined by (B.12) satisfies both (B.5) and (B.3). The fact that (B.12) is the *only* such solution follows from the uniqueness theorem, Theorem B.3 in Sec. B.2.4.

B.2.2 The Equation $\dot{x} = ax + be^{kt}$

One of the reasons for using the particular constructive technique of the preceding section is that it generalizes easily to nonhomogeneous equations. Indeed, proceeding as above, the equation to be studied can be written as

$$\frac{d}{dt}(xe^{-at}) = be^{(k-a)t}. \tag{B.13}$$

Thus, if x is any solution of

$$\dot{x} = ax + be^{kt}, \tag{B.14}$$

then the function

$$X(t) = x(t)e^{-at}$$

has a derivative which is equal to $be^{(k-a)t}$. However, if we recall the fundamental theorem of calculus, then we can write down another function with the same derivative. In fact,

$$\int_{t_0}^{t} be^{(k-a)\tau}\, d\tau$$

is such a function. Therefore, the derivative of the difference

$$X(t) - \int_{t_0}^{t} be^{(k-a)\tau}\, d\tau$$

is 0, and the difference is a constant. The integral can be evaluated explicitly, and taking note of the two different cases, we obtain

$$X(t) - \frac{b}{k-a}e^{(k-a)t} = c, \qquad k \neq a,$$

and

$$X(t) - bt = c, \qquad k = a,$$

where c is a constant. Writing this in terms of the function x, we have

$$x(t) = \frac{b}{k-a}e^{kt} + ce^{at}, \qquad k \neq a,$$

and

$$x(t) = (bt + c)e^{at}, \qquad k = a.$$

The reader should note that the above reasoning holds for any choice of the constant c, and for this reason the constant is referred to as an arbitrary constant. If we take $c = 0$, then we have

$$x_p(t) = \frac{b}{k-a}e^{kt}, \qquad k \neq a,$$

or (B.15)

$$x_p(t) = bte^{at}, \qquad k = a,$$

as a particular solution of (B.14). It is left to the reader to verify this assertion. It is a consequence of Theorem B.1 as well as the above construction that

$$x_p + ce^{at}$$

satisfies (B.14) for any constant c. We shall use this fact in finding a solution of (B.14) which satisfies an initial condition at the point t_0. If $x(t_0)$ is to be equal to x_0, then in the case $k \neq a$ we are to select c so that

$$x_0 = \frac{b}{k-a}e^{kt_0} + ce^{at_0};$$

that is,

$$c = \left[x_0 - \frac{b}{k-a}e^{kt_0}\right]e^{-at_0}.$$

Therefore a solution of the initial value problem is

$$x(t) = \frac{b}{k-a}e^{kt} + \left(x_0 - \frac{b}{k-a}e^{kt_0}\right)e^{a(t-t_0)}. \tag{B.16}$$

The details of the case $k = a$ are similar and left to the reader.

B.2.3 The Equation $\dot{x} = ax + b_1 e^{k_1 t} + \cdots + b_m e^{k_m t}$

The methods introduced above provide a means for finding solutions to the m equations

$$\dot{x} = ax + b_j e^{k_j t}, \qquad j = 1, 2, \ldots, m. \tag{B.17}$$

Denote these solutions by $x_1, \ldots, x_m$, respectively, and define a function x by

$$x = x_1 + \cdots + x_m. \tag{B.18}$$

Then x is indeed a solution of (B.4):

$$\begin{aligned}\dot{x} &= \dot{x}_1 + \cdots + \dot{x}_m \\ &= (ax_1 + b_1 e^{k_1 t}) + \cdots + (ax_m + b_m e^{k_m t}) \\ &= a(x_1 + \cdots + x_m) + b_1 e^{k_1 t} + \cdots + b_m e^{k_m t} \\ &= ax + b_1 e^{k_1 t} + \cdots + b_m e^{k_m t}.\end{aligned}$$

Again making use of Theorem B.1, we can find a solution of (B.4) which satisfies an initial condition (B.3). To this end, let x_p be any particular solution of (B.4) and x_h be the solution of $\dot{x} = ax$, constructed above; i.e., $x_h(t) = e^{at}$. Then

$$x = x_p + cx_h$$

is also a solution of (B.4) for any constant c. If we wish to have $x(t_0) = x_0$, then we select c so that

$$x_0 = x(t_0) = x_p(t_0) + cx_h(t_0),$$

or, equivalently,

$$c = \frac{x_0 - x_p(t_0)}{x_h(t_0)}.$$

It follows that the function

$$\begin{aligned}x &= x_p + \left[\frac{x_0 - x_p(t_0)}{x_h(t_0)}\right] x_h \\ &= x_p + [x_0 - x_p(t_0)] e^{a(t-t_0)}\end{aligned} \tag{B.19}$$

is a solution of (B.4) which satisfies the desired initial condition.

B.2.4 A Uniqueness Theorem

In this section we substantiate the claims made above concerning the uniqueness of solutions to initial value problems involving (B.4).

Theorem B.3: Given real numbers t_0 and x_0, there is exactly one solution of (B.4) which satisfies $x(t_0) = x_0$.

Proof: Equations (B.15) and (B.18) can be used to obtain a particular solution x_p of (B.4), and consequently (B.19) is one solution of (B.4) which satisfies $x(t_0) = x_0$. It remains to prove that this is the only one. Suppose that x and y are two such solutions. Then their difference $z = x - y$ is a solution of the initial value problem

$$\dot{z} = az \tag{B.20}$$

$$z(t_0) = 0. \tag{B.21}$$

We shall show that every solution of (B.20) and (B.21) is identically 0.

If $z \equiv 0$, then we are finished. If not, then there is a t_1 such that $z(t_1) \neq 0$. As above, (B.20) is equivalent to

$$\frac{d}{dt}(ze^{-at}) = 0,$$

and after integrating this expression from t_1 to τ, we obtain

$$\int_{t_1}^{\tau} \frac{d}{dt}(ze^{-at})\,dt = 0$$

or

$$z(\tau)e^{-a\tau} - z(t_1)e^{-at_1} = 0.$$

But this last relation can be written as

$$z(\tau) = z(t_1)e^{a(\tau - t_1)},$$

and this must hold for all values of τ. But, by (B.21) we have

$$0 = z(t_1)e^{a(t_0 - t_1)},$$

which is obviously false. This contradiction indicates that the existence of a value t_1 such that $z(t_1) \neq 0$ is impossible, and our proof is complete.

As an immediate corollary to this theorem we note that if x is a solution of $\dot{x} = ax$, then either $x \equiv 0$ or $x(t) \neq 0$ for every t. A solution x of $\dot{x} = ax$ which is not the zero function is said to be a *nontrivial solution* of that equation.

As our final result in this development, we shall show how this uniqueness theorem can be used to improve Theorem B.1.

Theorem B.4: Let x_p be any particular solution of (B.4), and let x_h be a nontrivial solution of the homogeneous equation (B.5). Then for each constant c the function

$$x = x_p + cx_h \tag{B.22}$$

is a solution of (B.4), and every solution of (B.4) is of the form (B.22) for some constant c.

Proof: The first assertion is contained in Theorem B.1. We turn to the second. Suppose that X is a given solution of (B.4), fixed but arbitrary, and suppose that X has the value X_0 at t_0. Then the function

$$\tilde{x} = x_p + \tilde{c}x_h$$

with $\tilde{c}$ determined by

$$x_p(t_0) + \tilde{c}x_h(t_0) = X_0,$$

that is, with

$$\tilde{c} = \frac{X_0 - x_p(t_0)}{x_h(t_0)},$$

is another solution of (B.4) which satisfies the same initial condition at t_0. The uniqueness theorem (Theorem B.3) now guarantees that $X = \tilde{x}$, and consequently X can be written in the desired form.

B.3 COMMENTS ON NONLINEAR EQUATIONS

The techniques introduced in the preceding section can easily be extended to handle equations of the form

$$\dot{x} = ax + g(t),$$

where the function g is more complicated than the sum of exponential functions which appear in (B.4). The same is true if the constant a is replaced by a function $a(t)$. However, if we admit a nonlinear dependence of the function f of (B.1) on x, then no such simple treatment is possible. The assumption that f is continuous is sufficient to guarantee that Eq. (B.1) has solutions and that there is at least one solution to the initial value problem (B.1) and (B.3). On the other hand, it may be very difficult to obtain an explicit representation for the solutions, and there may be many solutions to a single initial value problem.

For the reader to better appreciate the latter point, we turn to an example. Consider the initial value problem

$$\dot{x} = 2|x|^{1/2}, \tag{B.23}$$

$$x(0) = 0. \tag{B.24}$$

In this problem the function f of (B.1) is $f(t, x) = 2|x|^{1/2}$, and this is clearly a continuous function on the whole t-x plane. It is a straightforward exercise to check that if a is any nonnegative real number, then the function

$$x_a(t) = \begin{cases} 0, & 0 \leq t \leq a, \\ (t-a)^2, & a < t < \infty, \end{cases}$$

is a solution of the initial value problem (B.23)–(B.24) on the interval $0 \leq t < \infty$. Consequently, this problem has infinitely many solutions, a few of which are graphed in Fig. B–1.

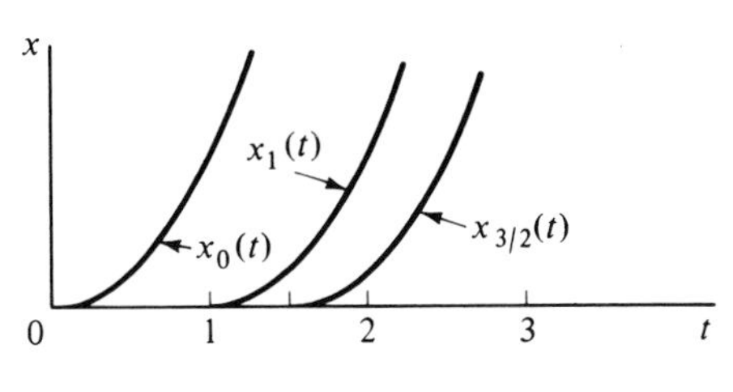

Fig. B–1

Most of the techniques which are useful in solving nonlinear equations are quite special and of little general interest. In this text we have considered relatively few models which involve nonlinear equations, and we have only attempted to solve some of them. As an example of a type of nonlinear equation which occurs sufficiently often to justify its inclusion in this appendix, we mention one known as a separable equation. An equation of the form (B.1) is said to be *separable* if the function f can be written as the product of two functions, one depending only on t and the other only on x. That is, if f can be written

$$f(t, x) = p(t)q(x), \tag{B.25}$$

where p and q are real-valued functions of a single variable and p is independent of x and q is independent of t. The term *separable* is intended to reflect the fact that if f is as in (B.25), then (B.1) can be written formally as

$$\frac{\dot{x}}{q(x)} = p(t), \tag{B.26}$$

where the left-hand side is independent of t and the right-hand side is independent of x. We proceed with our study by considering Eq. (B.26). This means, obviously, that we must avoid values of x for which $q(x) = 0$. Suppose that P is an antiderivative of p, that is, $dP/dt = p$, and Q is an antiderivative of $1/q$. Also, for t in an interval I and for c a constant, let the equation

$$Q(x) = P(t) + c \tag{B.27}$$

define implicitly a function $x = x(t)$, $t \in I$. Then the function x defined by (B.27) is a solution of (B.26) on I. Indeed, differentiating (B.27) with respect to t, one has

$$\frac{dQ}{dx}\frac{dx}{dt} = \frac{dP}{dt},$$

which, in view of the properties of P and Q, is another form of (B.26). Note that (B.27) provides at most an implicit definition of a function x. It may be difficult or even impossible to obtain a useful expression for x directly as a function of t.

C Real Linear Spaces

We shall collect here a few of the most basic results from linear algebra. It is our intention to provide enough background that students with no prior contact with linear spaces or matrices will be prepared for the discussion in the text, particularly Chap. 4. With this limited goal in mind, we do not always formulate definitions and theorems in the most general way, and much of what is done here admits substantial generalizations. There are many textbooks presenting this material at different levels of abstraction, and the reader will find them under titles involving linear algebra and matrices.

C.1 TERMINOLOGY

An *n vector* **x** is an ordered set of n real numbers $(\xi_1, \xi_2, \ldots, \xi_n)$. The number ξ_i is called the ith coordinate of the vector **x**. We systematically denote vectors by boldface type (**x**) and real numbers by italic type.

The set of all n vectors is called real n space and is denoted by R^n.

Certain vectors in R^n are particularly important. The *zero vector*, denoted by **0**, is the vector all of whose coordinates are 0. Note the distinction between the vector $\mathbf{0}$ and the real number 0. The kth unit vector in R^n is the vector whose coordinates other than the kth are 0, while the kth coordinate is 1. We ordinarily denote the kth unit vector by $\mathbf{u}_k$. Thus

$$\mathbf{u}_k = (0, 0, \ldots, 0, \underbrace{1}_{k\text{th coodinate}}, 0, \ldots, 0).$$

Much of the utility of the spaces R^n arises from the fact that certain very natural operations can be defined on them.

Let $\mathbf{x} = (\xi_1, \ldots, \xi_n)$, $\mathbf{y} = (\eta_1, \ldots, \eta_n)$ denote vectors in R^n. We say that $\mathbf{x} = \mathbf{y}$ if $\xi_i = \eta_i$, $i = 1, 2, \ldots, n$.

By the *sum*, $\mathbf{x} + \mathbf{y}$, of $\mathbf{x}$ and $\mathbf{y}$ we mean the vector in R^n whose coordinates are $\xi_i + \eta_i$, $i = 1, 2, \ldots, n$. That is,

$$\mathbf{x} + \mathbf{y} = (\xi_1 + \eta_1, \xi_2 + \eta_2, \ldots, \xi_n + \eta_n).$$

We do not define a product between two vectors whose result is a vector. Instead we define *scalar multiplication* as an operation between a real number and a vector whose result is again a vector and an *inner product* as an operation between two vectors which results in a real number. We postpone the definition of the latter to Sec. C.3. For the former, let λ be a real number and let $\mathbf{x}$ be a vector in R^n. The scalar product or product $\lambda\mathbf{x}$ is defined to be the vector whose coordinates are $\lambda\xi_i$, $i = 1, 2, \ldots, n$. That is,

$$\lambda\mathbf{x} = (\lambda\xi_1, \lambda\xi_2, \ldots, \lambda\xi_n).$$

The reader should consider in detail the cases $n = 2$ and $n = 3$ and verify the connections between these definitions and the familiar models for the plane and ordinary physical 3-space. If $n = 2$, for example, then we can identify R^2 with a plane having a coordinate system with perpendicular coordinate axes. The unit vectors $\mathbf{u}_1$ and $\mathbf{u}_2$ are the unit points in the positive direction on each of the coordinate axes. The vector $\mathbf{x} = (\xi_1, \xi_2)$ can be interpreted either as a *point* in the plane or as a *directed line segment* from (0, 0) to (ξ_1, ξ_2). Addition of vectors is found to be the familiar *parallelogram law* (for addition of forces). Scalar multiplication can be interpreted as the operation of stretching a vector by a factor λ. It is an actual stretching if $\lambda > 1$ and a contraction if $0 < \lambda < 1$. The direction of $\lambda\mathbf{x}$ is opposite to that of $\mathbf{x}$ if $\lambda < 0$.

Vector addition behaves much like ordinary addition of real numbers. Indeed, by using the definitions and the corresponding properties of the real numbers, which we take as known, one can prove that

$$\mathbf{x} + \mathbf{y} = \mathbf{y} + \mathbf{x} \qquad \text{for any } \mathbf{x}, \mathbf{y} \in R^n;$$

$$(\mathbf{x} + \mathbf{y}) + \mathbf{z} = \mathbf{x} + (\mathbf{y} + \mathbf{z}) \qquad \text{for any } \mathbf{x}, \mathbf{y}, \mathbf{z} \in R^n;$$

for any $\mathbf{x}, \mathbf{y} \in R^n$, there is a $\mathbf{z} \in R^n$ such that $\mathbf{x} + \mathbf{z} = \mathbf{y}$.

Likewise, it can be shown that scalar multiplication satisfies

$$\lambda(\mathbf{x} + \mathbf{y}) = \lambda\mathbf{x} + \lambda\mathbf{y},$$
$$(\lambda + \mu)\mathbf{x} = \lambda\mathbf{x} + \mu\mathbf{x},$$
$$\lambda(\mu\mathbf{x}) = (\lambda\mu)\mathbf{x},$$
$$1\mathbf{x} = \mathbf{x}.$$

When we refer to a vector space we mean a set of vectors together with these definitions for addition and scalar multiplication. The astute reader will observe that many of the results derived below depend only on the properties listed immediately above and not on the fact that the *scalars* are real numbers and the *vectors* are n-tuples of real numbers. Such results are valid for an arbitrary linear space—which could be defined as an abstract mathematical system satisfying conditions similar to those specified above.

C.2 SETS OF VECTORS

Frequently our concern will be more with a certain subset of vectors in R^n than with the entire space. In this section we shall introduce certain subsets with special properties.

A subset M of the space R^n is said to be a *subspace* (linear subspace, linear manifold) if it is closed under vector addition and scalar multiplication. That is, if

$$\mathbf{x}, \mathbf{y} \in M \text{ implies } \mathbf{x} + \mathbf{y} \in M,$$
$$\mathbf{x} \in M,\ \lambda \in R \text{ implies } \lambda\mathbf{x} \in M.$$

For example, $M_1 = \{\mathbf{x} \in R^n : \xi_1 = 0\}$ is a subspace of R^n, while $M_2 = \{\mathbf{x} \in R^n : \xi_1 = 1\}$ is not.

Subspaces play a fundamental role in our study of vector spaces. As a result of the definition, i.e., closure, and the properties of vector addition and scalar multiplication, it follows that these properties also hold in the subspace. That is, the subspace itself is a vector space embedded in a larger vector space.

For example, in R^3 (with the usual geometrical conventions) planes and straight lines through the origin are subspaces. Incidentally, planes can be identified with R^2 and lines with R^1. Two-dimensional surfaces other than planes and one-dimensional curves other than lines are not subspaces. Neither are planes nor lines which do not pass through the origin.

Given a finite set of vectors, $\mathbf{x}_1, \mathbf{x}_2, \ldots, \mathbf{x}_k$ in R^n, by a *linear combination* of $\mathbf{x}_1, \ldots, \mathbf{x}_k$ we mean a vector of the form $\lambda_1\mathbf{x}_1 + \lambda_2\mathbf{x}_2 + \cdots + \lambda_k\mathbf{x}_k$, $\lambda_j \in R$, $j = 1, 2, \ldots, k$. Let us denote by $S[\mathbf{x}_1, \ldots, \mathbf{x}_k]$ the set of all linear combinations of $\mathbf{x}_1, \ldots, \mathbf{x}_k$. We refer to $S[\mathbf{x}_1, \ldots, \mathbf{x}_k]$ as the *span* of $\mathbf{x}_1, \ldots, \mathbf{x}_k$ or the space generated by $\mathbf{x}_1, \ldots, \mathbf{x}_k$. The following result is clear from the definitions.

Theorem C.1: For any set $\mathbf{x}_1, \ldots, \mathbf{x}_k$ of vectors in R^n, $S[\mathbf{x}_1, \ldots, \mathbf{x}_k]$ is a subspace of R^n.

For example, it is clear that for the unit vectors $\mathbf{u}_1, \mathbf{u}_2, \mathbf{u}_3 \in R^3$, we have $S[\mathbf{u}_1, \mathbf{u}_2, \mathbf{u}_3] = R^3$. Indeed, $\mathbf{x} = (\xi_1, \xi_2, \xi_3) = \xi_1(1, 0, 0) + \xi_2(0, 1, 0) + \xi_3(0, 0, 1) = \xi_1\mathbf{u}_1 + \xi_2\mathbf{u}_2 + \xi_3\mathbf{u}_3$.

For a particular set $\mathbf{x}_1, \ldots, \mathbf{x}_k$ of vectors in R^n an important question is whether or not there is a proper subset $\mathbf{x}'_1, \mathbf{x}'_2, \ldots, \mathbf{x}'_j$ of $\mathbf{x}_1, \ldots, \mathbf{x}_k$ such that $S[\mathbf{x}'_1, \ldots, \mathbf{x}'_j] = S[\mathbf{x}_1, \ldots, \mathbf{x}_k]$. To answer this question, we shall find additional terminology useful.

A set $\mathbf{x}_1, \ldots, \mathbf{x}_k$ of vectors in R^n is said to be *linearly dependent* (or simply *dependent*) if there is a nontrivial linear combination in $S[\mathbf{x}_1, \ldots, \mathbf{x}_k]$ which is the zero vector, that is, if there are real numbers $\lambda_1, \ldots, \lambda_k$ *not all zero* such that

$$\lambda_1\mathbf{x}_1 + \lambda_2\mathbf{x}_2 + \cdots + \lambda_k\mathbf{x}_k = \mathbf{0}.$$

If a set of vectors is not dependent, then it is said to be independent.

For example, the vectors $(1, 2, -1), (-1, 0, 1), (1, 1, -1)$ are linearly dependent in R^3. This follows since $(1, 2, -1) - (-1, 0, 1) - 2(1, 1, -1) = (0, 0, 0)$. The vectors $(1, 2, -1), (-1, 0, 1), (1, 1, 0)$ are independent.

For some geometrical insight, consider again the case of R^3. Here four or more vectors are always linearly dependent; three vectors are linearly dependent if and only if they lie in the same plane; and two vectors are linearly dependent if and only if they lie on the same line. The planes and lines referred to here must pass through the origin.

Theorem C.2: Let $\mathbf{x}_1, \mathbf{x}_2, \ldots, \mathbf{x}_k$ be vectors in R^n and suppose that $\mathbf{y}_i \in S[\mathbf{x}_1, \ldots, \mathbf{x}_k]$, $i = 0, 1, 2, \ldots, k$. Then the vectors $\mathbf{y}_i$, $i = 0, 1, \ldots, k$, are linearly dependent.

Proof: The most direct proof of this result utilizes theorems on systems of linear equations. Since one of our goals is the derivation of such theorems, we must look for an alternative proof of this basic theorem. The method we select is a proof by induction.

Consider first the case $k = 1$. With the notation of the statement of the theorem we have $\mathbf{y}_0 \in S[\mathbf{x}_1]$, $\mathbf{y}_1 \in S[\mathbf{x}_1]$. Thus there are real numbers λ_{01} and λ_{11} such that $\mathbf{y}_0 = \lambda_{01}\mathbf{x}_1$, $\mathbf{y}_1 = \lambda_{11}\mathbf{x}_1$. If both λ_{01} and λ_{11} are 0, then $\mathbf{y}_0 = \mathbf{0} = \mathbf{y}_1$, and the conclusion is obvious. If not both λ_{01} and λ_{11} are 0, then $\lambda_{11}\mathbf{y}_0 - \lambda_{01}\mathbf{y}_1 = \mathbf{0}$, and consequently $\mathbf{y}_0$ and $\mathbf{y}_1$ are linearly dependent.

We assume next that the result holds for $k = m - 1$ and prove that it holds for $k = m$. We assume that

$$\begin{aligned} \mathbf{y}_0 &= \lambda_{01}\mathbf{x}_1 + \lambda_{02}\mathbf{x}_2 + \cdots + \lambda_{0m}\mathbf{x}_m = \sum_{j=1}^{m} \lambda_{0j}\mathbf{x}_j, \\ &\vdots \\ \mathbf{y}_m &= \lambda_{m1}\mathbf{x}_1 + \lambda_{m2}\mathbf{x}_2 + \cdots + \lambda_{mm}\mathbf{x}_m = \sum_{j=1}^{m} \lambda_{mj}\mathbf{x}_j. \end{aligned}$$

As above, if all the coefficients λ_{ij} are 0, then the conclusion is obvious. If, on the other hand, some coefficient is not 0, then, by relabeling the $\mathbf{x}$'s and $\mathbf{y}$'s if necessary, we can assume that $\lambda_{01} \neq 0$. Define $\mathbf{z}_1, \ldots, \mathbf{z}_m \in R^n$ by $\mathbf{z}_j = \mathbf{y}_j - \lambda_{j1}(\lambda_{01})^{-1}\mathbf{y}_0$, $j = 1, 2, \ldots, m$. Clearly, since all the $\mathbf{y}$'s are in $S[\mathbf{x}_1, \ldots, \mathbf{x}_m]$, so are all the $\mathbf{z}$'s. But more than this is true. Expanding the above definition of $\mathbf{z}_j$ and using the fact that each $\mathbf{y}_j$ can be written as a linear combination of the $\mathbf{x}$'s, we see that each $\mathbf{z}$ can be written as a linear combination of the $\mathbf{x}$'s in which the coefficient of $\mathbf{x}_1$ is 0. That is, actually $\mathbf{z}_j \in S[\mathbf{x}_2, \ldots, \mathbf{x}_m]$ for $j = 1, 2, \ldots, m$. Thus we have m vectors each of which can be written in terms of $m - 1$ vectors. By the induction hypothesis the vectors $\mathbf{z}_1, \ldots, \mathbf{z}_m$ are linearly dependent. Suppose that $\mu_1, \ldots, \mu_m$ are real numbers (not all 0) such that

$$\sum_{j=1}^{m} \mu_j \mathbf{z}_j = \mathbf{0}.$$

It follows that

$$\sum_{j=1}^{m} \mu_j(\mathbf{y}_j - \lambda_{j1}(\lambda_{01})^{-1}\mathbf{y}_0) = \mathbf{0},$$

or

$$\left[-(\lambda_{01})^{-1} \sum_{j=1}^{m} \mu_j \lambda_{j1}\right]\mathbf{y}_0 + \sum_{j=1}^{m} \mu_j \mathbf{y}_j = \mathbf{0},$$

and therefore the vectors $\mathbf{y}_0, \mathbf{y}_1, \ldots, \mathbf{y}_m$ are dependent. Q.E.D.

Theorem C.1 has some important corollaries. The first follows immediately from the observation that for any $\mathbf{x} \in R^n$ we have $\mathbf{x} \in S[\mathbf{u}_1, \ldots, \mathbf{u}_n]$.

Corollary 1: Any set of $n + 1$ vectors in R^n is dependent.

Corollary 2: Any system of n homogeneous linear equations in $n + 1$ variables has a nontrivial solution.

Proof: Suppose that the equations are given by

$$\sum_{j=0}^{n} a_{ij}\xi_j = 0, \qquad i = 1, 2, \ldots, n,$$

where the a_{ij}'s are known and the ξ_j's are the variables. The vectors $\mathbf{a}^j = (a_{1j}, a_{2j}, \ldots, a_{nj})$, $j = 0, 1, 2, \ldots, n$, constitute a set of $n+1$ vectors in R^n and as such are linearly dependent. Thus there are real numbers $\xi_0, \xi_1, \ldots, \xi_n$ not all 0 such that

$$\mathbf{a}^0\xi_0 + \mathbf{a}^1\xi_1 + \cdots + \mathbf{a}^n\xi_n = \mathbf{0}.$$

Writing out this vector equation in coordinate form, we have the desired result.

In answer to the question raised above in motivation for this discussion, we have the following:

Corollary 3: Given $\mathbf{x}_1, \ldots, \mathbf{x}_k \in R^n$, there is a proper subset $\mathbf{x}'_1, \ldots, \mathbf{x}'_j$ of $\mathbf{x}_1, \ldots, \mathbf{x}_k$ such that $S[\mathbf{x}_1, \ldots, \mathbf{x}_k] = S[\mathbf{x}'_1, \ldots, \mathbf{x}'_j]$ if and only if $\mathbf{x}_1, \ldots, \mathbf{x}_k$ are linearly dependent.

Proof: The *only if* part of the corollary follows at once from the theorem. For the *if* part, suppose that $\mathbf{x}_1, \ldots, \mathbf{x}_k$ are linearly dependent. Then there are real numbers $\lambda_1, \ldots, \lambda_k$, not all 0, such that $\sum_{j=1}^{k} \lambda_j \mathbf{x}_j = \mathbf{0}$. By relabeling, if necessary, we can assume that $\lambda_1 \neq 0$. Then

$$\mathbf{x}_1 = \frac{-\lambda_2}{\lambda_1}\mathbf{x}_2 + \cdots + \frac{-\lambda_k}{\lambda_1}\mathbf{x}_k,$$

and this is sufficient to imply that $S[\mathbf{x}_1, \ldots, \mathbf{x}_k] = S[\mathbf{x}_2, \ldots, \mathbf{x}_k]$. Q.E.D.

Continuing the line of thought indicated in Corollary 3, it is useful to ask for the *smallest* subset of $\mathbf{x}_1, \ldots, \mathbf{x}_k$ which has the same span. It is convenient to rephrase the question slightly.

Let T be a subset of a vector space V. The *rank* r of T is the maximum number of independent vectors which can be selected from T. If r is the rank of T, then a set of r independent vectors in T is a *basis* of T.

Note that if $V = R^n$, then Corollary 1 shows that the rank of R^n is no larger than n. In fact, since $\mathbf{u}_1, \mathbf{u}_2, \ldots, \mathbf{u}_n$ are independent, it follows that the rank of R^n is $\geq n$ and we have the following:

Theorem C.3: The rank of R^n is n.

Theorem C.4: The set $\mathbf{x}_1, \ldots, \mathbf{x}_k$ is a basis for V if and only if

1. $\mathbf{x}_1, \ldots, \mathbf{x}_k$ are linearly independent, and
2. $\mathbf{x} \in S[\mathbf{x}_1, \ldots, \mathbf{x}_k]$ for any $\mathbf{x} \in V$.

Proof: First we assume that $\mathbf{x}_1, \ldots, \mathbf{x}_k$ is a basis. Then condition 1 holds by definition. Also, if $\mathbf{x}$ is any vector in V, the vectors $\mathbf{x}, \mathbf{x}_1, \ldots, \mathbf{x}_k$ are linearly dependent. Thus there exist constants $b, c_1, c_2, \ldots, c_k$ (not all 0) such that

$$b\mathbf{x} + \sum_{j=1}^{k} c_j \mathbf{x}_j = \mathbf{0}.$$

But $b \neq 0$, for otherwise the $\mathbf{x}_i$'s would be dependent, and so

$$\mathbf{x} = -\sum_{j=1}^{k} (b)^{-1} c_j \mathbf{x}_j,$$

and $\mathbf{x} \in S[\mathbf{x}_1, \ldots, \mathbf{x}_k]$.

In the other direction, assuming that $\mathbf{x}_1, \ldots, \mathbf{x}_k$ are independent, condition 2 implies that the rank of V is k. Indeed, for any other $\mathbf{x}$ in $\mathbf{V}$ one can find constants c_j, $i = 1, \ldots, k$, such that

$$\mathbf{x} - \sum_{j=1}^{k} c_j \mathbf{x}_j = \mathbf{0}.$$

That is, $\mathbf{x}, \mathbf{x}_1, \ldots, \mathbf{x}_k$ are dependent.

C.3 INNER PRODUCTS

If $\mathbf{x} = (\xi_1, \xi_2, \ldots, \xi_n)$, $\mathbf{y} = (\eta_1, \eta_2, \ldots, \eta_n)$ are two vectors in R^n, then the *inner product* of $\mathbf{x}$ and $\mathbf{y}$ is

$$\mathbf{x} \cdot \mathbf{y} = \sum_{i=1}^{n} \xi_i \eta_i.$$

Thus the inner product is a function from $R^n \times R^n$ to R.

We have at once that $\mathbf{x} \cdot \mathbf{y} = \mathbf{y} \cdot \mathbf{x}$, $\lambda\mathbf{x} \cdot \mathbf{y} = \lambda(\mathbf{x} \cdot \mathbf{y})$ for all real numbers λ, and $(\mathbf{x} + \mathbf{y}) \cdot \mathbf{z} = \mathbf{x} \cdot \mathbf{z} + \mathbf{y} \cdot \mathbf{z}$.

The reader should recall the geometrical significance of the inner product in R^3. In this case the inner product of $\mathbf{x}$ and $\mathbf{y}$ is the length of the projection of $\mathbf{x}$ on a unit vector in the direction of $\mathbf{y}$ multiplied by the length of $\mathbf{y}$. Thus the inner product of two vectors in R^3 is positive, 0, or negative according as the *angle* between them (in the usual sense) is acute, right, or obtuse.

The following theorem is fundamental for our work with systems of equations and inequalities. Its proof is rather complicated though illuminating. The lemma is an important result in its own right.

Theorem C.5: Let $\mathbf{x}_1, \ldots, \mathbf{x}_k$ be k linearly independent vectors in R^n. Then for each j, $1 \leq j \leq k$, there is a vector $\mathbf{y}_j$ such that

$$\mathbf{x}_i \cdot \mathbf{y}_j = \begin{cases} 1, & \text{if } i = j, \\ 0, & \text{if } i \neq j. \end{cases}$$

We shall base our proof on the following lemma. The result as well as the technique of proof is useful beyond the application made here and admits an interesting geometric interpretation.

Lemma: Let $\mathbf{x}_1, \ldots, \mathbf{x}_k$ be an ordered set of k linearly independent vectors in R^n. Then there is an ordered set $\mathbf{v}_1, \ldots, \mathbf{v}_k$ of k vectors in R^n with the following properties:

1. The vectors $\mathbf{v}_1, \ldots, \mathbf{v}_k$ are linearly independent and $S[\mathbf{v}_1, \ldots, \mathbf{v}_j] = S[\mathbf{x}_1, \ldots, \mathbf{x}_j]$ for $1 \leq j \leq k$. In particular, for each j, $1 \leq j \leq k$, $\mathbf{v}_j \in S[\mathbf{x}_1, \ldots, \mathbf{x}_j]$ and $\mathbf{x}_j \in S[\mathbf{v}_1, \ldots, \mathbf{v}_j]$.
2. $\mathbf{v}_i \cdot \mathbf{v}_j = 0$ for $i \neq j$ and $\mathbf{v}_j \cdot \mathbf{v}_j \neq 0$ for $i, j = 1, 2, \ldots, k$. $\mathbf{v}_i \cdot \mathbf{x}_j = 0$ for $j < i$.

Proof: We define

$$\mathbf{v}_1 = \mathbf{x}_1$$

$$\mathbf{v}_2 = \mathbf{x}_2 - \frac{\mathbf{x}_2 \cdot \mathbf{v}_1}{\mathbf{v}_1 \cdot \mathbf{v}_1}\mathbf{v}_1,$$

- - - - - - - - - - - - - - - - - - - -

$$\mathbf{v}_j = \mathbf{x}_j - \sum_{i=1}^{j-1} \frac{\mathbf{x}_j \cdot \mathbf{v}_i}{\mathbf{v}_i \cdot \mathbf{v}_i}\mathbf{v}_i,$$

- - - - - - - - - - - - - - - - - - - -

$$\mathbf{v}_k = \mathbf{x}_k - \sum_{i=1}^{k-1} \frac{\mathbf{x}_k \cdot \mathbf{v}_i}{\mathbf{v}_i \cdot \mathbf{v}_i}\mathbf{v}_i.$$

Note that since the $\mathbf{x}_j$'s are independent, we have $\mathbf{v}_i \cdot \mathbf{v}_i \neq 0$ for each i and these definitions make sense.

It is clear from the definitions that $\mathbf{x}_j \in S[\mathbf{v}_1, \ldots, \mathbf{v}_j]$, $1 \leq j \leq k$. Also, $\mathbf{v}_1 = \mathbf{x}_1$ and consequently we have $\mathbf{v}_2 = \mathbf{x}_2 + \mathbf{w}_1$, where $\mathbf{w}_1 \in S[\mathbf{x}_1]$. An induction verifies that $\mathbf{v}_j = \mathbf{x}_j + \mathbf{w}_{j-1}$, where $\mathbf{w}_{j-1} \in S[\mathbf{x}_1, \ldots, \mathbf{x}_{j-1}]$, $2 \leq j \leq k$. It follows from this that if a nontrivial linear combination of the $\mathbf{v}_j$'s represents the zero vector, then there is a nontrivial linear combination of the $\mathbf{x}_j$'s accomplishing the same thing. Thus linear dependence of the $\mathbf{v}_j$'s implies linear dependence of the $\mathbf{x}_j$'s. This implication shows that $\mathbf{v}_1, \ldots, \mathbf{v}_k$ are linearly independent. Thus all of property 1 is established.

Also, by construction,

$$\mathbf{v}_2 \cdot \mathbf{v}_1 = 0,$$

$$\mathbf{v}_3 \cdot \mathbf{v}_1 = 0, \qquad \mathbf{v}_3 \cdot \mathbf{v}_2 = 0,$$

- -

$$\mathbf{v}_j \cdot \mathbf{v}_1 = 0, \qquad \mathbf{v}_j \cdot \mathbf{v}_2 = 0, \ldots, \mathbf{v}_j \cdot \mathbf{v}_{j-1} = 0,$$

- -

$$\mathbf{v}_k \cdot \mathbf{v}_1 = 0, \qquad \mathbf{v}_k \cdot \mathbf{v}_2 = 0, \ldots, \mathbf{v}_k \cdot \mathbf{v}_{k-1} = 0.$$

By property 1, for each j, $\mathbf{x}_j \in S[\mathbf{v}_1, \ldots, \mathbf{v}_j]$, and so

$$\mathbf{x}_1 \cdot \mathbf{v}_2 = 0, \mathbf{x}_1 \cdot \mathbf{v}_3 = 0, \ldots, \mathbf{x}_1 \cdot \mathbf{v}_k = 0,$$
$$\mathbf{x}_2 \cdot \mathbf{v}_3 = 0, \mathbf{x}_2 \cdot \mathbf{v}_4 = 0, \ldots, \mathbf{x}_2 \cdot \mathbf{v}_k = 0,$$
$$\mathbf{x}_j \cdot \mathbf{v}_{j+1} = 0, \ldots, \mathbf{x}_j \cdot \mathbf{v}_k = 0,$$
$$\mathbf{x}_{k-1} \cdot \mathbf{v}_k = 0.$$

Thus both the proof of property 2 and the proof of the lemma are complete.

Let us now turn to the proof of the theorem.

Proof of Theorem C.5: Consider a fixed integer j, $1 \leq j \leq k$, and the ordered set of vectors $\mathbf{x}_1, \mathbf{x}_2, \ldots, \mathbf{x}_{j-1}, \mathbf{x}_{j+1}, \ldots, \mathbf{x}_k, \mathbf{x}_j$. There exist vectors (which depend on j, but we suppress this dependence since we consider j fixed) $\mathbf{v}_1, \ldots, \mathbf{v}_k$ for which we have (among other things) that

$$\mathbf{v}_k \cdot \mathbf{v}_k \neq 0 \qquad \text{and}$$
$$\mathbf{v}_k \cdot \mathbf{x}_1 = 0, \mathbf{v}_k \cdot \mathbf{x}_2 = 0, \ldots, \mathbf{v}_k \cdot \mathbf{x}_{j-1} = 0,$$
$$\mathbf{v}_k \cdot \mathbf{x}_{j+1} = 0, \ldots, \mathbf{v}_k \cdot \mathbf{x}_k = 0.$$

Also $\mathbf{x}_j \in S[\mathbf{v}_1, \ldots, \mathbf{v}_k]$, and this implies that $\mathbf{v}_k \cdot \mathbf{x}_j \neq 0$. Indeed, recall that $\mathbf{v}_i \cdot \mathbf{v}_j = 0$, $i \neq j$.

If we set $\mathbf{y}_j = \mathbf{v}_k/(\mathbf{x}_j \cdot \mathbf{v}_k)$, then we have a vector which satisfies the assertion of the theorem.

Corollary: Let $\mathbf{x}_1, \ldots, \mathbf{x}_k$ be a linearly independent ordered set of vectors in R^n. Then for each ordered set of k real number $\lambda_1, \ldots, \lambda_k$ there is a vector $\mathbf{y} \in R^n$ such that $\mathbf{x}_j \cdot \mathbf{y} = \lambda_j$, $1 \leq j \leq k$.

Proof: Take $\mathbf{y} = \sum_{j=1}^{k} \lambda_j \mathbf{y}_j$, where the $\mathbf{y}_j$'s are the vectors of the conclusion of the theorem.

The reader should draw some pictures to aid in his understanding of these results in R^2 and R^3.

C.4 MATRICES

An $m \times n$ *matrix* is a rectangular array of real numbers with m rows and n columns. If the matrix is $\mathbb{A}$, then the element in the ith row and jth column will be denoted by a_{ij}, $i = 1, 2, \ldots, m$, $j = 1, 2, \ldots, n$. The vector $(a_{i1}, \ldots,$

a_{in}), the ith row of $\mathbb{A}$, is a vector in R^n and we denote it by $\mathbf{a}_i$. The vector $(a_{1j}, \ldots, a_{mj})$, the jth column of $\mathbb{A}$, is a vector in R^m and we denote it by $\mathbf{a}^j$. We write $\mathbb{A} = (a_{ij})$ for the matrix.

Given an $m \times n$ matrix $\mathbb{A}$, its *transpose* $\mathbb{A}^T$ is the matrix whose i–j element is a_{ji}. That is, if a^T_{ij} is the element in the ith row and jth column of $\mathbb{A}^T$, then

$$a^T_{ij} = a_{ji}, \qquad i = 1, 2, \ldots, n; j = 1, 2, \ldots, m.$$

Thus the operation $\mathbb{A} \longrightarrow \mathbb{A}^T$ interchanges the rows and columns of $\mathbb{A}$. If $\mathbb{A}$ is an $m \times n$ matrix, then $\mathbb{A}^T$ is an $n \times m$ matrix.

Let $\mathbb{A}$ be an $m \times n$ matrix and $\mathbb{B}$ an $n \times k$ matrix. Then $\mathbb{AB}$ is defined to be the matrix $\mathbb{C}$, where

$$c_{ij} = \mathbf{a}_i \cdot \mathbf{b}^j, \qquad 1 \leq i \leq m, 1 \leq j \leq k.$$

Note that $\mathbf{a}_i \in R^n$ and $\mathbf{b}^j \in R^n$, and thus this definition makes sense. In particular, if $\mathbf{x} \in R^n$, then

$$\mathbb{A}\mathbf{x} = (\mathbf{a}_1 \cdot \mathbf{x}, \mathbf{a}_2 \cdot \mathbf{x}, \ldots, \mathbf{a}_m \cdot \mathbf{x}).$$

It follows immediately from the definitions of matrix multiplication and inner product that

$$\mathbb{A}(\mathbf{x}_1 + \mathbf{x}_2) = \mathbb{A}\mathbf{x}_1 + \mathbb{A}\mathbf{x}_2 \qquad \text{and} \qquad \mathbb{A}(\lambda\mathbf{x}) = \lambda(\mathbb{A}\mathbf{x}).$$

Matrix notation is particularly convenient for discussing systems of linear equations. Indeed, the system of equations

$$\begin{aligned} a_{11}\xi_1 + a_{12}\xi_2 + \cdots + a_{1n}\xi_n &= \beta_1, \\ a_{21}\xi_1 + a_{22}\xi_2 + \cdots + a_{2n}\xi_n &= \beta_2, \\ \cdots\cdots\cdots\cdots&\cdots\cdots \\ a_{m1}\xi_1 + a_{m2}\xi_2 + \cdots + a_{mn}\xi_n &= \beta_m, \end{aligned}$$

where the a_{ij}'s are given and the ξ_i's are to be found, can be written compactly as $\mathbb{A}\mathbf{x} = \mathbf{b}$. Here $\mathbb{A}$ is the $m \times n$ matrix (a_{ij}), $\mathbf{x} = (\xi_1, \ldots, \xi_n)$ and $\mathbf{b} = (\beta_1, \ldots, \beta_m)$. We shall not usually distinguish notationally between vectors written in row or column form. It will ordinarily be clear from the context which is meant, and if situations arise in which misinterpretation is possible, we shall indicate just what is intended.

The $n \times n$ matrix with 1's on the main diagonal (upper left to lower right) and 0's elsewhere is known as the $n \times n$ *identity matrix*. We denote this matrix by $\mathbb{I}_n$ or simply by $\mathbb{I}$ when there is no ambiguity concerning its size. An $n \times n$ matrix $\mathbb{A}$ is said to be *invertible* if there is an $n \times n$ matrix $\mathbb{B}$ such that $\mathbb{AB} = \mathbb{BA} = \mathbb{I}$. Not all $n \times n$ matrices are invertible, but if $\mathbb{A}$

is invertible, then the matrix $\mathbb{B}$ is unique and it is called the *inverse* of $\mathbb{A}$. We write $\mathbb{B} = \mathbb{A}^{-1}$.

We are now in a position to take advantage of the terminology and results we have derived to obtain some very useful theorems on systems of linear equations.

C.5 LINEAR EQUATIONS

We first consider systems of homogeneous equations. We know from Corollary 2 of Theorem C.2 that if $\mathbb{A}$ has more columns than rows, then $\mathbb{A}\mathbf{x} = \mathbf{0}$ has a nontrivial solution. It is straightforward to check that the set of all solutions forms a linear space, the solution space of the equation $\mathbb{A}\mathbf{x} = \mathbf{0}$. Our first theorem gives additional information on the size of the solution space.

Theorem C.6: Let $\mathbb{A}$ be $m \times n$ and let the rank of $S[\mathbf{a}_1, \ldots, \mathbf{a}_m]$ be k. Then there are $n - k$ linearly independent solutions of $\mathbb{A}\mathbf{x} = \mathbf{0}$.

Proof: Renumbering the rows of $\mathbb{A}$ if necessary, we suppose that $\mathbf{a}_1, \ldots, \mathbf{a}_k$ are linearly independent and $S[\mathbf{a}_1, \ldots, \mathbf{a}_k] = S[\mathbf{a}_1, \ldots, \mathbf{a}_m]$.

To begin, suppose that $k = n$ and $\mathbf{x}$ is such that $\mathbf{a}_i \cdot \mathbf{x} = 0$, $i = 1, 2, \ldots, n$. Then, since $\mathbf{x} \in R^n = S[\mathbf{a}_1, \ldots, \mathbf{a}_n]$, we have $\mathbf{x} = \sum_{i=1}^{n} \lambda_i \mathbf{a}_i$ for suitable real numbers λ_i. It follows that $\mathbf{x} \cdot \mathbf{x} = \sum_{i=1}^{n} \lambda_i (\mathbf{x} \cdot \mathbf{a}_i) = 0$, which implies that $\mathbf{x} = \mathbf{0}$.

If $k < n$, then we can select $n - k$ vectors $\mathbf{v}_1, \ldots, \mathbf{v}_{n-k}$ from the set $\mathbf{u}_1, \ldots, \mathbf{u}_n$ in such a way that $\mathbf{a}_1, \ldots, \mathbf{a}_k, \mathbf{v}_1, \ldots, \mathbf{v}_{n-k}$ are linearly independent and span R^n. One way to accomplish this is to proceed as follows. Consider the set $\mathbf{a}_1, \ldots, \mathbf{a}_k, \mathbf{u}_1$. If this set is linearly independent, call it S_1: if not, discard $\mathbf{u}_1$ and call the result S_1. Next consider $S_1 \cup \{\mathbf{u}_2\}$. If this set is linearly independent, call it S_2; otherwise set $S_2 = S_1$. The set S_n is the desired set. To continue the argument, by Theorem C.5, there are vectors $\mathbf{y}_1, \mathbf{y}_2, \ldots, \mathbf{y}_{n-k}$ with the following properties: For each i, $i = 1, 2, \ldots, n - k$,

$$\mathbf{y}_i \cdot \mathbf{a}_j = 0, \qquad j = 1, 2, \ldots, k,$$
$$\mathbf{y}_i \cdot \mathbf{v}_i = 1,$$
$$\mathbf{y}_i \cdot \mathbf{v}_j = 0, \qquad i \neq j,\ j = 1, 2, \ldots, n - k.$$

We claim that the vectors y_i, $i = 1, 2, \ldots, n - k$, are linearly independent and that each is a solution of $\mathbb{A}\mathbf{x} = \mathbf{0}$.

Since each vector $\mathbf{a}_i$ can be written as a linear combination of $\mathbf{a}_1, \ldots,$

$\mathbf{a}_k$, the equations $\mathbf{y}_i \cdot \mathbf{a}_j = 0, j = 1, 2, \ldots, k$, imply that $\mathbf{y}_i \cdot \mathbf{a}_j = 0$ for all j and therefore $\mathbb{A}\mathbf{y}_i = \mathbf{0}$.

To show linear independence, suppose that there are constants $\lambda_1, \ldots, \lambda_{n-k}$ such that

$$\lambda_1 \mathbf{y}_1 + \cdots + \lambda_{n-k}\mathbf{y}_{n-k} = \mathbf{0}.$$

Taking the inner product of each side with $\mathbf{v}_1$, we conclude that $\lambda_1 = 0$. Similarly, taking the inner product of each side with $\mathbf{v}_i$, $i = 2, 3, \ldots, n-k$, we conclude that $\lambda_i = 0$, $i = 2, \ldots, n-k$. Thus the $\mathbf{y}_j$'s are linearly independent. Q.E.D.

Next, we turn to equations of the form $\mathbb{A}\mathbf{x} = \mathbf{b}$. Our basic theorem for such equations asserts that either the equation is solvable or else $\mathbb{A}$ and $\mathbf{b}$ are of a rather special form. Several applications of this result are given in Chap. 4.

Theorem C.7: For any $m \times n$ matrix $\mathbb{A}$ and $\mathbf{b}$ ($\neq \mathbf{0}$) in R^m, exactly one of the following holds:

1. There is an $\mathbf{x} \in R^n$ such that $\mathbb{A}\mathbf{x} = \mathbf{b}$, or
2. There is a $\mathbf{y} \in R^m$ such that $\mathbb{A}^T\mathbf{y} = \mathbf{0}$, $\mathbf{b} \cdot \mathbf{y} = 1$.

Proof: We show first that if statement 1 holds, then statement 2 cannot. Suppose that there is an $\mathbf{x} = (\xi_1, \ldots, \xi_n)$ such that $\mathbb{A}\mathbf{x} = \mathbf{b} = (\beta_1, \ldots, \beta_m)$. Then $\mathbf{a}_i \cdot \mathbf{x} = \beta_i$, $i = 1, 2, \ldots, m$. Also, for any vector $\mathbf{y} = (\eta_1, \ldots, \eta_m)$ satisfying $\mathbb{A}^T\mathbf{y} = \mathbf{0}$ we have

$$\begin{aligned} 0 = \mathbf{x} \cdot (\mathbf{a}^1 \cdot \mathbf{y}, \mathbf{a}^2 \cdot \mathbf{y}, \ldots, \mathbf{a}^n \cdot \mathbf{y}) &= \sum_{j=1}^{n}\Big(\sum_{i=1}^{m} a_{ij}\eta_i\Big)\xi_j \\ &= \sum_{j=1}^{n}\sum_{i=1}^{m} a_{ij}\xi_j\eta_i = \sum_{i=1}^{m}\Big(\sum_{j=1}^{n} a_{ij}\xi_j\Big)\eta_i \\ &= (\mathbf{a}_1 \cdot \mathbf{x}, \mathbf{a}_2 \cdot \mathbf{x}, \ldots, \mathbf{a}_m \cdot \mathbf{x}) \cdot \mathbf{y} = \mathbf{b} \cdot \mathbf{y}. \end{aligned}$$

Consequently, any $\mathbf{y}$ which satisfies $\mathbb{A}^T\mathbf{y} = \mathbf{0}$ must also satisfy $\mathbf{y} \cdot \mathbf{b} = 0$. This proves that if statement 1 is true, then statement 2 is false.

It remains to show that if statement 1 is false, then statement 2 is true. Suppose that $\mathbf{a}^1, \ldots, \mathbf{a}^k$ is a basis for $S[\mathbf{a}^1, \ldots, \mathbf{a}^n]$. Then the set $\mathbf{a}^1, \ldots, \mathbf{a}^k$, $\mathbf{b}$ is linearly independent. Otherwise $\mathbf{b} \in S[\mathbf{a}^1, \ldots, \mathbf{a}^k]$, in which case statement 1 is true. By Theorem C.5 there is a vector $\mathbf{y} \in R^m$ such that $\mathbf{a}^i \cdot \mathbf{y} = 0$, $i = 1, 2, \ldots, k$, $\mathbf{b} \cdot \mathbf{y} = 1$. But each column vector can be written as a linear combination of $\mathbf{a}^1, \ldots, \mathbf{a}^k$ so that $\mathbb{A}^T\mathbf{y} = \mathbf{0}$. Q.E.D.

Note that this theorem provides a positive criterion for asserting that equation $\mathbb{A}\mathbf{x} = \mathbf{b}$ has no solution. To prove that $\mathbb{A}\mathbf{x} = \mathbf{b}$ is unsolvable, one need only produce a solution to $\mathbb{A}^T\mathbf{y} = \mathbf{0}$, $\mathbf{b} \cdot \mathbf{y} = 1$.

The geometry of the situation is as follows: If $\mathbb{A}\mathbf{x} = \mathbf{b}$ has no solution, then $\mathbf{b}$ is not contained in $V = S[\mathbf{a}^1, \ldots, \mathbf{a}^m]$. In this case, the theorem asserts that there is a vector $\mathbf{y}$ which is perpendicular to every vector in V and not perpendicular to $\mathbf{b}$. We use the term *perpendicular* here to mean that the inner product of two vectors is 0. The usage is based on the geometry of R^2 and R^3.

Index

J

K

L

M